Sustainable Land Use Practices and Agricultural Innovations in India

Hasibur Rahaman • Sakil Ansari

Sustainable Land Use Practices and Agricultural Innovations in India

A Case Study from Murshidabad District of West Bengal

Springer

Hasibur Rahaman
Department of Geography, School of Earth Science
Hemwati Nandan Bahuguna Garhwal University (A Central University)
Srinagar Garhwal, Uttarakhand, India

Sakil Ansari
Department of Geography
Aligarh Muslim University
Aligarh, Uttar Pradesh, India

ISBN 978-3-031-83245-1 ISBN 978-3-031-83246-8 (eBook)
https://doi.org/10.1007/978-3-031-83246-8

This Springer imprint is published by the registered company Springer Nature Switzerland AG
The registered company address is: Gewerbestrasse 11, 6330 Cham, Switzerland

Preface

Achieving sustainable goals is not possible without understanding the aspirations of about one-fourth workforces of the world who are dependent on agricultural systems for foods and livelihood options. The evolving challenges of food security, environmental degradation, climate change, and rural development in India call for innovative approaches that harmonize agricultural productivity with sustainable land use practices. In this context, the district of Murshidabad, located in the heart of West Bengal, offers a rich case study that encapsulates both the challenges and the potential solutions faced by agrarian communities across the nation. This book, *Sustainable Land Use Practices and Agricultural Innovations in India: A Case Study from Murshidabad District of West Bengal*, is the culmination of years of research, fieldwork, and deep engagement with farmers, local stakeholders, academicians, and agricultural experts.

The seed for this work was planted during our early experiences and interactions with rural farming communities, where we were born, grew, and observed the resilience of the people in the face of changing environmental and economic conditions. The urgency to balance the pressures of population growth, food security, agricultural income, and environmental conservation such as soil erosion, pollution, floods, and drought became increasingly evident. This led us to explore innovative and sustainable land use practices that could empower farmers while ensuring long-term environmental stability.

Through this book, our aim is to provide a comprehensive examination of how sustainable agricultural innovations can be effectively implemented in rural India by land size classes. The case study of Murshidabad serves as a focal point to highlight the specific challenges faced by small and marginal farmers, as well as the potential solutions that can emerge through collaborative efforts between farmers, policymakers, academicians, researchers, policy think tank, and agricultural scientists.

At its core, this book seeks to address the critical question: How can we increase agricultural productivity and profitability without further straining the environment? Through a blend of theoretical analysis and practical insights, we hope to offer a road map for achieving sustainable agricultural development and suitable land use measures through innovation processes, not just in Murshidabad, but across India and other regions facing similar challenges.

This book is written for a diverse audience—students, researchers, policymakers, and anyone interested in sustainable land use, agricultural innovation practices, livelihoods, and rural development. I hope that readers will find it useful as a resource for understanding the nuances of sustainable land use and its practical application in real-world scenarios.

We would like to take this opportunity to express our gratitude to the many individuals and institutions who contributed to this project. The farmers of the Murshidabad generously shared their experiences and insights, which became the foundation of this work. We are also indebted to our colleagues and mentors who provided invaluable feedback and guidance during the research process.

In conclusion, this book is a reflection of our belief that the path to sustainable development lies in the intersection of innovation, community engagement, and environmental stewardship. It is our hope that this work will inspire further research, discussions, and actions towards building a sustainable future for agriculture in India.

Srinagar Garhwal, Uttarakhand, India — Dr. Hasibur Rahaman
Aligarh, Uttar Pradesh, India — Dr. Sakil Ansari

Contents

Introduction

Background of the Study

As per a recent estimation of World Bank, agricultural sector has engaged about 26% of global workforce and contributes 4.3% of world's gross domestic product (GDP). In 2023, the GDP share from agriculture in South Asia was 15.98% (World Bank) against the employed workforce of 41.61% (International Labour Organization, 2022). The agricultural income includes forestry, fishing, and agriculture. According to National Sample Survey Office (NSSO), during 2022–2023, about 45.76% of India's total workforce was involved in agriculture and shared 15% of GDP. At national and regional level, the contribution of agriculture in GDP and workforces is decreasing, but still this sector is important for poverty alleviation, reducing hunger, climate change mitigation, and the generation of employment horizontally and vertically at global stage irrespective of intensive and extensive nature of agricultural practices. The intensive subsistence form of agriculture does not require dedicated skills and training; therefore, Africa, South and Central Asia, and countries from Central and South Americas still depend on traditional farming practices. The technological, institutional, and infrastructural development are, no doubts, eliminating workforce directly from agriculture horizontally, but use of automation, precision agriculture, and high tech farming processes increases new avenue for employment vertically. To face global threats and challenges like climate change, disaster, desertification, soil erosion, land pollutions, and food security, the entire agricultural system should be rationalized; therefore, we need agricultural innovation (AI). The AI is a process of developing and applying convenient techniques, technologies, ideas, and practices to improve farm output, sustainability, efficiency, and adjust global challenges and threats. The basic tenet of AI is to use the land for present need without compromising the future fertility. And, from here sustainable land use (SLU) is the first essential premise of AI.

The SLU is not only key of AI rather sustainable development goals (SDGs) have also been focused on this. The SLU emphasizes balanced distribution and the preservation of environmental resources—including water and soil—along with efficient, responsible practices that integrate economic, social, and environmental considerations. The main and key components of SLU are agroforestry, crop rotation, crop diversification, conservation of water, management of tillage, urban greenery, organic farming, and soil health based farming practices. Conversely, the sustainable land use management (SLM) involves planning, governance, and long-term management of land resources which will be beneficial to people of today and tomorrow. The SLM also integrates and balances the social, economic, and environmental existence for well-being. The fundamental focus of SLM includes policies, regulations, and practices to use land in sustainable manner. As like SLU, the SLM also has various components such as zonation of land for planning, integrated land use systems, community participation, environmental impact assessment, and sustainable forestry and agriculture for productivity and efficiency. Keeping in mind of its aims, objectives, and focus, the AI in this study includes SLU and SLM as integral part of discussion under 'sustainable land use practices' (SLUP).

Sustainable Land Use Practice, Agricultural Innovation, and SDGs

The needs of this study also become relevant because of SDG objectives. AI and SLU are integral parts of this long-term goals to enhance economic growth and development, food security, climate resilient agriculture, and natural resources management. The following SDGs are important regarding SLUP:

1. SDG-1 focuses on sustainable land use through AI. To enhance agricultural productivity in any region, the crop rotation, crop diversification, precision agriculture, conservation of tillage, agroforestry, soil, and water efficiency are essential. The goal of zero hunger can be reached through increasing agricultural productivity.
2. SDG-6 stresses on clean water and sanitation through reducing pollutions and efficient water management in agriculture and households. Focusing on water efficiency, the land use management can be enhanced in semiarid and low rainfall regions.
3. SDG-8 focusses on decent work and economic growth by efficient use of land for economic development. The rural economy of small landholders may be enhanced through advancement in agriculture after technological innovation.
4. SDG-9 targets industry, innovation, and infrastructural expansion after vertical farming, hydroponics, aerophonic, smart agriculture, and market horticulture. Sustainable industrial growth may be possible when AI boosts digital platform (mobile application, software, etc.) making agricultural system more robust against climate change and land pollution.

5. SDG-12 aims at responsible consumption and production based on efficient and ensured agricultural system. Innovation in food storage, waste, consumption, and sustainable production system reduces carbon footprint from land use processes.
6. SDG-13 focuses on climate action only after sustainable land use and agricultural innovation like afforestation, agroforestry, and reforestation. The sustainable land use practices in cultivation process have reduced the emissions of greenhouse gases, resultantly an agricultural system become climate resilient.
7. SDG-14 points at life below water. This goal targets water-based agriculture, fishing and restricted water, and land pollution. The SLU avoids the use of chemical fertilizer, on the one hand, and the AI restricts the farming system from unsustainable chemical process along canal, river, and irrigation sources on the other hand.
8. SDG-15 explains about life on land. To make sustainable life practice of plants and animal including agriculture, it is required to halt desertification, land pollution, soil erosion, alkalinity, and salinity problem. The SLUP can boost economic prosperity if existing land resources get due concern from management, planning, attention, and focus.

The adaptation of SLU and leveraging of AI can promote agroforestry, precision agriculture, regenerative farming, and climate-resilient agriculture.

The title of this book can be looked under following two heads, namely sustainable land use in agriculture and agricultural innovation. In the following section, existing literatures from these fields have been analysed with sources and findings. The literatures from the world, India, and local levels are presented coherently.

Sustainable Land Use in Agriculture

World

Tanrivermis (2003) examined that the intensification of agricultural land use geared up by agricultural mechanization, use of agri-chemicals, rapid population growth, industrialization, urbanization leads to the deterioration of soil quality as well as the environment. Therefore, he suggested new legal and institutional regulations which integrate agricultural policies and environmental concerns for sustainable land use management.

A study by Zhao et al. (2008) discussed the concepts and aims of sustainable agricultural development in China with reference to land use practices. They pointed out some obstacles of sustainable agricultural innovation, i.e., water use shortage, cultivated land loss, inappropriate usage of fertilizers and pesticides, and environmental degradation which limit appropriate land use practices. To promote sustainable land use practice, the need of the hour is the requirement of long-term policy intervention along with positive public response towards appropriate measures.

A study on 'Multifunctional Urban Agriculture for Sustainable Land Use Planning in the United States' by Lovell (2010) discussed that how urban agriculture offers a viable alternative in city areas for improving sustainability of land use. The study looks forward for viable alternatives such as new strategies, policies, and program through government funding to protect and improve complex urban ecosystem though sustainable land use practices. To face climate change impact on urban life, climate-resilient agriculture with agricultural innovation will attribute positively.

The benefit of agroforestry practices for the sustainable land use system in Nigeria has been discussed by Alao and Shuaibu (2013). They concluded that optimal use of land for agriculture along with forestry production not only increases food and sustainable wood production but also improves the soil quality.

Mockel (2015) highlighted the importance of standard tested technologies and methods such as integrated pest protection, organic farming, precision farming, and conservation tillage to make sustainable land use in agricultural. The high level of emission from agriculture, environmental pollution, and productivity barrier makes European agricultural land use less fertile. Apart from positive check on pollution, introduction of efficient technologies is equally essential for sustainable land use.

India

Jayakumar and Arockiasamy (2003) analysed the land use land cover change of the Eastern Ghats in Tamil Nadu. They suggested some remedial measures, i.e., regular testing of soil, selection of crops based on soil properties and climatic condition, crop rotation, land management, and application of manure and pesticides for the optimum and sustainable utilization of land resources.

The consequence of agricultural expansion by overexploitation of forest causes serious ecological and environmental problems in Kerala (Kumar, 2005). Agroforestry holds promise in order to mitigate these problems. Therefore, it has been suggested that the state should provide attention for the adoption of agro-forestry practices to protect sustainable land use over irregular expansion of agricultural practices.

Srivastava et al. (2016) proposed a commercial ecological agriculture with least disturbing, resilience building, resource use efficient, labour and skill intensive, low input, and diversified system to achieve land use sustainability in existing agroecosystem. By doing this, it would improve and restore the soil quality as well as increase the productivity.

A study on 'Impact of Sustainable Land-Use Management Practices on Soil Carbon Storage and Soil Quality in Goa State, India' by Paramesh et al. (2021) recommended sustainable agriculture with alternative land use system in the hills and sequential cropping in coastal plains to enhance soil organic carbon, improve soil quality and, increase farmer's income at various capacity.

Crop diversification, an important dimension of agricultural development, benefits us ecologically, environmentally, economically, and sustainably over time and space (Rahaman, 2021). Crops diversification across land size classes provides sustainable land use practices across the blocks in Malda district of West Bengal and promotes regenerative agricultural system across the district.

West Bengal

A study by Jana et al. (2015) on 'Multiplier Effect of Home Gardening' in sustainable land use practice found that the food and nutritional security, employment generation, and maintaining soil health are the best possible outcome from annual and perennial cropping, shrub, and fisheries. The households with a high level of education using modern inputs generate more revenues from sustainable agricultural practices i.e., home gardening.

In a study, Ghosh and Chakma (2018) measured the sustainability of agricultural land use applying ecological footprint method. The study revealed that agricultural production relative to population demand of food represents a higher sustainable agricultural system, while crop-wise assessment explored some negative aspects with respect to land utilization and cropping pattern. Hence, the study urged necessary shifting of existing cropping pattern for agricultural sustainability.

Parihari et al. (2021) assessed the land suitability for appropriate crop suggestion using GIS-based sustainable agricultural production. In the study, they found that rice is a highly suitable crop in the given land followed by wheat, sugarcane, and cashew. For effective cropping measures and sustainable production, along with mapping, the proper and timely intervention is required keeping in mind temporal and spatial changes.

The pattern and impact of agricultural land use practices on livelihood assets of farmer in Malda district (Mandal & Saha, 2013) revealed that the socioeconomic conditions and regional economy have great impacts on changing land use pattern of local farmers irrespective of other essential criterion.

Agricultural Innovation

World

Feder and Umali (1993) conducted a review on 'adoption of agricultural innovation and the impact of policy intervention that promotes technology adoption'. The study revealed that all types of technology, market structures, and the nature and duration of the policy intervention under different agro-climatic regions directly impact technological adoption in green revolution technology diffusion at different stages of agriculture.

The role of agri-technological innovation for climate adaptation in agriculture in Ontario, Canada, was explored by Smithers and Blay-Palmer (2001). They concluded the study by recommending that technological innovation in agriculture be promoted, but only after accounting for climate adaptation.

Ariza et al. (2013) tried to develop a framework for measuring innovation and its key determinants in agriculture firms in Colombia. Based on three main tools, i.e., Innovation Matrix, Innovation Index, and Econometric Model, they have classified the technological level of every innovation as the major, intermediate, and minor level of advancement.

A study by Lapple et al. (2015), titled 'Measuring and Understanding the Drivers of Agricultural Innovation: Evidence from Ireland,' found that farm size, production intensity, access to credit, and agricultural education positively drive agricultural innovation. In contrast, older farmer age and off-farm employment act as barriers to innovation.

In a case study of Nigeria, Ogunniyi et al. (2017) revealed that the households implementing agricultural innovation led by agricultural research programs had a better livelihood, productivity, and more diversified income.

India

A study conducted on 'Agricultural Innovations Among Indian Farmers' by Roy (1968) discovered that the diffusion of agricultural innovation is strongly related to social status, education, and personal contact with the outside world, overshadowing the fragmentation, commercialisation, specialisation, and farm efficiency effects. Socioeconomic status and exposure of mass media also play vital roles to innovation practices in agriculture.

Ali (2016) found that smallholder vegetable growers having land size less than 2 ha are more inclined towards adaptation of innovative agricultural practices, i.e., crop rotation, usage of green manure, diversification crops, sorting and grading of production, and market and price analysis.

Tanti et al. (2022) discussed the importance of agricultural innovation in adopting institutional roles, such as subsidies on farm machinery, credit facilities, extension supports, and field demonstration of technologies by smallholder farmers for climate-smart agriculture.

A study on 'An Empirical Analysis on Adoption of Precision Agricultural Techniques Among Farmers of Punjab for Efficient Land Administration' conducted by Khanna and Kaur (2023) revealed that several constraints such as non-awareness of innovative techniques by framers, financial limitation, issue related to cost, etc., are the main hurdles for the non-adaptability of innovative agricultural techniques and practices.

West Bengal

Ghosh (2010) has claimed that the mechanization of farms is strongly determined by irrigation, access to institutional credits, size of landholdings, and proportion of HYV and fertilizer-intensive crops. The mechanized form of agriculture has the ability to promote innovation in which young age farmers are more adaptive than old block farmers.

The failure of three institutional factors, i.e., comprehension, customization, and generalization of Technology Transfer Training (TTT) is responsible for the lag in the process of technology adoption by farmers (Chandra et al., 2018). The AI is compromised when institutional barrier hinders the use and availability of modern and updated technologies.

Reddy et al. (2020) argued in a study that adopting innovative approaches in sustainable intensification, diversification, market reforms, and access to livelihood capitals can lead to the inclusive sustainable intensification of agriculture in West Bengal.

A study by Reddy et al. (2020), titled 'Inclusive Sustainable Intensification of Agriculture in West Bengal, India: Policy and Institutional Approaches,' highlights that the absence of coherent and comprehensive policies and programs is a key constraint contributing to unsustainable agricultural practices. To promote and adopt inclusive, sustainable agricultural innovations, synergy among various formal and informal institutions including panchayat Raj institutions is needed.

Nature of Problems

One of the major problems facing India today is the decreasing man–land ratio. In this situation, the land must be utilized properly and preciously to its full capacity. Another major problem of Indian agriculture is the small and fragmented landholding, as is the case of the Murshidabad district. It directly influences agricultural land use as the size of the landholding decides the degree of risk for the farmers. The identification of suitable cropping patterns for specific regions governs the overall success of agricultural innovation. The cropping pattern reveals that most of the farming land in the study region is covered by food crops. Rice, wheat, maize, gram, etc., are important staple crops that are found in the district. With the development of irrigation and other technological factors, commercial food crops, fruits, and vegetables are gradually occupying land under cash crops. These changing scenarios in cropping patterns indicate the changes in overall agricultural innovation.

Agriculture occupies an important place in India's economy, especially in West Bengal, where it contributes 21.87% (State Economy, Govt. of West Bengal, 2019–2020) to the state domestic product (SDP). In West Bengal, agriculture remains the backbone and crucial to the state economy as around 70% (Statistical Abstract of West Bengal, 2019) of rural lives, directly and indirectly, depend on it.

Besides the structural transformation of the state economy, the contribution of agriculture in SDP follows a declining trend from 41.16% in 1970–1971 to 21.87% in 2019–2020 (Rahaman, 2020). Therefore, the development of this sector is the primary concern for planners and geographers to find out means and ways for overall development.

In India, agricultural innovation was paved over for the first time when the first 5-year plan (1951–1956) was initiated. Agriculture was given as the first priority in the third 5-year plan. The primary aim was to increase food production and raw materials for different industries substantially. The government realizes that the nation needs to focus on rapid growth of agricultural sector to sustain and maintain a higher tempo of industrial development. But in true sense, agricultural innovation in India enlightens the academic discussion shortly after introducing the 'Green Revolution' in the country.

Green revolution brings high-yielding variety (HYV) seeds, new implements and machineries, extended irrigation facilities, and improved fertilizers and pesticides to Indian agriculture. Previously, Indian agriculture was characterized by a subsistence farming system, primitive cultivation techniques, and low yields. On the brighter side, the green revolution helps stabilize the country's food crisis and reaches beyond sufficient food production. At the cost of bumper production and productivity, soil sustainability drastically reduced; as a result, salinity, alkalinity, and loss of fertility have become the new reality of the green revolution in India. The introduction of HYV seeds in the country has doubled the yields of crops, especially the food grain crops, but this has led to inter-regional disparities in Indian agriculture (Chakravarti, 1973). The bumping drawback of the green revolution, the ever-growing population, and the new food security policy have slowed down agricultural contribution to the national economy (Misra & Puri, 2011).

The sudden innovation in agriculture due to the green revolution has created many problems in rural India. The rapid change in production caused economic imbalance among the farmers, thereby creating interregional agricultural disparities. In order to enable planning and execution, the government has started counting the agricultural situation of the country. This led to the initiation of the first-ever comprehensive Agriculture Census in 1970–71. The Agriculture Census has been conducted every 5 years since then to measure the extent and pattern of the agricultural situation in the country. The primary purpose of the first Agriculture Census was to collect statistics related to the various aspects of agriculture to improve farmers' income and living standards. Soon after the second Agriculture Census, it came to light that there was a horizontal expansion of agricultural land use as the barren lands, uncultivated lands, and culturable wastelands were brought back to cultivation.

Once again, in 1992, the agricultural innovation and land use pattern got reshaped as liberalization, privatization, and globalization (LPG) brought new market opportunities to the farmers. As a result, the vertical expansion of agriculture begins. Farmers started shifting from traditional to export-oriented crops, as evident from various research publications (Rahaman, 2020).

In the Murshidabad district, 80.28% of people (District Census Handbook, 2011) are still living in rural areas where agriculture is the principal occupation. About 75% of the total geographical area in this study region is under the net show area. However, the major problem is that more than 96% of the landholders are marginal and small farmers. Thus, it becomes essential to point out the major problems of the farmers regarding agricultural land use and how these problems can be solved through better agricultural land use planning. Therefore, this study will examine the changing pattern of sustainable agricultural land use intensity and highlight how these changing patterns from traditional to commercial crops lead to overall success in agricultural innovation.

Previously available literature suggests that an increase in agricultural land use through proper land utilization and remunerative cropping patterns significantly impacts the structural transformation of agricultural innovation. While judging these findings, it may be noted that these studies are done in various micro-locations in the country or at the district or state levels using either primary household data or secondary data relating to different time periods. Since agricultural land use is a continuing process that changes over time and across space, it is necessary to investigate such studies for different regions of the country using both aggregate and household data. This study is set in this broader framework in the Murshidabad district of West Bengal. It may be noted that there exists hardly any study on this subject for the district.

According to the 2015–16 Agriculture Census, Murshidabad district of West Bengal became home for more than 96% of farmers having a land size of fewer than 2 ha. Traditional low-intensive, subsistence, and low-commercial agriculture boost their income and livelihood at a plodding pace. The district identifies several constraints, such as decreasing groundwater table, climate change and variability, irregular monsoon rainfall, lack of inputs and technological adaptation, instability and volatility of prices of different crops, slowed down public investments, etc. In this context, this work on 'Sustainable Land Use Practices and Agricultural Innovations in India: A Case Study from Murshidabad District of West Bengal' has become a subject of exciting discussion.

Significance of the Study

The increasing pressure of population creates crucial problems on the limited land resources. Therefore, it is very significant to study the use and misuse of land, its exploitation, and conservation for the welfare of human beings, society, and the country. It can only get through proper planning of land resources which suggests that optimum utilization of land resources must contribute to the maximum of its capacity and must serve some valuable purpose to the society. It has been observed that rapid population growth during the last few decades creates changes to the utilization pattern of land by converting agricultural land into settlements, forests into cropland, and others. The scope of this study encompasses important elements of land use related to agricultural land use intensity, cropping pattern, cropping intensity, crop productivity, and innovations of agricultural and land use planning. Land use planning must emphasize the sustainable use of land for various purposes. Every piece of land should be checked and adequately utilized; otherwise, once it is changed into settlements, it will never return to the farmers. India supports more than 18% of the world's population, but it has only about 2.4% land area of the world. Thus, such a vast population is a matter of consideration which clearly indicates the need to enhance the agricultural innovation process to meet these challenges. Therefore, this study has attempted to focus on changes in land use patterns with the main focus on agricultural land use and how these changes lead to the innovation process of agriculture in the selected study area.

Choice of the Topic and Study Area

The scale of problem is fundamental in geographical studies. In agricultural geography, data is collected and observations are generalized. A district-level study would provide a frame at the micro-level on which further research can be based. Keeping this in mind, the Murshidabad district of West Bengal has been chosen as an area of investigation. Several considerations influence the choice. Firstly, little work has been done to assess the significance of various physio-socio-economic factors concerning agriculture in the Murshidabad district. It is felt that such a study at the micro-level would provide a valuable approach for obtaining a complete understanding of the problems of agriculture in the region. Secondly, Murshidabad district has a significant location with respect to agriculture in the state and country. The district is a good representation of geology, physiography, climate, drainage, natural vegetation, soils, and other socio-economic phenomena. Therefore, the study of agricultural land use will help to a certain extent to understand the agricultural geography of the district. Thirdly, the district has a varied physical base, i.e., it represents significant variations in topography. Thus, it is possible to evaluate the influence of various physical elements on agricultural land use. All these considerations have led to the choice of Murshidabad district as the region for this study in order to understand the agricultural land use of the region from a time–space perspective.

Research Question

Agricultural land use is an essential dimension of the agricultural innovation of a region. It is directly related to the pattern of the cropping system as well as the intensity of cropping. Hence, the improvement in the agricultural land use pattern assumes better agricultural output across different land size categories in the district. Here is the research question:

> In the present context, can changing agricultural innovation ensure sustainable land use practices in Murshidabad district?

Objectives

This study is committed to answering the research question raised by the researchers. For this purpose, five objectives and three hypotheses have been framed. This study aims to examine the pattern of change, various factors influencing it, and its contribution to the overall innovation of agriculture. The specific objectives of the study are the following:

1. To examine the status of sustainable agricultural land use and cropping pattern changes under different land size categories at the block level.
2. To measure the level of agricultural innovation across land size class categories at the block level.
3. To estimate the impact of input–output dimensions of agricultural innovation on agricultural land use across land size classes at the block level.
4. To find out the major factors influencing agricultural innovation and sustainable land use practice under different land size classes in the sampled villages.
5. To suggest some policy-based interventions for sustainable agricultural land use and agricultural innovation in Murshidabad district.

Hypotheses

The following three hypotheses set as null are to be tested for rejection under the assumption that they are true:

1. **Ho**: There is no significant difference in the amount of change in agricultural land use across different land size categories.
2. **Ho**: Agricultural input negatively correlates with agricultural output under different land size classes.
3. **Ho**: Agricultural land use is not significantly related to overall agricultural innovation across land size categories.

Database

This work is constituted from both primary and secondary sources of data.

Primary Data

Twenty-six (26) villages have been carefully chosen for micro-level study. Here one village is selected from each block. Only villages populated by the maximum number of cultivators in the 2011 Census of India have been selected for the study. The systematic random sampling is used to determine 30 agricultural households from each village. The representativeness (as used by NSSO, Agriculture Census, and others) is the only criterion for selecting 30 sample sizes. Thereby, a total of 780 households have been surveyed through scheduled questionnaires.

Secondary Data

Secondary data is obtained from the following sources:

1. Agriculture Census of India
2. Input Survey, Govt. of India
3. Census of India
4. National Sample Survey
5. Statistical Handbook of Murshidabad
6. Office of the DDA Admin., Murshidabad
7. Office of the Horticultural Department, Murshidabad
8. Office of the Soil Testing Laboratory, Murshidabad
9. Reports on Strategic Research and Extension Plan of Murshidabad (SREP)
10. Agricultural Technology Management Agency (ATMA), Murshidabad
11. District Irrigation Plan Book, Murshidabad

Data for a few variables which are not available for the census year 2015–16 have been extrapolated. For the extrapolation, first, the data of the Agriculture Census 2005–06 and 2011 were interpolated, and then the average of the last 3 years was taken as actual data for 2015–16. The same technique is used for the 2006–07 and 2011–12 Input Survey data to get the 2016–17 data.

$$\text{Interpolation} = \frac{\text{Base year data} + \left(\text{Recent year data} - \text{Base year data}\right)}{\text{Number of years}\left(\text{Here five}\right)}$$

Here, for Agriculture Census data, the base year is 2005–06, and the recent year is 2010–11. For Input Survey data, the base and recent years are 2006–07 and 2011–12, respectively.

For the data split, the number of households and gross cropped area are taken as the base of the block.

Methodology

Qualitative and quantitative methods have been used for data analysis and mining in the study.

Qualitative Techniques

Strengths, weaknesses, opportunities, and threats (SWOT) analysis is a purely qualitative technique used in Chap. 6.

Quantitative Techniques

1. The UNDP method is applied to compute index value and composite standardized score. We know that Human Development Index (HDI), Gender Inequality Index (GII), and Multidimensional Poverty Index by UNDP and agricultural development in India (ministerial report) have been prepared by using the UNDP method. The standardization technique of the UNDP method is mentioned below:
   - For positive variables

$$\frac{(\text{Actual Value} - \text{Minimum Value})}{(\text{Maximum Value} - \text{Minimum Value})}$$

   - For negative variables

$$\frac{(\text{Maximum Value} - \text{Actual Value})}{(\text{Maximum Value} - \text{Minimum Value})}$$

The minimum value would be 0, and the maximum would be 100 if the data is in percentage form. Otherwise, the maximum and minimum values are considered from the table value when the data is in absolute numbers.

2. Yang's Yield Index used to compute productivity indices

Name of the crops	The area under crop in the block	Yield in quintal/ha		Crop yield in the block as the percentage of the district	Percentage multiplied by area (in ha)
		Average yield in the block	Average yield in the district		
1	2	3	4	5 = col. 3/ col. 4 * 100	6 = col.5 * col. 2

3. Gibbs-Martin's technique is used to calculate the crop diversification index. The index is computed as follows:

$$\text{Crop Diversification Index} = 1 - \frac{\Sigma X^2}{(\Sigma X)^2}$$

Here, X is the percentage of the total cropped area occupied by an individual crop at a point of time. The index value of this technique ranges from 0 to 1. The crop diversification seems to be higher when the index value approaches 1, while the value reaching near 0 reflects a specialization of crop cultivation.

4. To calculate cropping intensity, the following formula is used:

$$\text{Cropping Intensity} = \frac{\text{Gross Cropped Area}}{\text{Net Area Sown}} \times 100$$

5. Karl Pearson's correlation of coefficient is used to check the degree of association between input and output dimensions of agricultural development. The equation for the calculation of Karl Pearson's coefficient of correlation is given below:

$$\mathbf{r} = \frac{\Sigma(\mathbf{X} - \bar{\mathbf{X}})(\mathbf{Y} - \bar{\mathbf{Y}})}{\sqrt{\Sigma(\mathbf{X} - \bar{\mathbf{X}})^2 \, \Sigma(\mathbf{Y} - \bar{\mathbf{Y}})^2}}$$

where $\bar{\mathbf{X}}$ is the mean of **X** variable, and $\bar{\mathbf{Y}}$ is the mean of **Y** variable

6. .The multiple linear regression model, focusing on the ordinary least square (OLS) method used for the estimation of the model's parameter, is statistically formalized as follows:

$$y_i = \beta_0 + \beta_1 x_{i1} + \beta_2 x_{i2} + _{\ldots} + \beta_{p-1} x_{i,p-1} + \varepsilon_i$$

where y_i represents the outcome or dependent variable, β_0 is the intercept (the mean of y when all $x_k = 0$), each β_k ($k = 1, 2, \ldots, p - 1$) represents a slope or estimated coefficient concerning the explanatory variables x_i ($i = 1, 2, \ldots, n$ unit of observations), and ε_i is the random error with a prior expectation of 0 and variance σ^2.

7. ANOVA, Karl Pearson's correlation coefficient and T-test have been used to test a significant level of framed hypotheses.

Data are processed in Excel and SPSS spreadsheets. Jamovi 2.3.2, IBM SPSS Statistics 20, and Excel have been used to calculate and run the statistical tests. Moreover, the maps and charts are prepared in ArcMap 10.5 and Excel.

Chapter Outline

This study comprises six chapters, excluding the introduction and conclusion.

The introductory section discusses the background, literatures, the statement of problem, the significance of the study, the choice of the topic and area, the research question, objectives and hypotheses, and the database and methodology for a meaningful shape of the study.

Chapter 1 deals with the overview of the study area. Here, different aspects of the physical and cultural set-up of the Murshidabad district are presented with various maps and diagrams and related data.

Chapter 2 contains the sustainable land use dynamics and cropping patterns under different land size categories in the Murshidabad district. After that, agricultural land use transformation is analysed and discussed in a spatial context across land size categories at two points of time at the block level. The last section deals with the reasons for agricultural land use changes through field survey data.

Chapter 3 analyses input and output determinants of agricultural innovation. A detailed account of the variables of input and output determinants of agricultural innovation is discussed first, and then the degree of association between them is analysed. The last section discusses the overall agricultural innovation concerning the input and output aspects at spatial context under different land size classes in two points of time.

Chapter 4 is devoted to understand the relationship between sustainable agricultural land use and agricultural innovation in the district. The association between agricultural land use and agricultural innovation is measured statistically through multiple regression analysis under different land size categories. Following this, a section on unit-wise change of these two domains is explained at the block level across land size classes, where the input and output dimensions of agricultural innovation are measured with comparisons of per unit change of sustainable agricultural land use.

Chapter 5 presents an analysis based on a household survey to check the sustainable agricultural land use and innovation relation further at the micro-level. The first section describes the physical, socio-economic, and demographic characteristics of the sample villages, and then the agricultural situation of these villages is explained in the next section.

Chapter 6 assesses the issues and challenges of agricultural innovation with respect to existing policies, programs, schemes, and projects introduced by the government for the district. Here, strengths, weaknesses, opportunities, and threats (SWOT) are discussed to identify key factors of agriculture in particular and the innovation process in general. A highlight of area-specific (village-level in this case) constraints faced by the farmers using a rating scale is discussed in the next section. Lastly, the evidence-based policy intervention has been suggested to sort out existing problems.

The study sums up with a summary of findings, suggestions, and conclusions.

References

- Alao, J. S., & Shuaibu, R. B. (2013). Agroforestry practices and concepts in sustainable land use systems in Nigeria. *Journal of Horticulture and Forestry*, *5*(10), 156–159.
- Ali, J. (2016). Adoption of innovative agricultural practices across the vegetable supply chain. *International Journal of Vegetable Science*, *22*(1), 14–23.
- Ariza, C., Rugeles, L., Saavedra, D., & Guaitero, B. (2013). Measuring innovation in agricultural firms: A methodological approach. *Electronic Journal of Knowledge Management, 11*, 185–198.
- Chakravarti, A. K. (1973). Green revolution in India. *Annals of the Association of American Geographers, 63*(3), 319–330.
- Chand, R. (1996). Diversification through high value crops in Western Himalayan region: Evidence from Himachal Pradesh. *Indian Journal of Agricultural Economics, 51*(4), 652–663.
- Chandra, P., Bhattacharjee, T., & Bhowmick, B. (2018). Does technology transfer training concern for agriculture output in India? A critical study on a lateritic zone in West Bengal. *Journal of Agribusiness in Developing and Emerging Economies*, *8*(2), 339–362.
- District Statistical Handbook of Murshidabad, Govt. of West Bengal.
- Feder, G., & Umali, D. L. (1993). The adoption of agricultural innovations: A review. *Technological Forecasting and Social Change, 43*(3–4), 215–239.
- Ghosh, B. K. (2010). Determinants of farm mechanization in modern agriculture: A case study of Burdwan districts of West Bengal. *International Journal of Agricultural Research, 5*(12), 1107–1115.
- Ghosh, B., & Chakma, N. (2018). Application of the ecological footprint method for measuring sustainability of agricultural land use at a micro level in Barddhaman district, West Bengal, India. *Current Science*, *115*(10), 1913–1920.
- GoI. (2006, December). *Towards faster and more inclusive growth: An approach to the 11th five year plan (2007–2012)*. Planning Commission, Govt. of India.
- GOI. (2018–2019). *Agricultural statistics at a glance*. Ministry of Agriculture. http://agricoop.nic.in/.
- Https://www.ilo.org/publications/major-publications/global-employment-trends-youth-2024. (2022). ILO. Retrieved September 10, 2024, from https://www.ilo.org/publications/major-publications/global-employment-trends-youth-2024
- Jana, S., Roy, J., Marambe, B., Weerahewa, J., Pushpakumara, G., Silva, P., Miah, Md. G., Punyawardena, R., & Premalal, S. (2015). Home gardens in the Paschim Medinipur District of West Bengal in India: A land use system with multiple benefits. *International Journal of Environment and Sustainable Development, 14*(2), 191. Inderscience Publishers.
- Jayakumar, S., & Arockiasamy, D. I. (2003). Land use/land cover mapping and change detection in part of Eastern Ghats of Tamil Nadu using remote sensing and GIS. *Journal of the Indian Society of Remote Sensing*, *31*, 251–260.

- Khanna, A., & Kaur, S. (2023). An empirical analysis on adoption of precision agricultural techniques among farmers of Punjab for efficient land administration. *Land Use Policy*, *126*, 106533.
- Kumar, B. M. (2005). Land use in Kerala: Changing scenarios and shifting paradigms. *Journal of Tropical Agriculture*, *43*, 1–12.
- Kumar, P., & Mathur, V.S. (1996). Structural changes in the demand for food in India. *Indian Journal of Agricultural Economics*, 51(4), 664-673.
- Läpple, D., Renwick, A., & Thorne, F. (2015). Measuring and understanding the drivers of agricultural innovation: Evidence from Ireland. *Food policy*, 51, 1-8.
- Lovell, S. T. (2010). Multifunctional urban agriculture for sustainable land use planning in the United States. *Sustainability*, *2*(8), 2499-2522.
- Mandal, R.B. (1982). *Land Utilization: Theory and Practice*. Concept Publishing House, New Delhi.
- Mandal, T., Saha, S. (2023). Agricultural Land Use Change and Its Impact on the Farmers' Livelihood Assets of Maldah District, West Bengal, India. In: Das, J., Halder, S. (eds) Advancement of GI-Science and Sustainable Agriculture. GIScience and Geo-environmental Modelling. Springer, Cham. https://doi.org/10.1007/978-3-031-36825-7_18
- Misra (2011), *"Indian Economy"*. 27th revised Edition, Himalaya Publishing House, Pvt. Ltd., Ramdoot, Dr Bhalerao Marg, Girgan, Mumbai-400004, pp59 to 64.
- Möckel, S. (2015). 'Best available techniques' as a mandatory basic standard for more sustainable agricultural land use in Europe? *Land use policy*, *47*, 342-351.
- National Sample Survey Organisation. (2023). Workforce Engaged in Agriculture. In *Ministry of Agriculture & Farmers Welfare Department of Agriculture & Farmers Welfare*. NSSO information through LOK SABHA. Retrieved September 26, 2024, from https://sansad.in/getFile/loksabhaquestions/annex/1714/AS228.pdf?source=pqals#:~:text=According%20to%20the%20Periodic%20Labour,allied%20sector%20during%202022%2D23.
- Ogunniyi, A., Oluseyi, O. K., Adeyemi, O., Kabir, S. K., & Philips, F. (2017). Scaling up agricultural innovation for inclusive livelihood and productivity outcomes in sub-Saharan Africa: The case of Nigeria. *African Development Review, 29*(S2), 121–134.
- Paramesh, V., Singh, S. K., Mohekar, D. S., Arunachalam, V., Misra, S. D., Jat, S. L., et al. (2022). Impact of sustainable land-use management practices on soil carbon storage and soil quality in Goa State, India. *Land Degradation & Development*, *33*(1), 28–40.
- Parihari, S., Das, K., & Chatterjee, N. D. (2021). Land suitability assessment for effective agricultural practices in Paschim Medinipur and Jhargram districts, West Bengal, India. In *Modern Cartography Series* (Vol. 10, pp. 285–311). Academic Press.
- Planning Commission. (1953). *First five-year plan.*
- Rahaman, H. (2020). *Diversified cropping pattern and agricultural development: A case study from Malda district, India* (p. 9). Springer.

- Rahaman, H. (2020). *Diversified cropping pattern and agricultural development: A case study from Malda district, India* (p. 1) Springer.
- Rahaman, H. (2021). Status of crop diversification. In: *Diversified cropping pattern and agricultural development*. Springer. https://doi.org/10.1007/978-3-030-55728-7_4
- Reddy, V. R., Chiranjeevi, T., & Syme, G. (2020). Inclusive sustainable intensification of agriculture in West Bengal, India: Policy and institutional approaches. *International Journal of Agricultural Sustainability*, *18*(1), 70–83.
- Roy, P. (1968). Agricultural innovation among Indian farmers. *Agricultural innovation among Indian farmers.*
- Satyasai, K. J. S., & Viswanathan, K. U. (1996). Diversification in Indian agriculture and food security. *Indian Journal of Agricultural Economics*, *51*(4), 674–679.
- Smithers, J., & Blay-Palmer, A. (2001). Technology innovation as a strategy for climate adaptation in agriculture. *Applied Geography, 21*(2), 175–197.
- Srivastava, P., Singh, R., Tripathi, S., & Raghubanshi, A. S. (2016). An urgent need for sustainable thinking in agriculture–An Indian scenario. *Ecological Indicators*, *67*, 611–622.
- Tanrivermis, H. (2003). Agricultural land use change and sustainable use of land resources in the Mediterranean region of Turkey. *Journal of Arid Environments*, *54*(3), 553–564.
- Tanti, P. C., Jena, P. R., & Aryal, J. P. (2022). Role of institutional factors in climate-smart technology adoption in agriculture: Evidence from an Eastern Indian state. *Environmental Challenges*, *7*, 100498.
- Zhao, J., Luo, Q., Deng, H., & Yan, Y. (2008). Opportunities and challenges of sustainable agricultural development in China. *Philosophical Transactions of the Royal Society B: Biological Sciences*, *363*(1492), 893–904.

Chapter 1
Overview of the Study Area

Contents

The preceding introductory section discusses the background, importance, nature of problem, research question, objectives and hypotheses, database, and methodology for a meaningful shape of the study. This chapter analyses the different human and

H. Rahaman, S. Ansari, *Sustainable Land Use Practices and Agricultural Innovations in India*, https://doi.org/10.1007/978-3-031-83246-8_1

physical aspects of the study area, i.e., Murshidabad district. The first section of this chapter deals with the background of the district, where the location of the study area, a brief history, and present administrative divisions are presented. The physical features such as physiography, geological structure, drainage pattern, beels, and soil of the study area are discussed in the second section. Subsequently, the aspects like climate, flora and fauna, demographic profile, economic elements, and transport have been interpreted in the following section. Various maps and diagrams have been put up to illustrate the data in the next sections.

1.1 Background of the District

1.1.1 Location of the Study Area

The Murshidabad district, the northernmost district of the presidency division, is located in the middle part of West Bengal, lying between 23° 43′ 30″ and 24° 50′ 20″ North latitudes and 87° 49′ 17″ and 88° 46′ 00″ East longitudes (District Census Handbook, 2011) (Fig. 1.1). It is entirely situated in the north of the Tropic of Cancer. The lower flow of river Ganga, bifurcated to Padma and Bhagirathi near

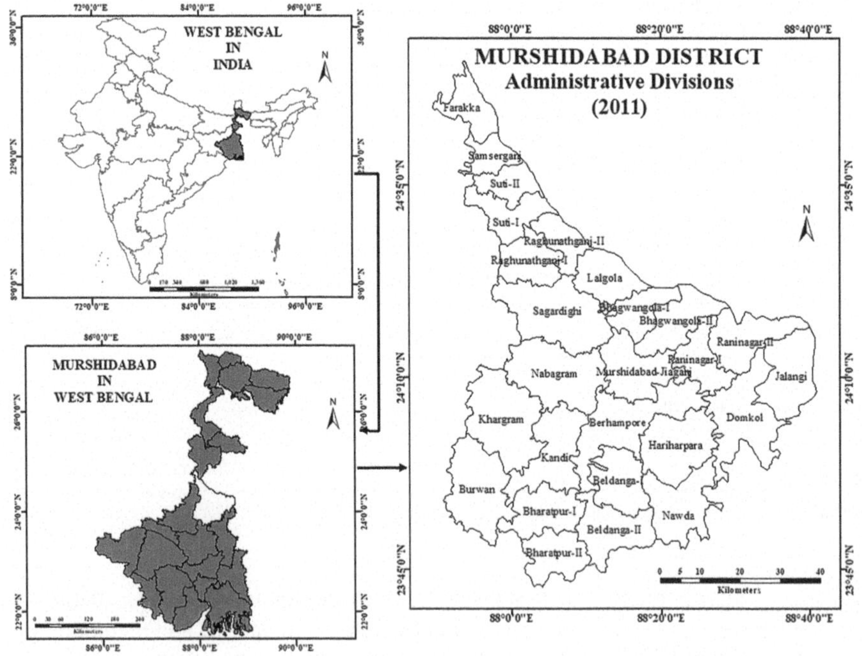

Fig. 1.1 Location of Murshidabad district. (Source: Base Map from Census of India)

Debipur of Suti-II block, creates a natural boundary of the district. The blossoming flow of the Padma River forms the northern and eastern borders of the district, separating the district from Malda district to the north and Bangladesh to the east. It is bounded by Burdwan and Nadia districts to the southern side. The state of Jharkhand (Pakur and Sahibganj districts) and Birbhum district form the western boundary of the district. The district covers 125.35 km of international borders with Bangladesh of 10 CD blocks, of which 42.35 km is on land. The remaining distance is covered by riverine boundaries (Padma River and her connected streams).

It is the seventh biggest district of West Bengal (covering 6.13% of the total geographical area of the state), spreading over an area of 5324 km^2, out of which rural and urban areas cover 5195.11 and 128.89 km^2, respectively (Census of India, 2011). The district's headquarters is Berhampore, situated at 24° 05′ 54″ North latitude and 88° 16′ 06″ East longitude.

1.1.2 Brief History of the District

The Murshidabad district has a long history that dates back to ancient India. During the fifth century AD, some portion of the present district belonged to the territory of the Gupta Empire. Then after, Sasanka ruled over almost the whole district during the seventh century AD. The capital city of his empire was Karnasubarna, which is presently located at the Kandi-Subdivision of the district. The historical literature pointed out that different dynasties ruled the district from seventh to seventeenth century AD. According to the Khalimpur copperplate inscription, the foundation of the Pal dynasty was led by Gopal around 750 AD. It lasted until 1144 AD, when Vijay Sen defeated Madan Pal, the last ruler of the Pal dynasty. The Sen dynasty was the last Hindu-ruled dynasty before the invasion of the Sultanate Empire by Bakhtiyar Khalji. After the termination of sultanate rule, Shahi dynasties and Sher Shah Suri ruled briefly over the district.

Thereafter the Mughal Empire's era began, which ruled till 1757 AD when Siraj-ud-Daulah, the last sovereign ruler of Bengal, lost the battle of Plassey to the East India Company. In 1765 AD, with the grant of Diwani, the East India Company took over the revenue management authority for the whole of Bengal. Soon after taking control, they shifted the centre of power from Murshidabad to Calcutta. It is interesting to mention that the district was initially known as Makhsudabad (meaning the selected city). In 1704 AD, Nawab Murshid Quli Khan renamed the town Murshidabad and transferred the capital city from Dacca (now Dhaka, Bangladesh) to Murshidabad. The unit and boundaries of the district were finalized in 1879. Since then, the present set-up of Murshidabad district has remained the same along with the boundaries.

1.1.3 *Administrative Divisions*

The Murshidabad district is comprised of 26 CD blocks and seven municipalities under five sub-divisions for the sake of smooth administrative functions. These five sub-divisions are Sadar, Kandi, Jangipur, Lalbagh, and Domkal. The Sadar sub-division comprises eight police stations responsible for maintaining law and order in five blocks and two municipalities, namely Berhampore, Beldanga-I, Beldanga-II, Nawda, Hariharpara, Berhampore (M), and Beldanga (M). Similarly, Kandi, Khargram, Burwan, Bharatpur-I, and Bharatpur-II are the five blocks along with Kandi municipality that come under the security of five police stations in Kandi sub-divisions. Jangipur sub-division comprises six police stations for seven blocks and two municipalities. These are Farakka, Samserganj, Suti-I, Suti-II, Raghunathganj-I, Raghunathganj-II, and Sagardighi. Jangipur and Dhuliyan are the two municipalities situated in this sub-division. The five blocks administered by six police stations in the Lalbagh sub-division are Lalgola, Bhagwangola-I, Bhagwangola-II, Murshidabad-Jiaganj, and Nabagram. Adding to this, two municipalities are also there: Murshidabad and Jiaganj-Azimganj. The Domkal sub-division comprises four police stations that bring law and order in four blocks, i.e., Domkal, Jalangi, Raninagar-I, and Raninagar-II (see Annexure A.1).

In the district, the highest institution for local governance is the Murshidabad Zilla Parishad or District Council. Every community development block corresponding to the Zilla Parishad is the Panchayat Samiti. A total of 254 gram panchayats and 4245 Gram Sansads fall under the jurisdiction of Zilla Parishad in the district, covering 1886 inhabited villages. There are 128 wards classified under seven different municipalities. The municipalities are responsible for the functioning of urban governance. Annexure A.1 reveals that the district has 72 census towns (according to the 2011 census) found in all the sub-divisions except for the Domkal sub-division.

1.2 Physical Setting

1.2.1 *Physiography*

The river Bhagirathi, flowing from northern to southern, plays a vital role in the study area, which divides the district into two topographical divisions, i.e., Rarh area (western part of Bhagirathi) and Bagri (east of Bhagirathi). The western part of the river reveals rugged terrain, locally known as Rarh, and a flat and rolling plain on the eastern side known as Bagri. These two physiographic divisions are almost equal in size but strikingly different in their relief, geology, soil, etc. The Rarh region is comprised of Farakka, Samserganj, Suti-I, Suti-II, Raghunathganj-I, Raghunathganj-II, Nabagram, Khargram, Burwan, Kandi, Bharatpur-I, and Bharatpur-II blocks, while the rest of the blocks are in the side of the Bagri region (Fig. 1.2).

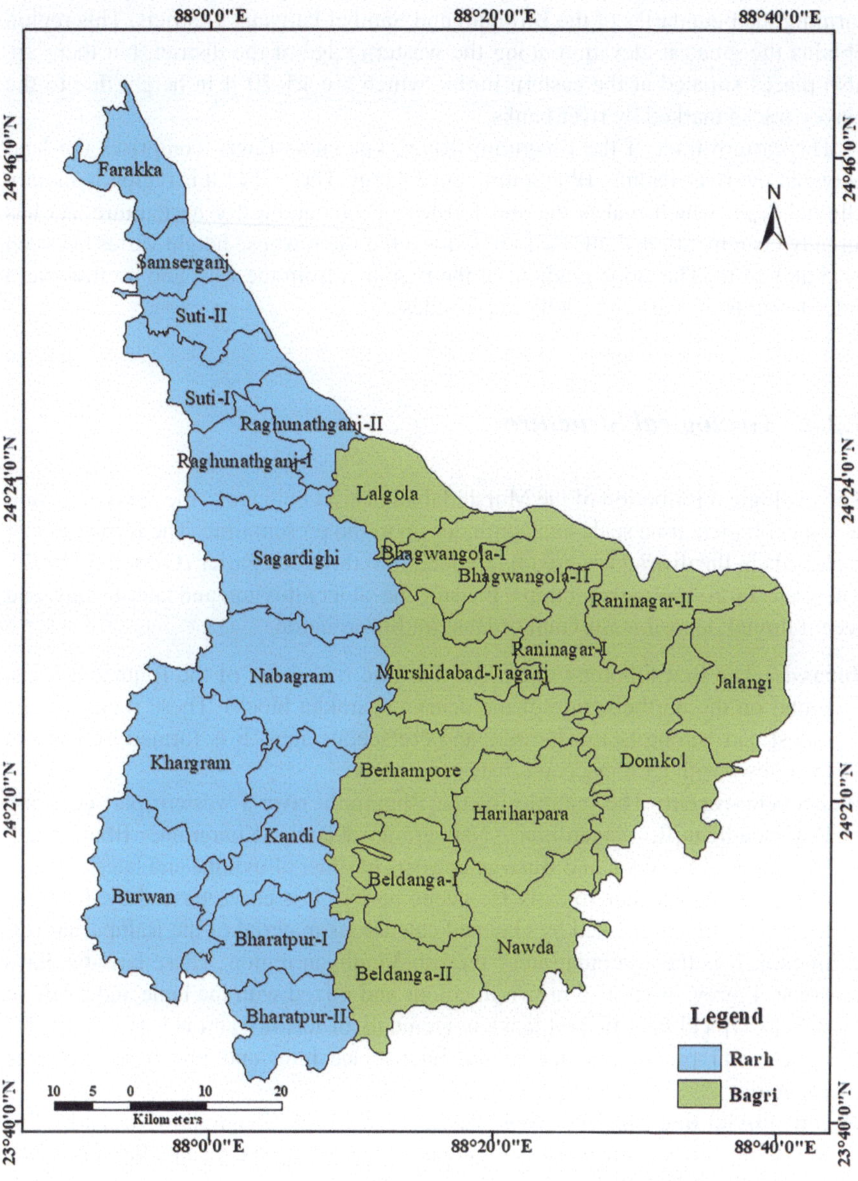

Fig. 1.2 Physiographic divisions of Murshidabad district. (Source: Based on Annexure A.1)

The western part of the study region, the Rarh, represents undulating and rugged terrain and is intercepted by several swamps and beds of old rivers. It is composed of mostly clay and laterite clay type of soil and is primarily a continuation of the sub-Vindhyan region. The altitude of this region is comparatively higher than the eastern region. As we move further, the elevation rises towards the west edge, forming the boundaries of the Birbhum and Santhal Parganas Districts. This region obtains the greatest elevation along the western edge of the district, but there are also places situated at the eastern limits, which are 15–20 ft in height due to the clayey tracks marked by river banks.

The eastern tract of the Bhagirathi River, known as Bagri, comprises the land between the river Padma, Bhagirathi, and Jalangi. The soil of this region is mainly alluvial type, which makes the soil fertile and favourable for agriculture. As has already been mentioned, the eastern part is a flat plain whose height varies between 12.5 and 24 m. The slope gradient of the region is from the west and northwestern sides towards the east and southeastern sides.

1.2.2 Geological Structure

The geological formation of the Murshidabad district belongs to the Jurassic period of the geological time scale and continues up to the present time. The formations of rock beds in the district are broadly divided into three categories (O'Malley, 1997). These are Jurassic-Rajmahal traps, Pleistocene-older alluvium and laterite clay, and recent fluvial deposit—alluvium of Bhagirathi formation.

Jurassic: The first lithology comprises basaltic rock belts of the Rajmahal traps, found on the northern side of the district (Farakka block). These rocks are the oldest and belong to the Jurassic and cretaceous ages. It is formed by basaltic lava flows with carbonaceous shales and clays.

Pleistocene-recent: The majority of the Bhagirathi river's western part covering Raghunathganj-I, Sagardighi, Nabagram, Kandi, Khargram, Bharatpur-I, Bharatpur-II Farakka, and Burwan consists of older alluviums and lateritic clays of the late Pleistocene to early Holocene period. It is characterized by the argillaceous sediment formed by clay and calcareous material of the Rampurhat formation. It is the continuation of the sub-Vindhyan region. These lateritic clays are stiff, grey, or reddish brown in colour and mixed with the lame and oxide of iron. A typical lateritic soil lacks in elements of fertility and is less suitable for agricultural production due to the intense leaching and low base exchange capacities.

Recent fluvial deposit: The whole eastern part of the Bhagirathi river and a few low-lying zones in the western region are occupied by the recent alluvium of the Bhagirathi and the Padma rivers. This formation comprises clays, silts, and sands brought by these rivers. Continuous deposition by the Bhagirathi and its distributaries leads to the thick alluvium depth of the soils in this region. That is why the soils of this region are more suitable for agriculture than the western part of the district, i.e., the Rarh region.

1.2.3 *Drainage Patterns*

It is a well-known fact that the drainage network of any region plays a vital role in agricultural activities and development. The principal river system of the district consists of the perennial streams of Bhagirathi, Jalangi, and Bhairab. Due to the district's geographical characteristics, most of the rivers are connected to the mighty Ganges flow and are primarily rain-fed. The streams cannot carry a larger quantity of silt received during the upper course. As a result, it leads to the shoal formation that ultimately obstructs navigation in these rivers. The river Ganges after entering the Murshidabad district bifurcates into two branches, one flowing towards Bangladesh as the Padma River. In contrast, the other flows through the Murshidabad district and is known as Bhagirathi. Besides these two major rivers, several small tributaries and distributaries also flow in the district. A detailed account of these river systems of the district is also presented in Fig. 1.1.

1.2.3.1 The Ganga and Bhagirathi Rivers

The Ganga river enters Murshidabad district at the extreme northern point. After flowing southeast direction for a few kilometres, the Ganga river bifurcates to Padma and Bhagirathi near Debipur of Suti-II blocks. The larger discourse of the Ganga river, taking the name Padma, enters Bangladesh just after the bifurcation. In contrast, the Bhagirathi river continues to flow in the eastern direction of the district. It divides the study region into two parts and significantly influences agriculture. To regulate the Bhagirathi's water, the Government of India has constructed a feeder canal to link Farakka to Bhagirathi near Raghunathganj-II Block. A few streams and rivers join the Bhagirathi from northwest to southwest, in which the Konloi and Bansloi are the most important tributaries, influencing agricultural activities there.

1.2.3.2 The Bhairab River

The Bhairab river, which originated from the Ganga river near Lalgola, is an old river whose earliest flow can be traced back as a continuation of river Mahananda. The Padma river feeds some water in it during the rainy season. After having a short journey across Murshidabad district, it meets river Jalangi in the northern part of Nadia district.

1.2.3.3 The Jalangi River

The river Jalangi has originated from the Ganga river and starts flowing along the west of the Hangordaobah. Then it turns eastward near the international boundary of Bangladesh. After that, it again changes direction towards the southern side and flows along the district boundary of Murshidabad with Nadia. Sialmari, one of the branches of the Padma river, meets the Jalangi river and increases the flow, which often results in floods during monsoon season. This river poses importance due to its influence on agricultural activities.

1.2.3.4 The Dwarka River

The Dwarka (locally known as Babla) originated in Santhal Parganas of Jharkhand has numerous backwaters and side channels that connect them with the Bhagirathi river. It enters the district near Morgram from Birbhum district and flows east and then in the southeastern direction, which forms the eastern boundary of the Kandi sub-division. This river has a rapid current and causes severe floods during the rainy season, adversely affecting agriculture.

1.2.3.5 The Brahmani, Mayurakshi, and Kuiya Rivers

The Brahmani, Mayurakshi, and Kuiya rivers are the tributaries of the Dwarka river and originate from the different hills of the Birbhum district. These rivers have strong currents and meet at the Hizal beel situated at the Kandi sub-division. These rivers have several small tributaries and influence agriculture. Due to the nature of stream current flow, these rivers often cause floods in the catchment areas.

1.2.3.6 The Bansloi and Pagla Rivers

The Bansloi and Pagla rivers are small but important tributaries of the Bhagirathi rivers, flowing parallel to each other in the west-east direction. Both rivers meet the Bhagirathi river near Jangipur town.

1.2.4 Beels (Ponds)

Apart from the rivers mentioned above, numerous beels (also known as Bils, Jhils), small lakes, ponds, etc. are also found in the district. These beels and lakes are the results of direct or indirect fluvial action. In the north, an area of 389 km^2 consists of a strip of low-lying plain land that becomes a vast lake during the rainy season. There is a plain of 129 km^2 at the confluence of the Mayurakshi and Dwarka rivers known as ‘Hizal Beel’. There is an area of a vast swampy tract of dark clay, popularly known as ‘Kalantar Beel’, in the southeastern corner of the district, which is about 129 km^2. Besides, many beels and tanks are found in the eastern part of the district. Among them, Damus Beel, connected with the Bhagirathi river, is a famous lake for producing a large variety of fish.

A canal of 26.5 km long, known as the Feeder Canal, takes off upstream of the Farakka Barrage on the right bank of the Ganga river. This canal conveys the Ganges water from Farakka Dam to the Bhagirathi river near Nurpur. This man-made project was taken up in 1960 to preserve and maintain the Kolkata port as well as the navigability of the Bhagirathi-Hooghly river.

1.2.5 Soil

The healthy growth of plants depends on the type of soil prevailing in a region. The nature and variety of soils decide the crops, ultimately ensuring better productivity. The clay soil supported by an adequate water supply is favourable for the cultivation of crops like paddy, jute, mesta, etc. Conversely, sandy soil rich in iron content supports crops like watermelon, cucumber, dry cactus trees, etc. The majority of the crops are grown in loamy soil with adequate irrigation and fertilizers support in the district. The study area is bestowed with sedimentary-based alluvial soils favourable for agricultural activities.

1.2.5.1 Soil Types

Several factors determine the development of various soils with various textures, colours, structures, depths, horizons, etc., in the district. Broadly, the district's soil is classified into five types, i.e., sandy, coarse loamy, fine loamy, and miscellaneous (Fig. 1.3). The coarse loamy and sandy texture soils cover 785.2 km^2 (14.8% of the district's total geographical area). Most of the district's soil is fine loamy, spreading over 2702 km^2. Around 30% of the total area (1590 km^2) is obtained by sandy clay, silt clay, and clayey soils.

Figure 1.4 reveals that the nature of the soil in the majority of the blocks is varied from clay loam to sandy loam. Burwan reports the maximum percentage of area under clay loam soil, i.e., 99%. It is immediately followed by Raghunathganj-I

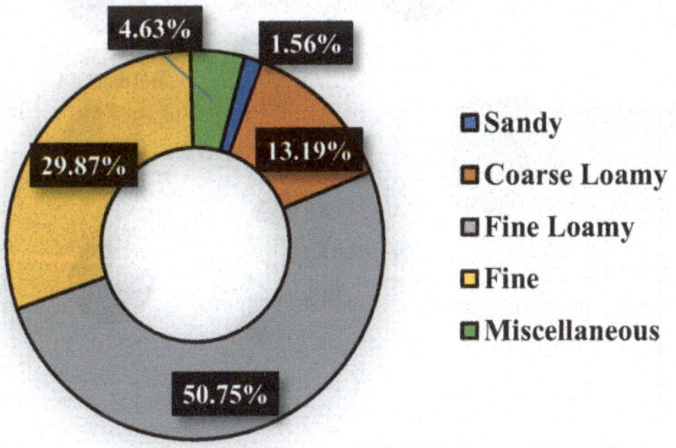

Fig. 1.3 Major soil types. (Source: Based on Table 1.1)

Table 1.1 Soil classifications in Murshidabad district, 2015–2016

Sl. no.	Texture class (surface)	Textural grouping	Area (km^2)	Percentage
1	Sandy	Sandy	82.9	1.6
2	Coarse loamy	Loamy	702.3	13.2
3	Fine loamy		2702	50.8
4	Fine	Clay	1590.1	29.8
5	Miscellaneous		246.7	4.6
Total			**5324**	**100**

Source: Deputy Director of Agriculture, Agriculture Department, Murshidabad

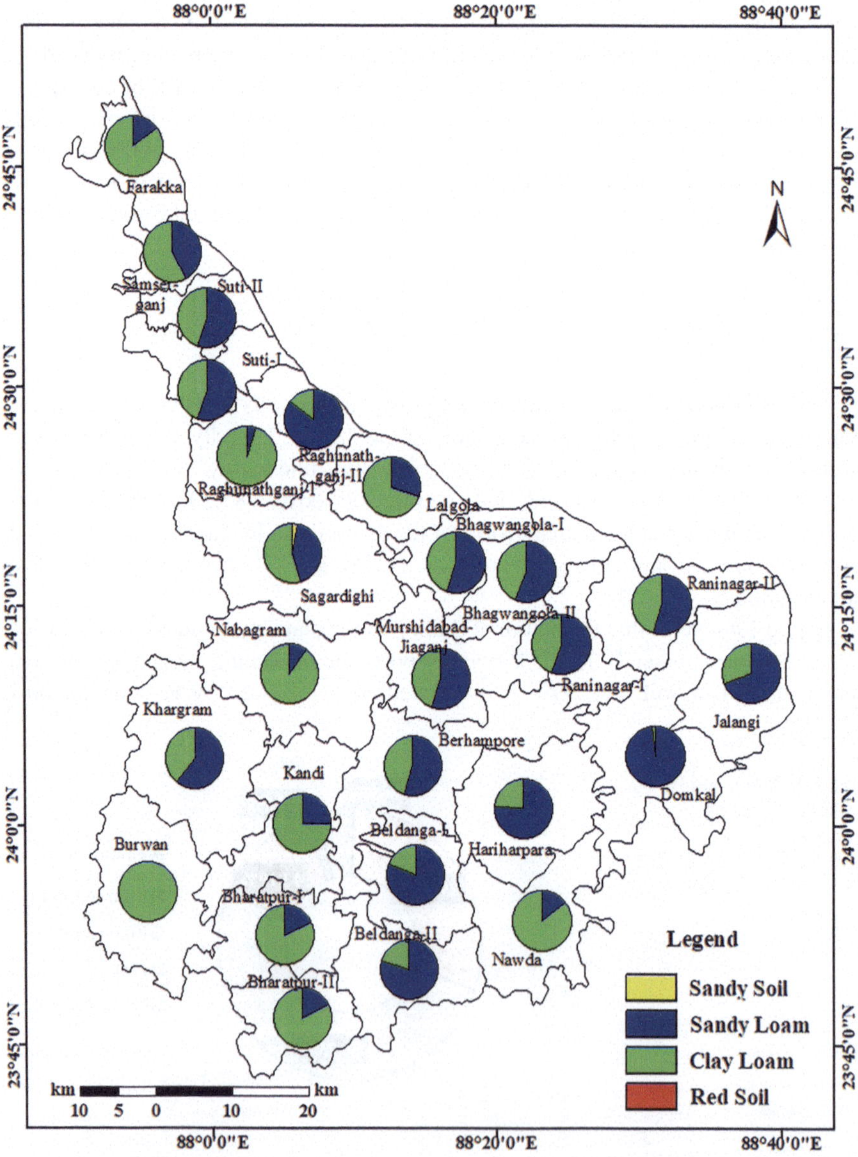

Fig. 1.4 Composition of major soil types. (Source: Based on Annexure A.2)

(95%), Nabagram (90%), and Farakka (85%). In contrast, the least area of clay loam soil is found in Domkal (2%). The highest and lowest percentages of sandy loam soil is indicated by Domkal (98%) and Burwan (01%), respectively. The district poses only 5.40% of the area covered by sandy soil. Only four blocks, namely Sagardighi (3%), Raninagar-I (1%), Raninagar-II (1%), and Bharatpur-I (0.40%), register sandy soil. Similarly, the red soil is found only in the Farakka block.

1.2.5.2 Soil Nutrient

The soil's nutrients can be categorized into two broader types based on the quality, chemical composition, and requirements. These are macronutrients and micronutrients. The correct proportions of macro- and micro-soil nutrients support healthy growth and production of crops. The nutrients that different plants and crops require in a large proportion are known as macronutrients of the soil. These are organic carbon (OC), nitrogen (N), phosphorus (P), and potassium (K). The elements of micronutrients include copper (Cu), iron (Fe), manganese (Mn), zinc (Zn), boron (B), and sulfur (S). These nutrients are required in a relatively small quantity than macronutrients in the soil. However, the role of these nutrients in plants' growth cannot be ignored.

The district is sufficient in terms of all four macro-soil nutrients (Fig. 1.5). Nitrogen in the soil is 83.04% sufficient, while the sufficiency of organic carbon (OC), phosphorus (P), and potassium (K) is 64.07%, 99.39%, and 83.14%, respectively. Every block in the district reports more than 50% sufficiency in terms of these four macro-soil nutrients (Table 1.4). Figure 1.6 demonstrates the sufficiency and deficiency of the six major micro-soil nutrients. Among these, iron (Fe), copper (Cu), manganese (Mn), and sulfur (S) are found to be sufficient in the Murshidabad district. Conversely, the soil lacks in terms of zinc (Zn) and boron (B).

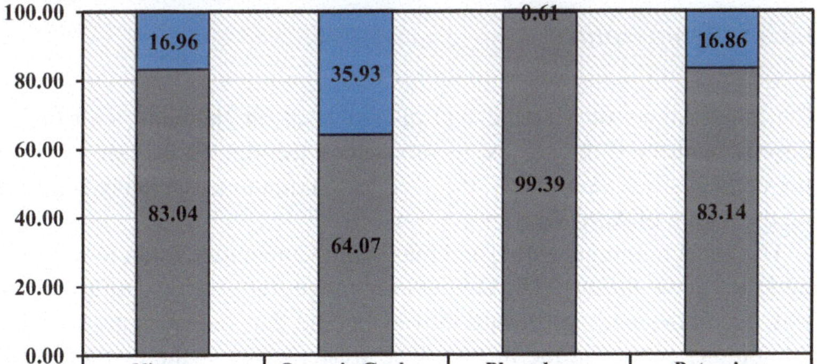

	Nitrogen	Organic Carbon	Phospherus	Potassium
Deficiency (%)	16.96	35.93	0.61	16.86
Sufficiency (%)	83.04	64.07	99.39	83.14

Fig. 1.5 Macro-soil nutrient status (2016–2017). (Source: Based on Annexure A.3)

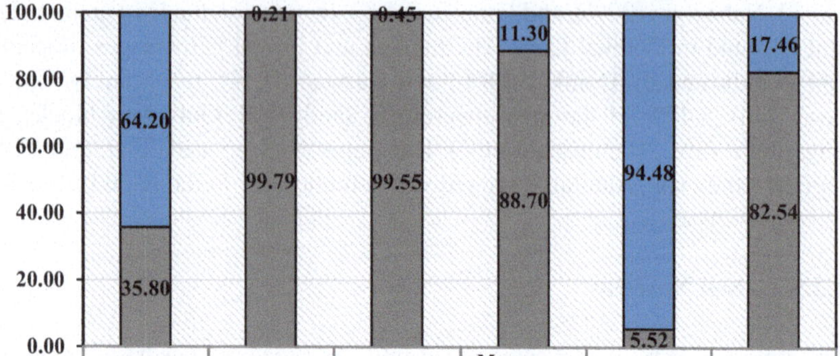

	Zinc (Zn)	Iron (Fe)	Copper (Cu)	Manganese (Mn)	Boron (B)	Sulphur (S)
Deficiency (%)	64.20	0.21	0.45	11.30	94.48	17.46
Sufficiency (%)	35.80	99.79	99.55	88.70	5.52	82.54

Fig. 1.6 Micro-soil nutrient status (2016–2017). (Source: Based on Annexure A.3)

1.3 Climate Condition

An oppressive hot summer climate with abundance humidity throughout the year prevails in the Murshidabad district. The district gets a fairly even distribution of rainfall due to the prevailing southwest monsoon. The precipitation of the monsoon season (June–September) contributes around 75% of the total annual rainfall. Average yearly rainfall stands at 1328 mm in the district. There are four seasons in the district: the relatively cold and dry winter season (from mid-November to the end of February), the summer season (March–May), the rainy or monsoon season (beginning of June to the end of September), and the post-monsoon season (October to mid-November).

1.3.1 Temperature

The temperatures of Murshidabad district keep on changing with the change of seasons. January is the coldest month, which records the average lowest temperature in the district. Conversely, the average maximum temperature is measured during the month of April–May.

The mean minimum (MM) and maximum temperatures start to fall from August until the end of January. The temperature range starts to increase from February to May due to the sun's northward movement and thus indicates the beginning of the hot summer season in the district. The western part of the district (Rarh region) records relatively higher temperatures throughout the year than the eastern part (Bagri region). As per the records, the highest temperature, 46.1 °C, was recorded in Berhampore in 1961 (District Census Handbook, 2011). Figure 1.7 reveals the mean annual maximum and minimum temperature of the district (from 1991 to

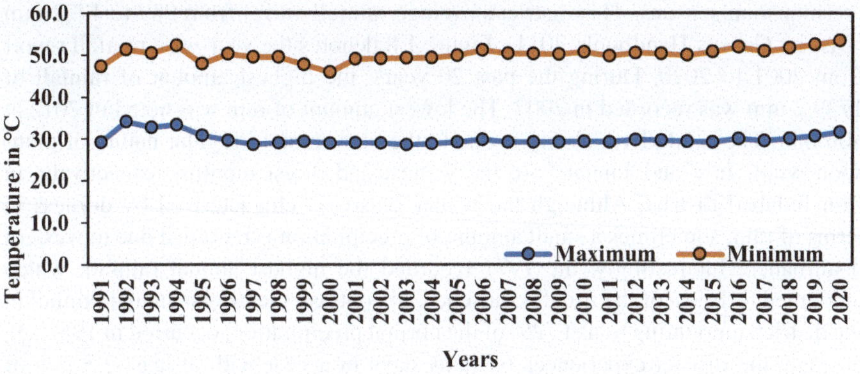

Fig. 1.7 Trends of mean annual temperature. (Source: Based on Table 1.2)

Table 1.2 Trends in mean annual minimum and maximum temperature in Murshidabad district

Year	M.M. temp.	M.M. temp.	Year	M.M. temp.	M.M. temp.
1991	29.2	18.2	2006	29.9	21.3
1992	34.2	17.2	2007	29.3	21.1
1993	32.8	17.7	2008	29.3	21.0
1994	33.4	18.9	2009	29.3	20.9
1995	30.9	17.1	2010	29.4	21.3
1996	29.7	20.6	2011	29.3	20.5
1997	28.7	20.9	2012	29.8	20.7
1998	29.1	20.5	2013	29.7	20.6
1999	29.4	18.4	2014	29.4	20.6
2000	29.0	17.0	2015	29.3	21.5
2001	29.1	20.0	2016	30.1	21.8
2002	29.0	20.2	2017	29.6	21.3
2003	28.6	20.7	2018	30.2	21.5
2004	28.9	20.5	2019	30.8	21.7
2005	29.3	20.6	2020	31.7	21.8

Source: Deputy Directorate of Agriculture (Admin.), Murshidabad (2021)

2020), which indicate that the district is experiencing a phenomenon of climate change. Global warming is affecting the climate change scenario worldwide and so is the Murshidabad district.

1.3.2 *Rainfall*

The district sees uneven and irregular rainfall distribution throughout the year. Rainfall mainly occurs from the southwest monsoon. The monsoon season brings 75–80% of the yearly rain; the rest happens during summer, winter, and pre-monsoon seasons. This district's average rainfall varies from 989 to 1722 mm (District Census Handbook, 2011). Figure 1.8 denotes the year-wise rainfall report from 2001 to 2020. During the past 20 years, the highest amount of rainfall of 1739.2 mm was recorded in 2007. The lowest amount of rain was noted in 2012 at 806 mm. The rainfall variability is due to the uneven and irregular nature of monsoon wind. July and January are the wettest and driest months, respectively, in Murshidabad district. Although the winter season is characterized by dryness in terms of rain, sometimes a small amount of precipitation is recorded due to western disturbance. Interestingly, the 1971 recorded the highest annual rainfall, which amounted to 169% of the usual occurrence. In contrast, the lowest annual rainfall in the district, amounting to just 72% of the normal precipitation, occurred in 1962. On average, the district experiences 64 rainy days in a year with at least 2.5 mm of rainfall.

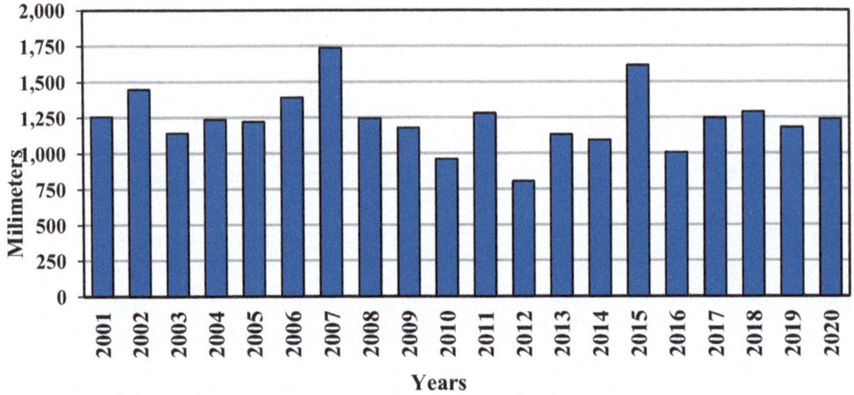

Fig. 1.8 Annual rainfall report. (Source: Based on Table 1.3)

Table 1.3 Year-wise rainfall report from 2001 to 2020

Year	Total rainfall (mm)	Year	Total rainfall (mm)
2001	1257.3	2011	1282
2002	1449.6	2012	806
2003	1141.6	2013	1131.9
2004	1236.6	2014	1093.4
2005	1221.4	2015	1616.1
2006	1392.9	2016	1006.3
2007	1739.2	2017	1248.4
2008	1245	2018	1290.3
2009	1178.5	2019	1181.7
2010	961	2020	1240.1

Source: Deputy Directorate of Agriculture (Admin.), Murshidabad (2021)

1.3.3 Humidity

The relative humidity is the percentage of water-vapour content in the atmosphere at a given temperature. It is the ratio of the actual content of water vapour to the maximum holding capacity in the atmosphere. Humidity plays a significant role in local weather conditions. The percentage of relative humidity remains high throughout the year. More specifically, the period of July–September records the highest relative humidity than the rest of the year due to the arrival of the monsoon season. This monsoon season also registers the least diurnal variation of relative humidity. The relative humidity during this season varies from 80% to 85%. Conversely, afternoons of the summer season have the lowest relative humidity (38–50%).

1.3.4 Cloudiness

The cloudiness condition in the district is maximum during the monsoon season, particularly in the month of July and August. In these 2 months, the sky remains overcast for 8–9 days per month, and generally, not a single day records a cloud-free sky. Cloudiness starts to decrease from October all over the district, and from December to March, the sky represents cloudless or lightly clouded.

1.3.5 Winds

During the summer and monsoon seasons, the winds are light, with a slight increase in force. The wind mostly blows from south to southeast directions during these seasons. While in the later part, winds are sometimes westerly in the morning. The winds are mainly calm in the afternoon and variable in the morning during the post-monsoon and winter seasons. The wind speeds are also higher during the monsoon season compared to the rest of the year.

1.3.6 Special Weather Phenomena

A few storms and depressions are created in the Bay of Bengal during the month of May and post-monsoon seasons almost every year. These storms and depressions often hit the district, causing high winds and widespread heavy rain. Depressions also originated during the southwest monsoon, which affects the district adversely with heavy rain. Thunderstorms are prevalent in the hot weather months of the summer season. These thunderstorms are sometimes violent and often accompanied by heavy showers, occasional hail, and storms coming from the northwest direction. Hence, these phenomena are known as nor'wester, locally called 'Kalbaisakhi'. The temperature suddenly falls down due to this phenomenon. Occasional patches of fog occur in the winter season.

1.4 Flora and Fauna

Just like the other deltaic districts of West Bengal, Murshidabad district also reports scattered flora. Numerous marshy species of plants are available in the swampy areas. Bamboo is commonly found all over the district. Similarly, the district is abundant with mangos, jackfruits, Segun, Sisoo trees, Babla, Pitali, Tentul, etc. Aswattha, Bot, Khend Sal, Mohua, and Palas are found in only a few parts of the district. Mulberry cultivations, popular for sericulture activities, are predominantly found in Beldanga, Berhampore, and Islampur areas. Wide varieties of mangos are available in the district, whose origin can be traced back to the days of Nawabs. These are Shadulla, Himsagar, Ranipasand, Mulayamjam, Kohitur, etc. The Murshidabad district contributes a small area under the forest covering only 770 ha, constituting 0.14% of the total geographical area (District Census Handbook, 2011). Therefore, there is good scope for forestry in the district. The district has approximately 153 sacred groves (small forest patches). Within which many trees, shrubs, bushes, herbs, insects, reptiles, birds, and mammals are found. About 30 medicinal plants, commonly used by traditional healers, are also noticed in these groves.

Although the bigger wild animals, such as tigers, rhinoceroses, a few wild buffalos, antelopes, and deers, were reported in 1857 by the Revenue Surveyor, they have entirely disappeared at present. Leopards are scarily found in Hariharpara. The other wild animals found in the district are wild cats, monkeys, jackals, wild ducks, snipes, pigeons, teals, etc. Cobra, dararash, karait, etc., are the varieties of snakes that are found all over the district and mostly appear after the occurrence of floods. The numerous water bodies in the district attract various families of birds and ducks. The district breeds different kinds of fish in rivers, ponds, beels, and lakes for local consumption as well as limited export. There is no wildlife sanctuary or national parks located in the district.

1.5 Demographic Profile

The potential and role of human resources are important for the growth and development of social and cultural development of a region. The following sections, therefore, briefly discuss population distributions, density, age-sex composition, and literacy.

1.5.1 Distribution

According to the 2011 census, 7,103,807 people live in the Murshidabad district. It contributes 7.78% of the total population of West Bengal and shares approximately 6% of the state. Annexure A.4 reveals that the Berhampore block contributes the maximum population, i.e., 642,110, among the blocks (9.04% district's population). With 158,024 number of peoples, the Bhagwangola-II block accounts for the least share of 2.22 of the district population. Sagardighi (345.42 km^2) and Samserganj (84.21 km^2) are the largest and smallest blocks, respectively, in terms of area coverage.

1.5.2 Density

Population density refers to the number of people living per 1 km^2 of area. According to the 2011 census, the density of the district stood at 1334 persons per km^2. Samserganj (3373) block appears to be the densest in the district, followed by Suti-II (2510), Raghunathganj-II (2182), and Farakka (2065). In Fig. 1.9, it is clear that the densest blocks are concentrated in the northern part of the district due to the fertile plain alluvial track of the Ganga and Bhagirathi rivers. Conversely, the Nabagram block (742) has the minimum population density, immediately followed by Khargram (858), Burwan (859), and Sagardighi (899).

1.5.3 Literacy

A person aged exactly seven or above, who can read and write with an understanding of any scheduled language, is treated as literate (Census of India). So, to calculate the literacy rate, first of all, we have to exclude the population below the age of seven from the entire population and then take the ratio of literate persons to the total population above the age of seven. In 2011, the district's literacy rate stood at 66.59%, in which the male and female literacy rates were 69.95% and 63.09%, respectively. The gender disparity in literacy is identically same with national pattern or neighbour districts like Malda (Rahaman & Rahaman, 2018; Rahaman, 2021). Figure 1.10 demonstrates the spatial pattern of literacy rates in the blocks where Berhampore (73.51%) and Samserganj (54.98%) are the most and least literate blocks, respectively, in the district.

Berhampore block once again ranks first in terms of the highest female as well as male literacy rates (70.34% and 76.52% correspondingly). In contrast, the minimum male and female literacy rates (60.42% and 49.57%, respectively) are noted in the Samserganj block. All the blocks (except for Berhampore) in the district lag behind the national (72.04%) and state average (77.08%) (Census of India, 2011).

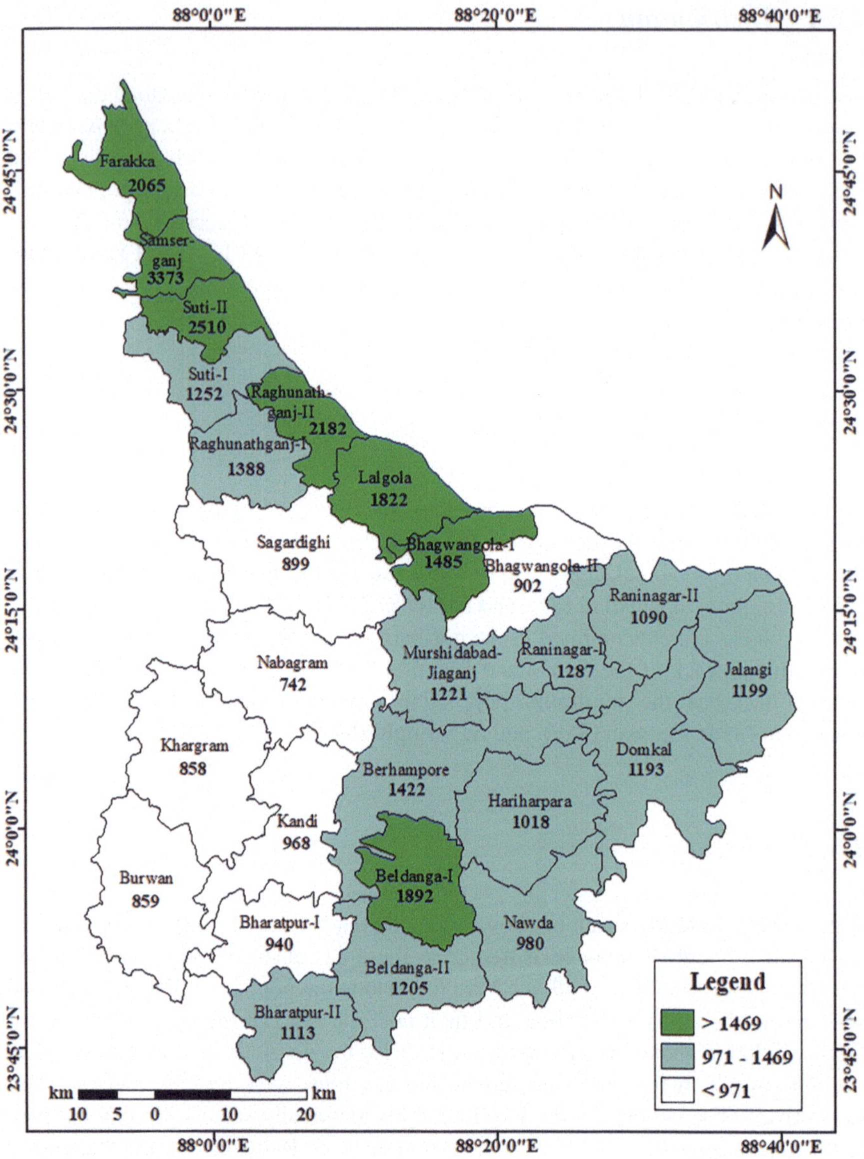

Fig. 1.9 Population density (2011). (Source: Based on Annexure A.4)

The district reports a 6.86% gender gap in literacy (Annexure A.4). Eight blocks, namely Kandi, Khargram, Burwan, Farakka, Nabagram, Samserganj, Suti-II, and Bharatpur-I, denote the male–female literacy gap of more than 10%, in which Kandi block notifies the maximum (13.41%). While the minimum gender gap is recorded by Bhagwangola-II (0.48%), followed by Domkal (1.16%), Hariharpara (1.58%), and Nawda (1.67%).

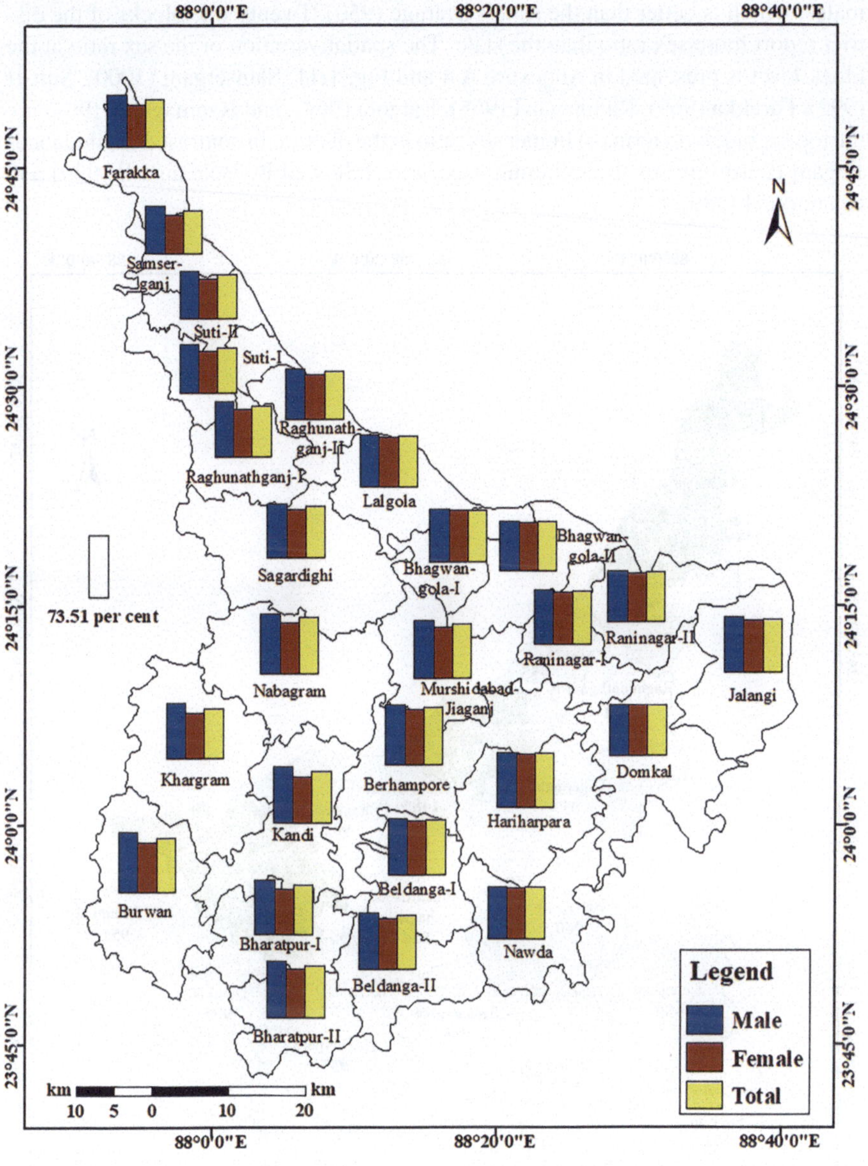

Fig. 1.10 Literacy rate (2011). (Source: Based on Annexure A.4)

1.5.4 Sex Ratio

The sex ratio is an essential demographic indicator of the population in a region. Generally, the number of females per 1000 males is considered as sex ratio. According to the 2011 census, the district has a sex ratio of 958 females per 1000 males, which is better than the state's average (950). Twenty-one blocks of the district report more sex ratio than the state. The spatial variation of the sex ratio at the block level is presented in Annexure A.4 and Fig. 1.11. Samserganj (1000), Suti-II (992), Farakka (969), Raninagar-I (965), Lalgola (964), and Raninagar-II (961) are the top six blocks in terms of higher sex ratio in the district. In contrast, Murshidabad-Jiaganj (936) informs the minimum sex ratio, followed by Beldanga-II (939) and Bharatpur-I (939).

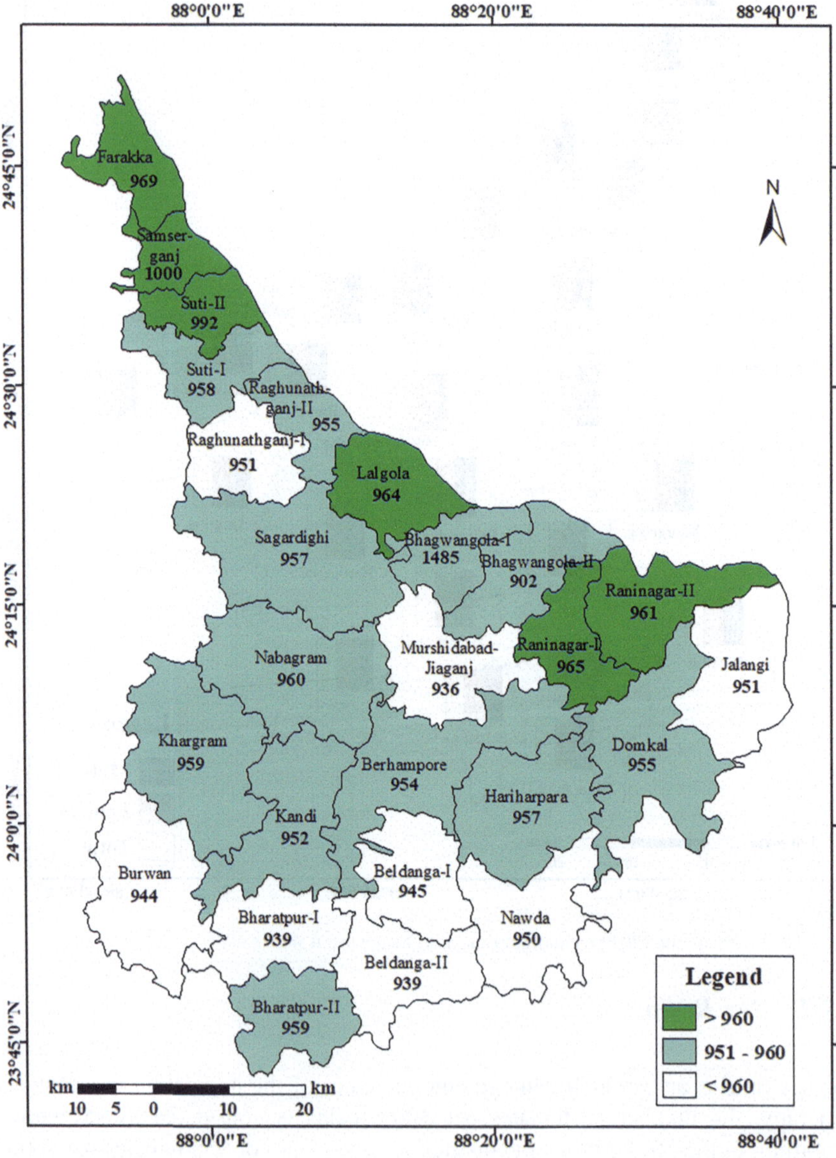

Fig. 1.11 Sex ratio (2011). (Source: Based on Annexure A.4)

1.5.5 Age-Sex Composition

Population characteristics such as age-sex compositions are essential to determine the planning and development of any region. The crucial aspects of the rate of fertility, mortality and migration play determinant roles in the age-sex composition of the population in an area.

Figure 1.12 displays the age-sex pyramid of the population for the Murshidabad district. Table 1.4 states that 60.18% of the district's population is in the working age group (15–59 years). In the same working-age population, the male population locates at 60.75% and females at 59.58%. The share of the female population surpasses the male population in terms of the 'aged group population' (>60 years). Interestingly, the same is the case with the below 15 years' age group. The young or juvenile population (below 18 years) shares 43.79% of the district's population.

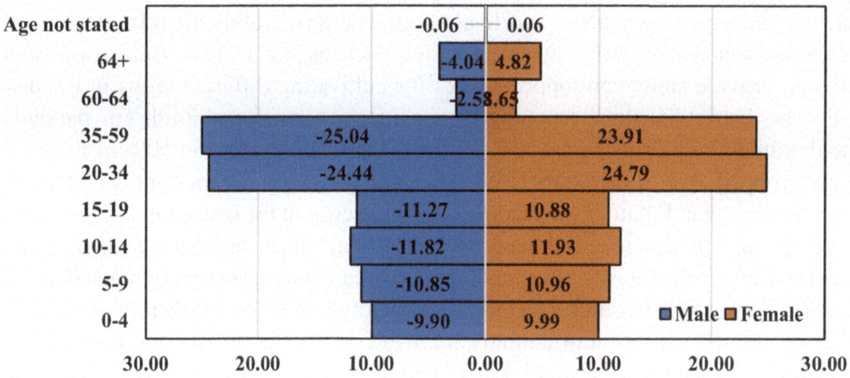

Fig. 1.12 Age-sex pyramid (2011)

Table 1.4 Age-sex composition of population in Murshidabad district (2011)

	Male		Female		Total	
Age groups (year)	Number	Percentage	Number	Percentage	Number	Percentage
0–4	359,130	9.90	347,405	9.99	706,535	9.95
5–9	393,757	10.85	381,098	10.96	774,855	10.91
10–14	428,602	11.82	414,886	11.93	843,488	11.87
15–19	408,705	11.27	378,087	10.88	786,792	11.08
20–34	886,469	24.44	861,856	24.79	1,748,325	24.61
35–59	908,417	25.04	831,193	23.91	1,739,610	24.49
60–64	93,551	2.58	92,278	2.65	185,829	2.62
64+	146,677	4.04	167,460	4.82	314,137	4.42
Age not stated	2256	0.06	1980	0.06	4236	0.06
All ages	3,627,564	100.00	3,476,243	100.00	7,103,807	100.00

Source: District Statistical Handbook, Murshidabad (2016)

Table 1.5 Percentage of landholders and area of holdings in Murshidabad district, 2016–2017

Size groups (HA)	Holders	Area	Average size of holdings
Marginal (below 1.0)	82.044	56.220	0.51
Small (1.0–1.99)	14.765	31.313	1.59
Semi-medium (2.0–3.99)	3.098	11.838	2.86
Medium (4.0–9.99)	0.090	0.580	4.85
Large (10 and above)	0.004	0.049	10.14
All groups	100.000	100.000	0.75

Source: Input Survey, Agriculture Census (2016–2017)

1.6 Economic Aspects

1.6.1 Agriculture

More than 81% of the rural population of Murshidabad district depend on agriculture for their earning sources and livelihood. The fertile alluvial soils and natural irrigation sources of different river plains, such as the Ganga, Bhagirathi, and Jalangi, provide significant opportunities for cultivating different crops in the district. The district has three cropping seasons: Rabi, Kharif, and Zaid. The paddy is the dominant crop growing twice a year, which covers more than 50% of the total gross cropped area in the district. Wheat and maize are the important cereal crops grown every year. Potato is the leading vegetable crop in the district in terms of area coverage and production, followed by brinjal, cabbage, cauliflower, tomato, etc. The top fibre crop in terms of production and area is jute. Most marginal and small farmers cultivate fruits and vegetables for income. Besides cultivation, livestock rearing and domestication of animals are also common in the district. Pisciculture in wetlands, ponds, beels, and lakes is another good source of income for some households.

The marginal and small land size holders account for more than 96% of the total farmers in the district, and their share of the cultivated area is around 87% (Table 1.5). The semi-medium, medium, and large land size classes hold 11.83%, 0.58%, and 0.05% of cultivable area, respectively, while the landholders belonging to these groups are 3.09%, 0.09%, and 0.04%, respectively. According to the 2016–2017 Input Survey, the average size of landholdings in Murshidabad district is 0.75 ha. It is 0.51 ha in the marginal, 1.59 ha in small, 2.86 ha in semi-medium, 4.85 ha in medium, and 10.14 ha in the large land classes (Table 1.5).

1.6.2 Industry

Heavy industries are not well-developed in the Murshidabad district. Here, only a few limited agriculture-based small industries are found, but these are also facing many institutional and infrastructural constraints. The main barrier to industrial development is the lack of mineral resources. These small-scale industries or cottage industries somewhat generate employment opportunities in the district. According to the 2011 census, there are 141 medium-type factories in the Murshidabad district where only 6053 persons are engaged, which is very low compared to the neighbouring districts (Table 1.6). The number of small industrial units in Murshidabad district was 2438, with an employment of 44,647 persons in 2010 (District Census Handbook, 2011). The tobacco industries, food processing industries, silk industries, beverage industries, manufacturing of paper and paper products, and chemical industries are prominent in the Murshidabad district.

1.6.3 Workforce

In the 2011 census, Murshidabad district reported a 2,589,907 and 4,513,900 persons as total workers and non-workers, respectively (Table 1.7). Thus, the work participation rate in the district is 36.46%, in which the male workers, 1,985,667 (76.67%), dominate the female workers, 604,240 (23.33%). The main workers, working for more than 180 days a year, share 78.07% of the total workers, where males and females constitute 83.73% and 59.49%, respectively. Conversely, the marginal workforce who works for less than 180 days a year accounts for only 21.93%. One of the common characteristics of the workforce is that the male denotes to the female workers in every aspect of the working categories. Only 14.71% of the total workforce are cultivators due to the limitation of cultivable land abridged by an ever-growing population. The agricultural labourers, household industries, and other workers report 32.52%, 17.99%, and 34.77%, respectively, of the total workforce.

Table 1.6 Number of factories and employees in Murshidabad district, 2011

Industries	No. of factories	No. of employees
Manufacture of food products	25	314
Manufacture of beverages	1	12
Manufacture of tobacco products	92	5368
Manufacture of textiles	1	63
Manufacture of paper and paper products	3	14
Manufacture of chemicals and chemical products	3	12
Manufacture of other non-metallic mineral products	3	80
Manufacture of basic metals	2	16
Manufacture of fabricated metal products, except machinery and equipment	4	23
Other manufacturing	2	78
Warehousing and support activities for transportation	5	73
All industry	**141**	**6053**

Source: District Statistical Handbook, Murshidabad (2016)

Table 1.7 Profession of working population in Murshidabad district, 2011

Description	Total population	Male	Female
Total workers	2,589,907	1,985,667	604,240
Main workers	2,022,083	1,662,624	359,459
Main workers—cultivators	346,049	336,521	9528
Main workers—agricultural labourers	627,126	597,348	29,778
Main workers—household industries	315,687	97,611	218,076
Main workers—others	733,221	631,144	102,077
Marginal workers	567,824	323,043	244,781
Marginal workers—cultivators	35,027	27,007	8020
Marginal workers—agricultural labourers	215,168	182,494	32,674
Marginal workers—household industries	150,320	19,900	130,420
Marginal workers—others	167,309	93,642	73,667
Non-workers	4,513,900	1,641,897	2,872,003

Source: Census of India (2011)

Table 1.8 Trends of road lengths in Murshidabad district

Year	Surface road (km)	Unsurfaced road (km)	Total lengths (km)
2006–2007	4060.68	4538.69	8599.37
2007–2008	4464.03	4577.45	9041.48
2008–2009	5075.58	4675.87	9751.45
2009–2010	5481.44	4675.52	10,156.96
2010–2011	5818.83	4747.24	10,566.07
2011–2012	6178.89	4626.61	10,805.50
2012–2013	5367.69	4666.70	10,034.39
2013–2014	6920.33	4504.46	11,424.79
2014–2015	7273.83	4593.38	11,867.21
2015–2016	7459.38	4712.86	12,172.24

Source: District Statistical Handbook, Murshidabad 2011 and 2016

1.7 Transport

Road and railways are the most crucial means and medium of the transportation system in Murshidabad district, as an almost flat terrain surface dominates the physiography of the district. NH 12 (previously known as NH34) is the only national highway that serves the district along with three other state highways, i.e., SH-07, SH-11, and SH-11A. The Farakka Barrage, situated at the northern point of the district, creates a vital link in road and rail communication. The significant railways, running from north–south, connect the district to North Bengal and Kolkata.

Table 1.8 and Fig. 1.13 represent the trends of road length (surfaced and unsurfaced) in the district from 2006–2007 to 2015–2016. The length of total roads has increased by 3572.87 km in that period. The surface road length has increased from 4060.68 km in 2006–2007 to 7459.38 km in 2015–2016. In the same time period, the unsurfaced road expanded from 4538.69 to 4712.86 km.

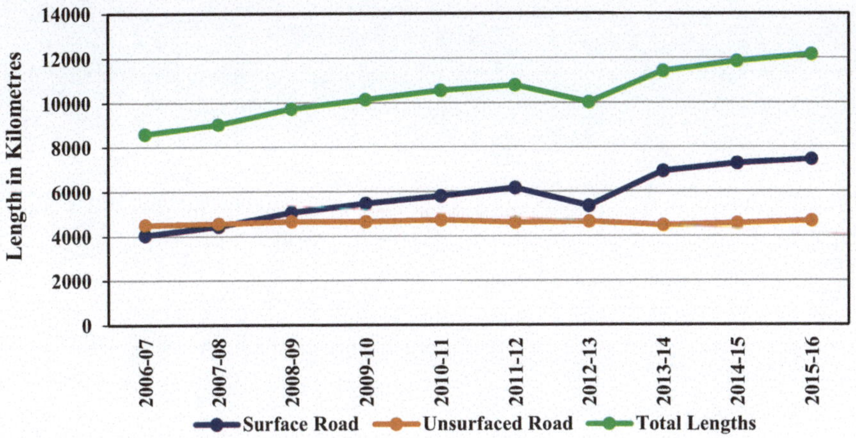

Fig. 1.13 Trends of road lengths

References

Census of India. (2011). District Census Handbook, Murshidabad, Directorate of Census Operations, West Bengal, series 20, Part XII-A.

O'Malley, L. S. S. (1997). Bengal district gazetteers: Murshidabad. Logos Press. (Original work published 1914).

Rahaman, H. (2021). Overview of the study area. In *Diversified cropping pattern and agricultural development*. Springer. https://doi.org/10.1007/978-3-030-55728-7_1

Rahaman, M., & Rahaman, H. (2018). Gender disparity in literacy in Malda District. *International Journal of Research in Social Sciences, 8*(10), 123–141.

Chapter 2
Sustainable Land Use Practices by Land Size Classes

Contents

H. Rahaman, S. Ansari, *Sustainable Land Use Practices and Agricultural Innovations in India*, https://doi.org/10.1007/978-3-031-83246-8_2

2.1 Land Use: Meaning and Concept

The word 'sustainable' is derived from Latin word 'sustinere', which means 'to hold up' or 'to endure'. As a part of speech, the word 'sustainable' is an adjective. This word emerged in the late twentieth century, focusing on the capacity to maintain something or endure over time, particularly in the context of environmental and resource management. Conversely, the word 'land' has its root in Proto-Germanic word 'landa', which means 'earth' or 'ground' and 'use' has derived from Latin word 'usus' meaning 'to use'. Altogether, the term sustainable land use refers to the practice of managing land resources in a way that assures the productivity as well as ecological balance over the long run. This involves using land in ways so that the resources and the environment do not get harmed, ensuring the continuous service for the current and future generations effectively. Land use is based on the functional dimension of land for different human purposes or economic activities (World Bank, 2005). The definitions and concepts of land use in the field of sustainability and development are numerous, but the most quoted and authentic definition is provided by Brundtland Report which defines 'humanity has the ability to make the development sustainable to ensure that it meets the needs of the present, without compromising the ability of future generations to meet their own needs' (Brundtland et al., 1987). The concept of sustainable land use has been widely accepted with the gradual extend of sustainable development in the field of land science. The notion of sustainable land use includes the rational development through proper use and protection of land resources on a particular space–time condition (Xie et al., 2020). Therefore, sustainable land use as a resource utilization method not only meets the needs of contemporary people but also secures the needs of the future generation without posing any risk.

The concept of land use has long been in the concerns of academic literature and moulded and revised its meaning since then. The term land use indicates the employment of land by humankind which includes settlement, pasture, cultivation, recreation, and so on (Meyer & Turner, 1994). Initially, the term land use is used in the language of city planning (Form, 1954; Guttenberg, 1959; Haar, 1957). With the passage of time, interest in the subject and knowledge of land use have expanded with more and more meanings added to it. After 1960s, the term 'land use' has been sometimes synonymously used with the term 'land cover' (Rahaman, 2021a). But there are several literatures on land use and land cover that clearly distinguish them in spatial and physical context. The term 'land cover' denotes the type of vegetation which covers the land surface including human structures, such as buildings or payment and other physical environmental aspects, i.e., soil, biodiversity, surfaces, and groundwater (Moser, 1996). While land use involves the manipulation of biophysical attributes of the land and the purpose and intension for the use of the land (Turner II et al., 1995). An international organization, Food and Agricultural Organization (1995) has provided a definition for land use. It stated 'land use concerns the function or purpose for which the land is used by the local human population and can be defined as the human activities which are directly related to the land, making use of its resources or having an impact on them'. The study of land use is a context driven in academic discussion, such as territorial scale, urban scale, agricultural, and so on. At the territorial scale, land is a resource involving larger land areas to yield raw materials for sustaining people and their activities. The urban scale focuses on the potentiality of land's surfaces in terms of location of various activities. In contrast, land use in agricultural scale is defined as the intensification of land through the inputs of capital, labour, and skills to gain greater concentration of production which is measured in terms of outputs per unit of land (Lambin et al., 2000). The rapid development of urbanization and industrialization, rapid growth of population, massive demand for resources and energy, and the ever-expanding urban scale land use have resulted in deterioration and degradation of ecological environment (Dadashpoor et al., 2019).

2.2 Models and Theories of Land Use

Modelling of land use change implies the use of artificial representation of the interaction within itself to unravel its dynamism and possible future development (Verburg et al., 2006).

1. Probably the first geographical model related to agricultural land use has been developed and published by Von Thünen (1966) in his book *The Isolated State*. In this model, he has explained that the net profit or economic rent as he prefers to say derived from any particular agricultural land use is the function of linear distance from any central market. The production of crops varies from the market according to their transport costs, land rent, perishability, and economic rationality. Therefore, different agricultural land use would form around a single central market with the fact that the intensity of cultivation will decrease with increasing distance from that market.
2. During the 1960s, an American scholar, Walter Isard (1952) has shown the relationship between industrial locational theories and agricultural locational theo-

ries in his work entitled 'A General Location Principle of an Optimum Space-Economy'. One of the minor differences he has indicated that while industrial location theory mainly deals with only one product, there is a set of products considered in agricultural locational theories. Another differentiation of both the locational theories is the cost of transporting raw materials. It is very important in case of industrial location, whereas it is usually equated to zero in the theories that deal with agricultural location.

3. Garrison and Marble (1957) developed an agricultural model regarding the spatial structure of agricultural activities where they proposed an equilibrium system for the optimum agricultural land use. Rather than analysing on an industry level, this model focuses on individual entrepreneur for optimum solution of maximum net returns from agricultural land use. This model further states that the intensity of cultivation increases continuously as we approach the market when a single land use and production function are under consideration. However, if either multiple land uses or multiple production functions are considered, there is a possibility of lower intensities of land use closer to the market than those further away.
4. A simple Markov-based conceptual model of agricultural land use was presented by Thornton and Jones (1998). They have designed this to construct a top-down land use model based on fewest processes that are useful for statistical analysis of land use change in a region. In this model, the cropping pattern and the effect of initial household size on wealth distribution appear to be a reasonable fashion within a simulated landscape.

2.3 Sustainable Development and Agricultural Land Use

Sustainable development is associated with the maintenance of development over time. Currently, there are several definitions of this terminology and hundreds of them are applied in practices (Elliott, 2012). The term 'sustainable development' has been extensively used and is firmly established in the political domain of international development thinking, as mentioned in the report of the World Commission on Environment and Development, entitled "Our Common Future" (WECD, 1987). It is defined as 'Development that meets the needs of the present without compromising the ability of future generations to meet their own needs' (Brundtland et al., 1987). In 2000, sustainable development as a policy goal was included as one of the eight 'Millennium Development Goals' (MDGs) that had to be achieved by 2015.

The objectives of providing income through the production of food and fibre in agricultural land use have been upgraded due to the increasing concerns of environment. The concept of sustainability is thus included in agricultural land use in the form of mixtures of long- and short-term goals (Zander & Kächele, 1999). The sustainable agricultural land use can be defined as the production of high-yielding crops while maintaining the preservation and well-being of environment and local peoples (Aznar-Sánchez et al., 2019). It deals with the social concerns for achieving sustainable development by reducing food waste, improving food accessibility, and changing diets (Borras Jr & Franco, 2012).

Tanrivermis (2003) have examined that intensification of agricultural land use geared up by agricultural mechanization, use of agri-chemicals, rapid population growth, industrialization, urbanization, etc., leads to the deterioration of soil quality as well as the environment. Therefore, he suggested new legal and institutional regulations which integrate agricultural policies and environmental concerns for sustainable land management. The consequence of agricultural expansion by overexploitation of forest causes serious ecological and environmental problems in Kerala (Kumar, 2005). Agroforestry holds promise in order to mitigate these problems. Therefore, it has been suggested that the state should provide attention for the adoption of agroforestry practices. A study on 'Multifunctional Urban Agriculture for Sustainable Land Use Planning in the United States' by Lovell (2010) discussed that urban agriculture offers a viable alternative in city areas for improving sustainability of land use. The benefit of agroforestry practices for the sustainable land use system in Nigeria has been discussed by Alao and Shuaibu (2013). They concluded that optimal use of land for agriculture along with forestry production not only increases food and sustainable wood production but also improves the soil quality. Srivastava et al. (2016) proposed a commercial ecological agriculture with least disturbing, resilience building, resource use efficient, labour and skill intensive, low input, and diversified to achieve land use sustainability of agro-ecosystem. It would improve and restore the soil quality as well as increase the productivity.

2.4 Land Use Classification in India

The statistics related to land utilization, providing the detailed distribution of the country's total geographical area into various uses, are essential as it determines the judicious use of available land resources in a region and finds out the existing land use problems prevailing in the country. The purpose of the land use classification is, therefore, to maximize the productivity of the land and its conservation for future generations (Mandal, 1990). The scientific and methodologically considerable classification of the present land use provides an understanding of the best use of each parcel of land. Thus, land use classification is not an end but a means of obtaining better use of land (Burley, 1961).

Land utilization statistics regarding land put to different uses have been more or less continuously available in India since 1884. In 1884–1885, the 'Returns of Agricultural Statistics of British India' was started to give details of cultivated and uncultivated areas and how much areas were irrigated. Besides, land under cropped, under current fallows, land available for cultivation, and the area under forests were also gathered.

The coverage of more geographical area under land use statistics has continuously increased from 1884–1885 to 1922–1923, but this remains stagnant during the follow-up years until after World War II. Owing to India's evolving territorial boundaries - including the separation of Burma, partition, and subsequent state reorganizations - consistent long-term data on land utilization patterns remain unavailable. In India, land utilization statistics were available under five categories till 1950–1951. They were:

1. Forest
2. Area not available for cultivation
3. Other uncultivated land excluding current fallows
4. Fallow land
5. Net sown area

It was found that this five-fold classification is not sufficient to meet the needs of agricultural planning. Therefore, the Technical Committee on Co-ordination of Agricultural Statistics was set up in 1948 by the Ministry of Food and Agriculture to remove the primacy and to break up the five-fold classification into smaller categories for better comprehension. In 1951, the Indian Council of Agricultural Research (ICAR) has recommended a 'nine-fold' land use classification and also put forward standard concepts and definitions for each land use class to be followed by all states in the country. This nine-fold classification was widely used in all scientific studies in the last century. They are:

1. Forest
2. Land put to non-agricultural uses
3. Barren and uncultivated land
4. Permanent pastures and other grazing lands
5. Land under miscellaneous tree crop
6. Cultivable waste
7. Fallow land other than current fallow
8. Current fallow
9. Net sown area

The ICAR land use classification is primarily based on whether a particular area is cultivated, grazed, or forested. The primary purpose here is to show the classification of the existing land in detail by its actual usage and not how a particular piece of land can potentially be utilized (Nath, 1969). Thus, the area under culturable wasteland is excluded from the actual culturable area, as it may not be possible to bring a large part of the area under cultivation, except at a huge cost. Therefore, the potential land use classifications are beset by several difficulties, as these classifications depend on the suitability of different regions for different uses, taking into account their natural as well as cultural endowments.

This 'nine-fold' classification of land use can be broadly classified into agricultural and non-agricultural uses. Non-agricultural use includes forest, land put to non-agricultural uses, and barren and uncultivable land. In contrast, permanent pastures and other grazing lands, land under miscellaneous tree crops, cultivable waste, fallow other than current fallow, current fallow, and net sown area are classified under agricultural land uses.

2.5 Types of Agricultural Land Use

The state governments of India have followed the nine-fold classification of land use provided by the Indian Council of Agricultural Research (ICAR) for land records. The present study has carefully adopted the 'nine-fold' classification of land use published by the Agriculture Census of India. The Agriculture Census land use classification has reached mainly three categories: the uncultivated area, the area under current fallow, and the net sown area. The fallow and uncultivated land are further classified into different categories. In brief, the nine-fold classification provided by the Agriculture Census is mainly focused on the agriculture perspective. Therefore, this adopted classification is more meaningful in the present analysis.

The 2015–2016th Agriculture Census denotes that around 88.73% of the total reporting area in India is under net sown area. Similarly, in West Bengal the net sown area accounts for 94.30% and in the district of Murshidabad, this figure goes up to 97.93%. This shows a more intensive cultivation of crops in the district throughout the year. It is then obvious that the percentage of current fallow land remains maximum in India (7.48%) than West Bengal (3%) and Murshidabad (0.85%). Again, India registers 2.79% of the total reporting area as total uncultivated land in the form of other uncultivated land excluding fallow land, fallow land other than current fallow, and culturable wasteland. In case of West Bengal and Murshidabad, the total uncultivated land reports 2.70% and 0.39%, respectively.

The total reporting area is almost equally distributed across different land size categories (24.03% in marginal, 22.91% in small, 22.84% in semi-medium, 20.16 in medium) except for large land class (9.07%) in India. This scenario is entirely different in West Bengal and Murshidabad district. More than half of the reporting area is cultivated by marginal farmers alone (52.39% in West Bengal and 52.87% in Murshidabad). Another 28.31% and 32.45% of reporting areas in West Bengal and Murshidabad district, respectively, are recorded by small land size category. This higher percentage of area share is because of the huge dominance of marginal and small farmers which is more than 96% of the total farmers.

The abovementioned nine-fold classification of agricultural land use by Agriculture Census across land size class category in Murshidabad district is individually explained below.

2.5.1 Total Area (TA)

According to Agriculture Census, the total area includes cultivated and uncultivated areas, with at least a part of it being put to agricultural practices during the reference period. The land occupied by the farm building, including the holder's house, is also added to the total area. It is essential to mention that government forest land, government wasteland, village common grazing land, and 'Abadi Area' (residential area) are entirely excluded from the total area of the holding.

Figure 2.1 reveals an increasing trend in the total reporting area in Murshidabad district because of the redefinition of the district boundary with Malda district in 2004. The Agriculture Census shows that the total reporting area was 441,779 ha in 1995–1996, and it increased to 442,677 ha in 2015–2016. Figure 2.2 displays that under land size class categories, the highest negative growth of the area is recorded in the semi-medium land size category (32,748 ha), followed by medium (12,482 ha) and large (241 ha) land size class categories. However, the marginal

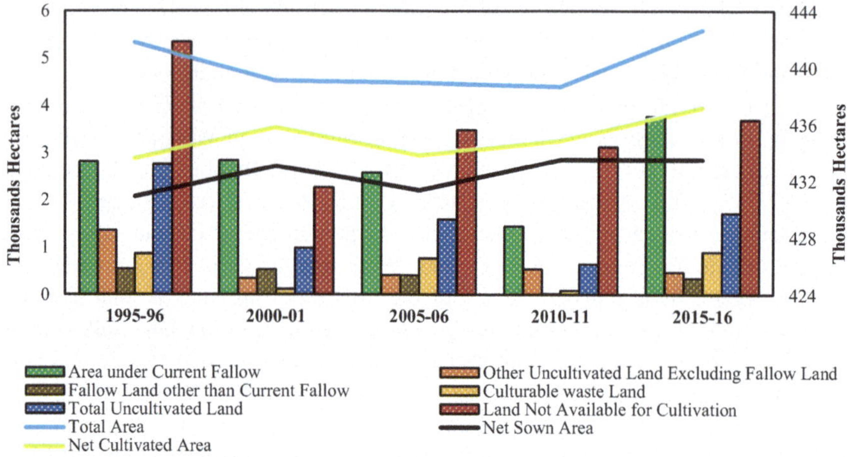

Fig. 2.1 Trend and pattern of land use in Murshidabad district from 1995–1996 to 2015–2016. (Source: Based on Annexure B.1)

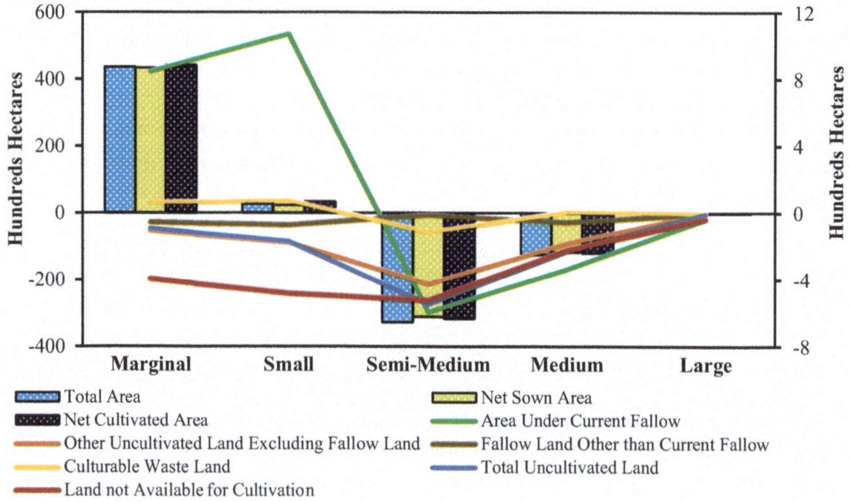

Fig. 2.2 Land size class-wise changes in land use in Murshidabad district from 1995–1996 to 2015–2016. (Source: Based on Annexure B.2)

and small land size class categories report a positive change with 43,670 and 2701 ha, respectively.

The block-wise data indicates that the top positive changes were observed in Sagardighi (8705 ha), Suti-II (3343 ha), Farakka (1088 ha), Bhagwangola-II (252 ha), Bhagwangola-I (70 ha), and Domkal (6589 ha) under marginal, small, semi-medium, large, and all land size class categories sequentially. Conversely, Khargram scores the highest but negative changes with 6299 ha area in marginal land class and 9778 ha in all land size class category from 1995–1996 to 2015–2016. The maximum negative changes under small and semi-medium land size categories are found in Sagardighi (5352 and 6069 ha, respectively). With a change of 1481 ha, Nabagram ranks top in negative change under the medium land class. Moreover, Berhampore (121 ha) posits the maximum negative change under the large land size class.

It is to be noticed that the data for 2020–2021 Agriculture Census are not available. Therefore, the data for 2020–2021 have been extrapolated. For the extrapolation, first, the data for 1995–1996 and 2015–2016 are interpolated, and then the average of the last 3 years was taken as actual data for 2020–2021. In the year 2020–2021, the total area in the district has slightly reduced (442,651 ha). The maximum and minimum areas are reported by Sagardighi (30,862 ha) and Raghunathganj-I (5257 ha), respectively (Fig. 2.3).

2.5.2 *Net Sown Area (NSA)*

Net sown area may be defined as the total cultivated area sown with crops and orchards during the reference year, irrespective of the fact that how many times it is being cultivated in a year. Thus, the area cultivated more than once during the same year will be counted once to calculate the net sown area. The NSA is the actual land for cultivation for agricultural purposes in a year of a region (Rahaman, 2021b).

It is seen from Fig. 2.1 that the changes of the net sown area remain positive in the district, where it has risen from 430,879 ha in 1995–1996 to 433,503 ha in 2015–2016. The main reason for increasing NSA is due to bringing of waste and fallow land for cultivation after reclamations during the past 20 years. Figure 2.2 highlights the significant increases in the area under marginal (43,306 ha) and small (2274 ha) land size class categories. The net sown area has declined in the remaining land classes, in which the semi-medium land size class indicates the maximum reduction, i.e., 31,090 ha from 1995–1996 to 2015–2016. This scenario happens due to the fragmentation of landholdings.

Figure 2.4 shows that Sagardighi posits the highest increase (8622 ha), while Khargram (6211 ha) block records maximum negative changes of NSA in marginal land size class during 1995–1996 to 2015–2016. Under the small land size class category, the top positive and negative changes are found in Suti-II (3624 ha) and Sagardighi (5399 ha), respectively. The highest decline of NSA in the semi-medium land class is again reported by Sagardighi (5979 ha), and Farakka (1113 ha) gains

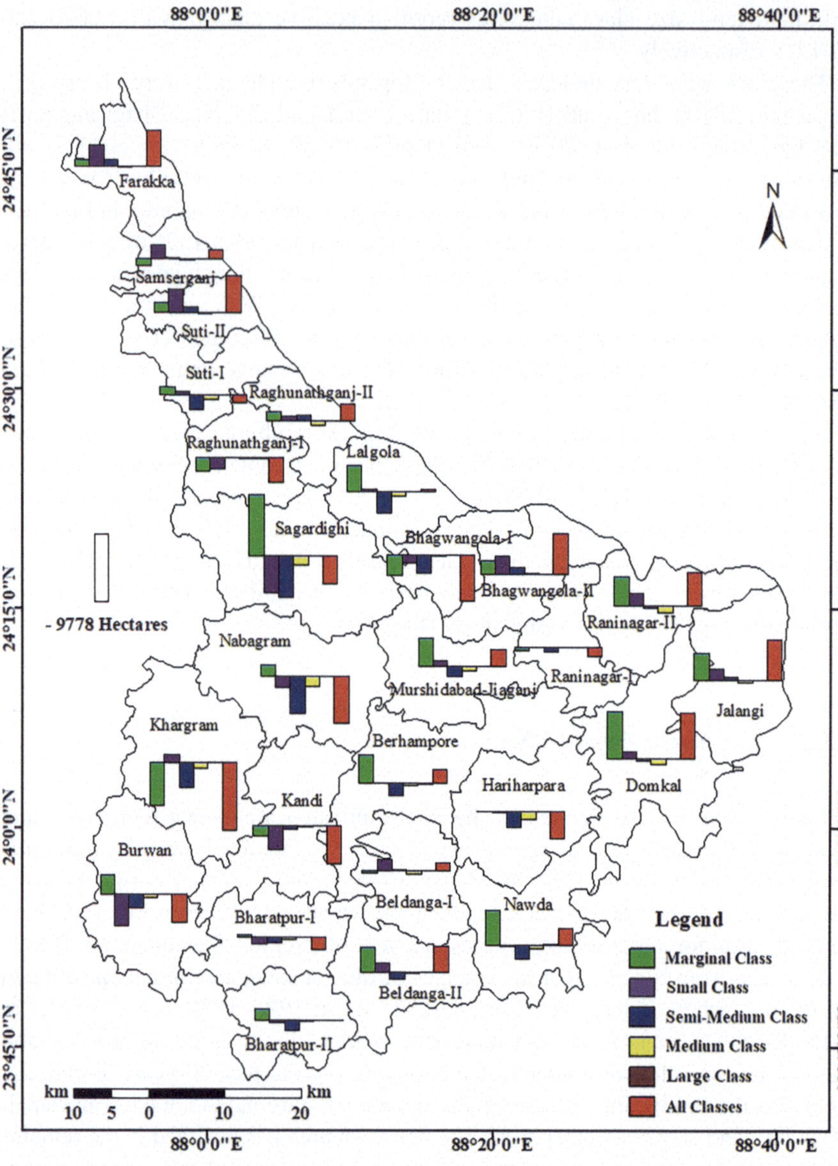

Fig. 2.3 Land size class-wise changes in total area in Murshidabad district, 1995–1996 to 2015–2016. (Source: Based on Annexure B.3)

maximum area. All the blocks except Bhagwangola-II (242 ha increased) experience negative changes of NSA under medium land class, out of which Nabagram reports the full deduction of 1494 ha. Bhagwangola-I (70 ha) gains the maximum area in the large land size class, and Berhampore (97 ha) denotes the biggest negative changes. Overall, 14 out of 26 blocks add area under this land classification

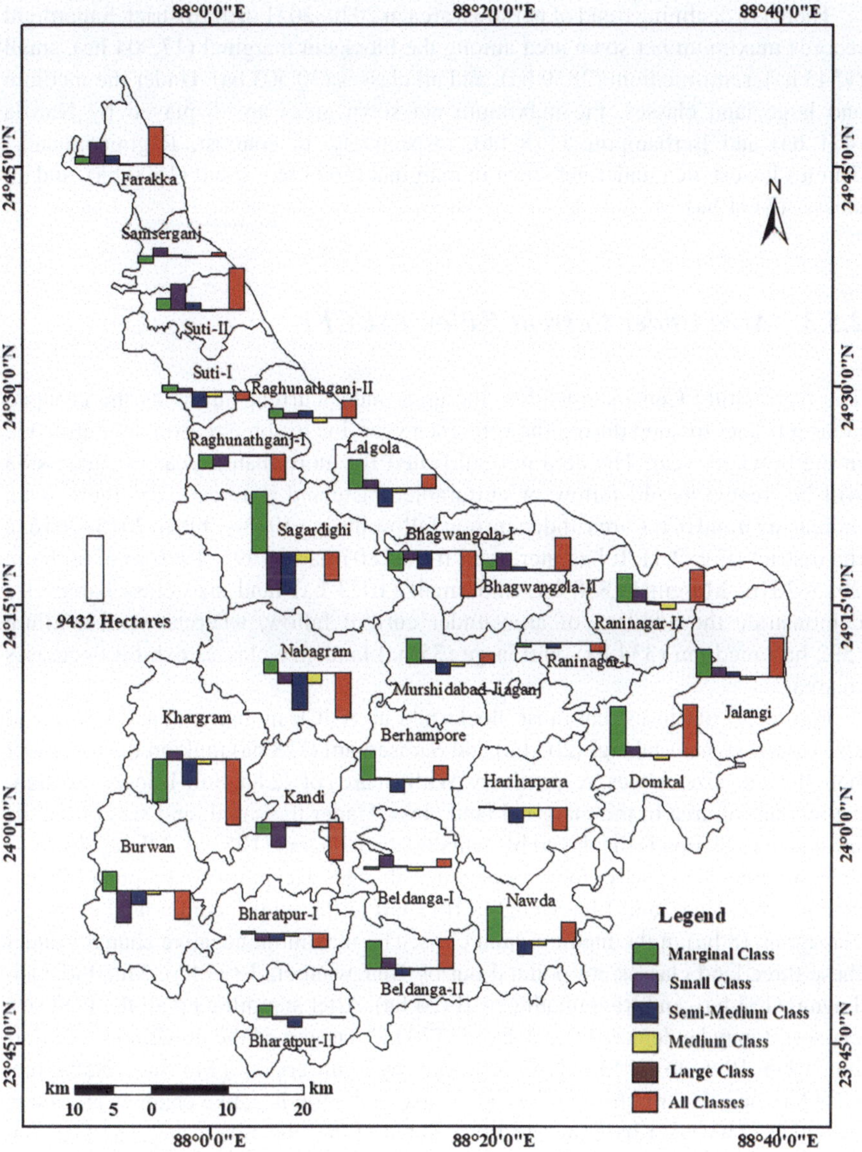

Fig. 2.4 Land size class-wise changes in net sown area in Murshidabad district, 1995–1996 to 2015–2016. (Source: Based on Annexure B.4)

from 1995–1996 to 2015–2016. Domkal (7179 ha) registers maximum gain, followed by Suti-II (5995 ha), Jalangi (5549 ha), Farakka (5313 ha), and Raninagar-II (4731 ha). Contrastingly, Khargram (9432 ha), Nabagram (6514 ha), Bhagwangola-I (6438 ha), and Kandi (5539 ha) have noted the maximum negative changes in ascending order.

There is a declining trend of net sown area in 2020–2021 in the district. Sagardighi records maximum net sown area among the blocks in marginal (17,904 ha), small (9543 ha), semi-medium (2839 ha), and all classes (30,503 ha). Under the medium and large land classes, the maximum net sown areas are displayed by Nawda (471 ha) and Berhampore (178 ha), respectively. In contrast, Raghunathganj-I denotes lowest area under net sown in marginal (1673 ha), small (1500 ha), and all classes (5141 ha).

2.5.3 Area Under Current Fallow (AUCF)

The Agriculture Census considers the area under current fallow as the cropped areas left uncultivated during the reference year due to some reason but cultivated in the previous year. The area not cultivated for more than 1 year in succession will be treated as old fallow or culturable wasteland. Interestingly, there is an increasing trend of the area under current fallow during 1995–1996 to 2015–2016 in the district (Fig. 2.1). It has increased from 2807 ha in 1995–1996 to 3766 ha in 2015–2016. Marginal (848 ha) and small (1073 ha) land size class categories demonstrate the addition of area under current fallow, whereas semi-medium (592 ha), medium (334 ha), and large (35 ha) land size classes exhibit decreases in area.

Figure 2.5 discloses that most blocks add up AUCF in marginal and small land size classes, where Jalangi (205 ha) and Samserganj (728 ha) rank on the top under both the land size classes, respectively. With an area of 72 ha, Suti-II loses the highest amount of area in the marginal land class. Under the small land size class, the most area reduction is displayed by Murshidabad-Jiaganj (105 ha). All the blocks in the remaining land classes indicate negative changes except for Bharatpur-I (38 ha), Nabagram (16 ha), and Bhagwangola-II (2 ha) in the semi-medium land class and Nabagram (8 ha) in the medium land class. The maximum negative changes under these three land classes are pointed out by Bhagwangola-I (93 ha), Murshidabad-Jiaganj (153 ha), and Raghunathganj-II (24 ha). After summing up all the land size classes, it may be depicted that Jalangi (770 ha) has gained the maximum area during 1995–1996 to 2015–2016, followed by Samserganj (744 ha), Nabagram (188 ha), and Sagardighi (149 ha). Conversely, the top negative changes are found in Suti-II (210 ha), Khargram (185 ha), Suti-I (179), and Bhagwangola-I (162 ha) sequentially.

In 2020–2021, a total of 3737 ha area has been left under current fallow in Murshidabad district, out of which marginal and small land classes record 1611 and 1993 ha, respectively. Jalangi block (816 ha) has maximum area under current fallow during this period, followed by Samserganj (722 ha). Interestingly, four blocks, namely Farakka, Beldanga-I, Kandi, and Suti-I, report less than 10 ha of area under current fallow due to the multiple cropping throughout the year.

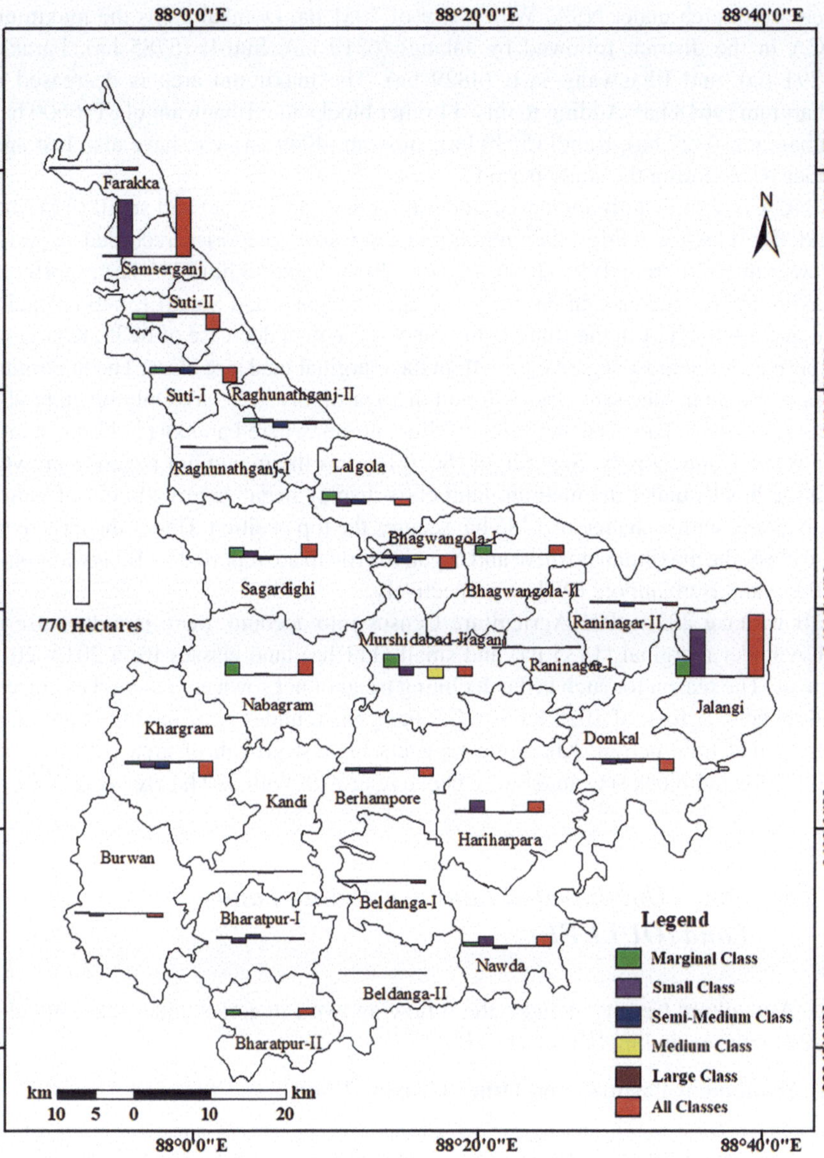

Fig. 2.5 Land size class-wise changes in area under current fallow in Murshidabad district, 1995–1996 to 2015–2016. (Source: Based on Annexure B.5)

2.5.4 *Net Cultivated Area (NCA)*

The net cultivated area is the actual cultivable land, including the net sown area and the area under current fallow land with reference year. The district added 3583 ha NCA in 2015–2016 from 1995–1996 (Fig. 2.1). This positive growth is due to the increase in area under NSA. With a rise of 7031 ha, Domkal gains the maximum NCA in the district, followed by Jalangi (6319 ha), Suti-II (5785 ha), Farakka (5291 ha), and Bhagwangola-II (4829 ha). The maximum area is decreased in Khargram (9618 ha). Adding to this, 11 other blocks like Bhagwangola-I (6600 ha), Nabagram (6327 ha), Kandi (5539 ha), Burwan (4049 ha), etc. have also lost area under NCA during the study period.

NCA is significantly increased under marginal (44,154 ha) and small (3347 ha) land size classes, while other remaining land size categories recorded negative growth in NCA from 1995–1996 to 2015–2016. Sagardighi (8735 ha) reports the maximum area increase in the marginal land category, and Suti-II (3540 ha) gains the maximum NCA in the small land size class. With a decrease of 6238, Khargram experiences the most negative growth in the marginal land size class. The maximum loss in the small land size class is found in Sagardighi (5325 ha). Only seven blocks get the positive changes in the semi-medium group, where Farakka (1112 ha) scores the most. Contrastingly, Sagardighi (6016 ha) has the maximum negative growth. All the blocks under the medium land class display negative growth, out of which Nabagram, with a change of 1486 ha, secures the top position. Under the large land size class, the maximum positive and negative growth are reported by Bhagwangola-I (70 ha) and Berhampore (97 ha), respectively.

If we take 2020–2021 Agriculture Census into account, there is a decrease in NCA under marginal (1435 ha) and small (111 ha) land classes from 2015–2016 census. The reason for such is the declining trend of net sown area as well as current fallow land. A total of 1028 ha of NCA has gained under semi-medium land class during that time period. Khargram registers highest growth of area with 313 ha, while Domkal block sees maximum negative growth with 229 ha area (Fig. 2.6).

2.5.5 *Other Uncultivated Land Excluding Fallow Land (OULEFL)*

The Agriculture Census includes the following two categories under other uncultivated lands, excluding fallow land (Fig. 2.7):

(a) **Permanent Pastures and Other Grazing Land**

This should include all grazing land, whether these are permanent pastures and meadows or not. Village common grazing land shall be excluded for purposes of census (Agriculture Census).

(b) **Land Under Miscellaneous Tree Crops**

This includes all cultivable land not included in the net sown area but is put to some agricultural use. Land under Casuarina trees, thatching grasses, bamboo bushes, and other grooves for fuel that is not included under 'Orchards' would be covered under this category (Agriculture Census).

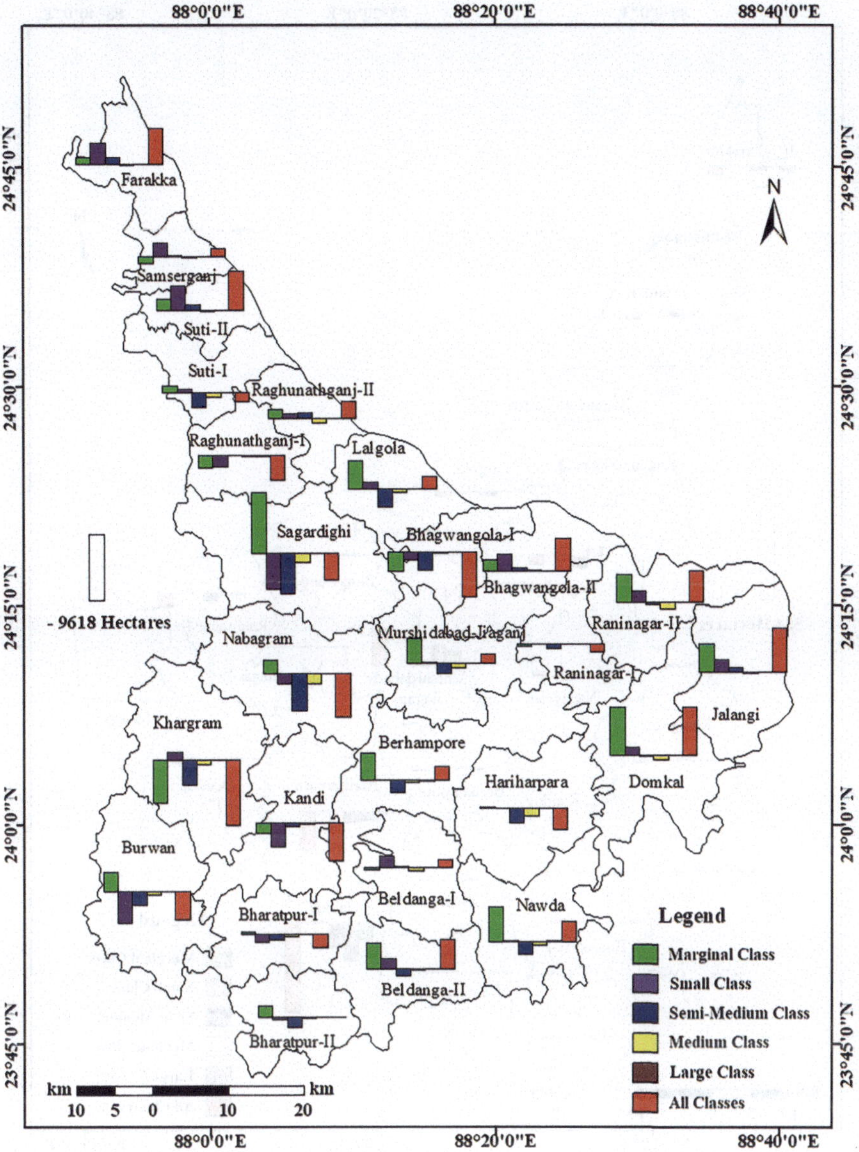

Fig. 2.6 Land size class-wise changes in net cultivated area in Murshidabad district, 1995–1996 to 2015–2016. (Source: Based on Annexure B.6)

Figure 2.1 shows that the area under OULEFL has minimized from 1351 ha in 1995–1996 to 475 ha in 2015–2016. Over time, the built-up area is encroaching on the permanent pastures and other grazing lands as well as land under miscellaneous tree crops due to the rapid population growth. All the land size classes face negative growth of OULEFL during 1995–1996 and 2015–2016. The block-level data reveals

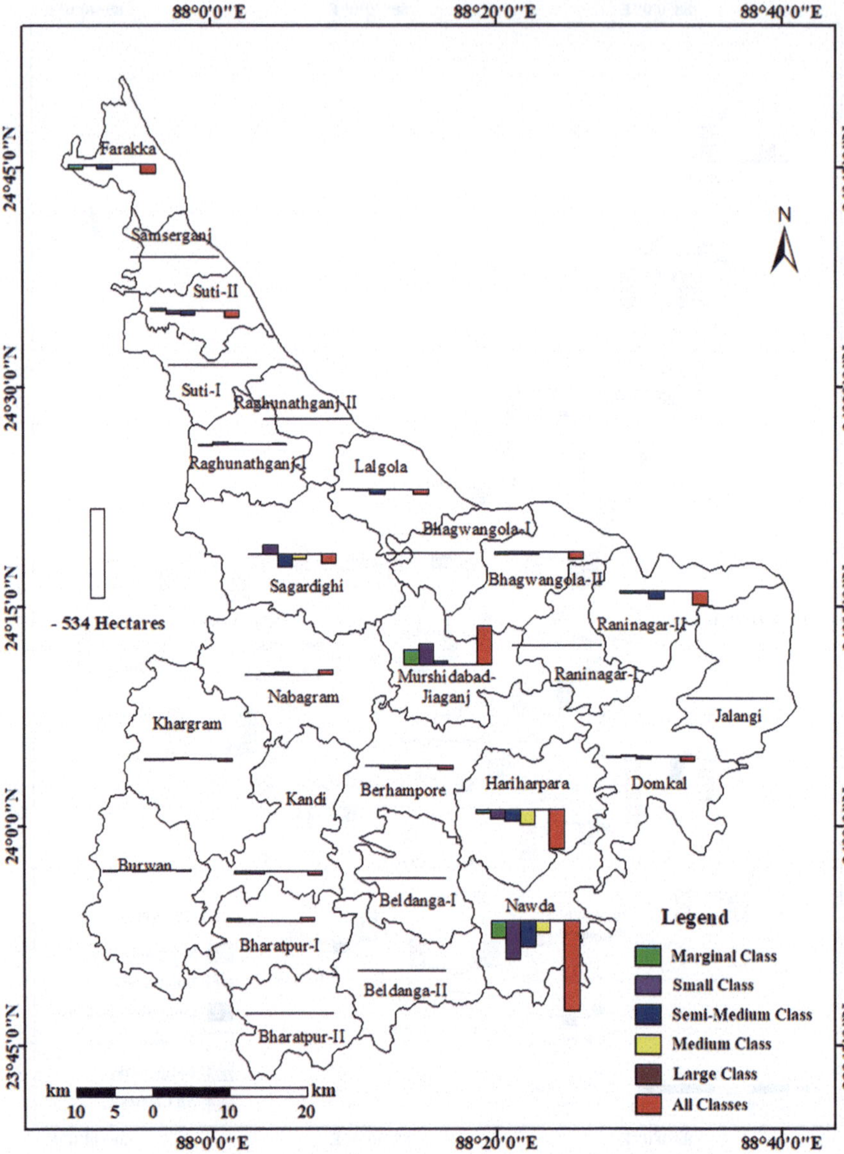

Fig. 2.7 Land size class-wise changes in other uncultivated land excluding fallow land in Murshidabad district, 1995–1996 to 2015–2016. (Source: Based on Annexure B.7)

that Murshidabad-Jiaganj is the only block that gains the maximum area under OULEFL in all land categories. A total of 222 ha area has increased in this block. Nawda registers the maximum decrease in marginal (95 ha), small (224 ha), and semi-medium (150 ha) land classes. The highest cut-off of OULEFL under medium land type is seen in Hariharpara (89 ha).

Again, area under OULEFL in the district has slightly increased to 503 ha in 2020–2021. Murshidabad-Jiaganj, with an area of 233 ha ranks top among the blocks in terms of area under OULEFL. As many as seven blocks, namely Burwan, Beldanga-II, Bharatpur-II, Jalangi, Raghunathganj-II, Samserganj, and Suti-I, report no area under OULEFL in 2020–2021.

2.5.6 *Fallow Land Other Than Current Fallow (FLOTCF)*

The census defines 'All the lands which were taken up for cultivation once but are temporarily out of cultivation for a period of not less than one year and not more than five years, i.e., greater than one year but less than or equal to five years are included under FLOTCF'. The possible reasons for keeping such lands under fallow are poverty of the cultivators, inadequate supply of water, variability of climate, unremunerative nature of farming, etc.

Figure 2.2 illustrates that the area under FLOTCF has decreased in all the land classes from 1995–1996 to 2015–2016. The total area of FLOTCF was 541 ha in 1995–1996 and reduced to 343 ha in 2015–2016. Seven of the total blocks add up the area under FLOTCF, where Suti-I (127 ha) gains the maximum. Contrary to this, Khargram (118 ha) loses the highest area during the study periods. The maximum reduction of FLOTCF under marginal, small, semi-medium, and medium land categories is reported by Khargram (61 ha), Beldanga-II (78 ha), Raninagar-II (12 ha), and Beldanga-II (5 ha), respectively (Fig. 2.8).

Out of the total FLOTCF in the district (348 ha) in 2020–2021, Suti-I alone records 123 ha area under FLOTCF, followed by Murshidabad-Jiaganj (45 ha), Bhagwangola-II (33 ha), and Burwan (31 ha).

2.5.7 *Culturable Wasteland (CWL)*

Culturable wasteland includes land available for cultivation, whether or not taken up for cultivation at any time but not cultivated during the current year and the last 5 years or more in succession for one reason or another. Such lands may be either fallow or covered with shrubs and jungles, which are not put to any use (Agriculture Census) (Fig. 2.9).

Figure 2.1 depicts an increasing trend of the area under culturable wasteland from 861 to 898 ha over 20 years in the district. All the land classes except semi-medium note a positive area growth under CWL (Fig. 2.2). None of the CWL areas is devoted to the large land group in 1995–1996 and 2015–2016. Raninagar-II reports the highest positive growth in all the land categories. A total of 441 ha area has increased in Raninagar-II. Similarly, the Jalangi block records the maximum negative change in all the land classes and the total amount accounts for 396 ha.

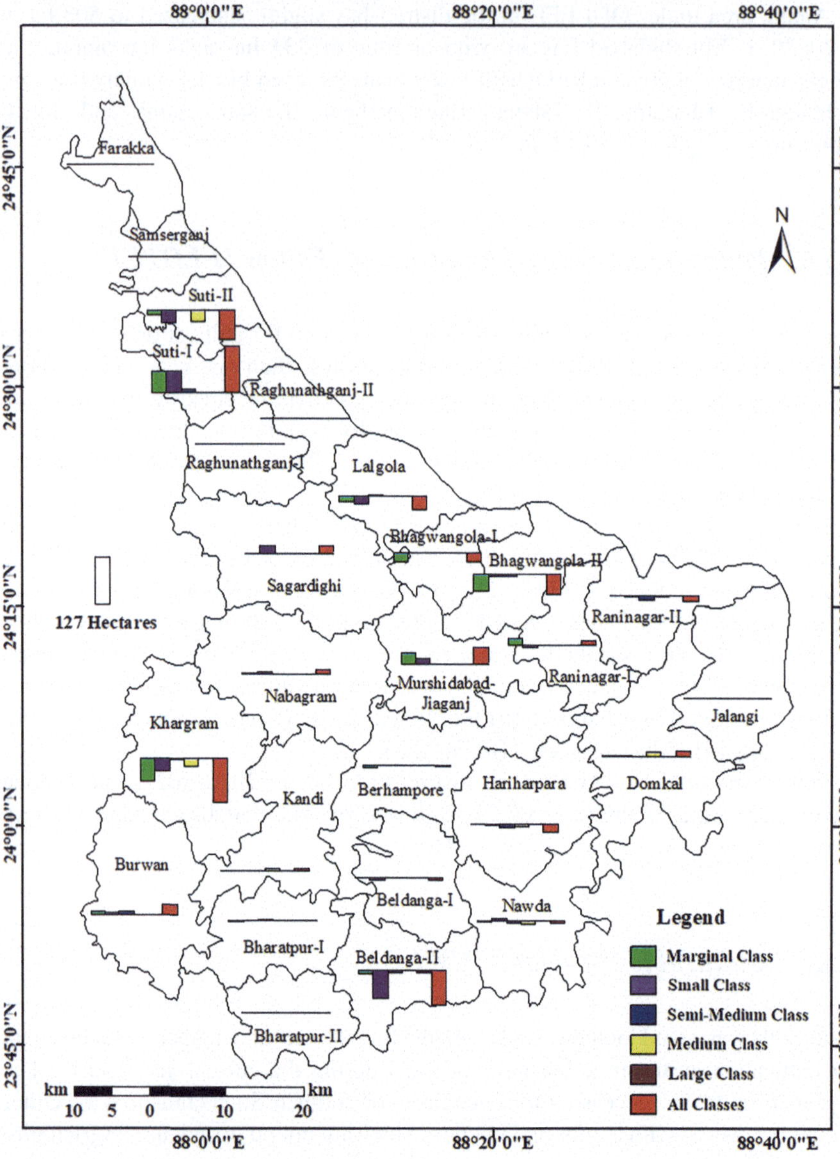

Fig. 2.8 Land size class-wise changes in fallow land other than current fallow in Murshidabad district, 1995–1996 to 2015–2016. (Source: Based on Annexure B.8)

The extrapolated data for 2020–2021 shows that Raninagar-II reports maximum area of CWL under all the land categories, i.e., 141 ha in marginal, 171 ha in small, 71 ha in semi-medium, 44 ha in medium, and a total of 427 ha under all classes.

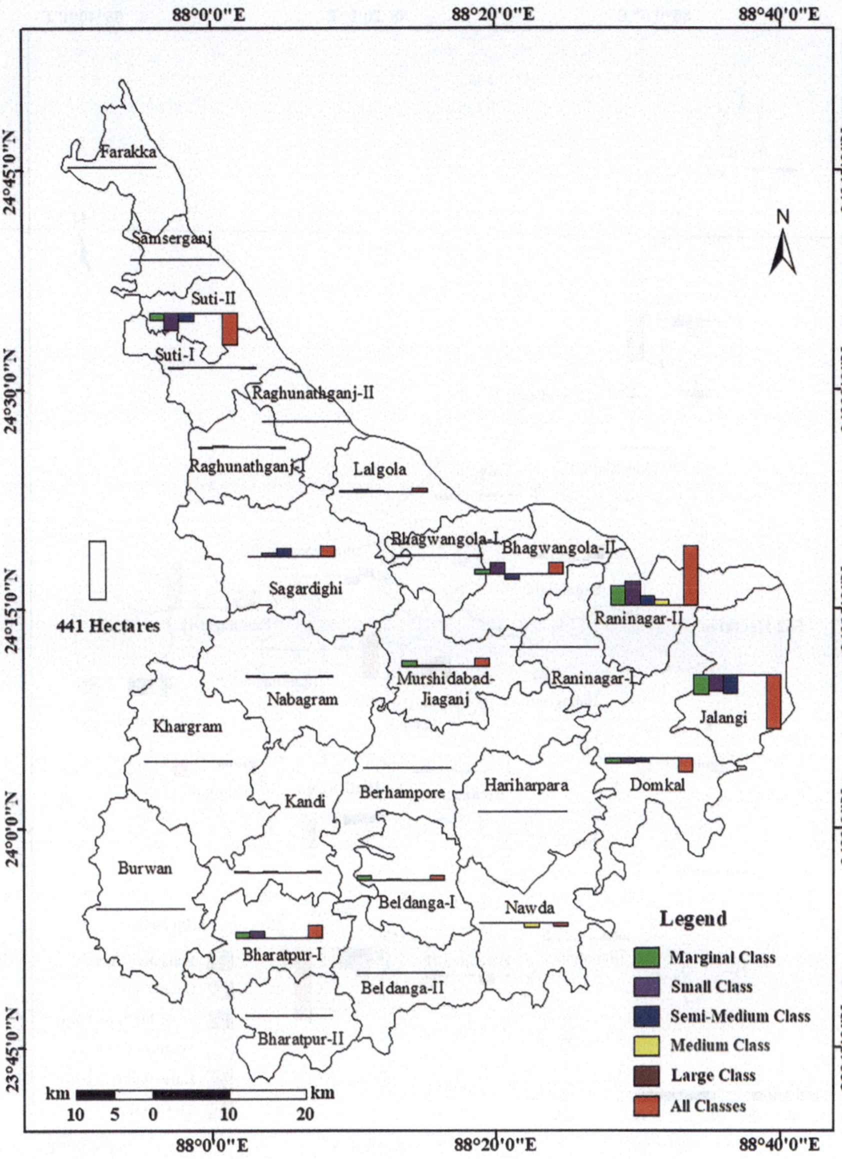

Fig. 2.9 Land size class-wise changes in culturable waste land in Murshidabad district, 1995–1996 to 2015–2016. (Source: Based on Annexure B.9)

2.5.8 Total Uncultivated Land (TUL)

The total uncultivated land is calculated by deducing the area devoted to total cultivable land from the total reported area in a particular region. All the land categories experience a negative growth of TUL in the district from 1995–1996 to 2015–2016

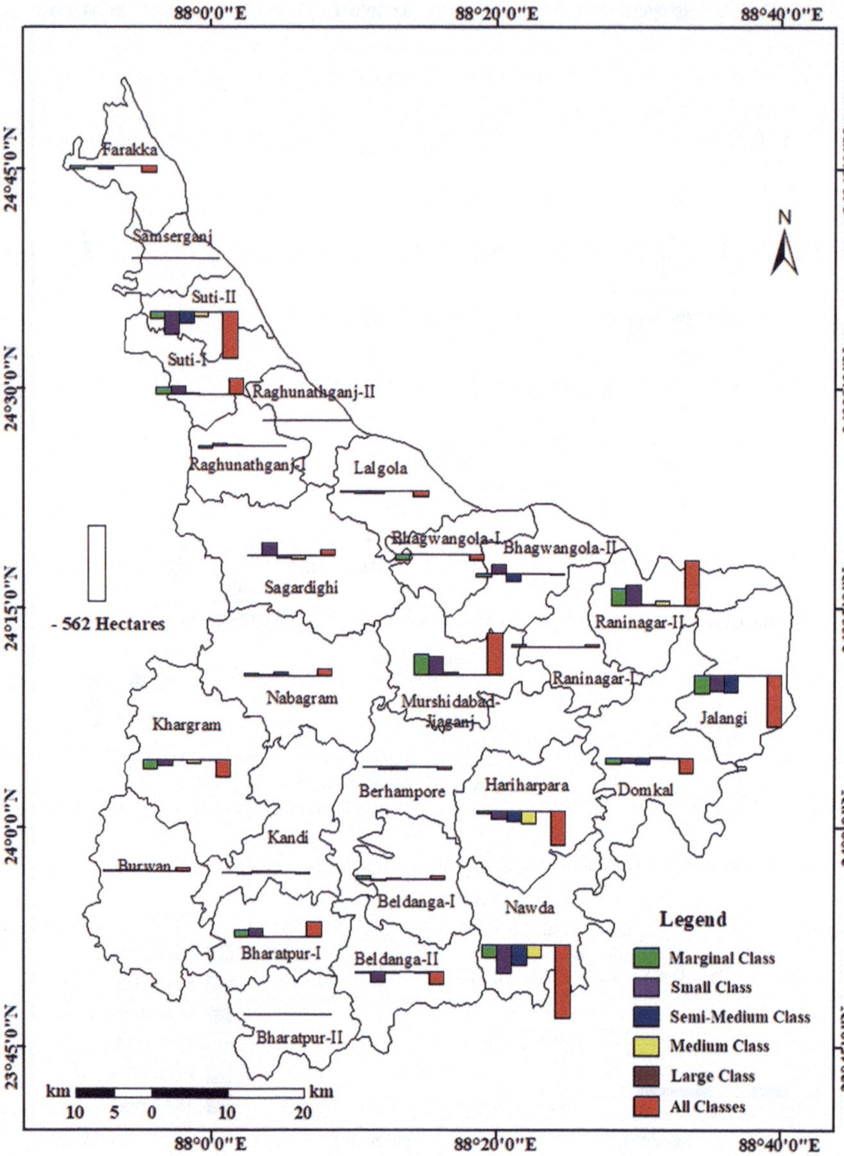

Fig. 2.10 Land size class-wise changes in total uncultivated land in Murshidabad district, 1995–1996 to 2015–2016. (Source: Based on Annexure B.10)

(Fig. 2.2). This land classification loses a total of 1031 ha area during the study period (Fig. 2.10).

Among blocks, Jalangi secures the top position in negative growth in the marginal land class. The maximum area decreases under small, semi-medium, and medium land categories are noted by Nawda (214, 154 and, 100 ha, respectively).

In reverse, Murshidabad-Jiaganj manages to gain maximum positive growth in marginal (158 ha) and semi-medium (23 ha) land groups. The highest area increases under small (158 ha) and medium (41 ha) land types are located in Raninagar-II. Overall, Nawda (562 ha) reduces the maximum area, followed by Jalangi (398 ha), Suti-II (351 ha), and Hariharpara (258 ha). There are still nine blocks in the district that report the addition of TUL during 1995–1996 and 2015–2016. Raninagar-I (345 ha) reveals the maximum positive growth, followed by Murshidabad-Jiaganj (325 ha) and Suti-I (122 ha).

The area under total uncultivated land in the district during 2020–2021 is 1753 ha. Raninagar-II displays maximum TUL that is 432 ha, followed by Murshidabad-Jiaganj (332 ha), Bhagwangola-II (181 ha), and Sagardighi (151 ha). There is no area under this land classification in Bharatpur-II, Raghunathganj-II, and Samserganj blocks.

2.5.9 Land Not Available for Cultivation (LNAC)

Forest area, area under non-agricultural use, and barren and uncultivated land are clubbed together and shown under LNAC.

(a) **Forest Area:** Only the private forest belonging to operational holders is covered for the purpose of the Agriculture Census. The area of crops raised in the forest and grazing lands or areas open for grazing within the forests should remain included under the forest area.
(b) **Area Under Non-agricultural Use:** 'This category encompasses all non-agricultural land uses—including buildings, tanks, and ponds—within the holdings of operational landholders.' (Agriculture Census).
(c) **Barren and Uncultivated Land:** This should include all barren and uncultivated land within the operational holding.

The LNAC is fallen to 3694 ha in 2015–2016 from 5340 ha in 1995–1996 (Fig. 2.1). All the land size categories display negative growth of LNAC due to the continued deforestation in the district. Figure 2.11 shows that Lalgola records the maximum negative change under all the land size classes except for the large category, with a total area of 1534 ha. Berhampore (24 ha) registers the maximum reduction in the large land type. Conversely, Bhagwangola-II and Murshidabad-Jiaganj gain a significant amount of area under LNAC under different land groups. These two blocks add 998 and 902 ha, respectively, from 1995–1996 to 2015–2016.

In 2020–2021, 3747 ha area is devoted to LNAC in the district, in which Bhagwangola-II adds maximum with 1163 ha. Across the land size categories, Bhagwangola-II denotes maximum area under marginal (371 ha) and semi-medium (499 ha) classes, while Murshidabad-Jiaganj reports highest under small class with 512 ha area of LNAC.

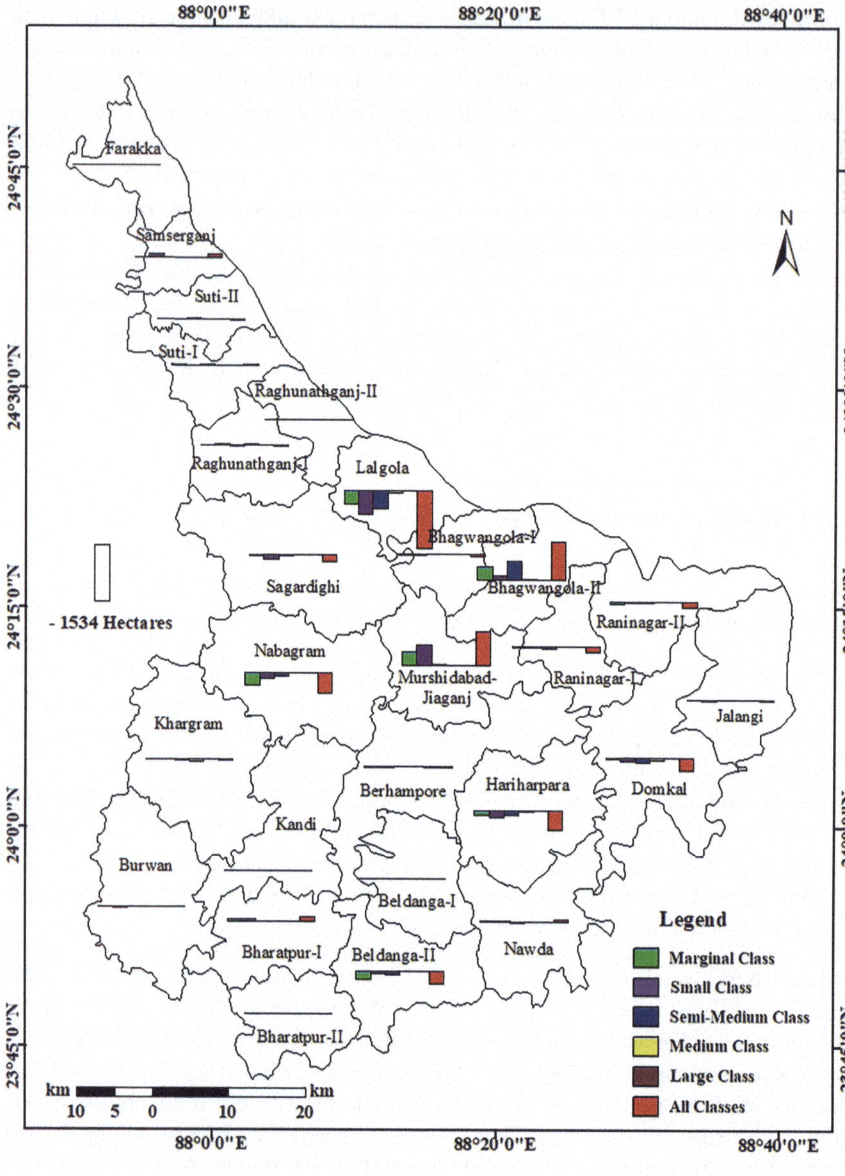

Fig. 2.11 Land size class-wise changes in land not available for cultivation in Murshidabad district, 1995–1996 to 2015–2016. (Source: Based on Annexure B.11)

2.6 Cropping Pattern

In simple words, a cropping pattern denotes the proportion of area under different crops in a particular region at a time. The proportion of area includes net sown area or gross cropped area depending upon the type of cropping practice in a region. It is a dynamic concept, as no crop arrangements remain permanently for a specific region. Thus, no cropping pattern is identically same across time and space (Siddiqui & Rahaman, 2019). However, the meaning and concept of cropping patterns vary from region to region depending on either the researcher's view or the objective of the study. Martinez-Casasnovas et al. (2005) have argued that cropping patterns can be defined differently based on either including or excluding the spatial location or distribution in the definition. For instance, Food and Agricultural Organization (FAO, 1996) defines the cropping pattern as the spatial representation of crops in a region and Bontkes and Van Keulen (2003) consider it as the spatial arrangement of different crops in a particular area as a result of climate, soil, available facilities, and socio-economic factors. According to N.J. Singh et al. (2011), the cropping pattern refers to the yearly sequence, spatial association as well as arrangement of crops in a given area. Conversely, the cropping pattern is considered as a list of all the crops produced in a region and their sequence within a year (Bigman & Srinivasan, 2002; Singh et al., 2001).

The cropping pattern study is essential because it is linked with the overall agricultural innovation of a region. In this context, Ranade (1980), in a study on 'Impact of Cropping Pattern on Agricultural Production' has discussed the effect of shifting of cropping pattern on agricultural production across 54 agro-climatic regions covering 16 major states of India from pre-green revolution to post-green revolution period. Changing cropping pattern in India is linked with crop diversification induced from income and livelihood option (Rahaman, 2021a). A diversified cropping pattern can also be used as a strategy to cope with the flood-related risk (Mandal & Bezbaruah, 2013).

The number of crops grown over time governs the type of cropping pattern prevailing in a region. It is called monoculture or mono-cropping pattern when only one crop is produced in the net sown area in a particular region. Contrary to this, if two or more than two crops are cultivated simultaneously, it is regarded as a mixed cropping pattern. Sometimes, more complex cropping systems exist, like intercropping systems, where on the same field, more than one crop is grown at the same time along rows containing a row of the main crop and three or four rows of other crops. Crop rotation is the cropping system practice to maintain the soil fertility of a unit of land. It may be defined as the repetitive cultivation of a planned succession of crops or a combination of crops and fallow on the same land.

The cropping pattern never remains the same in any region over time. Several important factors determine the cropping pattern of a region. The physical aspects like soil, temperature, rainfall, natural calamities, etc. always play an important role in the choice of cropping pattern in an area. Although with the advancement of modern technology, we can minimize the risk of the physical determinants but cannot completely ignore them. Morgan and Munton (1971) have argued that the physical environment actually limits the choice to grow different crops altogether. Economic issues such as the demands of certain crops, inflation, income opportunities, and government price policies play a critical role in a particular cropping pattern (Zandstra, 1981). Mahesh (1999) has done a location-specific study of Kerala regarding the causes and consequences of changing cropping patterns. He discusses that various social, demographic, and technological aspects have influenced cultivation practices and crop pattern. In most cases, changing cropping pattern encourages growth and development in agricultural output but sometimes measurement of such change-induced development becomes difficult (Usmani & Rahaman, 2021).

2.7 Crop Category-Wise Changing Cropping Pattern in Murshidabad District

The Agriculture Census defines the gross cropped area as the total area sown once plus more than once in a particular year. It means the area of a specific piece of land will be counted as many times as there will be sowing in a year. This total area is also known as the total cropped area or total area sown.

The overall GCA in the district has increased from 823,506 ha in 1995–1996 to 890,982 ha in 2015–2016. The district has gained GCA positively due to improved institutional, technological, and infrastructural innovation in agriculture. Figure 2.12 reveals that under land size classes, the maximum increase in GCA is found in marginal land type (11.27%) followed by small land category (1.01%) in two decades. The remaining land groups report negative growth in GCA in which semi-medium land type posits the maximum negative change (9.64%). All other land groups exhibit negative GCA growth, led by semi-medium types (-9.64%). The most severe declines occur in under (-32.80%) and Farakka (-9.49%). With a positive change of 18.48%, Farakka gains the highest GCA in the small land category, and Burwan (12.25%) reports the maximum negative change in the same land type. Raghunathganj-I (4.42%) and Bhagwangola-I (0.36) are the only blocks having positive GCA changes under semi-medium and medium land size classes, respectively. Nabagram (19.60%) indicates the maximum negative change in semi-medium size, while the same is located by Raghunathganj-II (10.45%) under medium class. The difference in GCA in the large category from 1995–1996 to 2015–2016 is almost negligible (0.04%).

It is to be noted that the data for 2020–2021 has been extrapolated based on the interpolated Agriculture Census data for 1995–1996 and 2015–2016. The maximum positive and negative deviations of GCA under marginal land category from 2015–2016 to 2020–2021 are recorded by Farakka (0.31%) and Sagardighi (1.07%), respectively. Under small land class, Burwan (0.43%) reports highest growth positively, while Farakka (0.60%) has seen maximum negative growth. Similarly, top positive growth under semi-medium class is found in Nabagram (0.64%), and under medium class, it is Raghunathganj-II (0.34%).

2.7.1 Changing GCA Under Cereals

Cereal crops are grasses cultivated for their grains' edible components (Arvanitoyannis & Tserkezou, 2008). These crops, also known as staple crops, are the primary diet source of essential calories, nutrition, and carbohydrates for

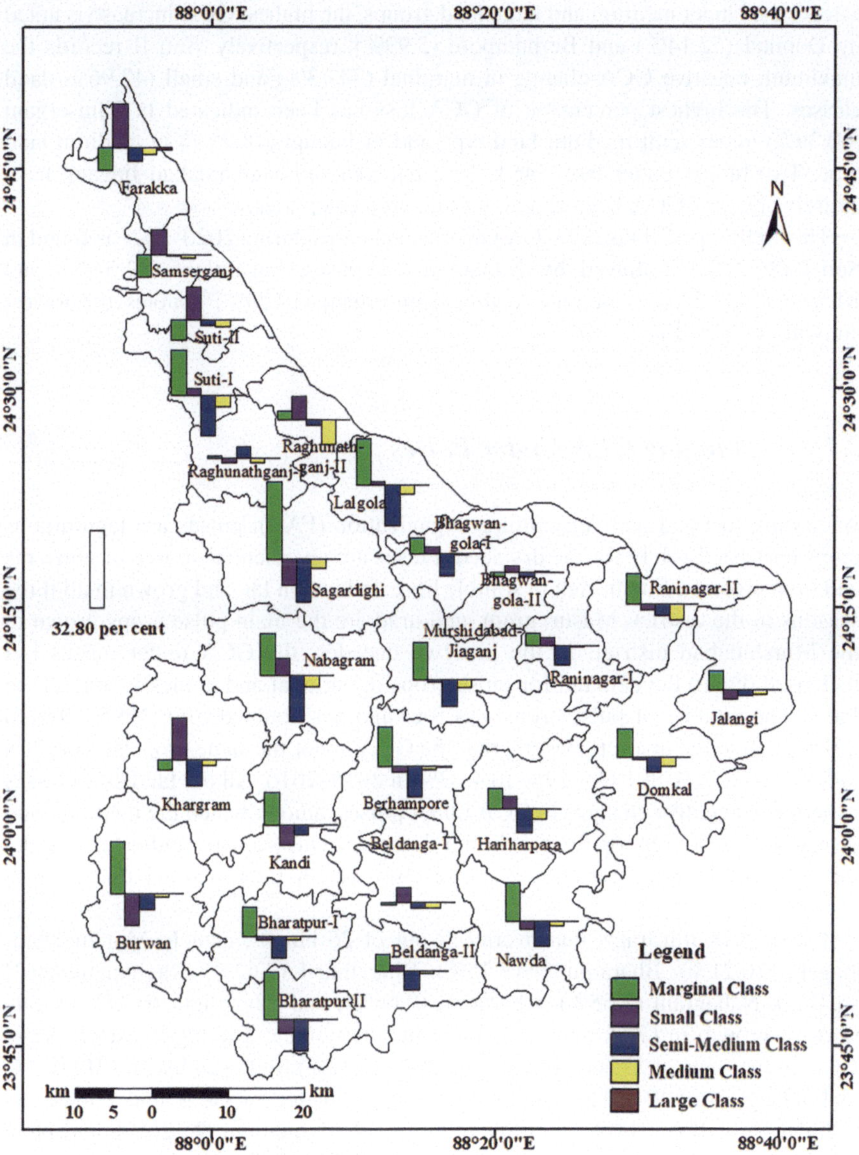

Fig. 2.12 Land size class-wise changes in gross cropped area in Murshidabad district, 1995–1996 to 2015–2016. (Source: Based on Annexure B.12)

billions of populations in the world. Wheat and rice together account for over 50% of cereal production worldwide. The GCA under cereal crops remains almost stable (only 0.35% positive change) in the district during 1995–1996 and 2015–2016, but the number of crops has increased from six to nine.

Nawda reports the maximum positive change of GCA under cereal crops in marginal (32.96%), small (32.22%), semi-medium (42.25), and all (32.72%) land class categories. Under medium and large land groups, the highest GCA increase is noted by Domkal (52.14%) and Berhampore (5.99%), respectively. Suti-II records the maximum negative GCA change in marginal (34.43%) and small (47.96%) land classes. The highest percentage of GCA loss has been indicated in Samserganj (60.39%) under semi-medium land type and in Jalangi (78.95%) in medium land type. Two blocks under large land class, Farakka and Murshidabad-Jiaganj, have entirely changed GCA from cereal to other crop categories.

The highest percentage of GCA under cereal crops during 2020–2021 is found in Suti-I (86.21%), followed by Nabagram (82.95%), Sagardighi (82.54%), and Khargram (82.32%). Contrary to this, Samserganj (12.29%) denotes the lowest amount of GCA (Fig. 2.13).

2.7.2 *Changing GCA Under Pulses*

According to Food and Agriculture Organization (FAO), pulses are leguminous crops harvested solely for the dry seed. Pulses are an essential source of nutrients and high protein content, rich in soluble fibre and low in fat, and grown in all three seasons in the district. Masur, gram, and urad are the main pulse crops grown in the Murshidabad district. In the last two decades, the GCA under pulses has increased due to the demand for pulse crops in national and domestic markets in India. The number of pulse crops was four and has doubled from 1995–1996 to 2015–2016 in the district. Despite that the GCA under the same crop category has fallen from 6.77% in 1995–1996 to 4.23% in 2015–2016. All the land size classes experience negative changes in GCA under pulses, among which the medium land category (4.53%) reports the maximum decrease. This scenario indicates that the farmers tend to grow varieties of pulse crops rather than quantity due to market demand.

Figure 2.14 illustrates that overall, 7 out of 26 blocks, namely Murshidabad-Jiaganj (10.21%), Bharatpur-II (7.75%), Bharatpur-I (7.62%), Raghunathganj-II (6.71%), Nabagram (2.65%), Khargram (0.66%), and Sagardighi (0.37%) locate positive growth in GCA under pulses in the study area. The highest decrease in GCA in the same class is recorded by Suti-I (21.32%), followed by Suti-II (19.14) and Jalangi (16.80%). The maximum negative change of GCA under marginal (22.65%), small (22.48%), semi-medium (27.42%), medium (25.26%), and large (32.69%) land categories is found in Samserganj, Suti-I, Farakka, Suti-I, and Raghunathganj-II, respectively.

A total of 4.86% of GCA is cultivated under pulse crops in 2020–2021. Raghunathganj-II reports an area of 24.86% GCA under this crop category which is the highest among the blocks in the district, while the lowest percentage of GCA is recorded in Raninagar-I block (0.45%).

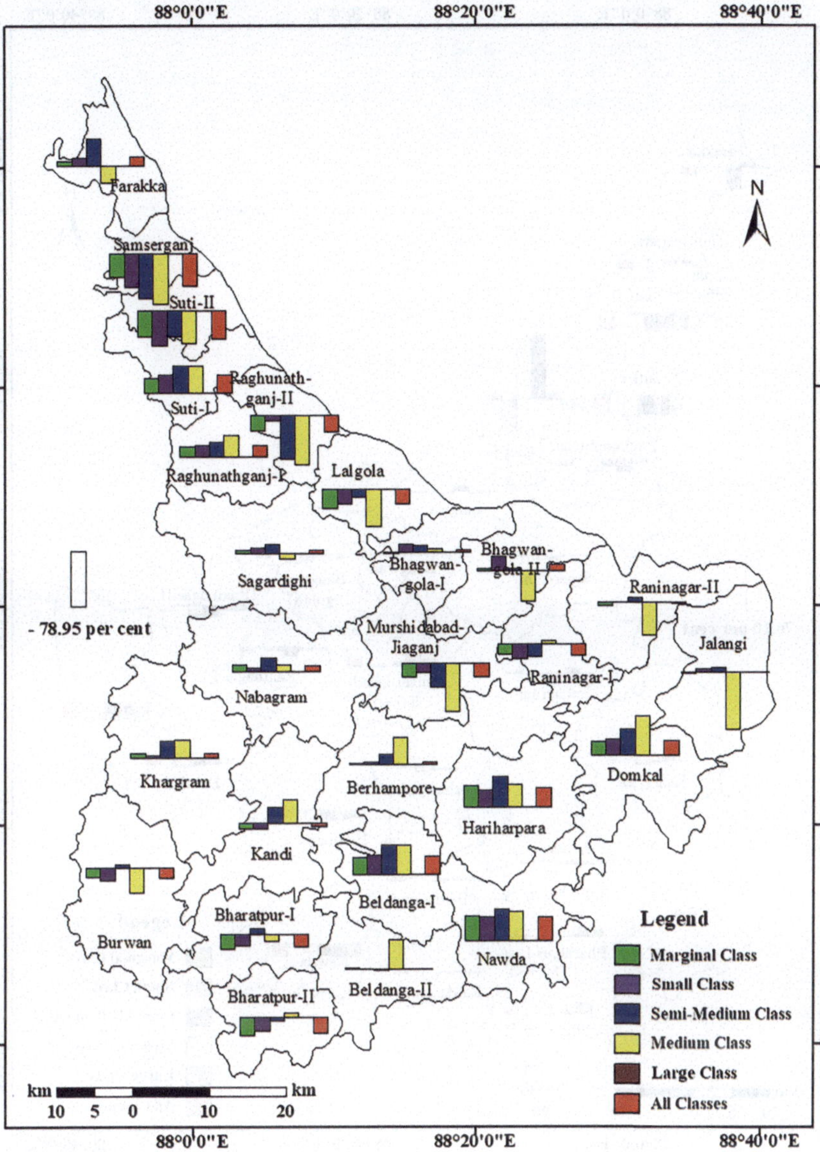

Fig. 2.13 Land size class-wise changes in GCA under cereals in Murshidabad district, 1995–1996 to 2015–2016. (Source: Based on Annexure B.13)

2.7.3 Changing GCA Under Spices

Spices are used for the colour, flavour, and preservation of food and beverages. These are used in different food dishes daily, making them a valuable substance. In the Murshidabad district, chillies, ginger, cardamom, turmeric, etc. are the major

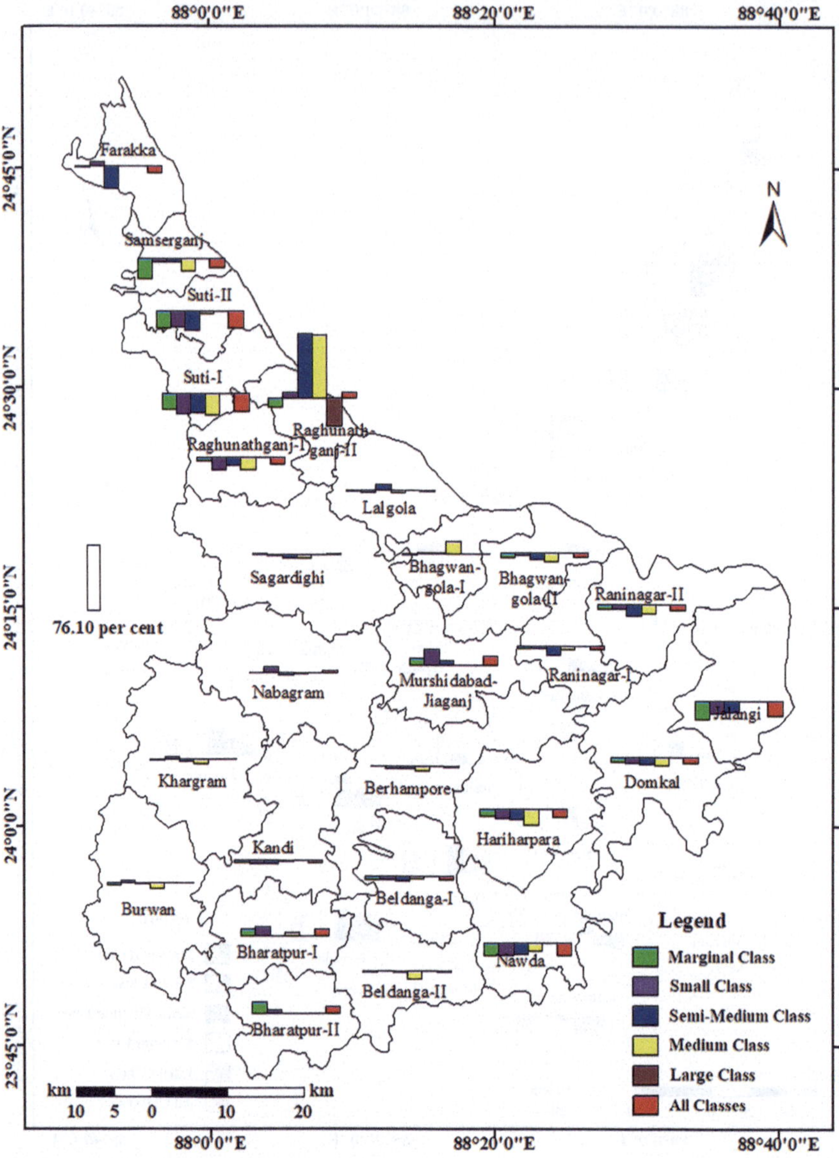

Fig. 2.14 Land size class-wise changes in GCA under pulses in Murshidabad district, 1995–1996 to 2015–2016. (Source: Based on Annexure B.14)

spice crops. The number of crops increased to 8 in 2015–2016 from 5 in 1995–1996. However, the GCA under spices has been reduced (Fig. 2.15).

All the land classes experience negative growth in GCA, in which the medium land class (0.41%) records the most. In the marginal land category, the most significant positive and negative change in GCA is noted by Suti-II (1.08%) and Samserganj

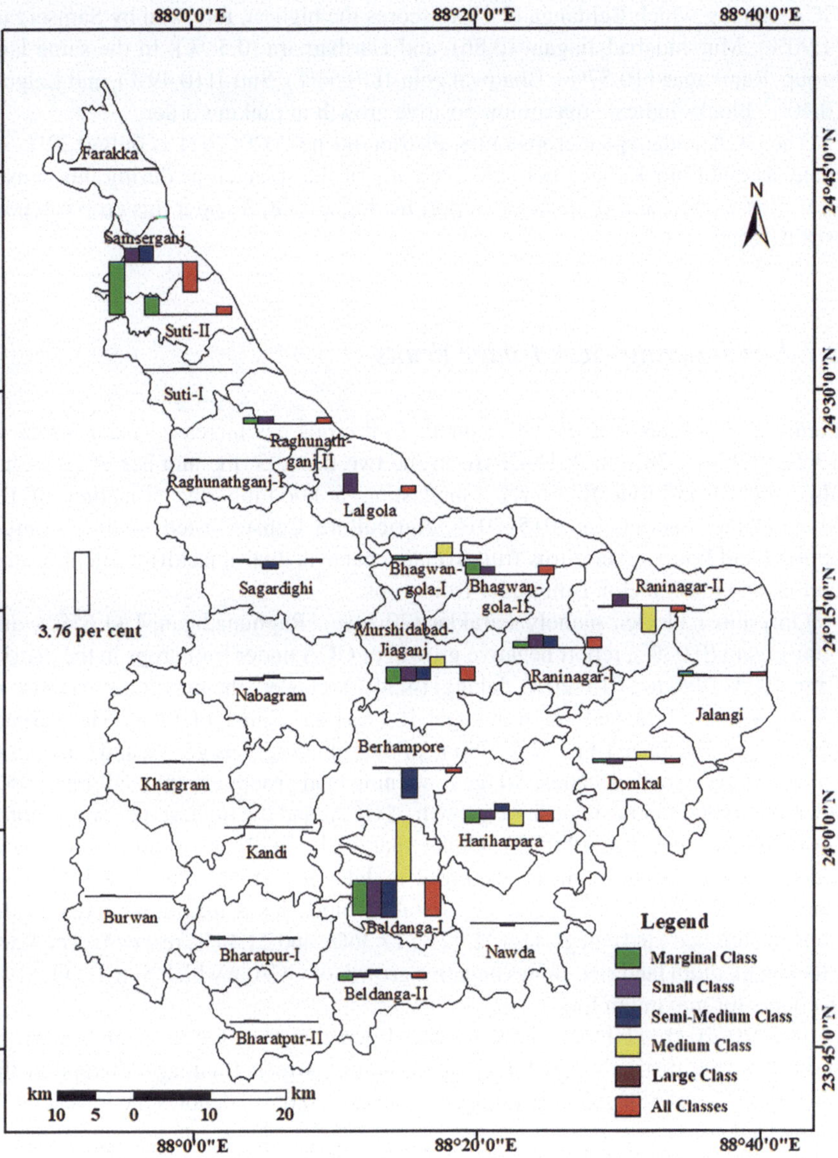

Fig. 2.15 Land size class-wise changes in GCA under spices in Murshidabad district, 1995–1996 to 2015–2016. (Source: Based on Annexure B.15)

(2.24%), respectively. Beldanga-I represents the maximum decrease in GCA under small (2.13%), semi-medium (2.24%), and medium (2.76%) land groups. Lalgola (1.15%) and Raninagar-I (0.64%) show the maximum increase in GCA under small and semi-medium classes. None of the blocks in medium land size has positive gains of GCA under spices. Overall, most of the blocks show negative growth of

GCA, among which Beldanga (2.17%) scores the highest, followed by Samserganj (1.90%), Murshidabad-Jiaganj (0.86), and Hariharpara (0.58%). In the same land group, Raninagar-I (0.57%), Bhagwangola-II (0.54%), Suti-II (0.49%), and Lalgola (0.40%) blocks indicate maximum positive growth in ranking order.

The GCA under spice crops in the district during 2020–2021 is only 0.22%. As many as eight blocks have not cultivated any of the spice crops during this census year. Raninagar-I and II, however, report the highest GCA under this crop category (0.83% each).

2.7.4 *Changing GCA Under Fruits*

Figure 2.16 reveals that the GCA under fruit crops has increased from 0.54% in 1995–1996 to 2.28% in 2015–2016. In the two decades, the number of crops has also risen from four to eleven, suggesting a booming diversification in the Murshidabad district. In 2015–2016, Agriculture Census listed mango, orange, lemon (acid lime), other citrus fruit, banana, papaya, guava, jackfruit, litchi, watermelon, and mask melon fruit under fruit crops.

Only three blocks, namely Farakka (20.24%), Raghunathganj-I (1.65%), and Samserganj (0.03%), report negative growth of GCA under fruit crops in the district from 1995–1996 to 2015–2016. Jalangi (4.34%) registers the maximum increase in GCA in total, followed by Raninagar-II (4.31%), Suti-I (4.12%), Hariharpara (2.61%), and Beldanga-I (2.41%). None of the fruit crops has grown under the large land category in both decades. All the remaining land groups except medium (1.06% negative change) indicate positive growth of GCA, and the highest increase is noted by marginal land class (2.03%). Under marginal, small, and semi-medium land classes, the maximum gain is displayed in Jalangi (6.13%), Suti-I (8.57%), and Raninagar-II (11.10%), respectively. Farakka block loses maximum GCA in the abovementioned land categories (22.92%, 17.36%, and 21.44%, respectively). Each block in medium land size shows negative growth of GCA in which Suti-II (11.51%) accounts for maximum loss.

In 2020–2021, 11 blocks have recorded percentage of GCA under fruit crops more than the district's percentage share among which Raninagar-II reports the maximum, i.e., 4.88%. Contrastingly, Samserganj (0.05%) denotes the lowest GCA under fruit crop category.

2.7.5 *Changing GCA under Vegetables*

The GCA under vegetable crops has increased from 7.88% in 1995–1996 to 11.10% in 2015–2016. There were only three crops listed in the 1995–1996 census. In 2015–2016, 23 crops were listed in vegetable crops in the Murshidabad district. The emerging demands of different vegetable crops in the local and national markets

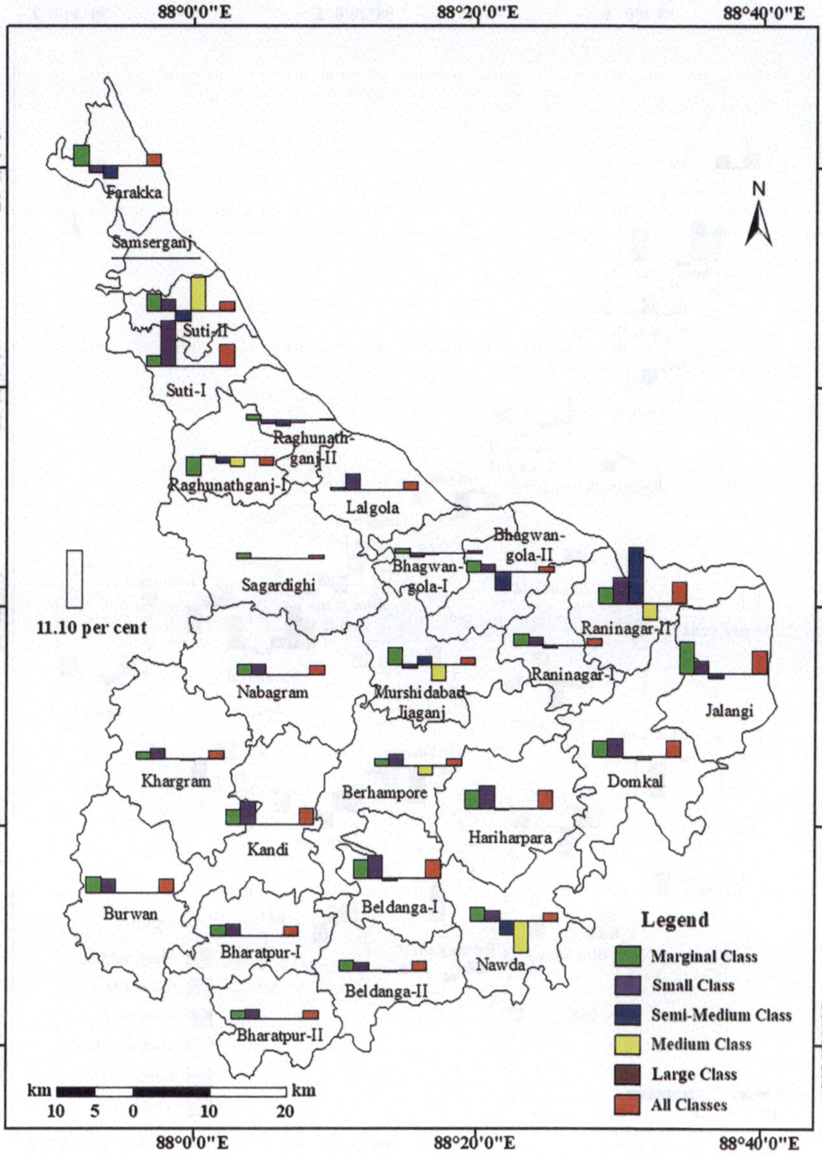

Fig. 2.16 Land size class-wise changes in GCA under fruits in Murshidabad district, 1995–1996 to 2015–2016. (Source: Based on Annexure B.16)

have led to higher diversification in vegetable crops. The district's significant crops cultivated under vegetables are brinjal, potato, onion, cauliflower, cabbage, tomato, and radish in the ranking ladder (Fig. 2.17).

Across land size classes, positive growth is recorded by marginal (4.38%) and small (2.29%) land groups. On the contrary, semi-medium (4.91%), medium

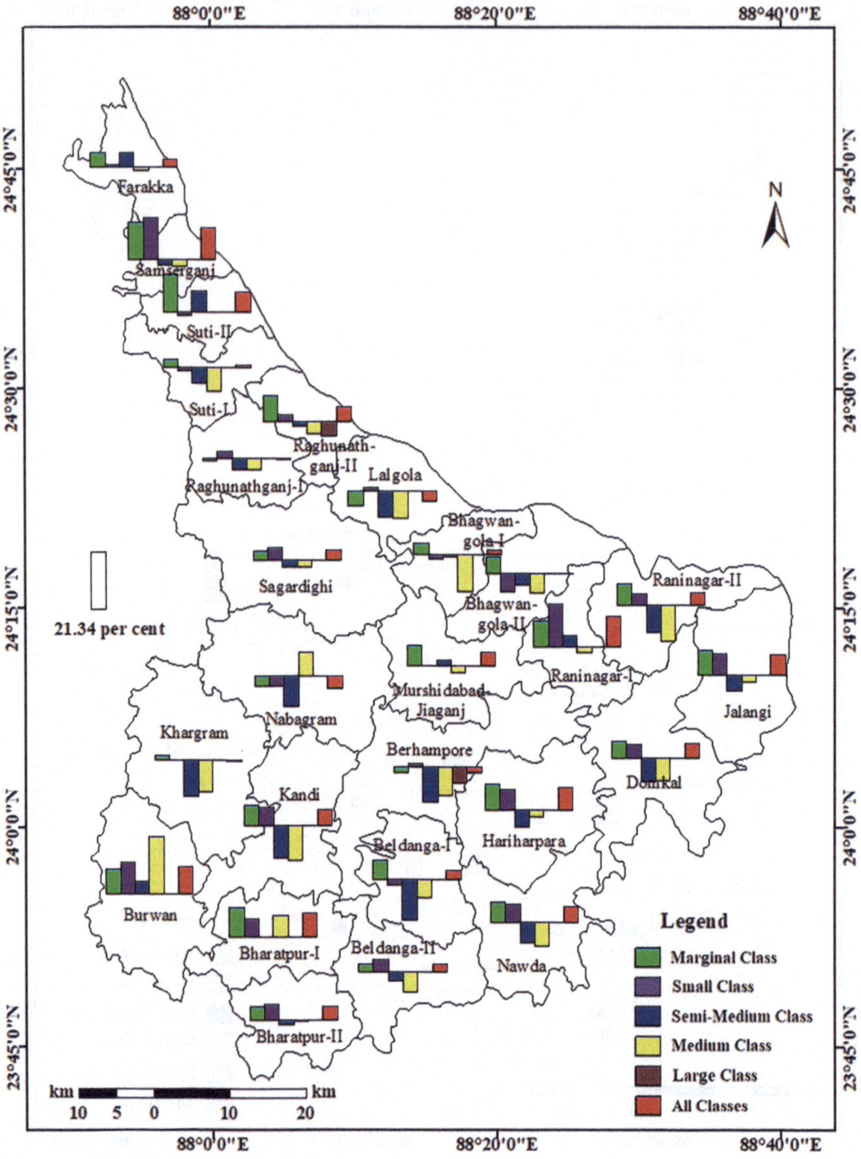

Fig. 2.17 Land size class-wise changes in GCA under vegetables in Murshidabad district, 1995–1996 to 2015–2016. (Source: Based on Annexure B.17)

(4.46%), and large (2.75%) land categories lose their share of GCA under vegetable crops. Under marginal land type, the maximum positive and negative changes are reported by Suti-II (14.46%) and Lalgola (5.48%), respectively. Raninagar-I (15.92%) in small, Suti-II (8.11) in semi-medium, and Burwan (21.34%) in medium class posit maximum positive growth of GCA. In comparison, the top negative change under the same categories is registered by Bhagwangola-II (6.62%), Beldanga-I (15.26%), and Bhagwangola-I (12.66%) sequentially. Overall, only four blocks depict negative growth of GCA, among which Nabagram (4.57%) ranks at the top. In the remaining blocks, Samserganj (11.64%), Raninagar-I (11.20%), and Burwan (9.84%) have gained the maximum GCA under vegetable crops.

One-tenth (10.87%) of the total GCA is allocated under this crop category in Murshidabad district during 2020–2021. The cultivation of vegetable crops is lowest in Farakka block (2.15%), while it is maximum in Jalangi (17.61%).

2.7.6 *Changing GCA Under Oilseeds*

The oilseed crops, one of the necessary cash crops, are grown in each block under different land categories in the Murshidabad district. The number of oilseed crops has increased from six in 1995–1996 to nine in 2015–2016. Mustard, rapeseed, sesamum, groundnut, coconut, linseed, etc. are the major oilseed crops in the district. A total of 0.51% GCA under oilseed has fallen from 1995–1996 to 2015–2016 in the district. The maximum GCA decrease is noted by Hariharpara (10.11%), followed by Khargram (7.38%), Nawda (5.29%), and Beldanga-I (5.06). Conversely, Suti-II (14.87) records the maximum positive growth.

In different land size classes, marginal (0.27%) and small (1.60%) highlight the negative change of GCA, while other remaining land classes add area under oilseed crops (Fig. 2.18). Hariharpara alone reports the maximum negative change in marginal (10.20%), small (9.44%), and semi-medium (11.30%) land categories. In a similar pattern, Suti-II registers the highest negative growth in marginal (12.03%), small (17.83%), and semi-medium (18.32%) land types. Under medium land classes, the maximum positive and negative differences are shown in Burwan (19.91%) and Kandi (16.83%), respectively.

The extrapolated data for 2020–2021 shows that GCA under oilseed crops has increased from 2015–2016 in the district. Across the blocks, Suti-II poses highest GCA with 21% area under this crop category, followed by Raninagar-I (12.67%). A few blocks, namely Suti-I, Nawda, Hariharpara, and Beldanga, report less than 1% of GCA under oilseed crops.

2.7.7 *Changing GCA Under Fibres*

Fibre, a non-food crop, is considered a cash crop and is grown for its fibres to make papers, cloths, and ropes. The share of GCA under fibre crops has been reduced from 12.63% in 1995–1996 to 11.85% in 2015–2016 in the Murshidabad district. There are four crops, namely jute, sunhemp, cotton, and mesta, grown in both the decades in the study area. Jute is the single dominant crop under fibre, accounting for more than 98% of the district's total fibre coverage in 1995–1996 and 2015–2016 census.

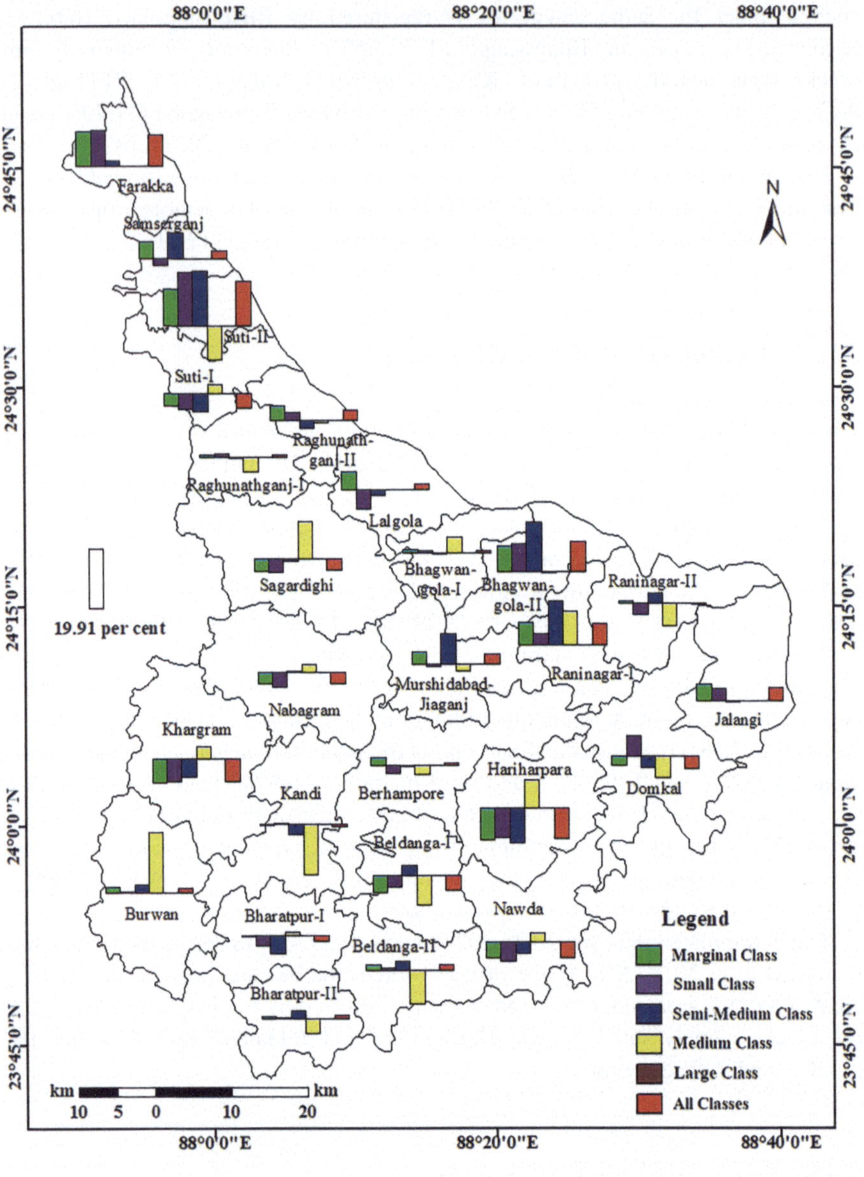

Fig. 2.18 Land size class-wise changes in GCA under oilseeds in Murshidabad district, 1995–1996 to 2015–2016. (Source: Based on Annexure B.18)

Marginal (2.82%) and small (1.00%) land classes locate a negative GCA change under fibre crops, while semi-medium (4.33%) and medium (2.63%) land groups report positive growth. Figure 2.19 reveals that Samserganj leads in terms of GCA increase in marginal (38.57%), semi-medium (60.78%), medium (87.10%), and all classes (44.19%) categories. The same in the small land category is seen in Suti-II

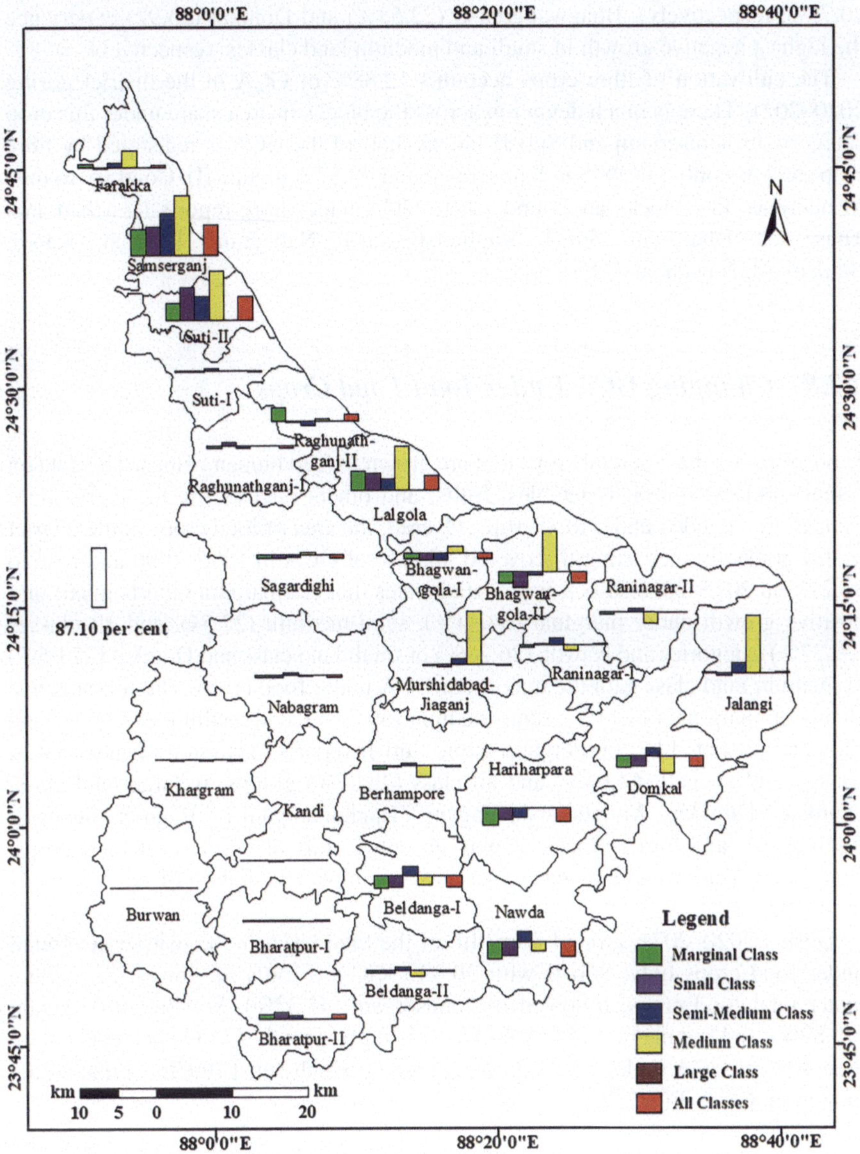

Fig. 2.19 Land size class-wise changes in GCA under fibres in Murshidabad district, 1995–1996 to 2015–2016. (Source: Based on Annexure B.19)

(46.92). There are two other blocks, i.e., Lalgola and Bharatpur-I, which indicate the positive growth of GCA under all the different land size categories during 1995–1996 and 2015–2016. The maximum negative change of GCA in marginal, semi-medium, and all class land types is recorded by Nawda (24.28%, 15.72%, and

20.75%, respectively). Bhagwangola-II (22.53%) and Domkal (25.92) experience the highest negative growth in small and medium land classes, respectively.

The cultivation of fibre crops accounts 12.88% of GCA in the district during 2020–2021. There is much deviation across the blocks in area share under this crop category. In Samserganj and Suti-II blocks, half of the GCA is registered by fibre crop category only (48.98% in Samserganj and 48.57% in Suti-II). Contrary to this, as many as nine blocks are found where GCA under fibre reports less than 1%. These are Khargram, Suti-I, Raghunathganj-I, Nabagram, Burwan, Kandi, Beldanga-I, Nawda, and Hariharpara.

2.7.8 Changing GCA Under Total Food Crops

Food crops are the essential crops that are grown to feed human beings. The data on cereals, pulses, spices, vegetables, fruits, and others released by the Agriculture Census are included under food crops. Overall, the area of food crops in the district is still gradually increasing. It covered 80.02% of GCA in 1995–1996 and rose to 82.21% in 2015–2016. Block-wise data locates that Hariharpara reports maximum positive growth under marginal (34.41%), semi-medium (22.54), and all classes (29.27%) categories and Nawda (26.26%) in small land class and Domkal (32.05%) in medium land class gain the maximum GCA under food crops. The top negative change in marginal (44.11%), semi-medium (69.24%), and medium (87.10%) land categories is noted by Samserganj, while Suti-II reports maximum negative GCA change under small (64.68%) and all class (48.22%) groups. In large land class, Domkal, Farakka, Murshidabad-Jiaganj, Raghunathganj-I, Raghunathganj-II, Raninagar-I, and Suti-I blocks display an entire shift of GCA from food crops. Conversely, Burwan and Nawda experience a complete alteration of GCA towards food crops.

During 2020–2021, around four-fifth of the total GCA in the dsitrict are found under food crops only. Suti-II with 30.38% GCA records the lowest area share under total food crops, followed by Samserganj (45.32%), Murshidabad-Jiaganj (50.89%), and Raninagar-I (54.09%). Contrastingly, a total of 11 blocks report more than 90% of GCA under all food crops, among which suti-I (99.75%) denots the maximum GCA (Fig. 2.20).

2.7.9 Changing GCA Under Total Non-food Crops

Non-food crops, also called 'industrial crops', include fibre crops, oilseeds, fodder and green manures, floriculture, bamboo, mulberry, rubber, and narcotics crops which are not directly used for human consumption. These crops are instead processed for more valuable and alternative products. The district has experienced a loss of 2.19% GCA under total non-food crops during 1995–1996 and 2015–2016,

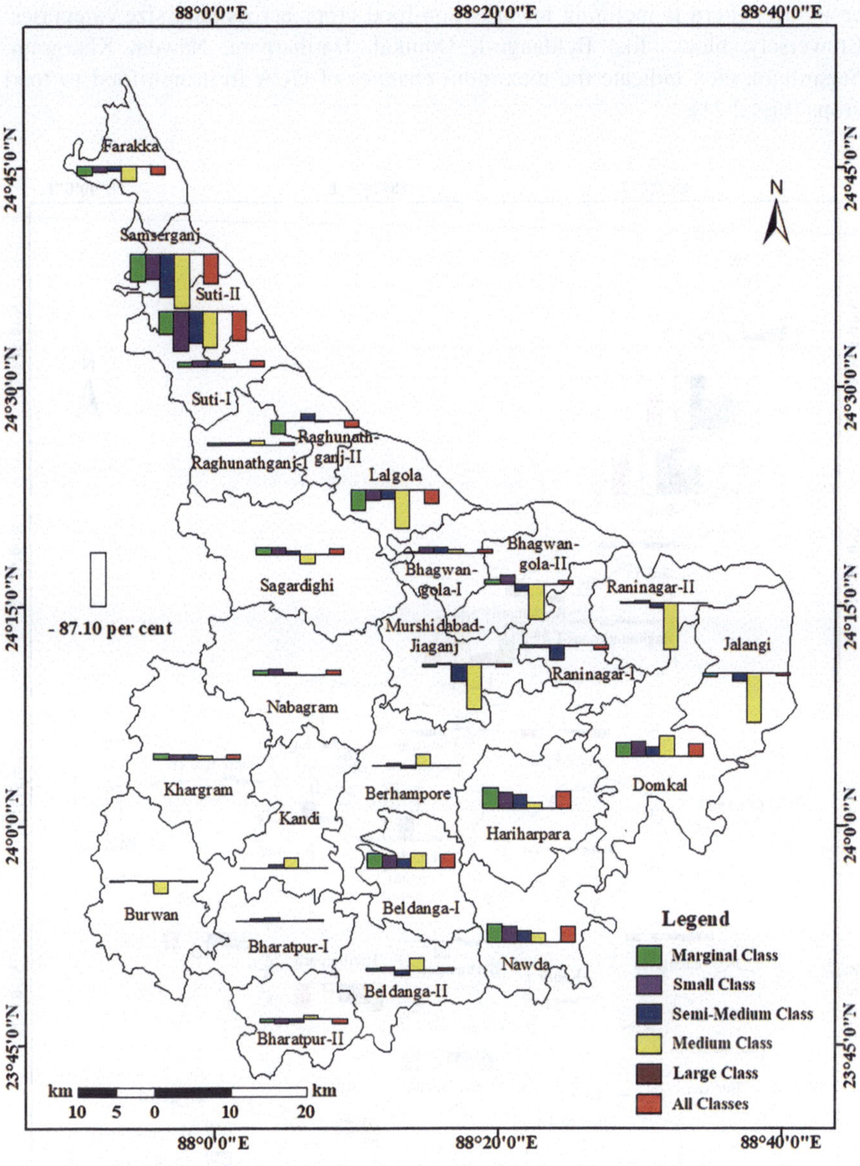

Fig. 2.20 Land size class-wise changes in GCA under total food crops in Murshidabad district, 1995–1996 to 2015–2016. (Source: Based on Annexure B.20)

which is not a good sign for overall agricultural innovation. Increases of 6.12% and 2.68% of GCA are registered under semi-medium and medium landholdings, respectively. In contrast, the marginal (4.02%) and small (2.60%) land classes report a reduction in the area under non-food crops. Most blocks of the northern part of the district, such as Farakka, Suti-II, Samserganj, and Raghunathganj-II, reveal that the cropping pattern is inclining towards non-food crops across land size categories. Conversely, blocks like Beldanga-I, Domkal, Hariharpara, Nawda, Khargram, Sagardighi, etc., indicate the maximum changes of GCA from non-food to food crops (Fig. 2.21).

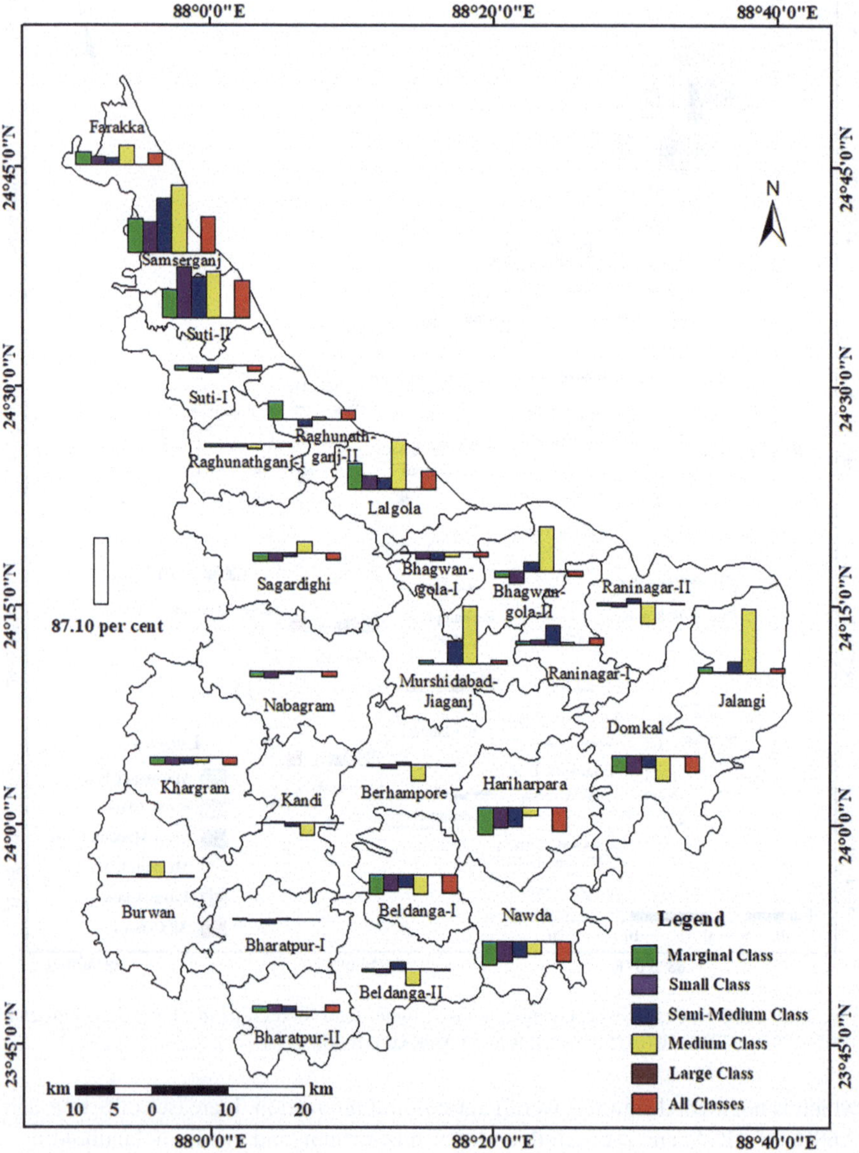

Fig. 2.21 Land size class-wise changes in GCA under total non-food crops in Murshidabad district, 1995–1996 to 2015–2016. (Source: Based on Annexure B.21)

2.8 Changing Cropping Pattern Under Major Crops in Murshidabad District

There are three cropping seasons noted in India. These are Kharif crops sown in June–July and harvested in September–October, Rabi crops sown in October–November and harvested in April–May, and Zaid crops in between Kharif and Rabi crops. In 2015–2016, the Agriculture Census listed 191 types of crops and divided them into 12 major crop categories in India. Similarly, the state of West Bengal records as many as 91 crops under 11 crop groups in 2015–2016. Cereal crops continue to account for the largest share of cultivated area and total agricultural output across both the state and India as a whole. During the same Agriculture Census, Murshidabad district displayed 82 crops under nine major crop categories. However, due to their negligible area share, floriculture, fodder, green manures, and other non-food crops are omitted from the present discussion. A list containing the name of these crops under the different categories is attached in the annexure section. According to their share in total GCA and listed in District Statistical Handbook, the principal crops are paddy, wheat, jute, mango, potato, mustard, gram, masur, maize, and others in the district. The details of these crops are further discussed across different land size class categories from 1995–1996 to 2015–2016 (Fig. 2.22).

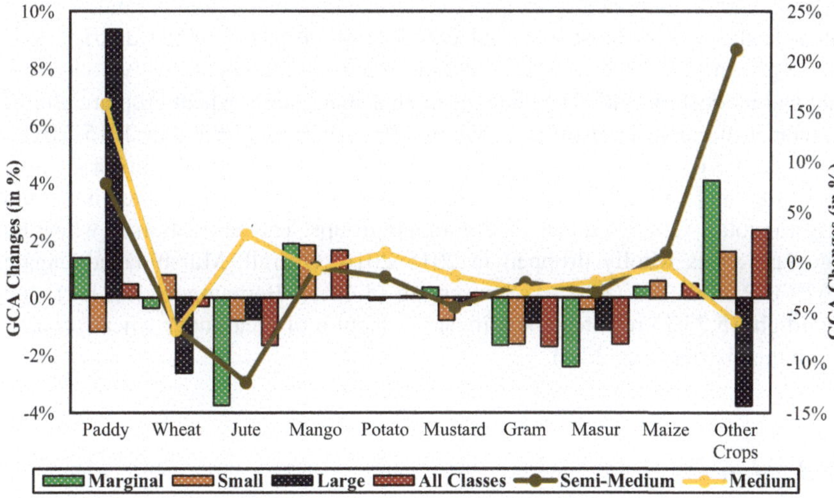

Fig. 2.22 Land size class-wise changes of gross cropped area under major crops in Murshidabad district, 1995–1996 to 2015–2016. (Source: Based on Annexure B.22)

2.8.1 *Paddy*

Paddy is the most dominant crop, occupying more than half the total gross cropped area in the Murshidabad district. The area under paddy in the district was 51.09% in 1995–1996, and it increased to 51.58% in 2015–2016. Besides the fact that this is the main food crop for the district's population, other factors such as fertile alluvial track, constant price, and availability of monsoon rainfall determine the dominance of the paddy crop. Surprisingly, all the land size classes experience an increase in area share except for the small land class. Nawda block notifies the maximum gain of the area under marginal (37.30%), small (32.32%), semi-medium (64.19%), and all (36.61%) landholdings and Domkal (67.47) leads in medium land type. Despite that, several blocks report a reduction in the area devoted to paddy during the same period, which is, in fact, a good sign for the betterment of the agriculture system. The maximum conversion of areas away from paddy under marginal, small, semi-medium, medium, and all land categories is noticed in Lalgola (29.46%), Suti-II (38.42%), Raghunathganj-II (54.22%), Jalangi (77.69%), and Lalgola (31.07%), respectively (Fig. 2.23).

2.8.2 *Wheat*

Wheat is the second most essential cereal crop cultivated in the district. It is mainly grown in the rabi cropping season, requiring a comparatively cooler climate and modest rainfall. The changes in area share under wheat crop are almost stagnant in the district (from 12.25% in 1995–1996 to 11.95% in 2015–2016). The maximum increase is recorded in Kandi under marginal (8.63%), small (16.30%), and all (9.94%), Lalgola (24.22%) under semi-medium, and Bhagwangola-I (8.84%) under medium land groups. The cultivation of wheat by large farmers is wholly dropped in 2015–2016. Overall, Murshidabad-Jiaganj (8.67%), Raninagar-I (8.64%), Samserganj (8.42%), Berhampore (5.81%), and Sagardighi (5.23%) register a significant reduction of area under wheat production in the district (Fig. 2.24).

2.8.3 *Jute*

Jute, a fibre crop often regarded as the 'golden fibre', is cultivated significantly throughout the district. Jute crop cultivation is continuously falling from 1995–1996 to 2015–2016. The marginal (2.74%), small (0.79%), and all (1.65%) land size classes have noted a reduction of area under jute crop, while semi-medium (4.45%) and medium (2.77%) land categories gain more area. The impact of variability in climatic factors, specifically the noticeable increase in temperature and

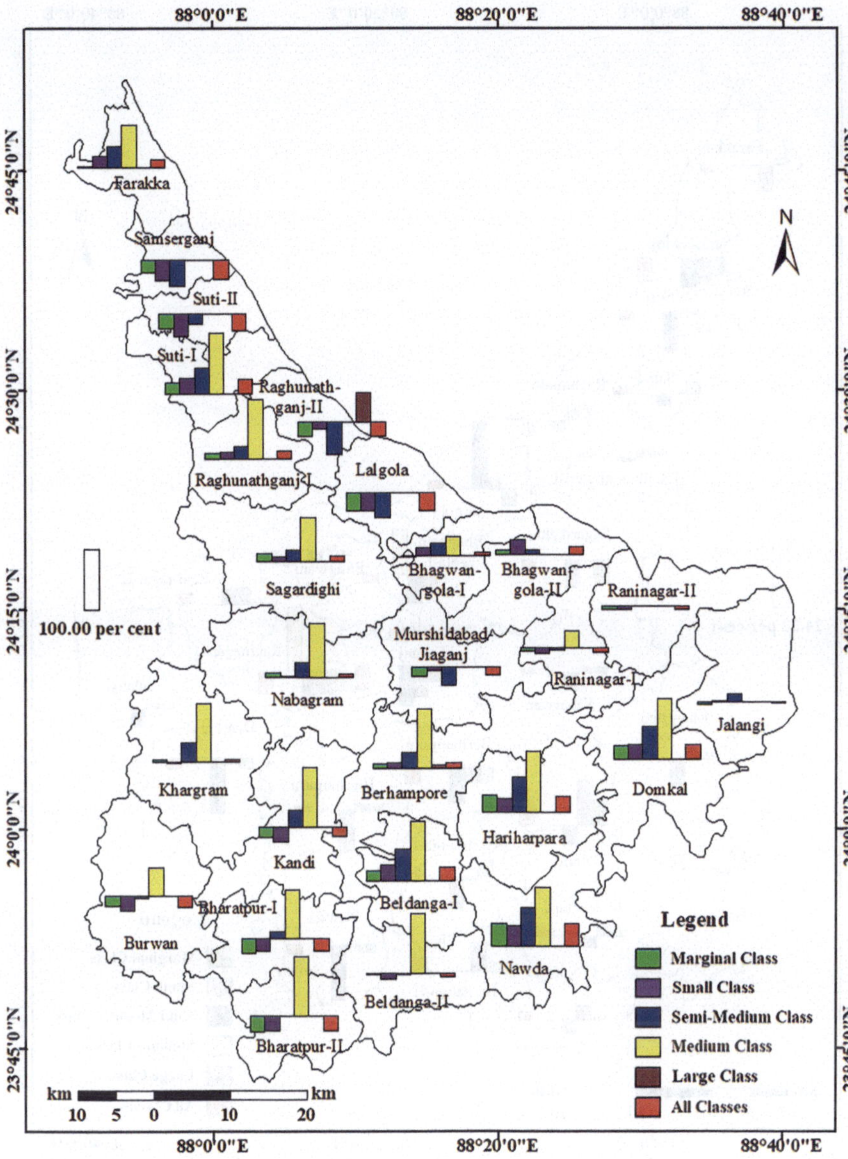

Fig. 2.23 Land size class-wise changes in GCA under paddy crop in Murshidabad district, 1995–1996 to 2015–2016. (Source: Based on Annexure B.23)

considerable variation in monsoon rainfall, causes significant fluctuation in jute production (Singh et al., 2019). Samserganj block alone gains the maximum area under marginal (38.95%), semi-medium (60.79%), medium (87.10%), and all (44.63%) landholdings. In small land type, the highest increase is recorded in Suti-II (47.08%). In contrast, Nawda in marginal (24.09%), semi-medium

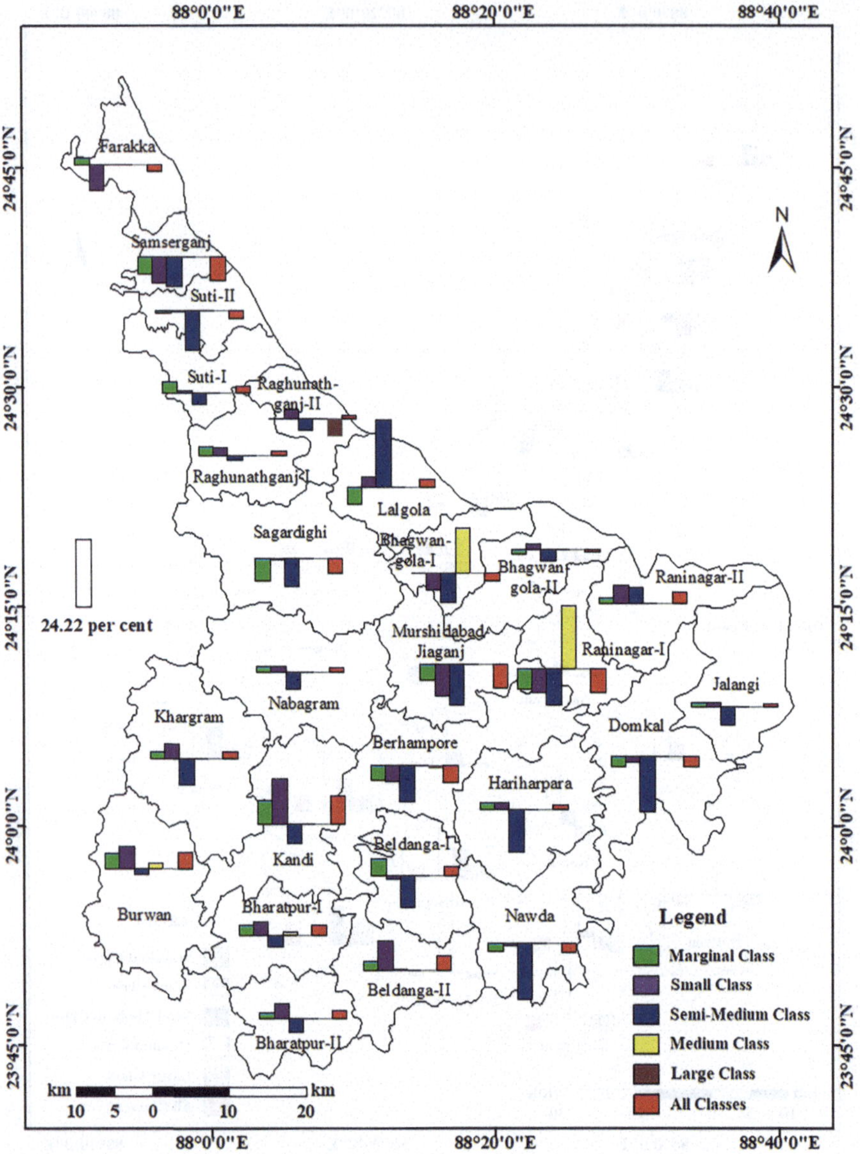

Fig. 2.24 Land size class-wise changes in GCA under wheat crop in Murshidabad district, 1995–1996 to 2015–2016. (Source: Based on Annexure B.24)

(15.72%), and all (20.59%), Bhagwangola-II in small (20.35%), and Domkal in medium (25.85%) land categories report maximum area loss in the district. It is essential to mention that in 1995–1996 only the Beldanga-I block had an area of 11.11% under jute cultivation in the large land group and was cut off in 2015–2016 (Fig. 2.25).

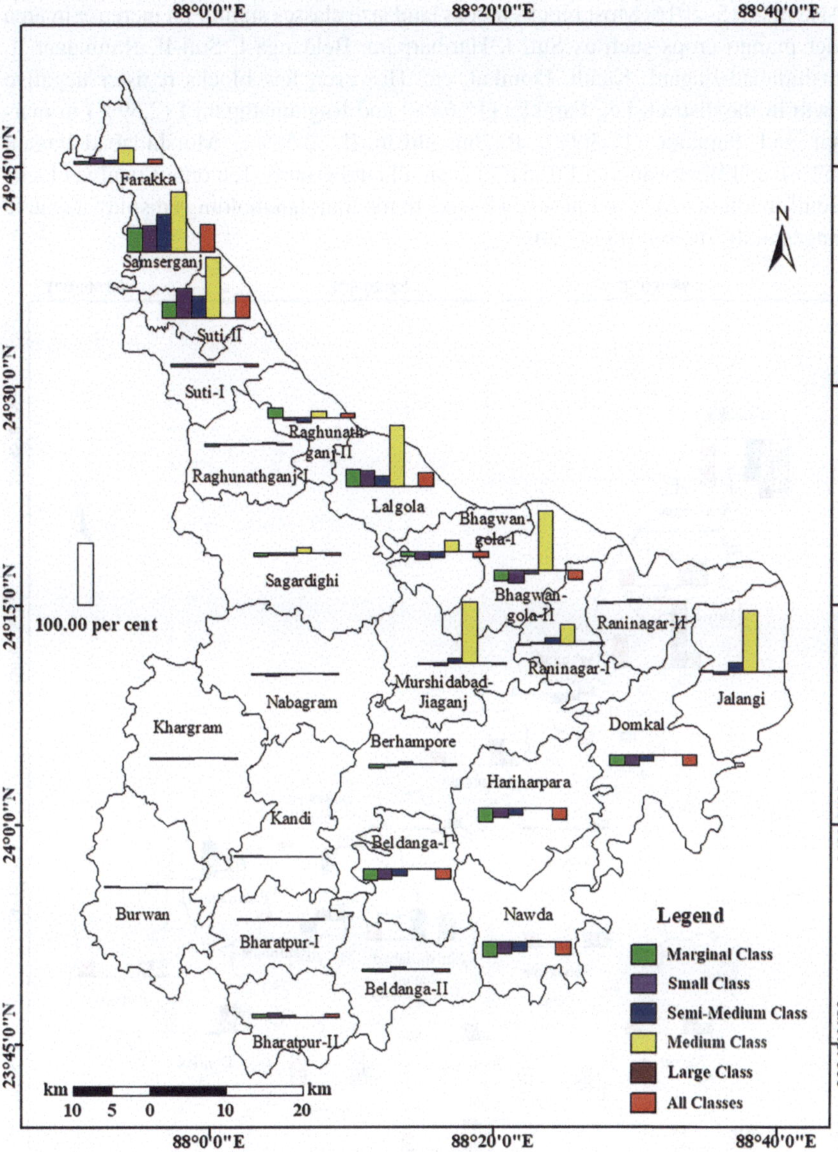

Fig. 2.25 Land size class-wise changes in GCA under jute crop in Murshidabad district, 1995–1996 to 2015–2016. (Source: Based on Annexure B.25)

2.8.4 *Mango*

Mango is the principal fruit crop and is the prime source of income for a large number of farmers in the Murshidabad district. The cropping pattern is diversifying from traditional cereal crops to fruits and vegetables. As a result, the area under mango in the district has increased more than five times, i.e., from 0.38% in 1995–1996 to

2.04% in 2015–2016. Most blocks across land size classes signify an increase in area under mango crops such as Suti-I, Hariharpara, Beldanga-I, Suti-II, Raninagar-II, Murshidabad-Jiaganj, Kandi, Domkal, etc. However, few blocks register negative growth in the district, i.e., Farakka (16.64%) and Raghunathganj-I (2.59%) in marginal and Farakka (17.36%), Raghunathganj-II (0.68%), Murshidabad-Jiaganj (0.52%), and Bhagwangola-I (0.51%) in small land groups. Ten out of twelve blocks in semi-medium and seven in seven blocks in medium landholdings display negative changes in the district (Fig. 2.26).

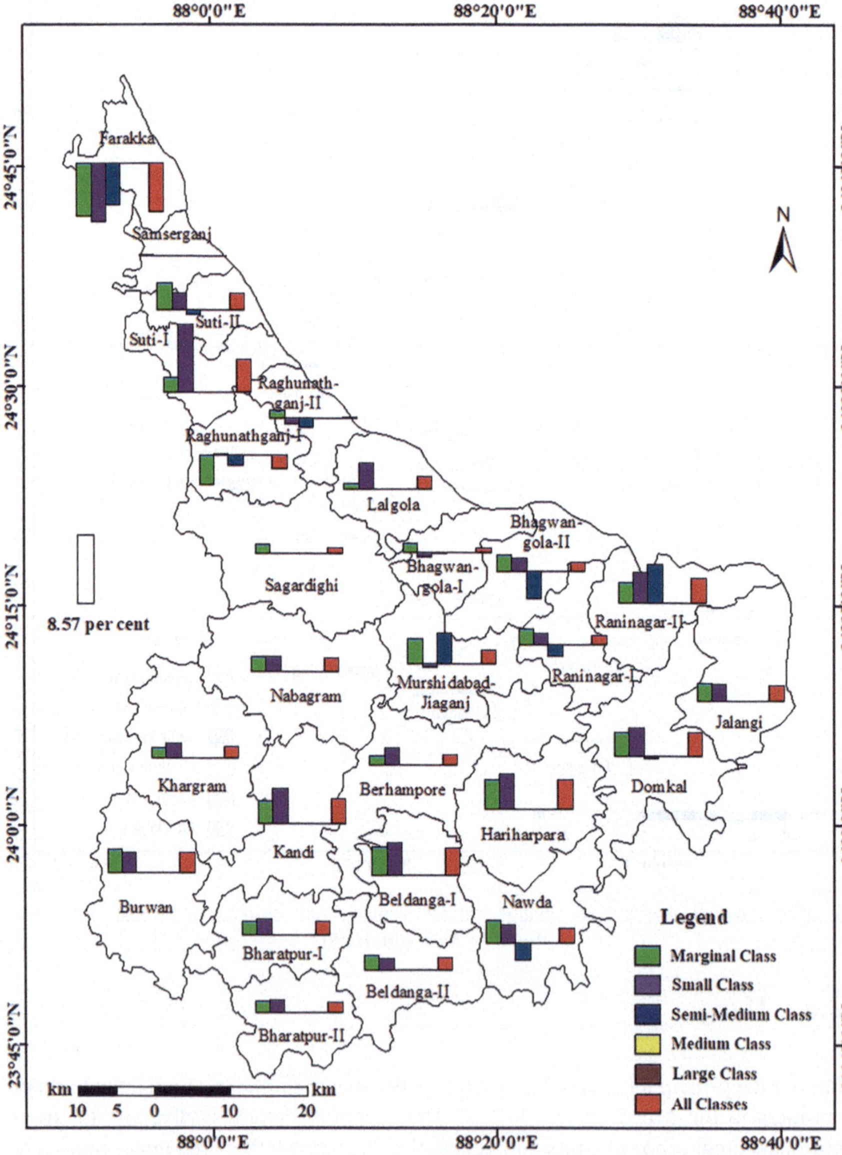

Fig. 2.26 Land size class-wise changes in GCA under mango crop in Murshidabad district, 1995–1996 to 2015–2016. (Source: Based on Annexure B.26)

2.8.5 Potato

In the district, the area under potato has increased by 0.05% from 1995–1996 to 2015–2016. The maximum area gain is noticed under the medium land class (0.99%), followed by semi-medium (0.69%) and marginal (0.02%) land categories. Conversely, the small landholding (0.10%) shows an adverse change in the study area. Suti-I (2.93%) and Farakka (2.71%) report the maximum positive growth, and Beldanga-I (2.05%) and Bharatpur-II (1.92%) reduce the highest area under marginal land class. A total of 18 blocks in the small land category depict negative differences, out of which Raninagar-II (1.86%) indicates the most. The maximum positive area change under semi-medium and medium landholdings is recorded by Suti-II (10.58%) and Burwan (22.28%), respectively. Overall, a significant increase in areas is noticed in Suti-II (4.14%), Raninagar-I (2.51%), Burwan (2.50%), and Beldanga-II (2.16%) (Fig. 2.27).

2.8.6 Mustard and Rapeseed

Agriculture Census listed mustard as an oilseed crop, occupying 4.96% and 5.13% of the district's gross cropped area in 1995–1996 and 2015–2016, respectively. The land size class-wise data reveals that the marginal (0.37%) and semi-medium (1.87) land groups report a positive change, while small (0.79%) and medium (1.36%) land types record a reduction of GCA under the mustard crop. Suti-II locates the maximum positive gain under marginal (12.66%), small (18.86%), semi-medium (20.03%), and all (15.85%) land categories. Medium landholding sees the maximum change of GCA in Sagardighi (12.61%). Conversely, the highest loss of GCA in marginal (9.57%), small (8.60%), semi-medium (9.60%), and all (9.24%) land classes are demonstrated by Hariharpara. In the medium land category, it is notified by Kandi (16.83%) (Fig. 2.28).

2.8.7 Gram

Gram, also known as chickpea or Bengal gram, is one of the district's most important pulse crops, mainly growing in the rabi cropping season. It is cultivated for human consumption as well as to feed domestic animals. The district is losing the area devoted to gram year after year. It was 1.95% of the total GCA in 1995–1996 and then reduced to only 0.25% of GCA in 2015–2016. Each block across the land classes has dropped the area under gram crop cultivation except for Raninagar-I (0.19%) in marginal, Farakka (1.03%) and Murshidabad-Jiaganj (0.89%) in small, Beldanga-II (0.12%) in semi-medium, Bharatpur-I (1.13%) in medium, and Bhagwangola-I (0.48%), Farakka (0.40%), and Murshidabad-Jiaganj (0.18%) in all land size categories. Overall, the maximum reduction in all the landholdings combined is found in Raghunathganj-II (7.75%) and Raghunathganj-I (6.55%) (Fig. 2.29).

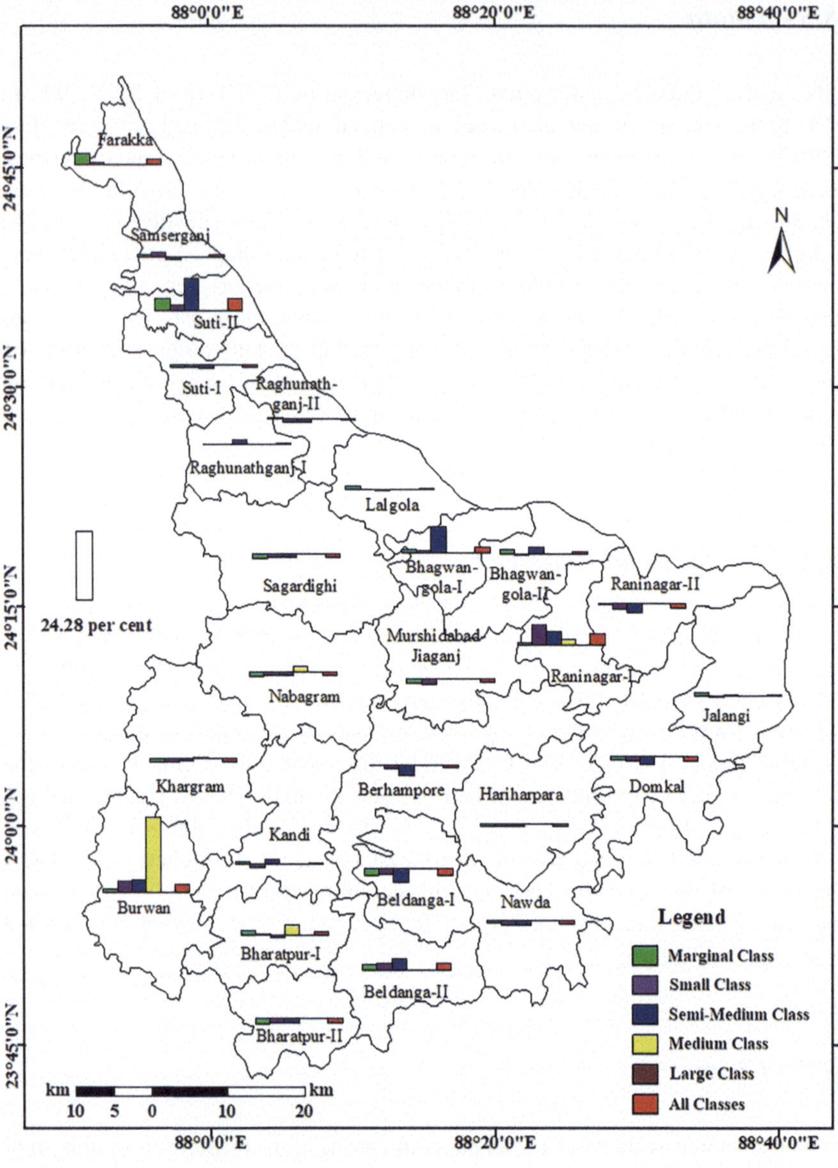

Fig. 2.27 Land size class-wise changes in GCA under potato crop in Murshidabad district, 1995–1996 to 2015–2016. (Source: Based on Annexure B.27)

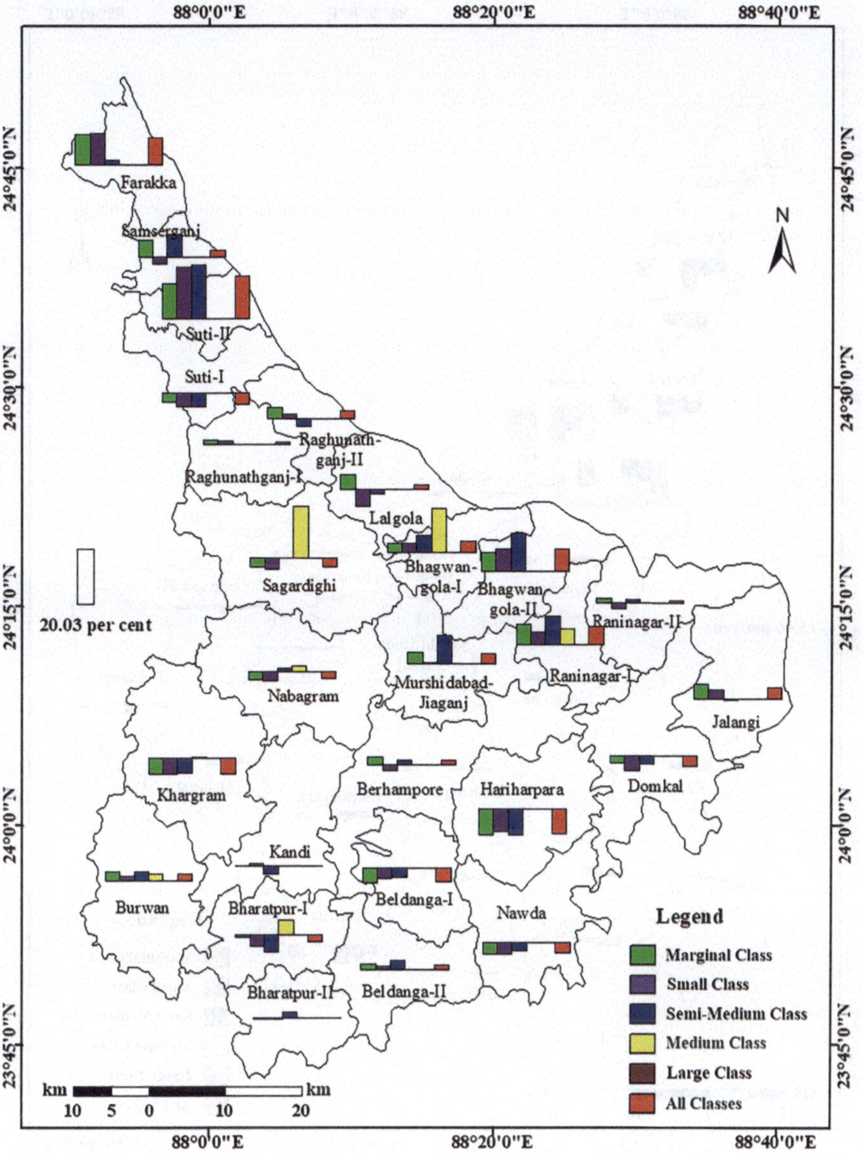

Fig. 2.28 Land size class-wise changes in GCA under mustard and rapeseed crop in Murshidabad district, 1995–1996 to 2015–2016. (Source: Based on Annexure B.28)

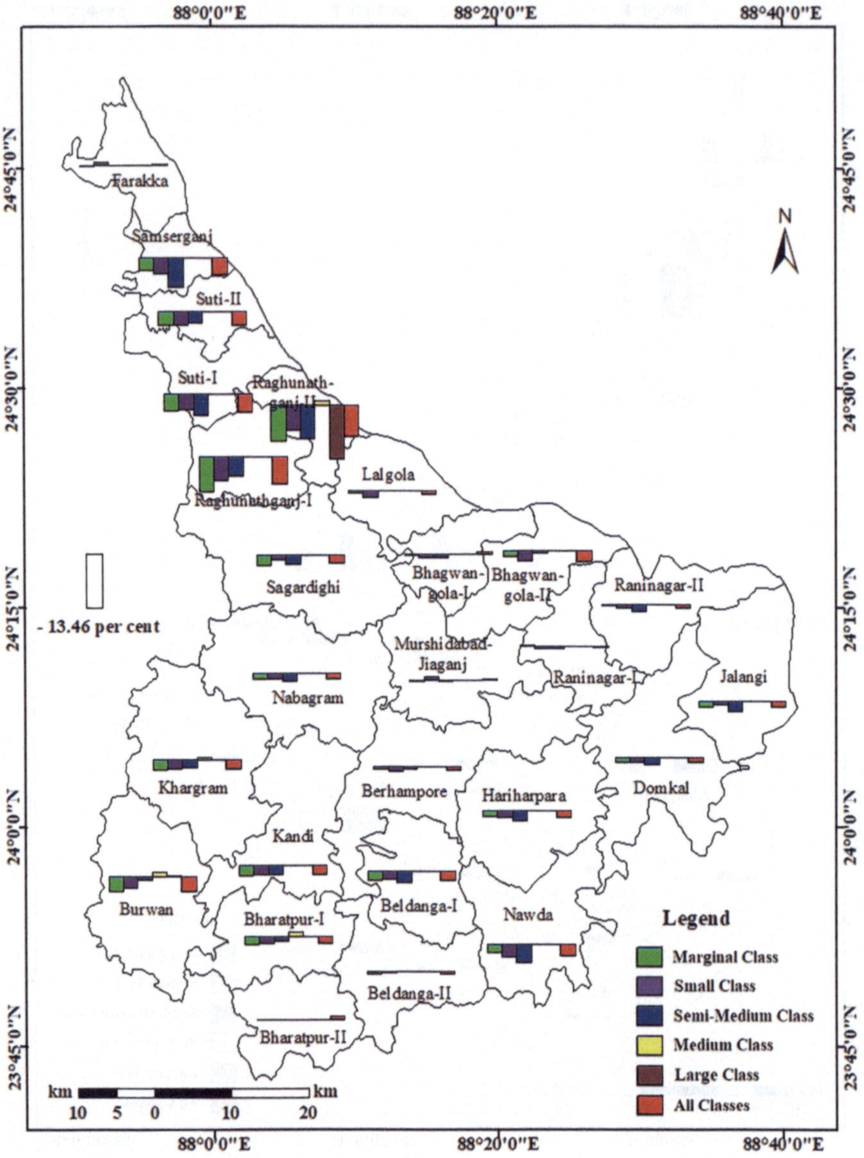

Fig. 2.29 Land size class-wise changes in GCA under gram crop in Murshidabad district, 1995–1996 to 2015–2016. (Source: Based on Annexure B.29)

2.8.8 *Masur*

Masur or lentil is the main pulse crop of Murshidabad district, mostly consumed as dry seeds. The district has recently lost the area under masur across land size categories, i.e., from 2.31% in 1995–1996 to 1.70% in 2015–2016, because the farmers are diverted towards mainly fruits and vegetable crops. Jalangi records the maximum reduction in marginal (19.20%), small (12.81%), and all (15.73%) land-holdings and the same under semi-medium and medium is recorded by Suti-II (12.46%) and Suti-I (10.17%), respectively. Conversely, some blocks gain GCA under masur cultivation, out of which Bharatpur-II (5.23%) leads in marginal, Samserganj in small (16.90%), semi-medium (7.59%), and all (7.04%), and Bhagwangola-I (15.18%) in medium land groups (Fig. 2.30).

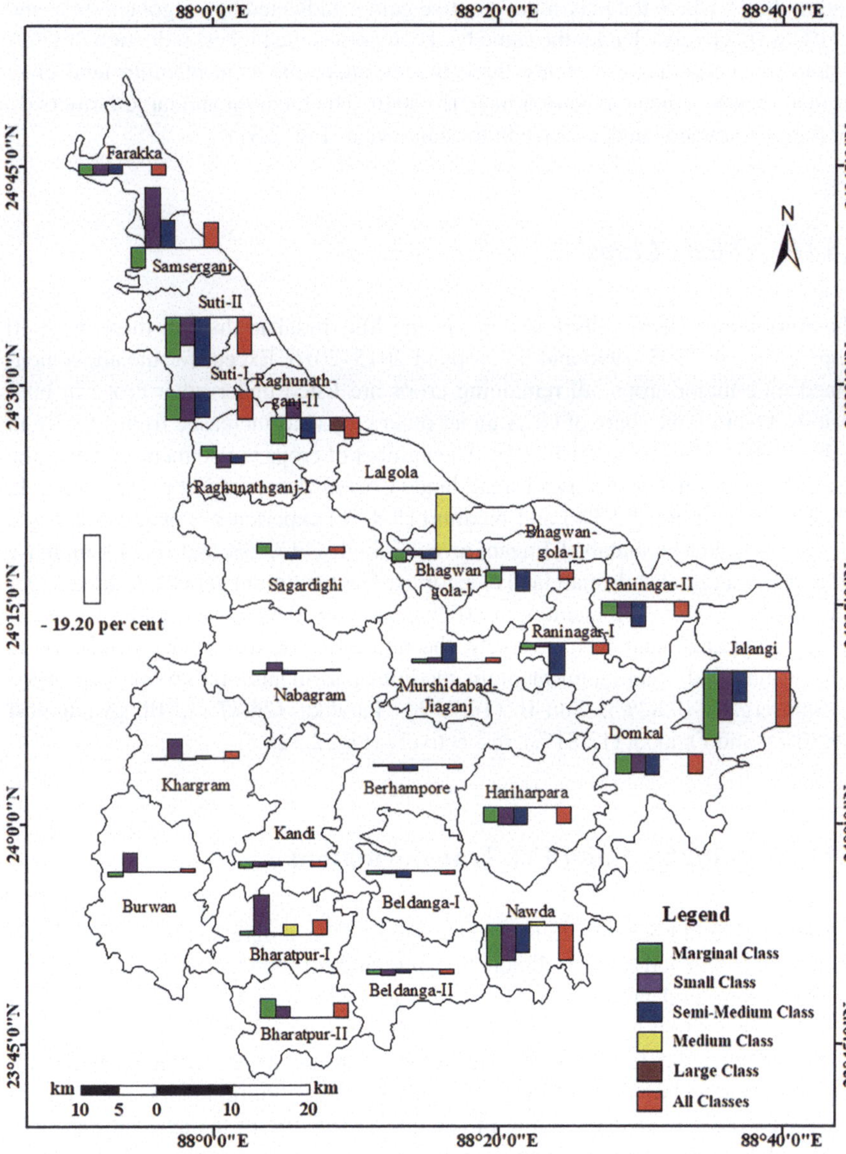

Fig. 2.30 Land size class-wise changes in GCA under Masur crop in Murshidabad district, 1995–1996 to 2015–2016. (Source: Based on Annexure B.30)

2.8.9 Maize

Maize is another significant cereal crop grown mainly in Kharif season in the district. It is often regarded as the 'Queen of cereals' due to its potential to produce the highest yield among cereal crops. The district has seen an increase in GCA under maize across the land size categories except for the medium class, which reports a loss of 0.20% GCA. Maize cultivation is mainly dominated in marginal and small landholdings, where the maximum positive gain is indicated by Lalgola (7.34% and 7.94%, respectively). Under the same land categories, the highest reduction of GCA is noticed in Raninagar-II. Only three blocks under the semi-medium land class devoted area to maize cultivation in 2015–2016. The medium and large farmers do not register any area under maize in the same year (Fig. 2.31).

2.8.10 Others Crops

The Agriculture Census discloses that in the Murshidabad district, there were 40 crops grown in 1995–1996 and 82 crops in 2015–2016. Except for the abovementioned nine major crops, all remaining crops are listed under other crops in both periods. Overall, the share of GCA under other crops has increased from 11.05% in 1995–1996 to 12.43% in 2015–2016. The number of crops is also increased over the years. The marginal (4.09%) and small land (1.62%) classes report positive growth, while semi-medium (5.54%) and medium (5.87%) experience a negative change. Bharatpur-I and Murshidabad-Jiaganj reveal as high as 17.60% and 16.74% increase in GCA in marginal and small land categories. The maximum gain of the area under semi-medium (82.58%), medium (72.07%), and all (20.44%) landholdings is notified by Raghunathganj-II. Conversely, the highest reduction of GCA under other crops in marginal, small, semi-medium, medium, and all land size groups is recorded in Samserganj (9.82%), Suti-II (19.75%), Farakka (26.87%), Bhagwangola-I (19.80%), and Farakka (12.81%), respectively (Fig. 2.32).

2.9 Sustainable Land Use Transformation

Agricultural land use refers to the primary use of the geographical area for different purposes and activities. It directs to using land for the primary purpose of producing agricultural products for human or animal consumption. Thus, it results from the interaction between man and the environment. Besides physical factors such as relief, climate, and soils, agricultural land use is also affected by socio-economic and technological factors. The study of agricultural land use of a region provides a base for understanding the complex structure of the agricultural landscape, which helps in better planning of that region. The net sown area, fallow land, and uncultivable land, excluding fallow land, form the agricultural land. The term

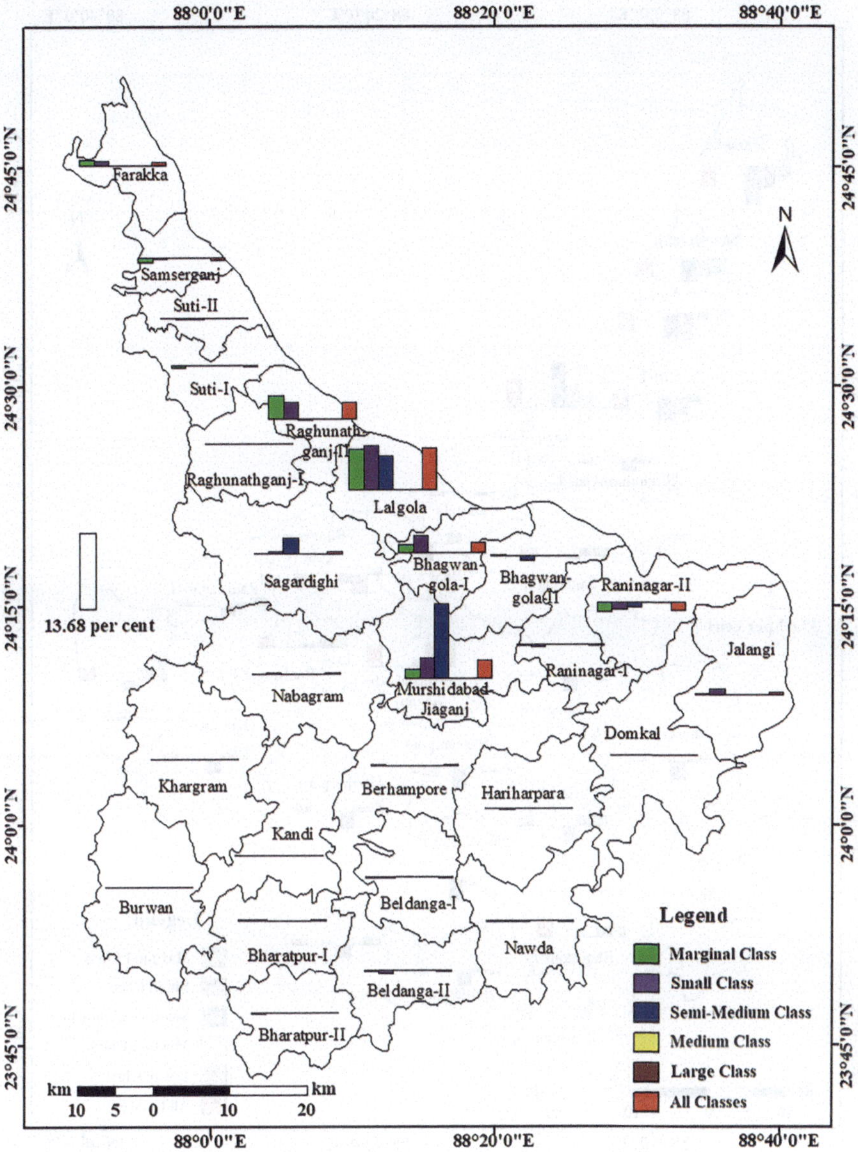

Fig. 2.31 Land size class-wise changes in GCA under maize crop in Murshidabad district, 1995–1996 to 2015–2016. (Source: Based on Annexure B.31)

'agricultural land use' denotes the extent of the gross cropped area under various crops during the reference year. In short, agricultural land use means a cropping pattern of an agriculture system, which is the proportion of area under multiple crops at a point in time and the spatial arrangement of crops and fallow on a given area. A cropping pattern is thus a dynamic concept that changes over space and time.

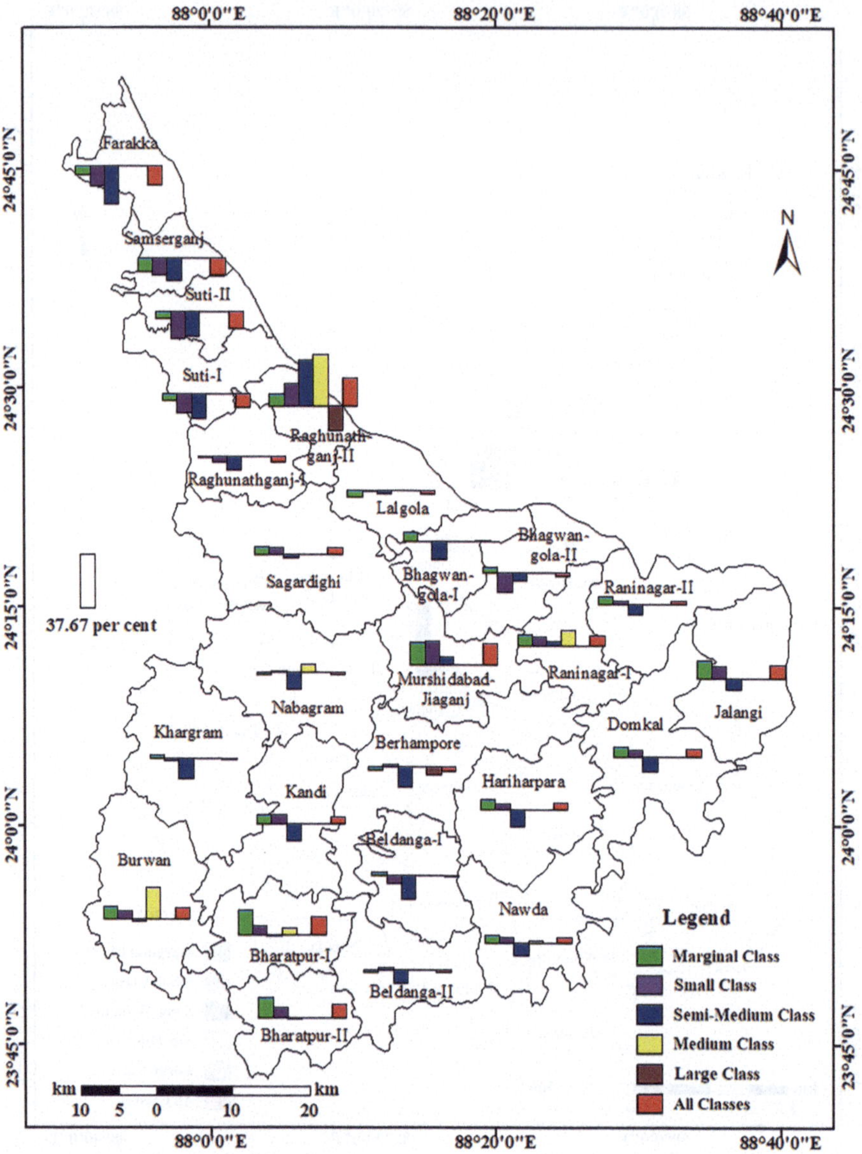

Fig. 2.32 Land size class-wise changes in GCA under other crops in Murshidabad district, 1995–1996 to 2015–2016. (Source: Based on Annexure B.32)

The agricultural land use pattern reflects the overall agricultural development of a particular region. Therefore, it is essential to show the agricultural land use transformation of the Murshidabad district from 1995–1996 to 2015–2016. Twenty-four variables from the ongoing discussion of land use and cropping patterns have been selected for the said purpose in the study area (see Annexure B.33).

Table 2.1 Common breaking points for agricultural land use transformation

Quartile division	1995–1996	2015–2016
Lower quartile	0.369	0.445
Upper quartile	0.425	0.510

Source: Calculated by authors

Data on the 24 variables described above are standardized first with the help of the UNDP method. Then a composite index of these variables has been constructed to show the spatial variation of sustainable land use intensity in the district. The calculated composite scores are categorized into three classes based on the quartile technique, where the first and third quartiles are taken as the breaking points. The common breaking point of composite score across the land groups in 1995–1996 is 0.369 for the lower quartile and 0.425 for the upper quartile. The same becomes 0.445 and 0.510, respectively, in 2015–2016 (see Table 2.1).

2.9.1 Sustainable Land Use Transformation Under Marginal Land Size Class

Figure 2.33 displays the agricultural land use intensity in the marginal land category for 1995–1996 and 2015–2016 in the district. Nine blocks have represented the index value above the upper quartile (0.425) in 1995–1996. Berhampore (0.475) emerges as the most intense block in terms of agricultural land use in the district, followed by Domkal (0.466) and Raghunathganj-II (0.462). Apart from the cereal crops, the dominant distribution of the gross cropped area under various crop categories like vegetables, oilseeds, fibres, etc. is the primary reason for such a higher index in those blocks. In the same year, five blocks are placed below the lower quartile (0.369), in which Raghunathganj-I (0.283) is located at the bottom. The remaining blocks are moderately intense regarding agricultural land use and lie between the upper and lower quartile index. In 2015–2016, the number of blocks situated below the lower quartile was reduced to three, where Raghunathganj-I (0.411) found itself at the bottom of the index list. Eight blocks have enjoyed better agricultural land use intensity (>0.510) in the said year, of which Domkal has the maximum index value, i.e., 0.578.

2.9.2 Sustainable Land Use Transformation Under Small Land Size Class

It is seen in Fig. 2.34 that under the small land size class, eight blocks lie above the upper quartile (0.425) in 1995–1996, whereas nine blocks are found above the third quartile (0.510) in the district during 2015–2016. Berhampore reports the maximum index value under this category in 1995–1996 (0.469) as well as in 2015–2016

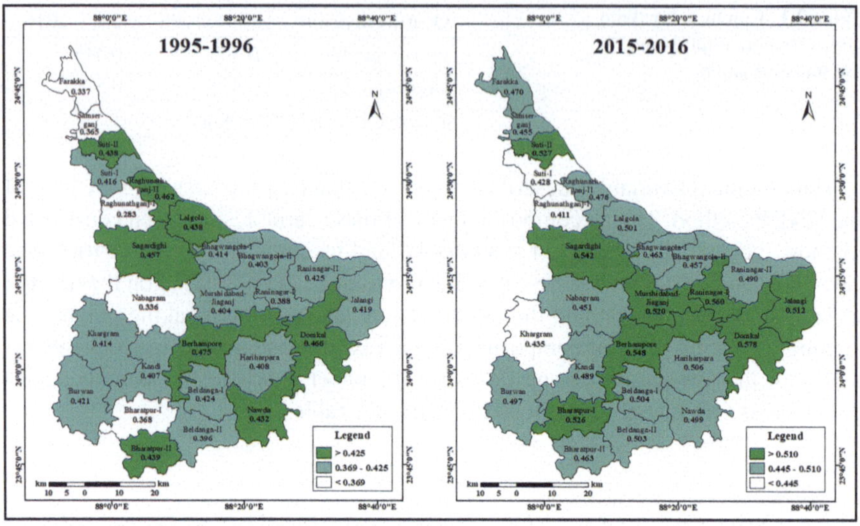

Fig. 2.33 Sustainable land use index in marginal land size category in Murshidabad district, 1995–1996 and 2015–2016. (Source: Based on Annexure B.34)

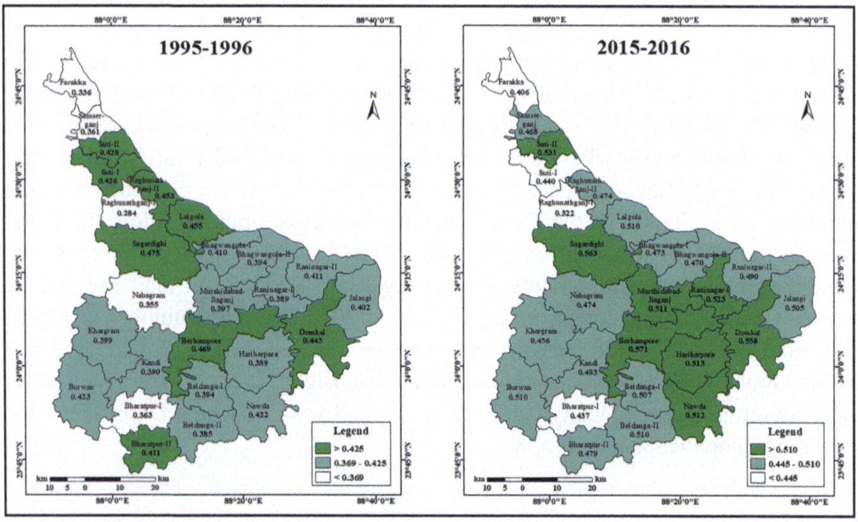

Fig. 2.34 Sustainable land use index in small land size category in Murshidabad district, 1995–1996 and 2015–2016. (Source: Based on Annexure B.34)

(0.571). Sagardighi, Domkal, Bharatpur-II, Suti-II, Burwan, etc. are the other blocks that have seen better indexes during the study periods. Five blocks registered index values below the lower quartile (0.369) in 1995–1996, and these were Raghunathganj-I (0.284), Farakka (0.336), Nabagram (0.355), Samserganj (0.361), and Bharatpur-I (0.363). However, four blocks were situated below the lower quartile (0.445) in 2015–2016. Nabagram has managed to upgrade itself to the moderate index category, while the rest of the four blocks have remained in the list of lower index categories in 2015–2016. The reasons for such a lower index in these blocks are the dominance of cereal crops in the cropping pattern and the existence of a large share of fallow and culturable wasteland in the total area.

2.9.3 Sustainable Land Use Transformation Under Semi-Medium Land Size Class

A total of five and four blocks under the semi-medium land category are located above the upper quartile in 1995–1996 and 2015–2016, respectively (Fig. 2.35). The five blocks that were found in 1995–1996 were Berhampore, Sagardighi, Lalgola, Domkal, and Bharatpur-II, in which Berhampore remained at the top with an index value of 0.458. Interestingly, in 2015–2016,Berhampore, Sagardighi, and Domkal have managed to hold their position above the upper quartile, and the highest index value in the said year is once again scored by Berhampore (0.542). This significant improvement in agricultural land use intensity is mainly because of the bringing back of uncultivated and fallow land areas to the net sown areas that were kept unused previously. Most of the blocks in 1995–1996 (15) as well as in 2015–2016 (12) are found in the medium range of agricultural land use. Six blocks, namely Raghunathganj-I (0.273), Farakka (0.325), Raghunathganj-II (0.342), Nabagram (0.344), Samserganj (0.350), and Bharatpur-I (0.352) were situated below the lower quartile (0.369) in 1995–1996. In 2015–2016, the number of blocks lying below the first quartile increased to ten, and Raghunathganj-I (0.352) sits at the bottom among them.

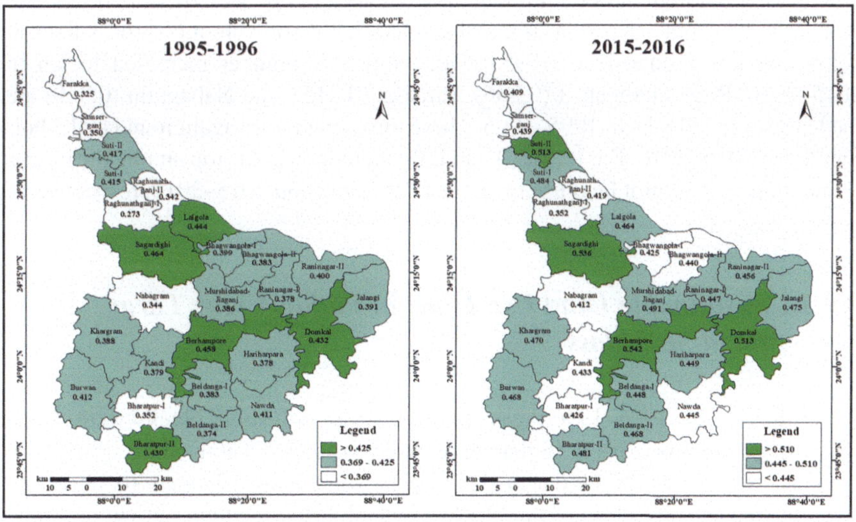

Fig. 2.35 Sustainable land use index in semi-medium land size category in Murshidabad district, 1995–1996 and 2015–2016. (Source: Based on Annexure B.34)

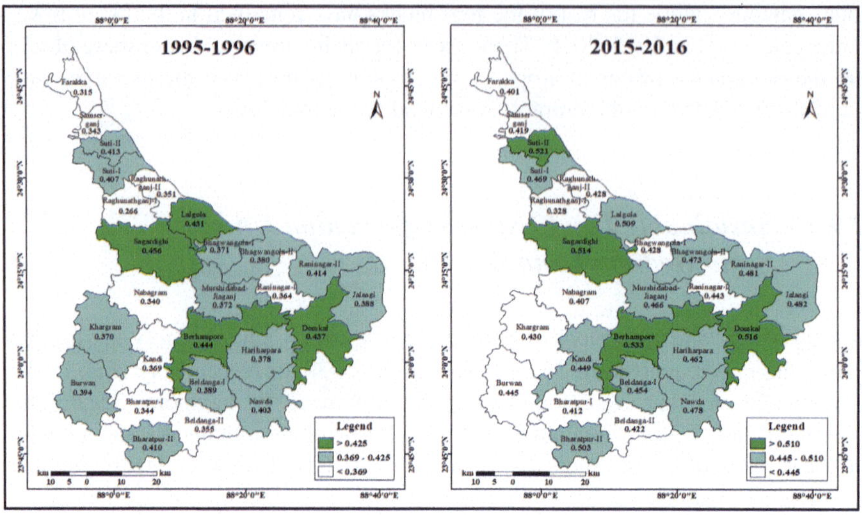

Fig. 2.36 Sustainable land use index in medium land size category in Murshidabad district, 1995–1996 and 2015–2016. (Source: Based on Annexure B.34)

2.9.4 *Sustainable Land Use Transformation Under Medium Land Size Class*

Figure 2.36 reveals four blocks that reported index values above the upper quartile (0.425) in 1995–1996. These were Berhampore (0.444), Sagardighi (0.456), Domkal (0.437), and Lalgola (0.431) sequentially. Except for Lalgola, all these blocks remained in the high agricultural land use category in 2015–2016. Suti-II has also joined the list in the said year. It may be noted that there were nine blocks located below the lower quartile in 1995–1996, while this number increased to ten in 2015–2016. Raghunathganj-I (0.266), Farakka (0.315), and Nabagram (0.340) are the least scored blocks in 1995–1996. These three blocks are again found at the bottom list in 2015–2016. Raninagar-II and Lalgola remain at the top among the blocks of medium agricultural land use range in 1995–1996 and 2015–2016, respectively.

2.9.5 *Sustainable Land Use Transformation Under Large Land Size Class*

According to the 1995–1996 Agricultural Census, only nine blocks were found where large farmers existed. Berhampore (0.491) is situated at the high agricultural land use range, while Beldanga-II (0.407), Murshidabad-Jiaganj (0.389), and Raghunathganj-II (0.375) are the blocks situated at the medium agricultural land use range. The remaining five blocks, i.e., Domkal, Farakka, Raghunathganj-I,

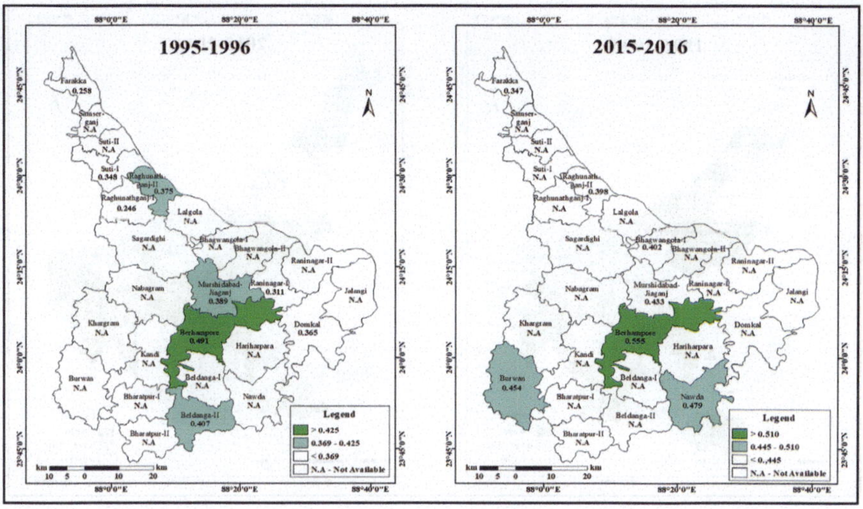

Fig. 2.37 Sustainable land use index in large land size category in Murshidabad district, 1995–1996 and 2015–2016. (Source: Based on Annexure B.34)

Raninagar-I, and Suti-I, had a low intensity of agricultural land use, of which Raghunathganj-I placed at the bottom with an index value of 0.246. In 2015–2016, large farmers were registered in only seven blocks, namely Berhampore, Burwan, Bhagwangola-I, Farakka, Murshidabad-Jiaganj, Nawda, and Raghunathganj-II. Berhampore (0.555) emerges as the most intense block in agricultural land use among the blocks mentioned above, followed by Nawda (0.479). Four blocks display index values lower than the first quartile and two between the upper and lower quartile during the same year. Due to the dominance of cereal crops, the large land size category reports lower intensity of agricultural land use compared to the other land size classes (Fig. 2.37).

2.9.6 Sustainable Land Use Transformation Under All Land Size Classes

Figure 2.38 represents the agricultural land use transformation under all land size class category in 1995–1996 and 2015–2016. As many as eight blocks were positioned above the upper quartile index (0.425) in 1995–1996, in which Berhampore (0.476) secured the maximum, followed by Sagardighi (0.474), Domkal (0.460), Raghunathganj-II (0.455), and Lalgola (0.445). There are five blocks located at the low intensity of land use under this category. Raghunathganj-I (0.286) has seen the lowest index value in the said year. In contrast, only three blocks were placed below the lower quartile in 2015–2016. The 2015–2016 Agricultural Census reports the same number of blocks located at the high agricultural land use intensity as found

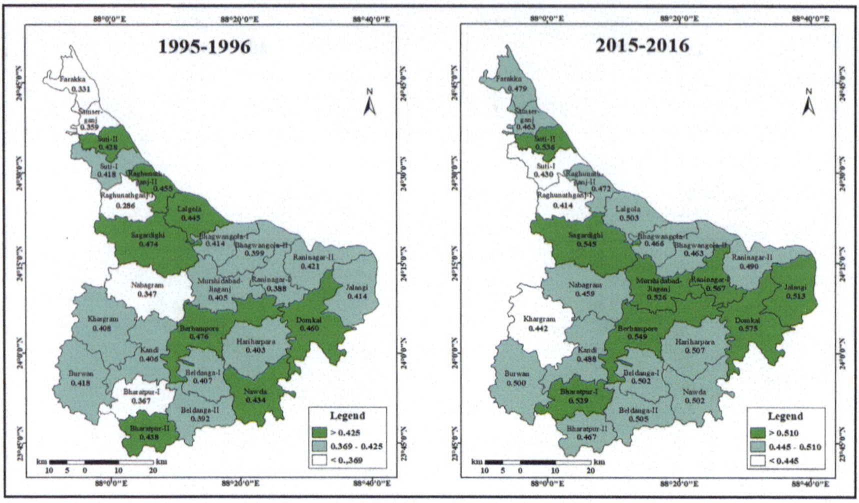

Fig. 2.38 Sustainable land use index in all land size category in Murshidabad district, 1995–1996 and 2015–2016. (Source: Based on Annexure B.34)

in 1995–1996. The maximum index value is reported by Domkal (0.575). Fifteen blocks are found in the moderate intensity of agricultural land use in the same year. High diversification in cropping pattern, increase in the gross cropped area due to the improved irrigation facilities, putting the fallow and uncultivated land into agricultural uses, etc. are the primary reasons for the higher land use of agricultural activities in the district.

2.10 An Insight from Field Survey

During the primary survey, it was found that more than 78% of the selected farmers in the district change their agricultural land use pattern very often, means in less than 3 years. In the different land size classes, the maximum percentage of respondents who react positively is seen in the marginal land category (81.25%), and the minimum rate of the farmers who change their crop pattern is recorded in the medium land class (52.85%). A significant proportion of farmers, 78.21% and 61.54% under small and semi-medium land groups, respectively, respond that they change their land use pattern at least once every 3 years both season-wise and crop-wise (Fig. 2.39). Most respondents agree that they change the cropping pattern from low-subsistence crops to high-value crops such as horticulture. The marginal and small farmers prefer to shift their cropping pattern to vegetables, pulses, and oilseed. In contrast, semi-medium and medium land farmers mainly change their crops from traditional cereals to fibre crops and, in some cases, to fruits and oilseeds.

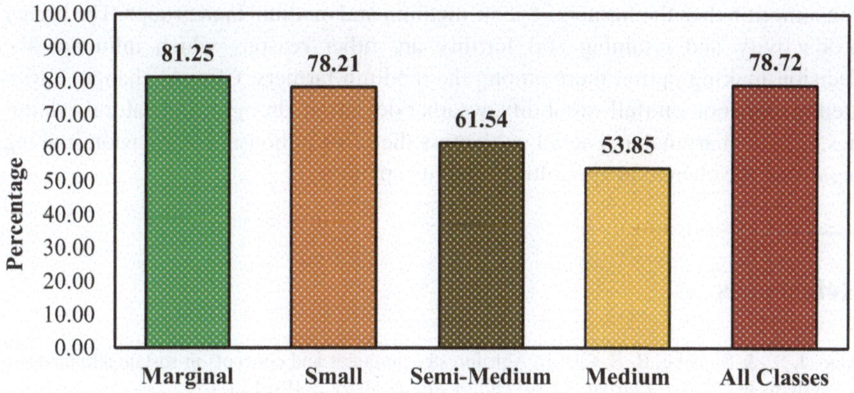

Fig. 2.39 Percentage of farmers who change agricultural land use very often in Murshidabad district, 2021. (Source: Based on Field Survey Data)

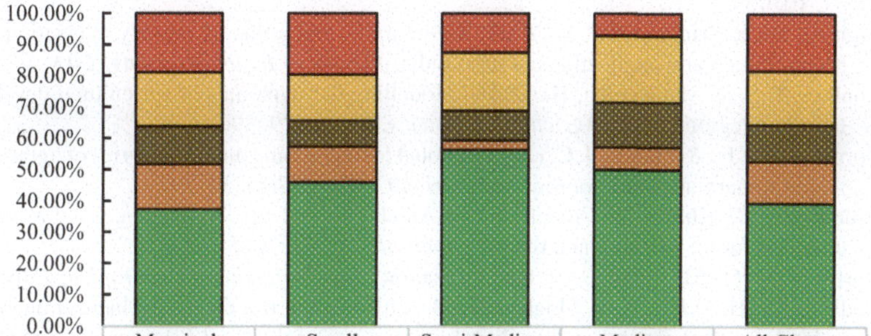

	Marginal	Small	Semi-Medium	Medium	All Classes
Consumption	18.93	19.67	12.50	7.14	18.40
Higher Productivity	17.36	14.75	18.75	21.43	17.26
Soil Fertility	12.03	8.20	9.38	14.29	11.56
Climate Change	14.40	11.48	3.13	7.14	13.36
Profit Opportunities	37.28	45.90	56.25	50.00	39.41

Fig. 2.40 Reasons for agricultural land use pattern changes in Murshidabad district, 2021. (Source: Based on Field Survey Data)

Figure 2.40 illustrates the various reasons for changing agricultural land use patterns in different land size classifications. An extensive literature survey has found that the primary reasons for changing agricultural patterns are profit opportunities, consumption, higher productivity, climate change, and soil fertility. Therefore, the nature of options in the field survey was kept close-ended with these options to see which reasons play a crucial role in decision-making to change agricultural patterns under the different land categories. The profit opportunity is the most significant reason in decision-making across all the land size classes. Beyond profit opportunity, cultivating food crops for consumption remains another vital factor in household decision-making. The marginal and small farmers are more inclined towards consumption than the farmers of semi-medium and medium land groups. The higher productivity and retaining soil fertility are other reasons which influence the decision-making option more among the medium farmers. Climate change, attributed to monsoon rainfall variability, weather deviation, drought, and natural calamities, affects marginal and small cultivators the most in household decision-making regarding the change in agricultural land use pattern.

References

Alao, J. S., & Shuaibu, R. B. (2013). Agroforestry practices and concepts in sustainable land use systems in Nigeria. *Journal of Horticulture and Forestry, 5*(10), 156–159.

Arvanitoyannis, I. S., & Tserkezou, P. (2008). Cereal waste management: Treatment methods and potential uses of treated waste. In *Waste management for the food industries* (pp. 629–702).

Aznar-Sánchez, J. A., Piquer-Rodríguez, M., Velasco-Muñoz, J. F., & Manzano-Agugliaro, F. (2019). Worldwide research trends on sustainable land use in agriculture. *Land Use Policy, 87*, 104069.

Bigman, D., & Srinivasan, P. V. (2002). Geographical targeting of poverty alleviation programs: Methodology and applications in rural India. *Journal of Policy Modeling, 24*(3), 237–255.

Bontkes, T. S., & van Keulen, H. (2003). Modelling the dynamics of agricultural development at farm and regional level. *Agricultural Systems, 76*(1), 379–396.

Borras, S. M., Jr., & Franco, J. C. (2012). Global land grabbing and trajectories of agrarian change: A preliminary analysis. *Journal of Agrarian Change, 12*(1), 34–59.

Brundtland, G., Khalid, M., Agnelli, S., Al-Athel, S., Chidzero, B., Fadika, L., et al. (1987). Our common future ('Brundtland report'). *Journal Medicine and War, 4*(1).

Burley, T. M. (1961). Land use or land utilization? *The Professional Geographer, 13*(6), 18–20.

Dadashpoor, H., Azizi, P., & Moghadasi, M. (2019). Land use change, urbanization, and change in landscape pattern in a metropolitan area. *Science of the Total Environment, 655*, 707–719.

Elliott, J. (2012). *An introduction to sustainable development*. Routledge.

FAO. (1996). *Agro-ecological zoning guidelines* (FAO Soils Bulletin 73). FAO.

Food and Agriculture Organization of the United Nations, Land, Water Development Division, & Agriculture Organization of the United Nations, Interdepartmental Working Group on Land Use Planning. (1995). *Planning for sustainable use of land resources: Towards a new approach* (No. 2). Food & Agriculture Organization of the UN (FAO).

Form, W. H. (1954). The place of social structure in the determination of land use: Some implications for a theory of urban ecology. *Social Forces, 32*, 317–322.

Garrison, W. L., & Marble, D. F. (1957). The spatial structure of agricultural activities1. *Annals of the Association of American Geographers, 47*(2), 137–144.

Guttenberg, A. Z. (1959). A multiple land use classification system. *Journal of the American Institute of Planners, 25*(3), 143–150.

Haar, C. M. (1957). Regionalism and realism in land-use planning. *University of Pennsylvania Law Review, 105*(4), 515–537.

Isard, W. (1952). A general location principle of an optimum space-economy. *Econometrica: Journal of the Econometric Society, 20*, 406–430.

Kumar, B. M. (2005). Land use in Kerala: Changing scenarios and shifting paradigms. *Journal of Tropical Agriculture, 43*, 1–12.

Lambin, E. F., Rounsevell, M. D., & Geist, H. J. (2000). Are agricultural land-use models able to predict changes in land-use intensity. *Agriculture, Ecosystems & Environment, 82*(1–3), 321–331.

Lovell, S. T. (2010). Multifunctional urban agriculture for sustainable land use planning in the United States. *Sustainability, 2*(8), 2499–2522.

Mahesh, R. (1999). *Causes and consequences of change in cropping pattern: A location-specific study* (p. 56). Kerala Research Programme on Local Level Development, Centre for Development Studies.

Mandal, R. B. (1990). *Land utilization: Theory and practice*. Concept Publishing Company.

Mandal, R., & Bezbaruah, M. P. (2013). Diversification of cropping pattern: Its determinants and role in flood affected agriculture of Assam Plains. *Indian Journal Of Agricultural Economics, 68*(2), 169–181.

Martínez-Casasnovas, J. A., Sánchez-Bosch, I., & Recasens, J. (2005). Impact assessment of land use change on soil erosion risk in Mediterranean vineyards. Soil Use and Management, 21(3), 312-321. https://doi.org/10.1079/SUM2005321.

Meyer, W. B., & Turner, B. L. (Eds.). (1994). *Changes in land use and land cover: A global perspective* (Vol. 4). Cambridge University Press.

Morgan, W. B., & Munton, R. J. C. (1971). *Agricultural geography*. Routledge.

Moser, S. C. (1996). A partial instructional module on global and regional land use/cover change: Assessing the data and searching for general relationships. *GeoJournal, 39*, 241–282.

Nath, V. (1969). The growth of Indian agriculture: A regional analysis. *Geographical Review, 59*, 348–372.

Rahaman, H. (2021a). Land use and cropping pattern dynamic. In *Diversified cropping pattern and agricultural development: A case study from Malda district, India* (pp. 35–69). Springer.

Rahaman, H. (2021b). Status of crop diversification. In *Diversified cropping pattern and agricultural development*. Springer. https://doi.org/10.1007/978-3-030-55728-7_4

Ranade, C. G. (1980). Impact of cropping pattern on agricultural production. *Indian Journal Of Agricultural Economics, 35*(2), 85–92.

Siddiqui, S. H., & Rahaman, H. (2019). Land size class wise change in cropping pattern in Malda district—A block level analysis. *Journal of Geography, Environment and Earth Science International, 21*(4), 1–19.

Singh, D. K., Jaiswal, C. S., Reddy, K. S., Singh, R. M., & Bhandarkar, D. M. (2001). Optimal cropping pattern in a canal command area. *Agricultural Water Management, 50*(1), 1–8.

Singh, N. J., Kudrat, M., Jain, K., & Pandey, K. (2011). Cropping pattern of Uttar Pradesh using IRS-P6 (AWiFS) data. *International Journal of Remote Sensing, 32*(16), 4511–4526.

Singh, R., Singh, G. S., & Bhattacharya, P. (2019). Sustainable land use planning through participatory approaches. *Journal of Environmental Management*, 243, 496–507. https://doi.org/10.1016/j.jenvman.2019.05.023.

Srivastava, P., Singh, R., Tripathi, S., & Raghubanshi, A. S. (2016). An urgent need for sustainable thinking in agriculture—An Indian scenario. *Ecological Indicators, 67*, 611–622.

Tanrivermis, H. (2003). Agricultural land use change and sustainable use of land resources in the Mediterranean region of Turkey. *Journal of Arid Environments, 54*(3), 553–564.

Thornton, P. K., & Jones, P. G. (1998). A conceptual approach to dynamic agricultural land-use modelling. *Agricultural Systems, 57*(4), 505–521.

Turner, B. L., II, Skole, D., Sanderson, S., Fischer, G., Fresco, L., & Leemans, R. (1995). *Land-use and land-cover change: Science/research plan* (IGBP Report No. 35, HDP Report No. 7). IGBP and HDP. Ch. 22.

Usmani, T. M., & Rahaman, H. (2021). *Identification of cropping pattern dynamics hotspots through ranking method: A geographic analysis from Malda district of West Bengal, India*. Asian Profile.

Verburg, P. H., Kok, K., Pontius, R. G., Jr., & Veldkamp, A. (2006). Modeling land-use and land-cover change. In *Land-use and land-cover change: Local processes and global impacts* (pp. 117–135). Springer.

Von Thünen, J. H. (1966). The isolated state (C. M. Wartenberg, Trans.). Pergamon Press. (Original work published 1826).

World Bank. (2005). Land policies for growth and poverty reduction. World Bank. https://doi.org/10.1596/0-8213-5863-4.

World Commission on Environment and Development (WCED). (1987). Our common future. Oxford University Press.

Xie, H., Zhang, Y., Zeng, X., & He, Y. (2020). Sustainable land use and management research: A scientometric review. *Landscape Ecology, 35*, 2381–2411.

Zander, P., & Kächele, H. (1999). Modelling multiple objectives of land use for sustainable development. *Agricultural Systems, 59*(3), 311–325.

Zandstra, H. G. (1981). *A methodology for on-farm cropping systems research*. Int. Rice Res. Inst.

Chapter 3
Status of Agricultural Innovation: Input–Output Dimensions

Contents

H. Rahaman, S. Ansari, *Sustainable Land Use Practices and Agricultural Innovations in India*, https://doi.org/10.1007/978-3-031-83246-8_3

From the previous chapter, it is found that the intensity of sustainable land use and the cropping pattern dynamics differ across land size class categories at the block level in the district. The land is the basic input of agriculture, so if it gets changed at spatiotemporal context, then the other inputs such as technologies, infrastructure, government policies, etc. also amplify the change positively in any agricultural systems. In this context, both input and output dimensions of agriculture have to be taken into consideration for the measurement of the overall agricultural innovation situation of a region. Therefore, the input and output aspects of agricultural innovation have been examined in this chapter. In the following sequence, the degree of association between input and output dimensions of agricultural innovation has been analysed. The overall agricultural innovation concerning the input and output aspects at spatial context under different land size classes in two points of time has been discussed in the last section.

3.1 Concept and Importance of Agricultural Innovation

Agricultural innovation refers to the overall development of agriculture induced by input and output parameters supported by modern practices of sustainability and management. The innovation can involve ideas and methods that improve efficiency and productivity in various aspects, such as crop production, cultivation, sustainable farming, and farm management. The objective of agricultural innovation revolves around food security, generating livelihood options and meeting the exiting challenges like climate change and population growth. In the era of sustainable development, horizontal and vertical agricultural innovations are the need of the hour.

Epistemologically, agricultural innovation is closely linked with agricultural modernization in terms of scope, focus, and sustainability. Agricultural modernization includes overall transformation of traditional agricultural system at large scale, mostly aiming at efficient agricultural systems. The emphasis remains at widespread mechanization, technological innovation, institutional mobility, fertilizer use, and infrastructural efficiency. The scope of this type of system targets the transition from traditional to industrial farming. The main focus is on scaling up productivity for sake of higher profit margin across time and space. Although both are similar in objectives, they significantly differ in target, scope, and focus.

The main concern of agricultural innovation in developing countries is to boost economic development by increasing productivity, making income, generating abundant foods, and reducing the demand on natural stacks (Alston & Pardey, 2021). The ongoing innovation system will lead to environmental threat if the pace of development is not linked with doubly green revolution principles. To do so, the sustainable agricultural innovation must always consider biophysical and environmental limits (Sayer & Cassman, 2013). And the limits can only be defined through structural policy impetus from governments and NGOs. In the last few decades, such attempts have been made by the United Nations.

The importance of agricultural innovation can be noted in achieving Sustainable Development Goals (SDGs) set by United Nations to improve present challenges arising from agricultural system. The issues include food security, reduction in poverty level, economic growth, improvement in food quality, environmental challenges, and climate change. Although, to make a better living world, the contribution of agricultural innovation is crucial and critical, there are specific goals which align with SDGs in direct and indirect ways:

1. SDG-1 advocates for poverty reduction, especially in developing countries based on available practices and economic opportunities. Target 1.1 and 1.a can be fulfilled through impact innovation of agriculture after introducing new technologies, practices, methods, and productivity in rural market areas. At present, mobile networking provides financial services to farmer regarding improved seeds.
2. SDG-2 focuses on zero hunger through sustainable agriculture and food security measures. The innovation practice in agriculture will directly impact and will be linked with target indicators of 2.3, 2.4, 2.a, and 2.b through mechanized farming, improved crop variety, and enhanced food productions.
3. SDG-6 aims at clean water and sanitation after bringing drip irrigation, sprinkle irrigation, tank irrigation, and water recycling methods in agricultural innovation. In future, solar power based irrigation methods for small-scale agriculture could be viable options.
4. SDG-8 emphasizes decent work and economic growth through innovation impact of mechanization of agriculture which can create new employment opportunities. The future digital platform to labours for crop monitoring, auto-

mated machine running, e-commerce, and reduced manual labours can be lucrative labour market job only after agricultural innovations.
5. SDG-12 stresses on responsible consumption and production by making circular and regenerative agricultural practices. The agricultural innovation reduces waste and encourages better recycle and storage solution.
6. SDG-13 highlights climate action through climate-resilient agriculture. To mitigate and adopt climate change effect, agricultural innovation at micro-level is the need of hour.
7. SDG-15 promotes life on land after bringing natural and ecological farming practices. Through the adoption of natural farming techniques, it supports life on land while combining agro-ecology and biodiversity conservation with modern agricultural innovations for comprehensive benefits.

3.2 Agricultural Innovation and Agricultural Development

Agriculture is defined as the science and art of cultivating soil, growing and harvesting crops, domestication of animals, and raising livestock. Agriculture as an independent discipline is closely associated with geography, sociology, economics, biological science, statistics, and other fields. Conversely, agricultural innovation focuses on overall growth and development of agriculture, such as productivity, income, upscaling techniques, and efficiency. On the other side, agricultural development denotes a process that creates growth, progress, and positive change in any particular component without damaging or degrading the resources of the environment. It is inevitably a normative concept almost synonymous with improvement or agricultural modernizations (Seers, 1969). Therefore, it can be said that the agricultural development is a subset of agricultural innovation in the conceptual domain of the discipline. As a policy target of a welfare state, agricultural development is a multidimensional concept, where crop productivity, diversification, and commercialization are the different dimensions of it (Krishna, 1992). The development of agriculture in a region signifies changes in the entire agricultural system as well as overall progress in the existing agricultural activities. This becomes only possible due to the proper utilization of modern inputs such as HYV seeds, advanced machinery, improved irrigation facilities, various fertilizers and pesticides, etc. in agriculture.

The importance of agricultural innovation cannot be ignored as this sector provides the basic need of society and humankind and, thus, contributes to the overall socio-economic growth in a region. Agricultural innovation helps to improve social life by providing food supply and promoting agro-based industries (Jathar & Beri, 1949). Undoubtedly, agricultural innovation plays a significant role in uplifting a nation's economy. Johnston and Mellor (1961) believed that the contribution of agriculture is substantial for development capital, especially in the initial stages of the process of growth. In a study, Witt (1965) critically reviewed the role of agriculture in economic development. After exploring the relationship of agriculture with economic attitude, migration, population growth, and food supply, he suggests that agricultural productivity must be increased to achieve economic development. In this regard, Winters et al. (1998) examined that the rate of growth of an economy is greatly determined by the degree of agricultural innovation, especially in developing nations. It contributes to economic development in numerous ways, such as increasing food supply, agricultural exports, transfer of labour resources, additional capital formation, and additional purchasing power as a result of an increasing level of income. Agricultural innovation aims to reduce the regional disparities existing in a region and find possible ways to minimize the gap of that region as a whole.

3.3 Determinants of Agricultural Innovation

The construction of the agricultural innovation index is necessary as it helps to identify comparative development status at unit level and the extent of disparities in a broader region. Therefore, to show the regional variation in agricultural innovation, the very first step is the selection of indicators which determine the process of development of agriculture. It is essential to distinguish between the indicators of inputs and outputs in the process of composition of an index. An application of various inputs of agriculture gives us output. So, there will be a duplication of information if we take these together. The indicators need to be chosen carefully and rationally, considering their impact in the growth of agriculture. The physical factors such as temperature, rainfall, humidity, and soil nutrients and non-physical factors such as land tenancy, irrigation, machinery, seeds, credit facilities, and socio-economic aspects influence agriculture innovation and its variation. Alston and Pardey (2021) measured the economics of agricultural innovation using some indicators in the study. These include the intensity of cultivation, induced by land, labour, and production which signify commercialization of agriculture along with development and diversification.

Nath (1969) constructed a composite index based on three factors, i.e., growth rate of agricultural outputs, use of modern input, and productivity per hectare to compare the level of agriculture innovation in various states of India. His paper was confined to the productivity dimension of agricultural innovation. Using a detailed district-wise data, Sharma (1971) constructed the index of agricultural innovation where he opined that it should not only be assessed by levels of productivity but also with consideration of various inputs such as irrigation, fertilizers, adoption of seeds, and extent of cultivated area. Rahaman et al. (2019) analysed the agricultural innovation of Malda district of West Bengal based on input–output dimensions considering the spirit of agricultural innovation for two decades.

To show the regional disparities of development and innovation in Andhra Pradesh, Alam (1974) employed six indicators, namely agricultural output per agricultural worker, agricultural output per acre, percentage of gross irrigated area, percentage of gross cropped area, canal irrigated, and doubled-cropped area. Tewari and Singh (1985) conducted a cross section temporal analysis of development and productivity of Indian agriculture. They have selected eight indicators to show the regional variation and these are value of produce, fertilizer consumption, irrigation coverage, irrigation intensity, number of agricultural workers per hectare net sown area, cropping intensity, percentage of commercial crops to gross cropped area, and consumption of power.

Bhadrapur and Naregal (1992) selected 18 indicators to show the regional disparities of agricultural innovation. These indicators include the various aspects of land tenancy, irrigation, fertilizer, farm technologies, credit facilities, price, cropping intensity, productivity, nature of workers, and land use. A study by Adams and Bumb (1978) recognized as many as 18 indicators to determine the patterns of agricultural innovation in Maharashtra. They concluded that human and animal power, rainfall, fertilizers, and mechanization supported by bank and cooperative credit facilities have helped to boost land productivity in Maharashtra. A total of 21 variables are selected by Ohlan (2013) to show the pattern of disparities development in India. He covers the broader aspects of production and productivity, mechanization and implements, animal husbandry, market and cold storage facilities, application of fertilizers, etc.

3.4 Input Variables of Agricultural Innovation

Based on an extensive literature survey from the ground of sustainable land use pattern and innovation of agriculture, a total of 32 variables of input dimension have been selected in this study to identify different levels of innovation in the block under six land size categories concerning the agricultural innovation and the extent of disparities. These variables are heads under the seven indicators. The output dimension is studied through four major outputs of agriculture, i.e., crop productivity, crop diversification, cropping intensity, and cropping pattern. Two types of analyses are done in this section. Firstly, the diversity and changes in the different aspects of agricultural innovation are analysed, and secondly, the composite index is applied to measure the level of agricultural innovation.

3.4.1 Land Size Holder

Agriculture Census defines an operational holder as ‘A person who holds the responsibility for the operation of an agricultural holding, meaning thereby, exercise technical initiative and responsibility for the operation of holding and may have the full economic responsibility for it (i.e., as owner) or share this with others (as a tenant)’. When two or more persons belonging to different households jointly share the responsibility for the operation of an agricultural holding, each one of them will be regarded as a holder or an operator. Only the marginal landholder has increased (4.47%) in the district from 1995–1996 to 2015–2016. The percentage share of landholders in the remaining land classes is decreased due to the fragmentation of

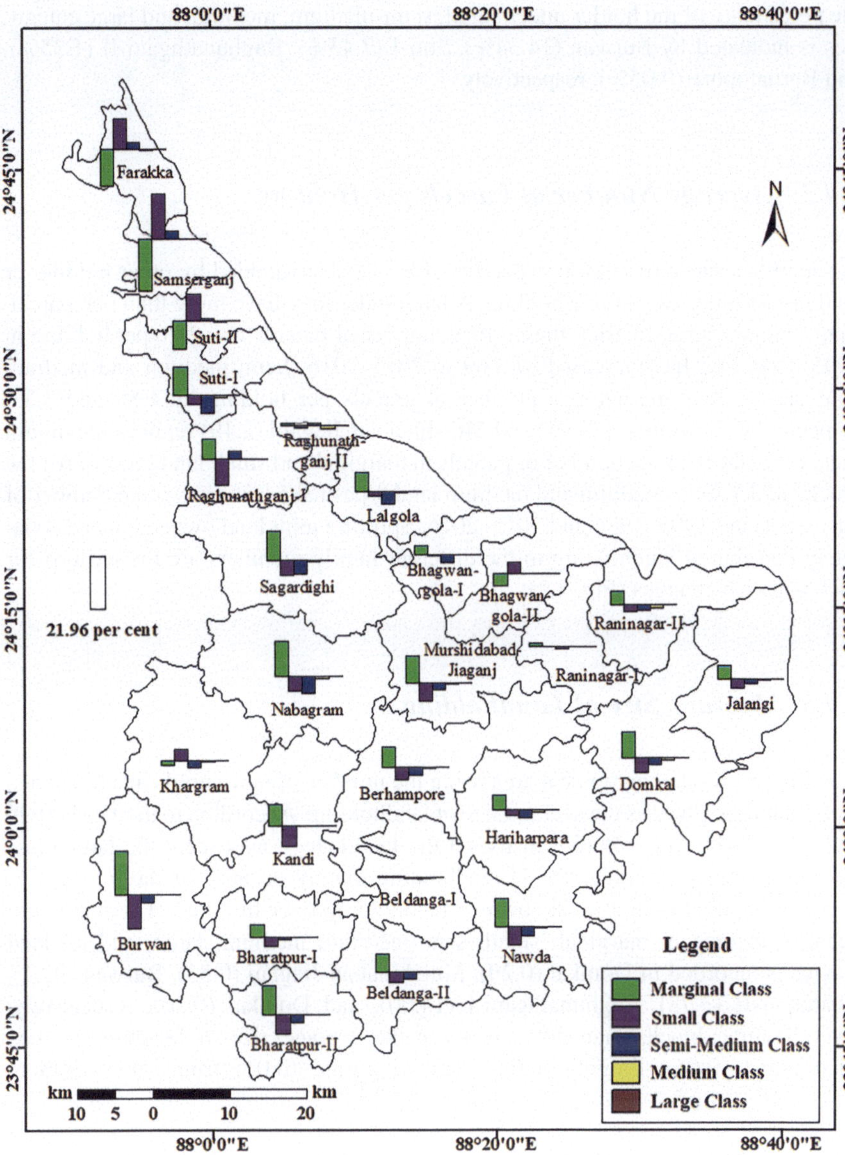

Fig. 3.1 Land size class-wise changes in percentage of landholders, 1995–1996 to 2015–2016. (Source: Based on Annexure C.2)

landholdings. Figure 3.1 shows that the maximum positive change in the marginal category is reported in Burwan (18.12%), while the top negative growth is recorded by Samserganj (21.96%). The maximum holder increase is found in Samserganj under small (18.84%) and semi-medium (3.21%), in Bhagwangola-II under medium (0.25%), and Bhagwangola-I in large (0.03%) land classes. The highest reduction in the percentage of the holder under small, semi-medium, medium, and large categories is indicated by Burwan (14.34%), Suti-I (7.43%), Raghunathganj-II (1.35%), and Berhampore (0.05%), respectively.

3.4.2 Average Number of Parcels per Holding

A parcel is a piece of land, irrespective of its size, surrounded by other holding or land not forming part of any holding. A landholder may have more than one parcel. In the Murshidabad district, the average number of parcels was 2.76 per holding in 1995–1996 and has increased to 3.69 in 2015–2016. Semi-medium and medium land classes lose the average number of parcels per holding by 1.80 and 3.33, respectively. Samserganj (1.93) and Murshidabad-Jiaganj (2.19) report a maximum increase in the average number of parcels in marginal and small land groups. All the blocks under semi-medium and medium land types have lost an average number of parcels during 1995–1996 and 2015–2016. Splitting joint land ownership and separately distributed landholding to the different family members are the reasons for such negative changes (Fig. 3.2).

3.4.3 Average Size of Landholding

Dividing the total geographical area using the number of households in a block at a particular time gives us the average size of landholding. According to the 1995–1996 and 2015–2016 Agriculture Census, all the land categories except the large class have dropped the average size of landholding mainly because of the pressure of growing population and breakdown of landholding over the time. The maximum negative change in marginal, small, semi-medium, medium, large, and all land classes is recorded in Domkal (0.20), Murshidabad-Jiaganj (0.53), Burwan (0.51), Bharatpur-II (3.78), Raghunathganj-I (14.00), and Domkal (0.30), respectively. Overall, three blocks manage to increase the average size of landholding, i.e., Bharatpur-I (0.06), Lalgola (0.04), and Khargram (0.01) from 1995–1996 to 2015–2016 (Fig. 3.3).

3.4.4 Wholly Owned and Self-Operated Area

The land for which the cultivator has the permanent heritable possession with or without the right of transfer and land operated under perpetual leased are defined as the own and self-operated area (Agriculture Census). Marginal (15.10%), small (6.31%), and all class (6.82%) categories register decreases in own and self-operated

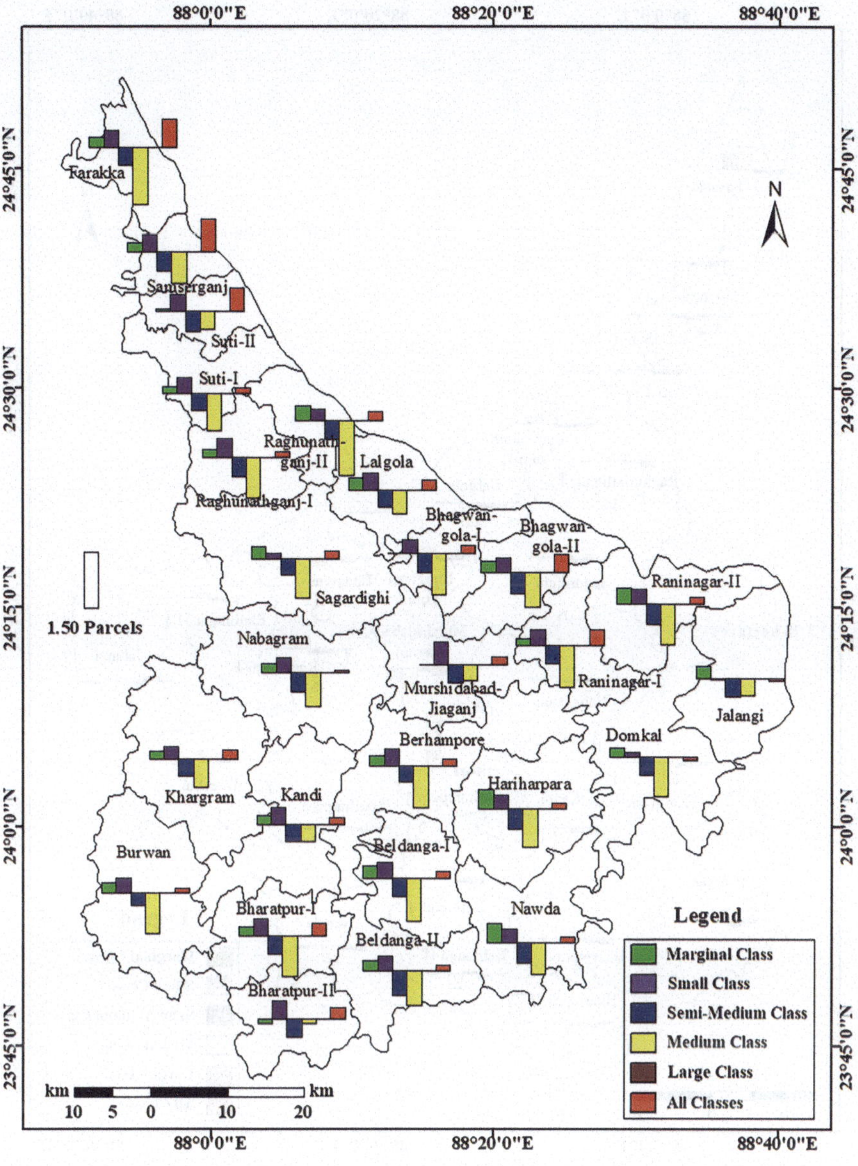

Fig. 3.2 Land size class-wise changes in average number of parcels per holdings, 1995–1996 to 2015–2016. (Source: Based on Annexure C.3)

area from 1995–1996 to 2015–2016. There is an increase in area under semi-medium (14.56%) and medium (40.28%) land classes because these farmers take their land back once they lease out. Only Suti-II (12.80) records the negative change in area in these two land categories. All the blocks under marginal land type except

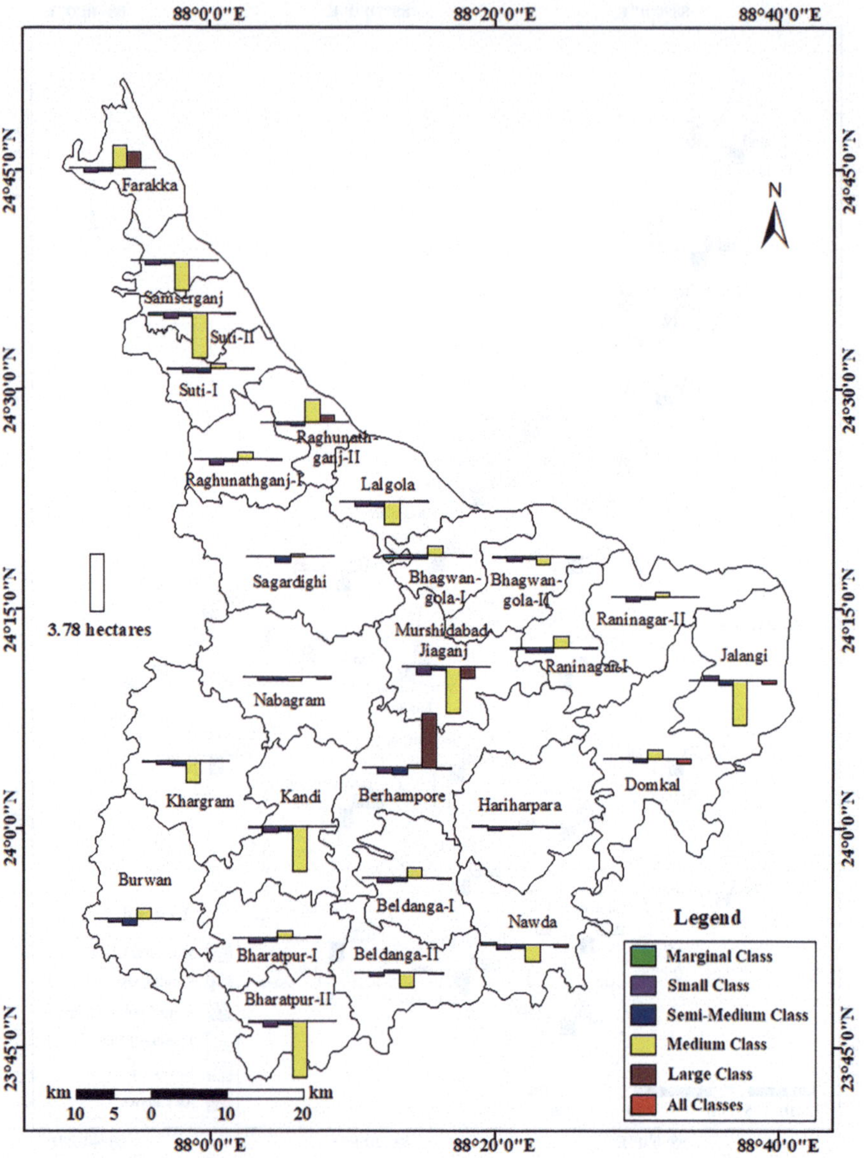

Fig. 3.3 Land size class-wise changes in average size of landholdings, 1995–1996 to 2015–2016. (Source: Based on Annexure C.4)

Farakka and Lalgola report a negative change, among which Bhagwangola (29.92%) shows the maximum. Lalgola (30.79%) and Jalangi (31.75%) explain the highest positive and negative growth in small land classes, respectively. Overall, six blocks in the district gained owned and self-operated area, in which Farakka takes the lead

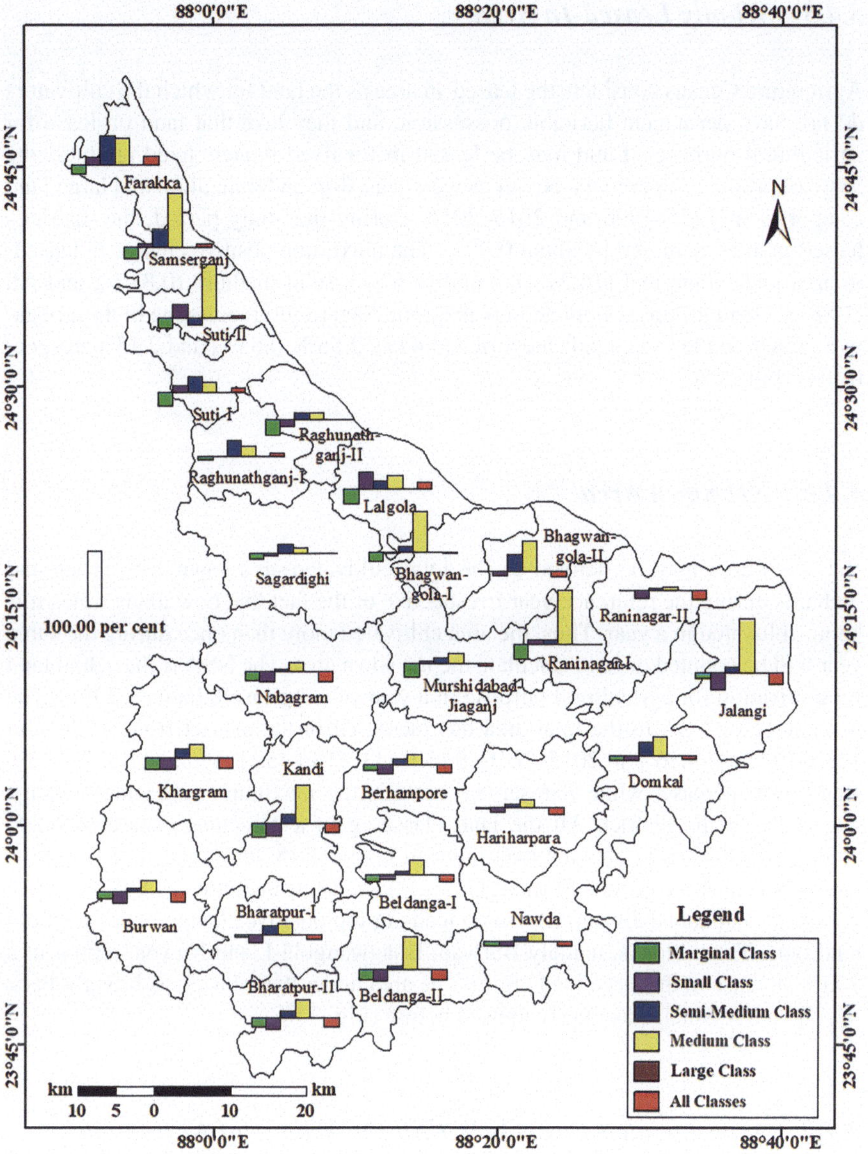

Fig. 3.4 Land size class-wise changes in wholly own and self-operated area, 1995–1996 to 2015–2016. (Source: Based on Annexure C.5)

with a 16.97% increase. The top negative change in area is secured by Jalangi (21.78%), followed by Nabagram (19.61%), Khargram (18.86%), Burwan (18.78%), and Beldanga-II (18.64%) (Fig. 3.4).

3.4.5 Wholly Leased-In Area

Agriculture Census considers the leased-in area as the land for which the cultivators do not have permanent heritable possession, and they take that land on lease for agricultural purposes. Land may be leased-in for fixed money, fixed produce, the share of produce, and others. No areas were leased-in under medium and large land categories in 1995–1996 and 2015–2016. During the study period, the district's leased-in area increased by about 0.95%. The maximum positive change in leased-in area under marginal (14.29%), small (8.62%), semi-medium (0.83%), and All (10.87%) land groups is noticed in Nabagram. Sagardighi reports the highest negative growth in the same land categories (3.44%, 2.56%, 0.83%, and 2.15%, respectively) (Fig. 3.5).

3.4.6 Net Sown Area

Net sown area may be defined as the total cultivated area sown with crops and orchards during the reference year irrespective of the fact that how many times it is being cultivated in a year. Thus, the area cultivated more than once during the same year will be counted once to calculate the net sown area. The NSA is the actual land for cultivation for agricultural purpose in a year of a region (Rahaman, 2021). The percentage of NSA to the total area has increased in the district from 97.26% in 1995–1996 to 98.76% in 2015–2016. Suti-II (11.32%) and Lalgola (11.06%) see a significant increase, while Samserganj records the maximum percentage change negatively in the district. All the land classes gain a percentage share of NSA, among which Lalgola gets the maximum in marginal (8.36%), and Suti-II reports the highest positive growth in small (17.73%) and semi-medium (13.80) land categories. Murshidabad-Jiaganj (20.86%) leads in top positive change under medium land type. Three blocks, namely Burwan, Bhagwangola-I, and Nawda, represent a 100% increase in the large land group. The area under these blocks is brought back to cultivation which was utterly unused before (Fig. 3.6).

3.4.7 Area Located Entirely Within the Village of Residence

If a cultivator's operating area is located within the village of their residence, it will be considered as the area located entirely within the village of residence. It is worth mentioning that if some of the operated areas fall outside the village of residence, those will not be part of this category. The percentage share of area located entirely within the village of residence has increased under all the land classes in the district from 1995–1996 to 2015–2016. Raghunathganj-II (48.77%), Khargram (58.47%), Kandi (75.73%), Raninagar-I (97.35%), and Raghunathganj-II (51.17%) report the

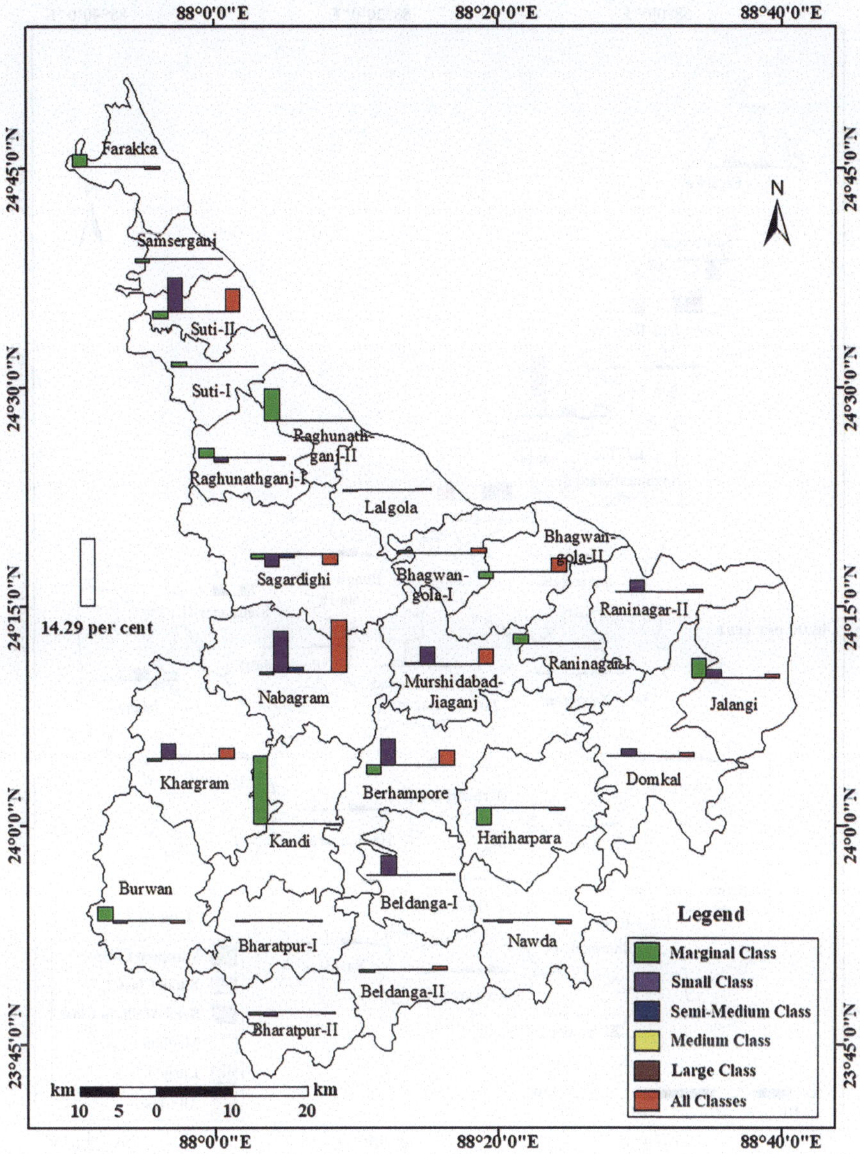

Fig. 3.5 Land size class-wise changes in wholly leased-in area, 1995–1996 to 2015–2016. (Source: Based on Annexure C.1)

maximum positive change under marginal, small, semi-medium, medium, and all class land categories, respectively. Farakka and Suti-I blocks decreased 100% of the area under large land size class either because of the shifting of cultivators to the other villages or selling their land to the cultivators who belong to other villages (Fig. 3.7).

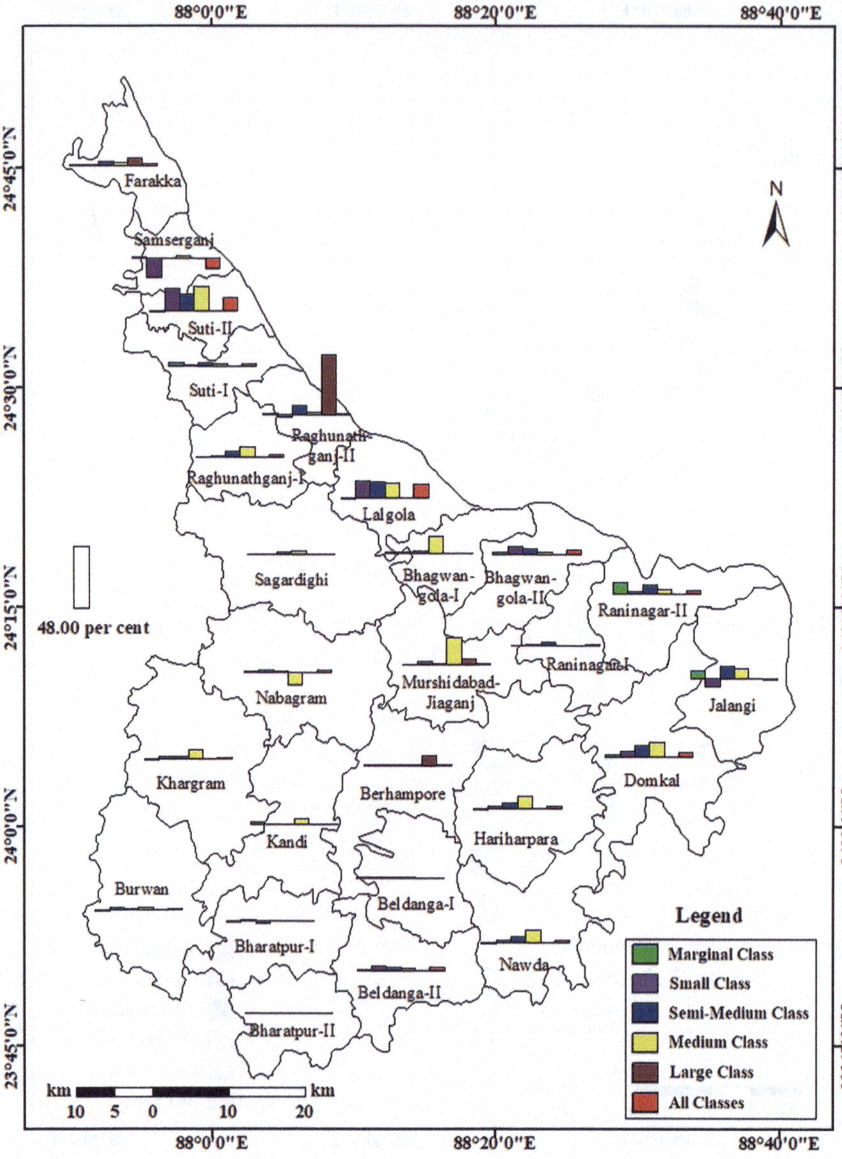

Fig. 3.6 Land size class-wise changes in net sown area in total area, 1995–1996 to 2015–2016. (Source: Based on Annexure C.7)

3.4.8 Area Located Entirely Outside the Village of Residence

An operated area is treated as an area located entirely outside the village of residence when 100% of the holding is found in the villages other than the cultivator's residential village. All the land size classes record a negative change in area in the

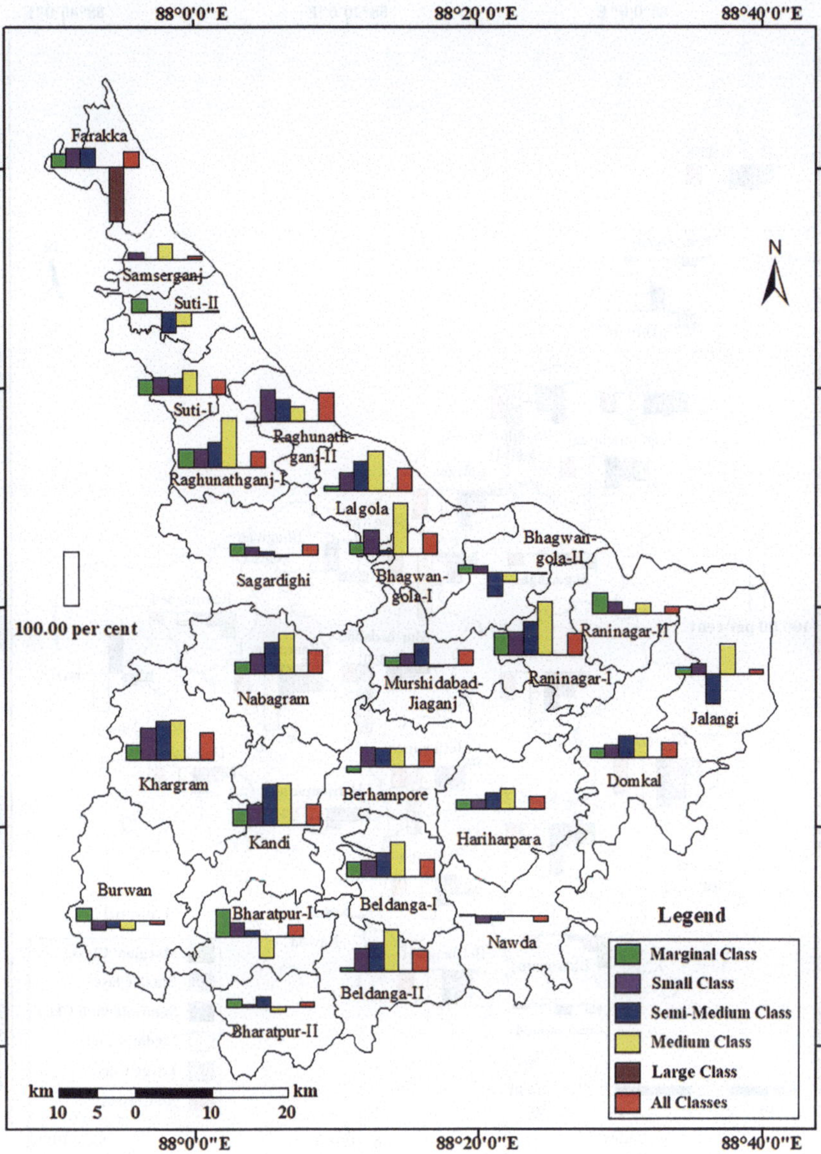

Fig. 3.7 Land size class-wise changes in area located entirely within the village of residence, 1995–1996 to 2015–2016. (Source: Based on Annexure C.8)

district, among which medium (34.41%) land type reduces the maximum, followed by small (25.27%), all (23.78%), semi-medium (22.73%), and marginal (19.01%), respectively during 1995–1996 and 2015–2016. Overall, only three blocks, i.e., Nawda (8.76), Burwan (7.67%), and Bhagwangola-II (1.40%), experience positive growth in the area located entirely outside the village of residence (Fig. 3.8).

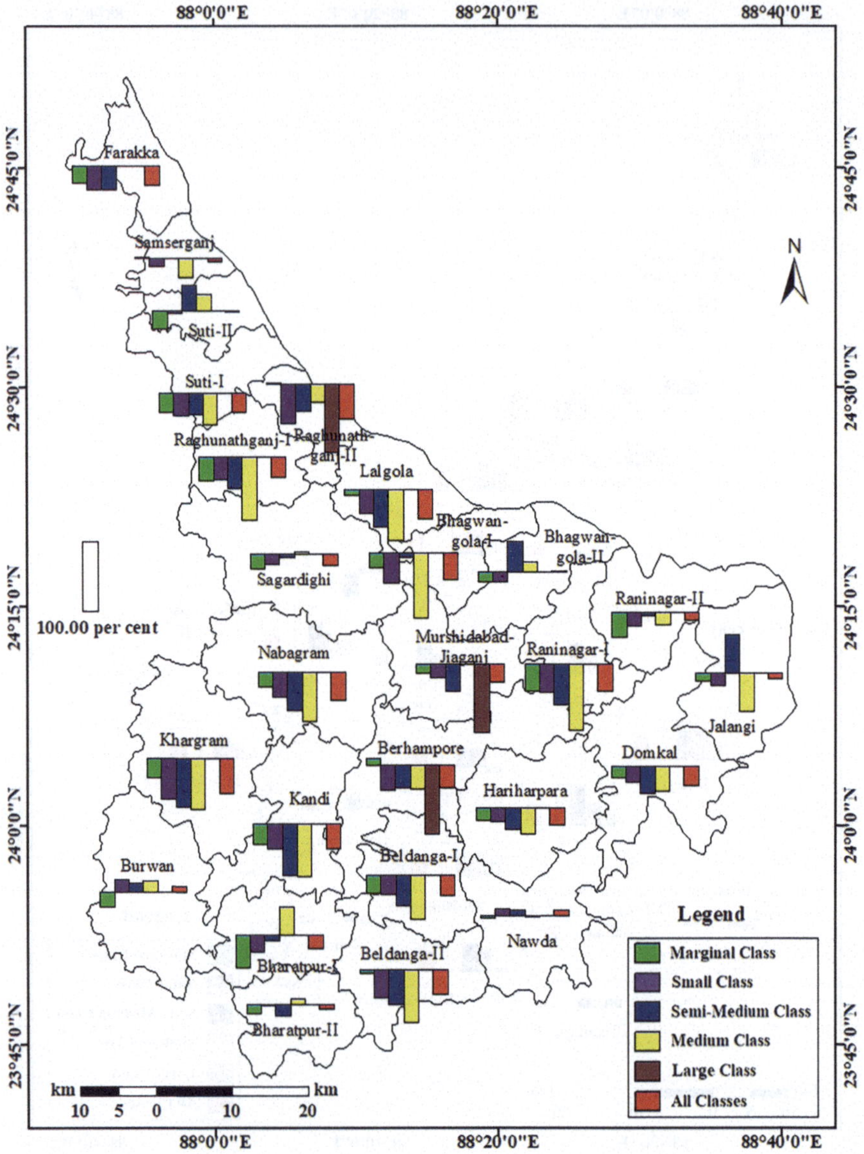

Fig. 3.8 Land size class-wise changes in area located entirely outside the village of residence, 1995–1996 to 2015–2016. (Source: Based on Annexure C.9)

3.4.9 Net Irrigated Area in Total Gross Cropped Area

Suppose the net sown area in an operational holding has received irrigation either a portion or wholly during the reference period, in that case, such areas are considered

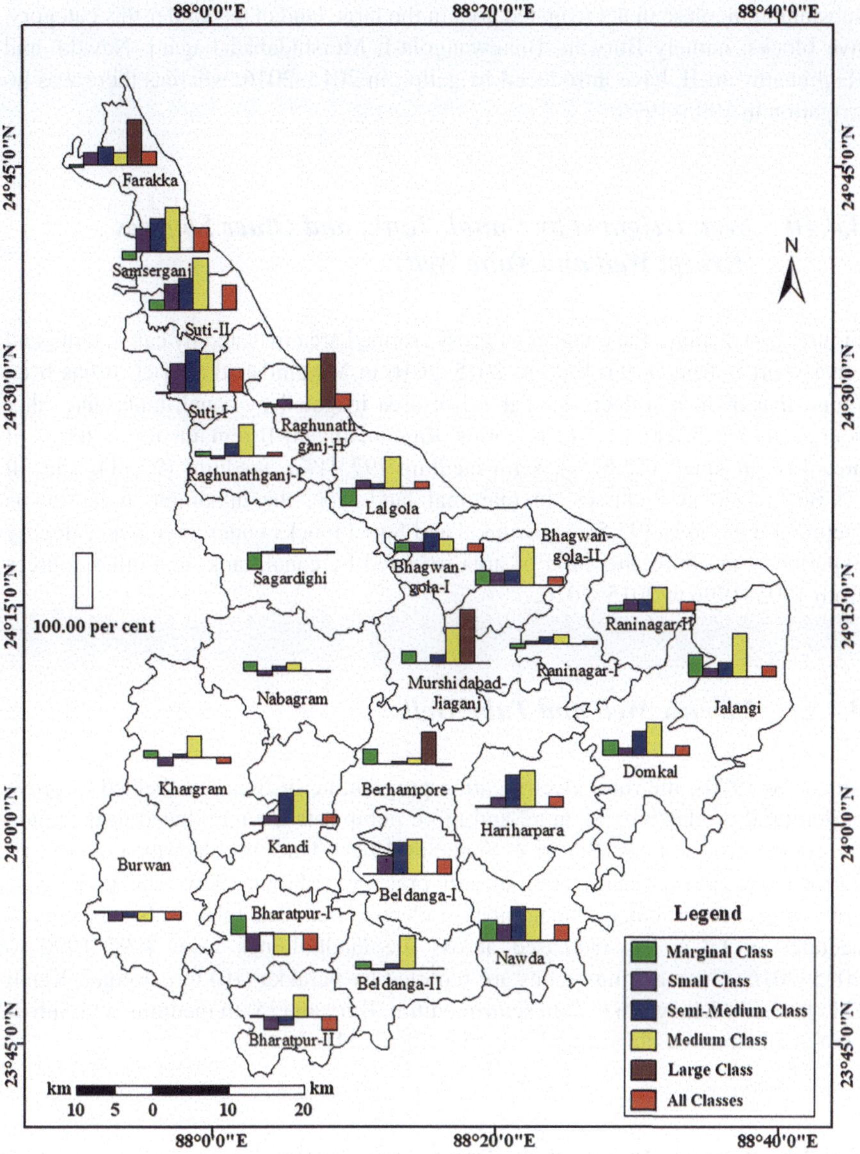

Fig. 3.9 Land size class-wise changes in net irrigated area in total GCA, 1995–1996 to 2015–2016. (Source: Based on Annexure C.10)

as net irrigated area. The percentage of net irrigated area in total GCA has increased under all the land categories in the district from 1995–1996 to 2015–2016. The maximum positive increase of net irrigated area under marginal (40.72%), medium (100%), and all (48.52%) land classes are noticed in Suti-II (Fig. 3.9). In the remaining land groups, Suti-I block scores the maximum positive growth. There is a substantial increase in net irrigated area in the large land class. Under this category, five blocks, namely Burwan, Bhagwangola-I, Murshidabad-Jiaganj, Nawda, and Raghunathganj-II, have introduced irrigation in 2015–2016, whereas there was no irrigation in 1995–1996.

3.4.10 Area Irrigated by Canal, Tank, and Other Sources (Except Well and Tube Well)

Figure 3.10 displays the changes of gross cropped area irrigated by canal, tank, and other sources from 1995–1996 to 2015–2016 in Murshidabad district. It has been found that there is a decreasing trend of area irrigated by canal, tank, and other sources under different land categories. Raghunathganj-II remains top in terms of area loss in small (52.67%), semi-medium (97.94%), medium (90.11), and all (79.06%) land size classes. In marginal land type, the maximum reduction is recorded in Farakka (95.54%). Domkal and Suti-I blocks under large land category experience an entire alteration of area irrigated by canal, tank, and other sources from 1995–1996 to 2015–2016.

3.4.11 Electric Well and Tube Wells

Since the 1970s, the rural electrification programme in India has helped improve agricultural productivity as more and more pump sets are introduced to irrigation through electric sources (Barnes & Samanta, 1984). Thus, the electrification of irrigation leads to the modernization of farm practices (Mellor, 1976). Due to the government's encouragement, the number of electric wells and tube wells per hundred hectares of GCA has increased across the landholdings from 1995–1996 to 2015–2016. The maximum gains are reported by Farakka (16) in marginal, Kandi (21) in small, Khargram (17) in semi-medium, Burwan (25) in medium, and Suti-II (15) in all land size classes in the district (Fig. 3.11).

3.4.12 Diesel-Operated Well and Tube Wells

Figure 3.12 illustrates that the number of diesel-operated wells and tube wells per hundred hectares of GCA is decreased under different land classes except for the large class from 1995–1996 to 2015–2016. Marginal landholding faces the loss of 22 diesel-operated wells and tube wells per hundred hectares GCA, followed by all (17), small (16), semi-medium (8), and medium (2) land groups. Nawda block alone

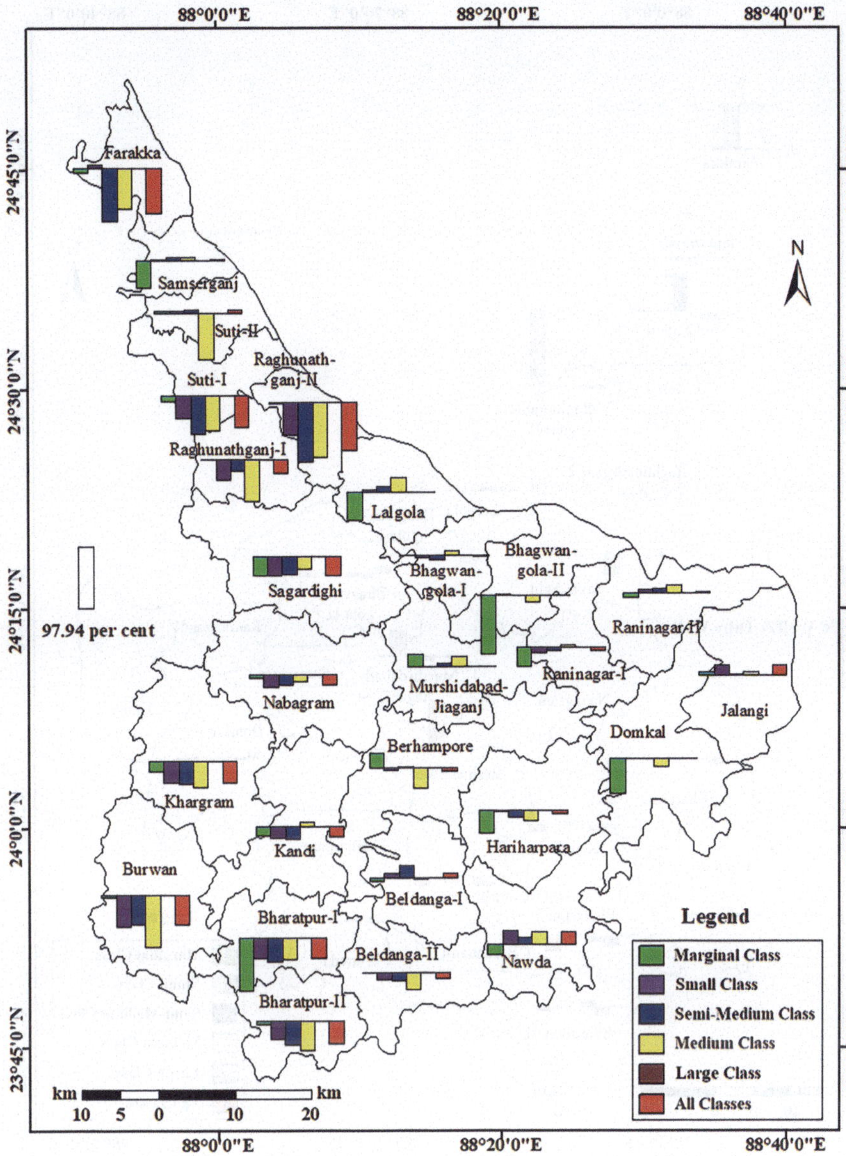

Fig. 3.10 Land size class-wise changes in area irrigated by tank and other sources, 1995–1996 to 2015–2016. (Source: Based on Annexure C.11)

depicts the maximum reduction under marginal (109), small (143), medium (28), and all (100) land classes. In the remaining semi-medium and large land categories, the highest negative growths are recorded in Beldanga-I (49) and Beldanga-II (20), respectively.

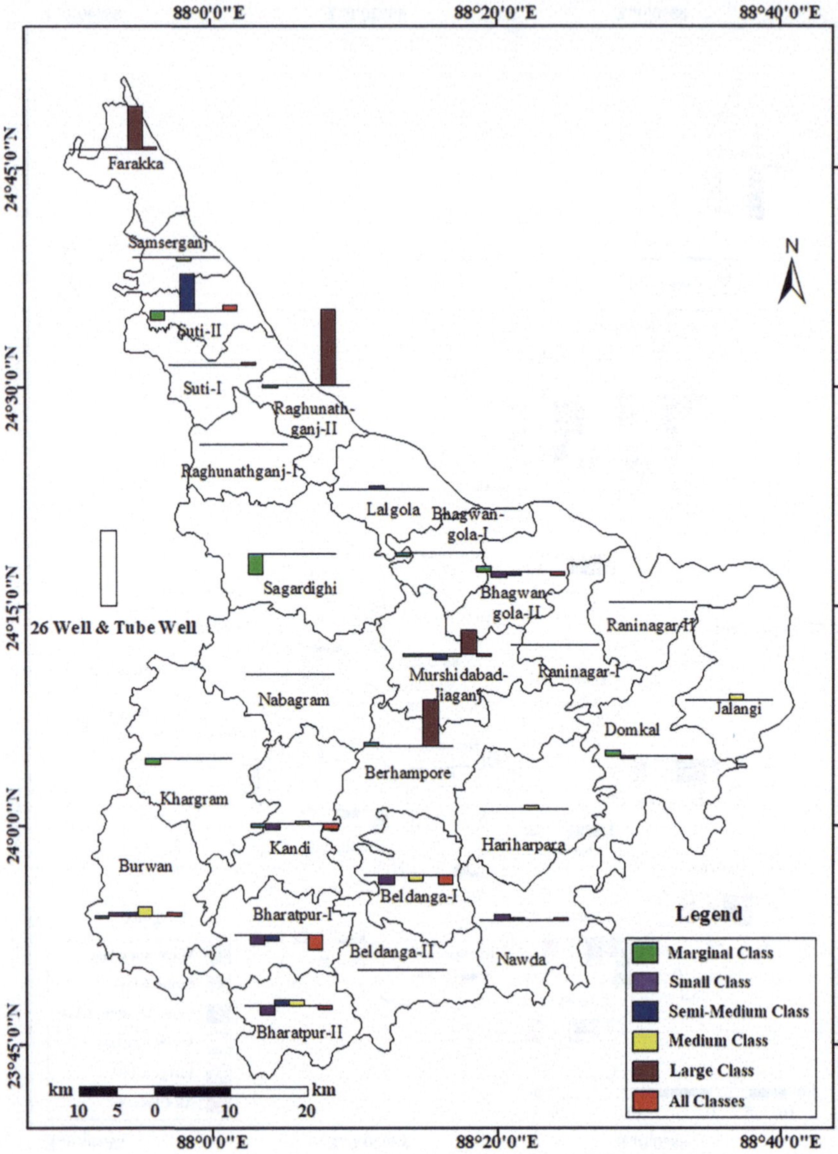

Fig. 3.11 Land size class-wise changes in number of electric well and tube well per hundred hectares of GCA, 1995–1996 to 2015–2016. (Source: Based on Annexure C.12)

3.4.13 Area Irrigated Under All Crops

The percentage of irrigated area under all crops has increased through different land categories in the district over the years. Figure 3.13 reveals that the blocks of

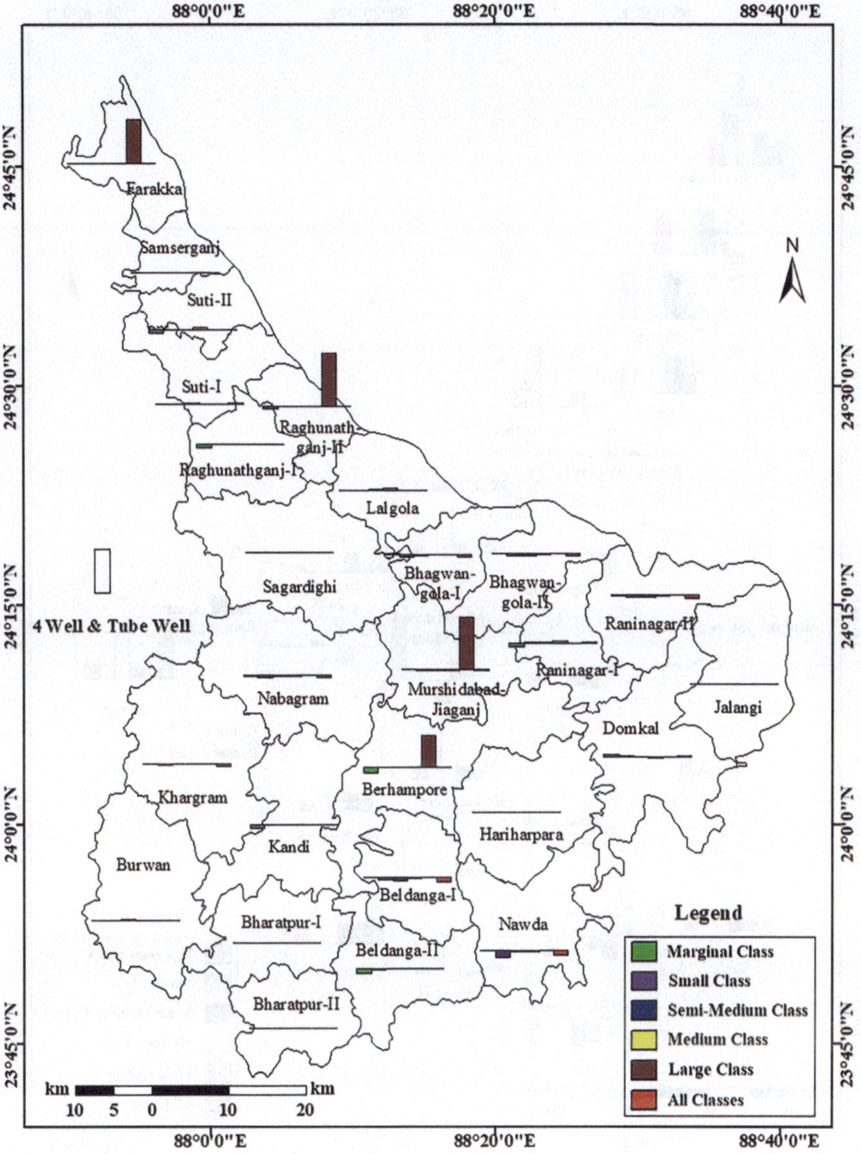

Fig. 3.12 Land size class-wise changes in number of diesel-operated well and tube well per hundred hectares of GCA, 1995–1996 to 2015–216. (Source: Based on Annexure C.13)

northern part of the district, i.e., Farakka, Samserganj, Suti-I, Suti-II, Raghunathganj-I, and Raghunathganj-II, report positive growth in all the land size classes. With an increase of 80.18% in marginal, 73.29% in small, 79.96% in semi-medium, 98.56% in medium, and 79.44% in all land classes, Suti-II blocks gain

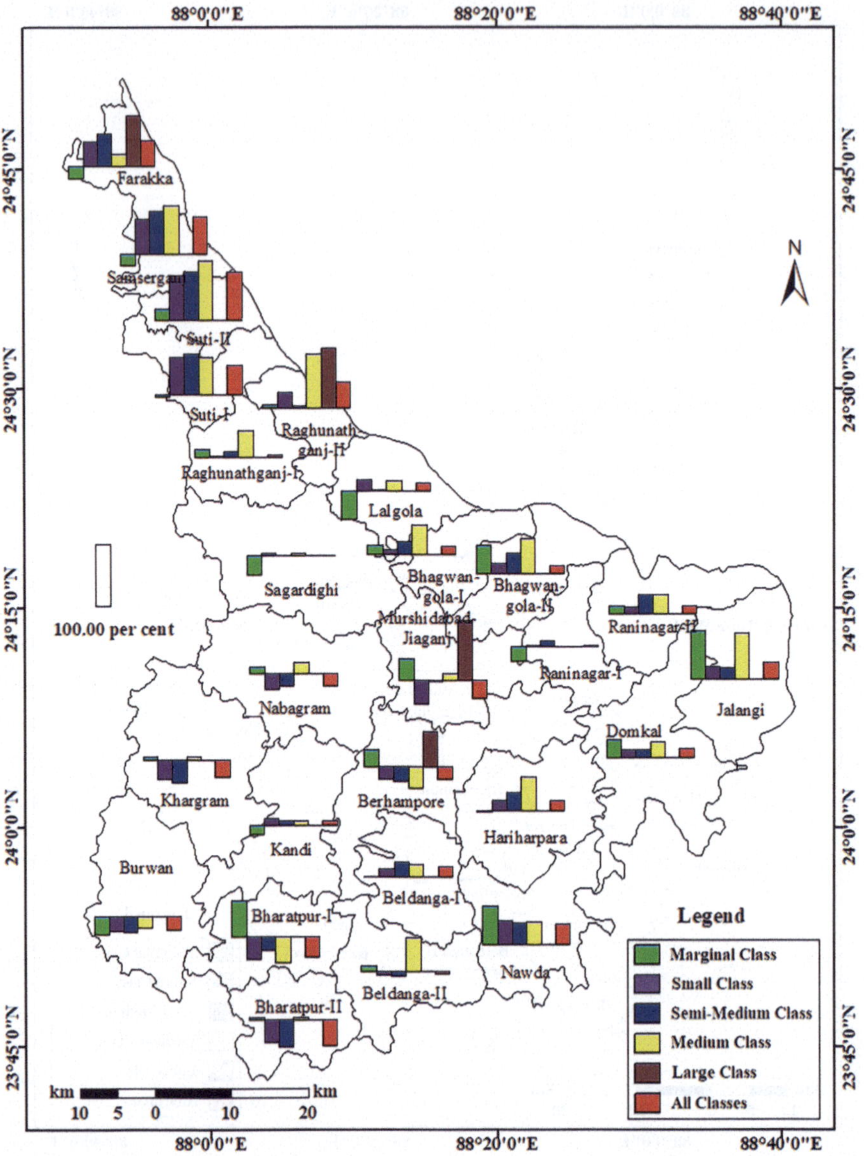

Fig. 3.13 Land size class-wise changes in irrigated area in GCA under all crops, 1995–1996 to 2015–2016. (Source: Based on Annexure C.14)

maximum irrigated area from 1995–1996 to 2015–2016. Area irrigated under all crops has significantly reduced in Bharatpur-II under marginal (46.40%), semi-medium (44.25%), and all (42.60%), in Murshidabad-Jiaganj under small (39.52%), and in Bharatpur-I under medium (43.22%) land groups.

3.4.14 *Operational Holder Using Certified Seeds*

The certified seeds, produced by registered seed growers and approved and certified by an official seed certifying agency, are used to maintain satisfactory genetic identity and purity, proper germination, and freedom from weeds, insects, and pests (Input Survey). The requirements for certified seeds have increased in the Murshidabad district over time. About 25% of operational holders used certified seeds in 1996–1997, while the percentage of operational holders increased to 45.96% in 2016–2017. All the land categories experience a positive change in certified seed users, in which medium land type (52.69%) accounts for the most. The maximum positive changes of the operational holder using certified seeds under marginal, small, semi-medium, medium, large, and all land classes are reported in Burwan (35.39%), Kandi (56.08%), Burwan (54.68%), Lalgola (64.15%), Nawda (77.35%), and Samserganj (40.41%), respectively (Fig. 3.14).

3.4.15 *Operational Holder Using Hybrid Seeds*

Hybrid seeds are regarded as one of the main contributors to the dramatic change in agricultural output (Rahaman, 2021). In agriculture and gardening, hybrid seeds are produced by crossing two different varieties of the same plant. These are disease resistant and give better yield and uniformity. Figure 3.15 depicts that the district gains a significant percentage of hybrid seed users across land classes from 1996–1997 to 2016–2017. Suti-II reports the maximum change under marginal (39.72%) and all (44.82%) land categories. In the remaining small, semi-medium, medium, and large land groups, the highest positive growths are recorded by Domkal (36.78%), Sagardighi (52.40%), Bharatpur-II (53.73%), and Bhagwangola-I (71.23%), respectively.

3.4.16 *Institutional Credits*

Agricultural credits that are provided by the government institutions such as regional rural banks (RRB), scheduled ommercial banks (SCB), National Bank for Agriculture and Rural Development (NABARD), cooperative banks, microfinance institutions, etc. are regarded as the institutional credit. The percentage of operational holders who took institutional credits has increased under different land categories in the district from 1996–1997 to 2016–2017. No households took institutional credit under large land class in both time periods. Raninagar-II (28.28%), Domkal (24.69%), Sagardighi (25.47%), and Murshidabad-Jiaganj (4.49%) display the highest gain of institutional credit takers under marginal, small, semi-medium, and medium land groups (Fig. 3.16).

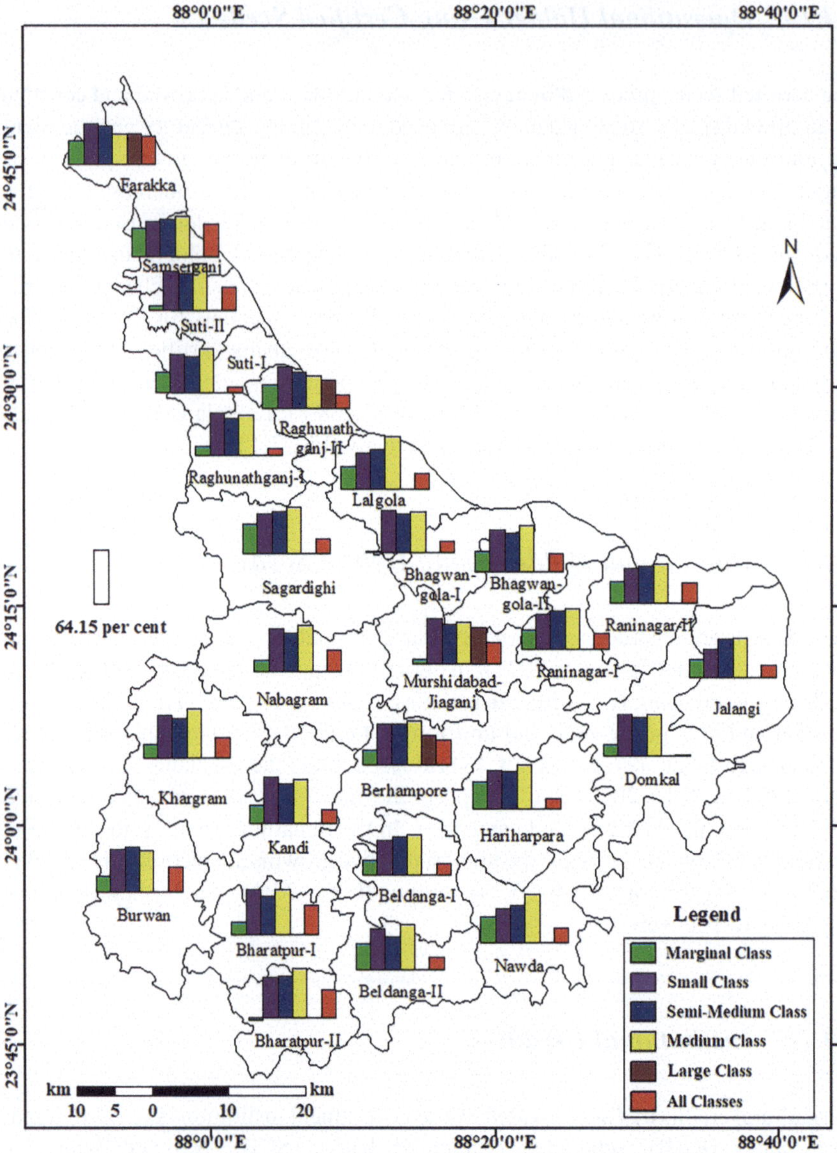

Fig. 3.14 Land size class-wise changes in operational holder used certified seeds, 1996–1997 to 2016–2017. (Source: Based on Annexure C.15)

3.4.17 Credit Taken from Primary Agricultural Credit Society (PACS)

Primary Agricultural Credit Society is a small cooperative credit institution that plays a vital role at the grass-root level (gram panchayat and village level) for the

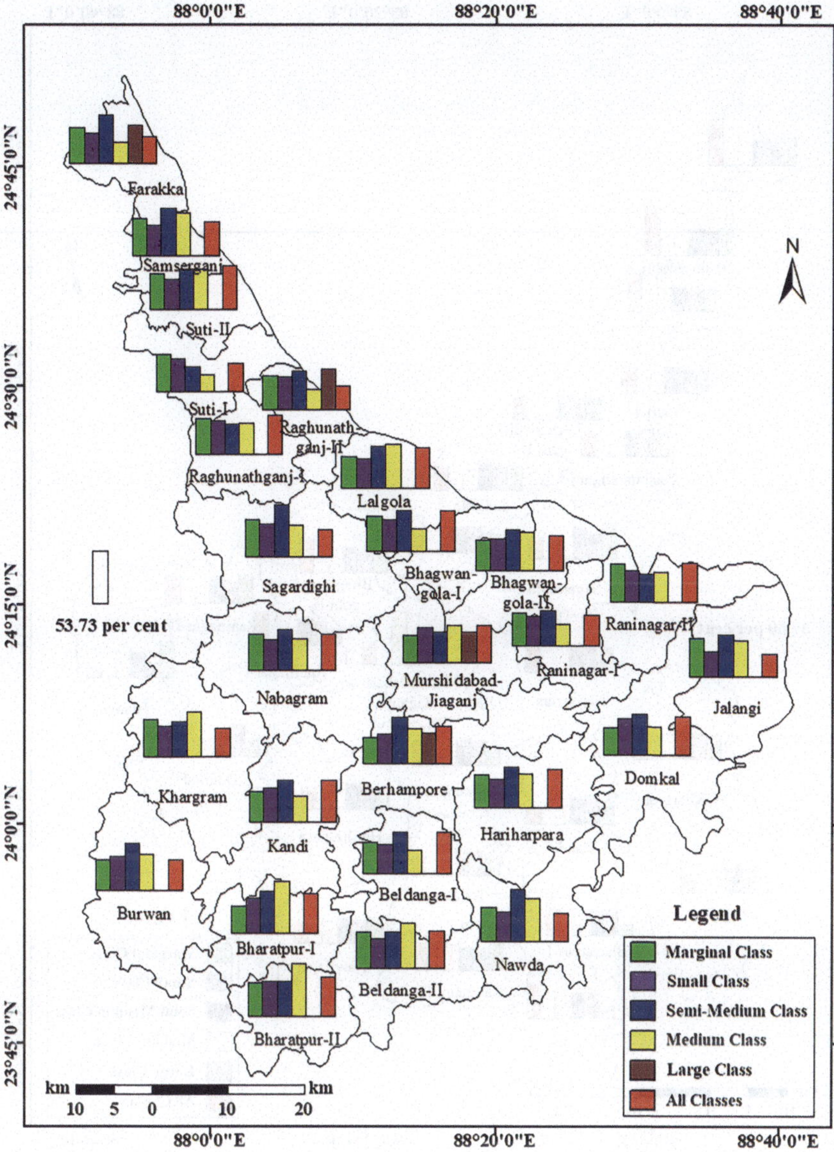

Fig. 3.15 Land size class-wise changes in operational holder used HYV seeds, 1996–1997 to 2016–2017. (Source: Based on Annexure C.16)

development of local areas by providing short-term and medium-term credit to the farmers. The district experienced a 7.05% increase in the percentage of operational holders taking credit from PACS. Marginal, small, semi-medium, and medium land size classes gain 6.07%, 7.43%, 11.81%, and 8.50%, respectively. Overall, Samserganj (10.94%) reports the maximum increase, while the minimum increase is noted by Jalangi (3.76%).

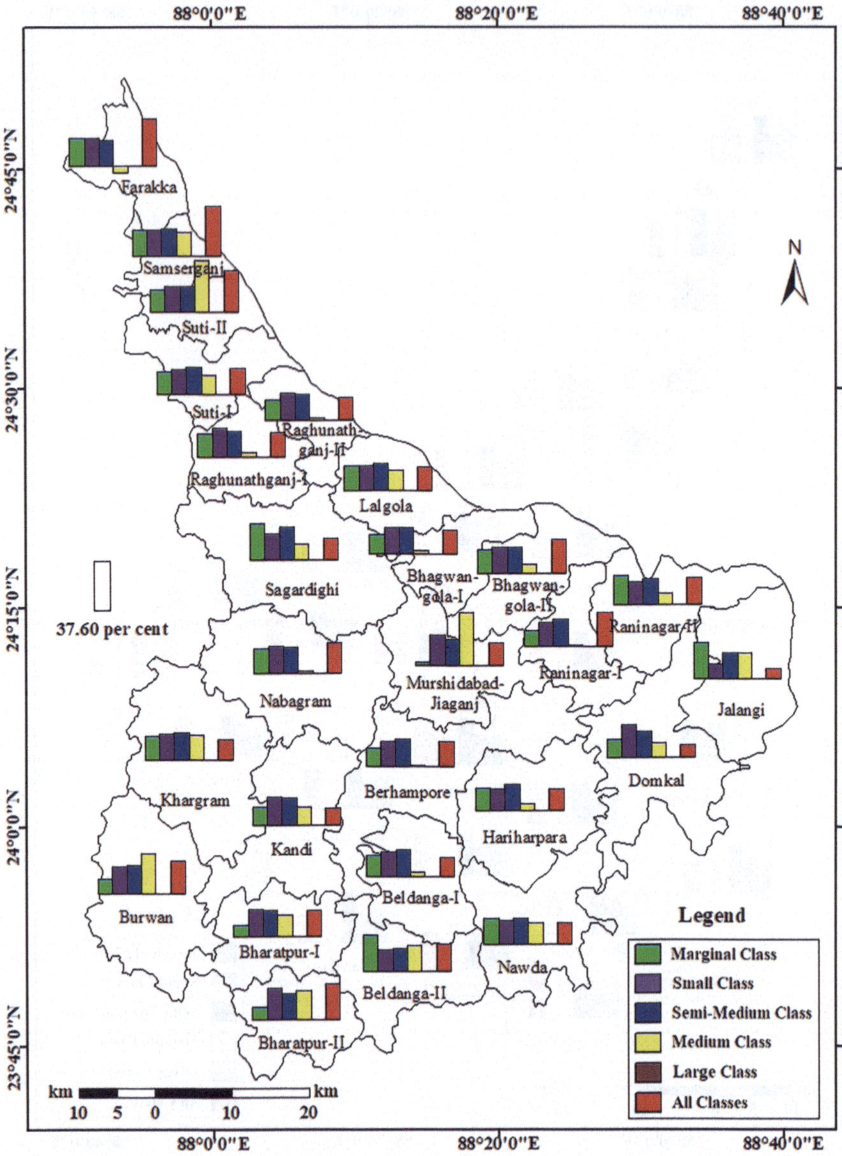

Fig. 3.16 Land size class-wise changes in operational holder took institutional credits, 1996–1997 to 2016–2017. (Source: Based on Annexure C.17)

3.4.18 Credit Taken from Primary Land Development Bank (PLDB)

Primary land development banks are the particular banks that provide long-term loans to the farmers assisting in land development and agricultural production.

These banks are organized at the district level with having their branches at the block level. PLDB credit takers in the district have a positive growth from 1996–1997 to 2016–2017. All the land classes register a positive gain, in which the medium land class gets the maximum (3.42%). However, a few blocks under different land categories report negative growth, i.e., Raghunathganj-I in marginal Bhagwangola-I, Bharatpur-II, Farakka, Raghunathganj-I, and Suti-I in semi-medium, and Farakka, Jalangi, and Raghunathganj-II in medium land classes (Figs. 3.17 and 3.18).

3.4.19 Credit Taken from Commercial Bank Branch (CBB)

The commercial bank branch is a financial banking institution which performs the function of providing financial services to the people. These banks accept the deposit from the general public and lend money as a loan for investments with a higher interest rate. Figure 3.19 reveals that the percentage of CBB loan beneficiaries has increased among all the blocks under different land categories from 1996–1997 to 2016–2017. Medium land type (29.58%) represents the maximum gain of loan takers, followed by semi-medium (18.73%), small (8.52%), and marginal (6.65%) land groups.

3.4.20 Credit Taken from Regional Rural Bank (RRB)

Regional rural banks were set up in 1975 to operate at the regional level in different states of India to primarily serve the rural areas with basic banking and financial services. These banks help the farmers by providing loans and credit facilities for agriculture innovation. The households under RRB loan beneficiaries have increased under marginal (0.27%), medium (6.48%), and all (0.28) land size categories. In contrast, small (0.10%) and semi-medium (0.96%) land classes record negative growth in the district from 1996–1997 to 2016–2017. With an increase of 1.47% and 2.00%, Sagardighi displays the highest growth in marginal and semi-medium land types, respectively. Under small, medium, and all land groups, the maximum gains are seen in Murshidabad-Jiaganj (1.23%), Bharatpur-II (13.99%), and Burwan (1.73%), correspondingly (Fig. 3.20).

3.4.21 Hand-Operated Machines

Agricultural mechanization is regarded as an essential input in the modern agricultural system. Efficient machinery helps increase agricultural production and productivity. It improves the utilization efficiencies of other inputs, the safety and comfort of the agricultural workers, and the quality and value of the produce. The

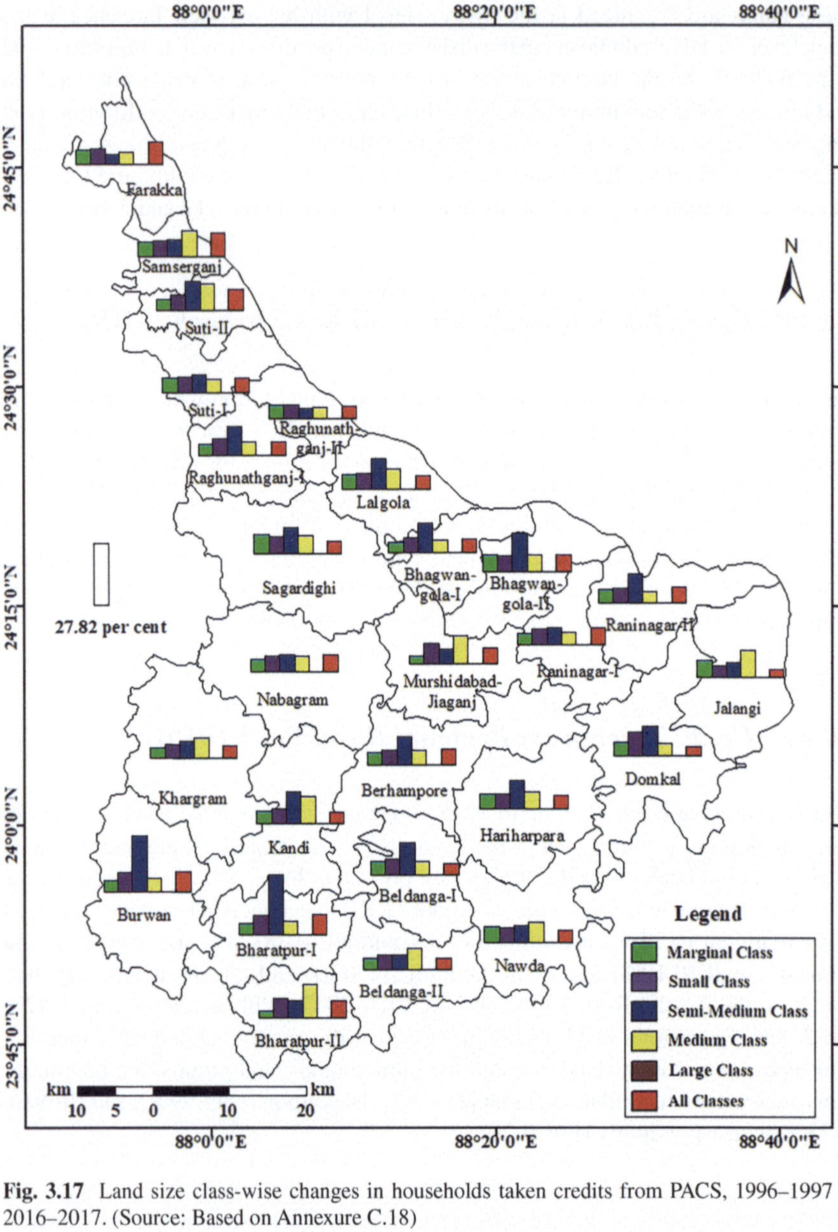

Fig. 3.17 Land size class-wise changes in households taken credits from PACS, 1996–1997 to 2016–2017. (Source: Based on Annexure C.18)

agriculture system is assumed to be moving from traditional to more advanced farming when the use of hand- and animal-operated machinery is losing its dominance and the use of the power-operated machines accelerates. All the land classes except the large group in the district display a reduction in the number of

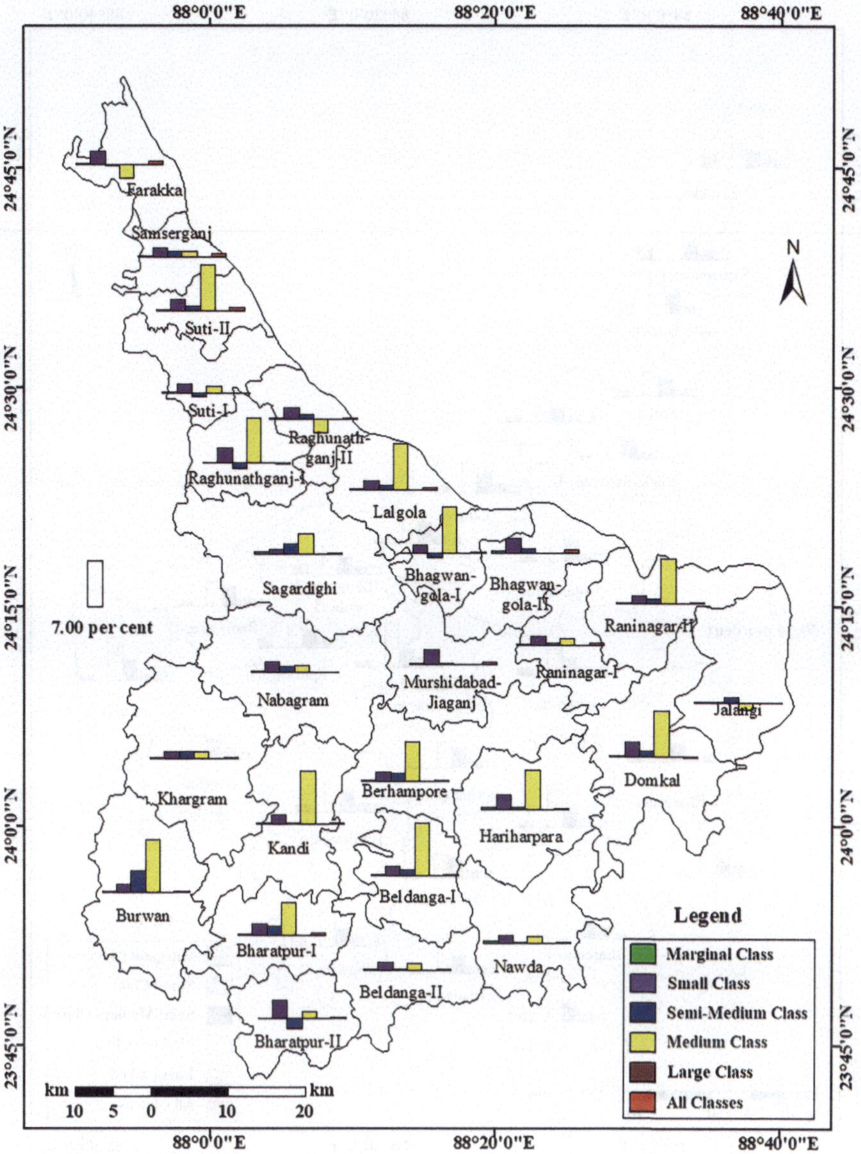

Fig. 3.18 Land size class-wise changes in households taken credits from PLDB, 1996–1997 to 2016–2017. (Source: Based on Annexure C.19)

hand-operated machines per 10 ha GCA during 1996–1997 and 2016–2017. Samserganj in marginal (57), Lalgola in small (25) and semi-medium (19), Jalangi (19) in medium, and Farakka (27) in all land size categories record the maximum change in the number of hand-operated machinery (Fig. 3.21).

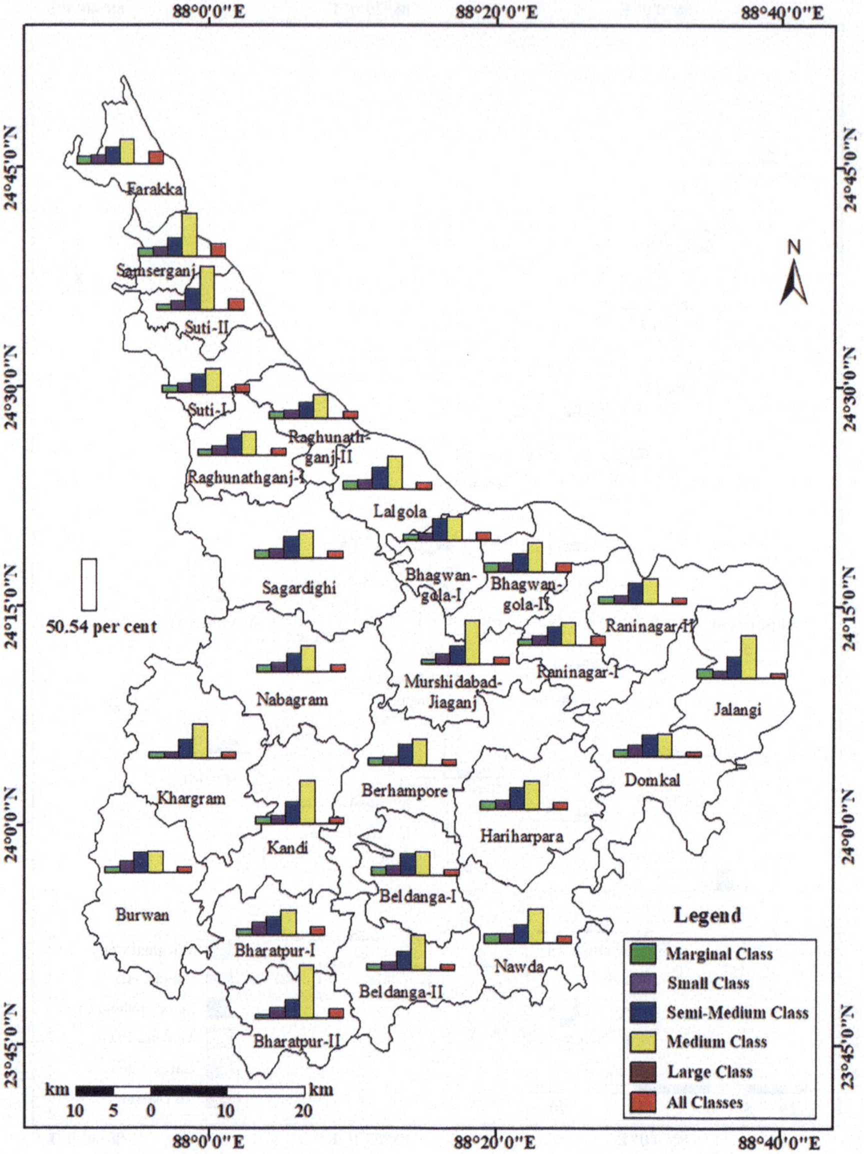

Fig. 3.19 Land size class-wise changes in households taken credits from CBB, 1996–1997 to 2016–2017. (Source: Based on Annexure C.20)

3.4.22 *Animal-Operated Machines*

In the district, the number of animal-operated machines has increased from 11 per 10 ha GCA in 1996–1997 to 15 per 10 ha GCA. Marginal (5), large (6), and all (3)

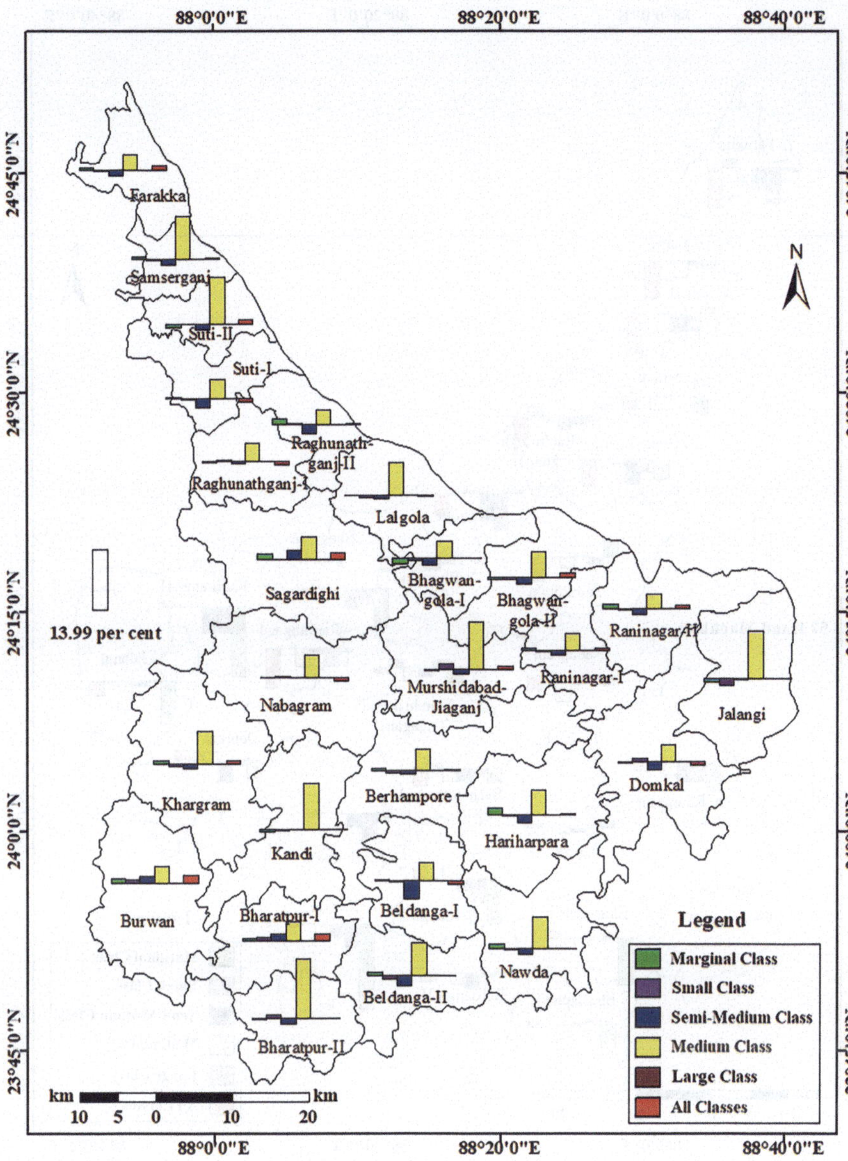

Fig. 3.20 Land size class-wise changes in households taken credits from RRB, 1996–1997 to 2016–2017. (Source: Based on Annexure C.21)

land groups report positive change, while the semi-medium (3) class records a decrease in animal-operated machines. No change has been found under small and medium land categories. Figure 3.22 reveals that overall, Suti-I (12) gains the maximum number, followed by Jalangi (11), Bharatpur-II (9), Kandi (9), and Murshidabad-Jiaganj (9).

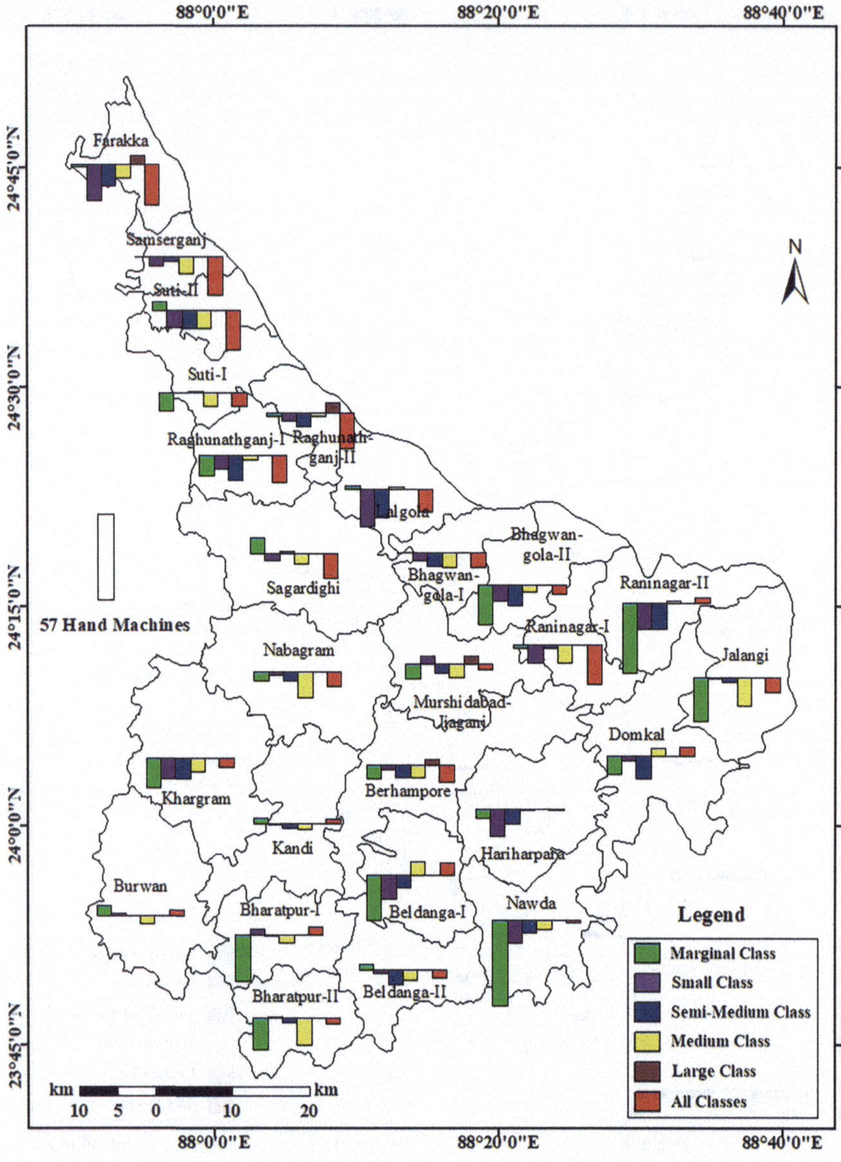

Fig. 3.21 Land size class-wise changes in number of hand-operated machines per 10 ha of GCA, 1996–1997 to 2016–2017. (Source: Based on Annexure C.22)

3.4.23 *Power-Operated Machines*

The district has seen a substantial increase in power-operated machinery such as tractors, threshers, harvesters, power chaff cutters, power cane crushers, electric pump sets, power tillers, etc. per hectare of the gross cropped area from 1996–1997

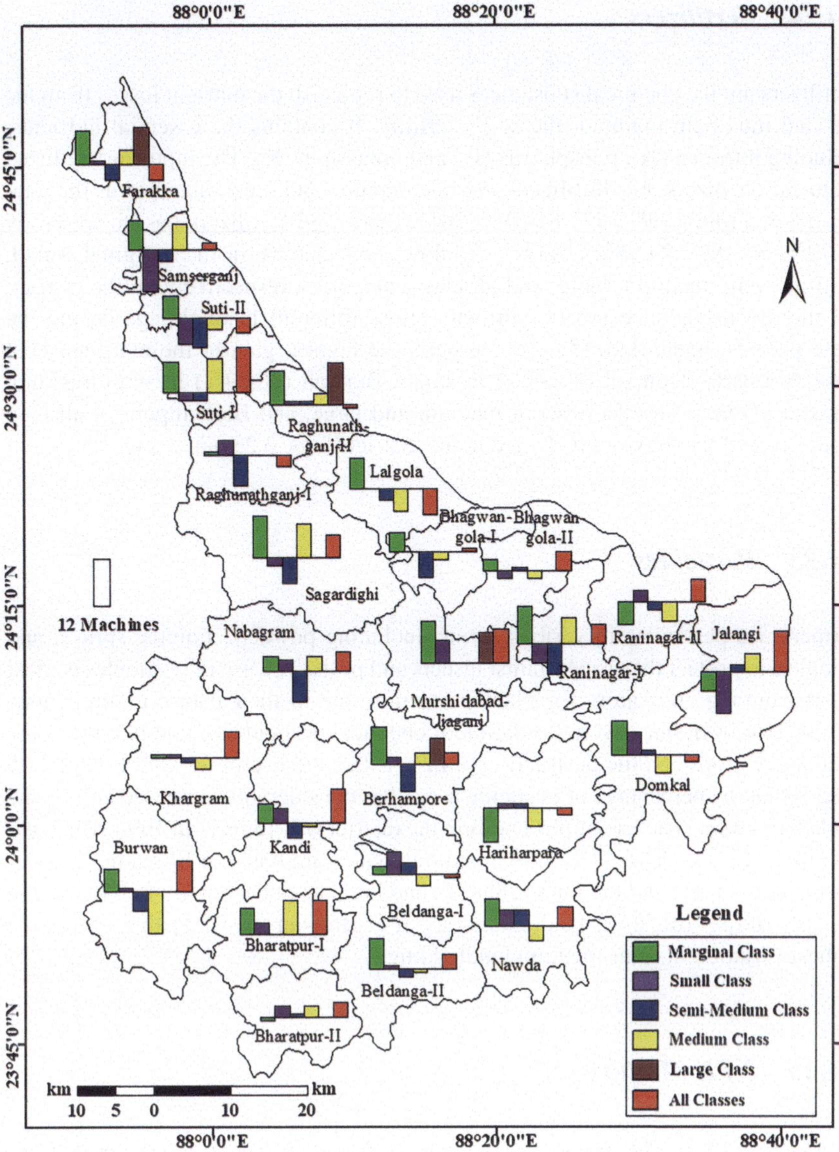

Fig. 3.22 Land size class-wise changes in number of animal-operated machines per 10 ha of GCA, 1996–1997 to 2016–2017. (Source: Based on Annexure C.23)

to 2016–2017. The maximum increase is noticed in marginal land class (19), followed by all (14) and large (10) land categories. Nawda (25), Suti-I (10), Kandi (11), Domkal (11), Burwan (11), and Kandi (20) are the blocks that have the highest increase under marginal, small, semi-medium, medium, large, and all land groups, respectively.

3.4.24 *Fertilizers*

Fertilizers are the chemical substances used to replenish the nutrient losses from the soil and thus help maintain the soil's fertility. It contains the essential nutrients, including nitrogen (N), phosphorus (P), and potassium (K). The input survey listed chemical fertilizers, bio-fertilizers, organic manure, and green manures as the total fertilizers. From 1996–1997 to 2016–2017, the use of fertilizers has increased by 32.72%, 33.48%, 24.98%, 24.08%, 74.86%, and 28.62% in the marginal, small, semi-medium, medium, large, and all class categories, respectively, in the district. All the blocks indicate positive growth across different landholdings during the same periods. Sagardighi (57.75%) reports the highest gain in the marginal land class. Similarly, Domkal (47.14%) in small, Burwan (38.89%) in semi-medium, Farakka (37.00% and 84.19%) in medium and large, and Berhampore in all land groups record the maximum change in the district (Figs. 3.23 and 3.24).

3.4.25 *Pesticides*

Farmers use pesticides in various forms, including powders, liquids, sprays, and granules, to protect the crops against insects and pests. The use of pesticides is most popular among cultivators for plant protection due to their nature of immediate results. However, one severe limitation of chemical pesticides is that it leads to the residues of crops and the destruction of insects that are useful for soil. Figure 3.25 reveals that the percentage of pesticide usage has considerably increased among the land size classes. The use of pesticides in the district was 18.01% in 1996–1997 and rose to 49.27% in 2016–2017. The maximum gains are noted in Bharatpur-I under marginal (54.52%) and medium (58.93%) and Farakka under small (45.04%), large (70.35), and all (62.76%) land categories. It is Burwan (49.01%) that scores the highest variation in semi-medium landholding.

3.4.26 *NPK Mixture*

The Agriculture Census considers nitrogen (N), phosphorus (P), and potassium (K) as the primary fertilizers used to provide soil nutrients. In the district, the use of NPK mixture has grown substantially under different land categories from 1996–1997 to 2016–2017. The small land class sees the maximum increase of 156 kg/ha among the land size groups, followed by medium (131 kg/ha), marginal (130 kg/ha), all (129 kg/ha), semi-medium (118 kg/ha), and large (73 kg/ha) land size holdings in decreasing order. Raninagar-II with 222 kg, Suti-I with 261 kg, Nawda with 296 kg, Domkal with 275 kg, Bhagwangola with 403 kg, and Nawda with 234 kg reports the highest gain under marginal, small, semi-medium, medium,

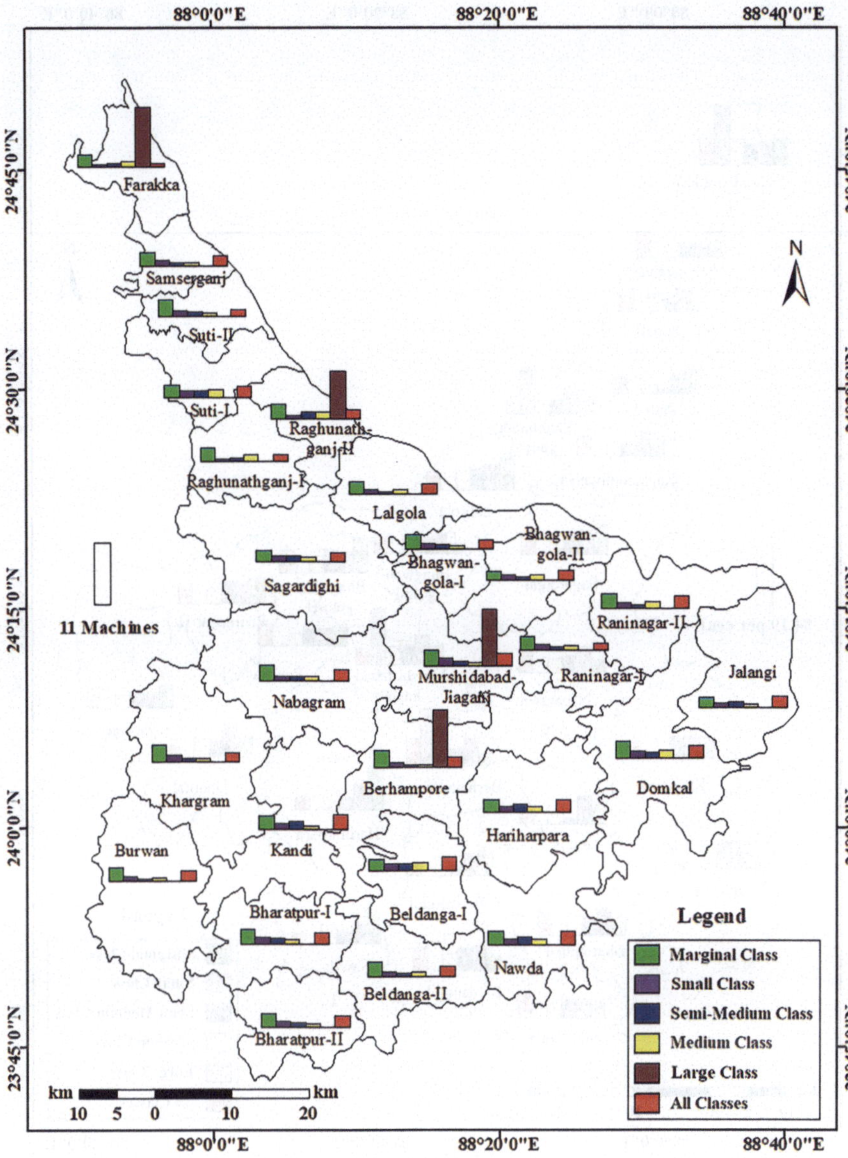

Fig. 3.23 Land size class-wise changes in number of power-operated machines per 10 ha of GCA, 1996–1997 to 2016–2017. (Source: Based on Annexure C.24)

large, and all land size classes, respectively. However, Berhampore, Burwan, Lalgola, Raninagar-I, and Raninagar-II register a reduction in NPK mixture consumption under semi-medium land type (Fig. 3.26).

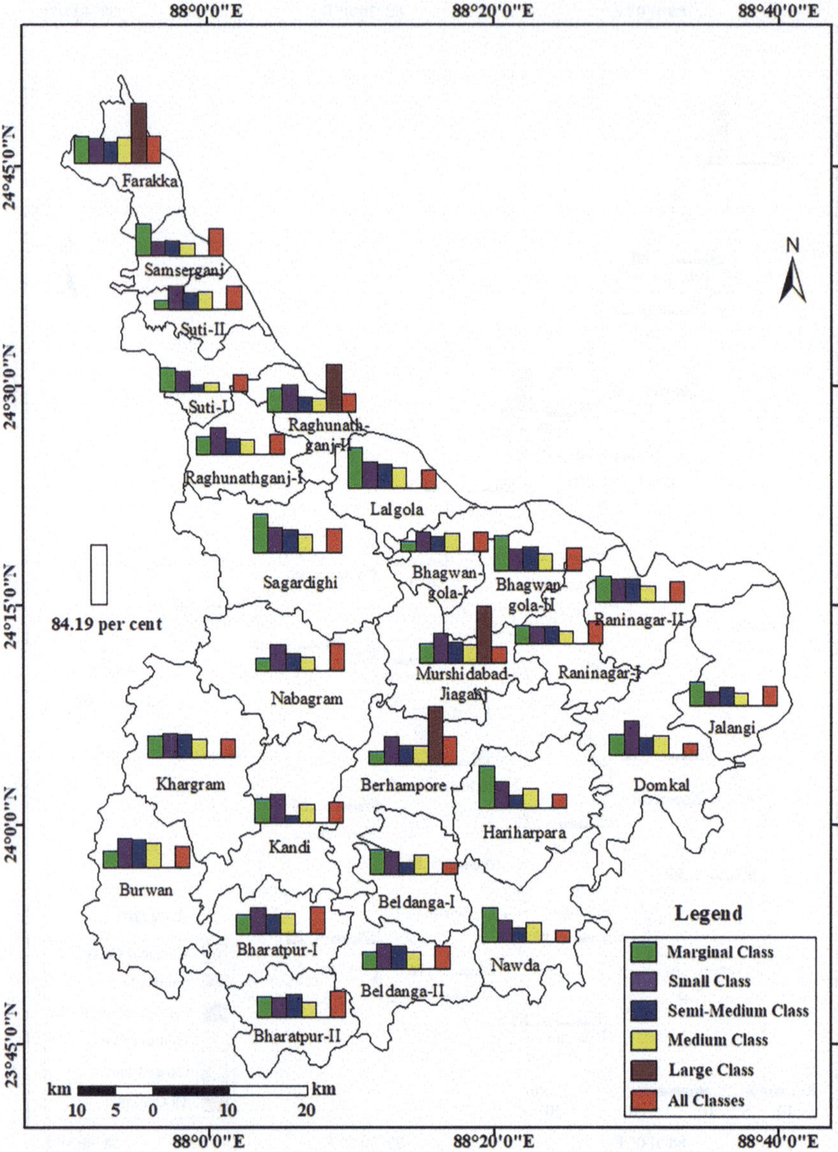

Fig. 3.24 Land size class-wise changes in GCA treated with fertilizers, 1996–1997 to 2016–2017. (Source: Based on Annexure C.25)

3.4.27 Urea, DAP, and Superphosphate Mixture

According to Agriculture Census, the popular macro-soil nutrients are urea, diammonium phosphate (DAP), and superphosphate, which are essential to raising the

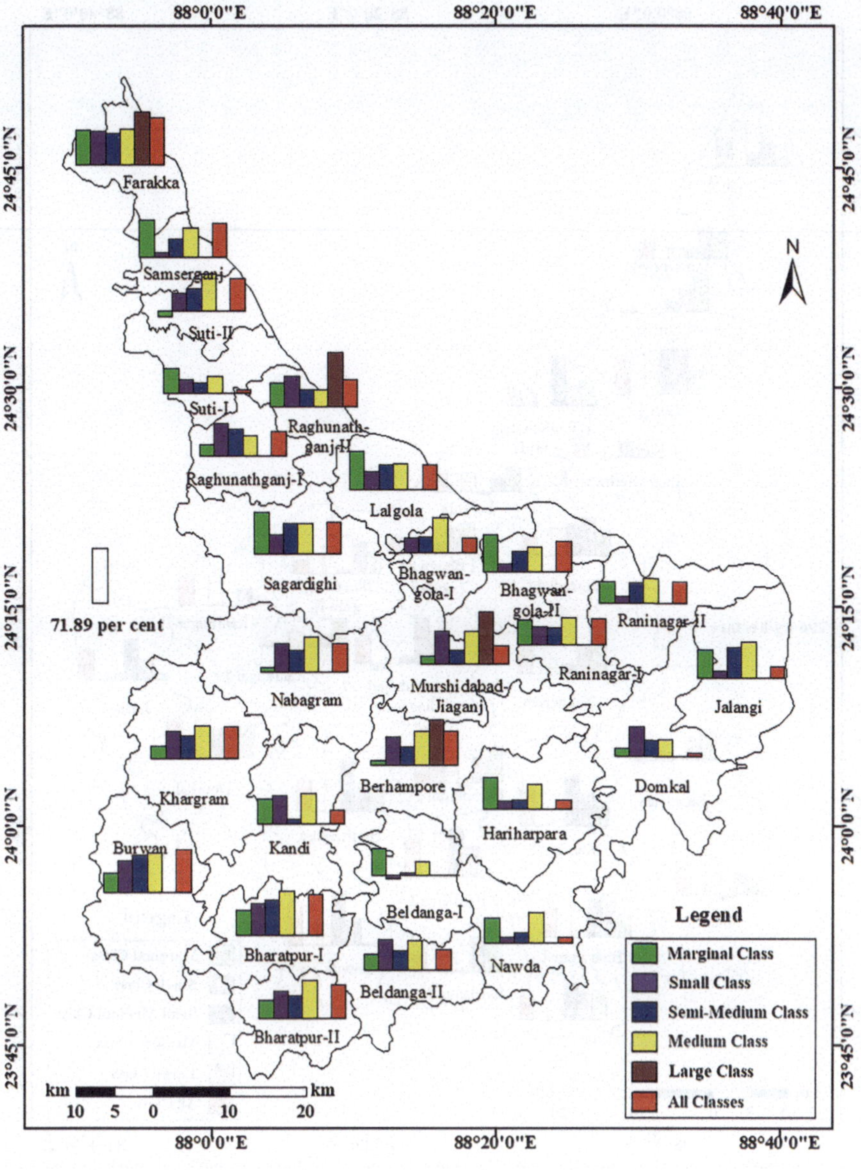

Fig. 3.25 Land size class-wise changes in GCA treated with pesticides, 1996–1997 to 2016–2017. (Source: Based on Annexure C.26)

productivity of various crops. Land size class-wise data locate an increase in consumption of urea, DAP, and superphosphate mixtures kilograms per hectare from 1996–1997 to 2016–2017 across the landholdings in the district, i.e., 96 kg/ha in marginal, 152 kg/ha in small, 124 kg/ha in semi-medium, 110 kg/ha in medium, 46 kg/ha in large, and 84 kg/ha in all land categories. The maximum change under

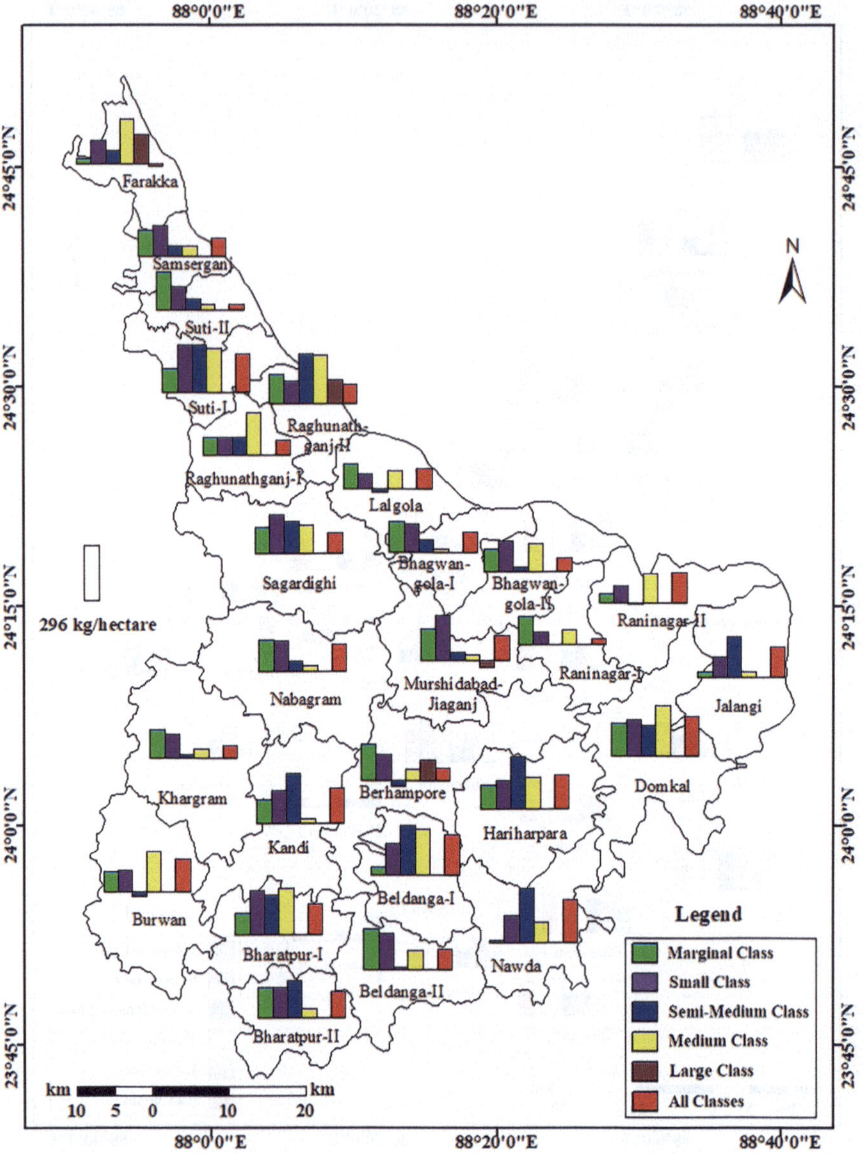

Fig. 3.26 Land size class-wise changes in use of NPK mixtures, 1996–1997 to 2016–2017. (Source: Based on Annexure C.27)

marginal land class is spotted in Raninagar-II (170 kg/ha), while Suti-I (236 kg/ha) tops the list under the small land group. Nawda reports the highest gain in semi-medium (284 kg/ha) and all (150 kg/ha) land size categories. The maximum positive difference under medium landholding is recorded by Domkal (242 kg/ha) (Fig. 3.27).

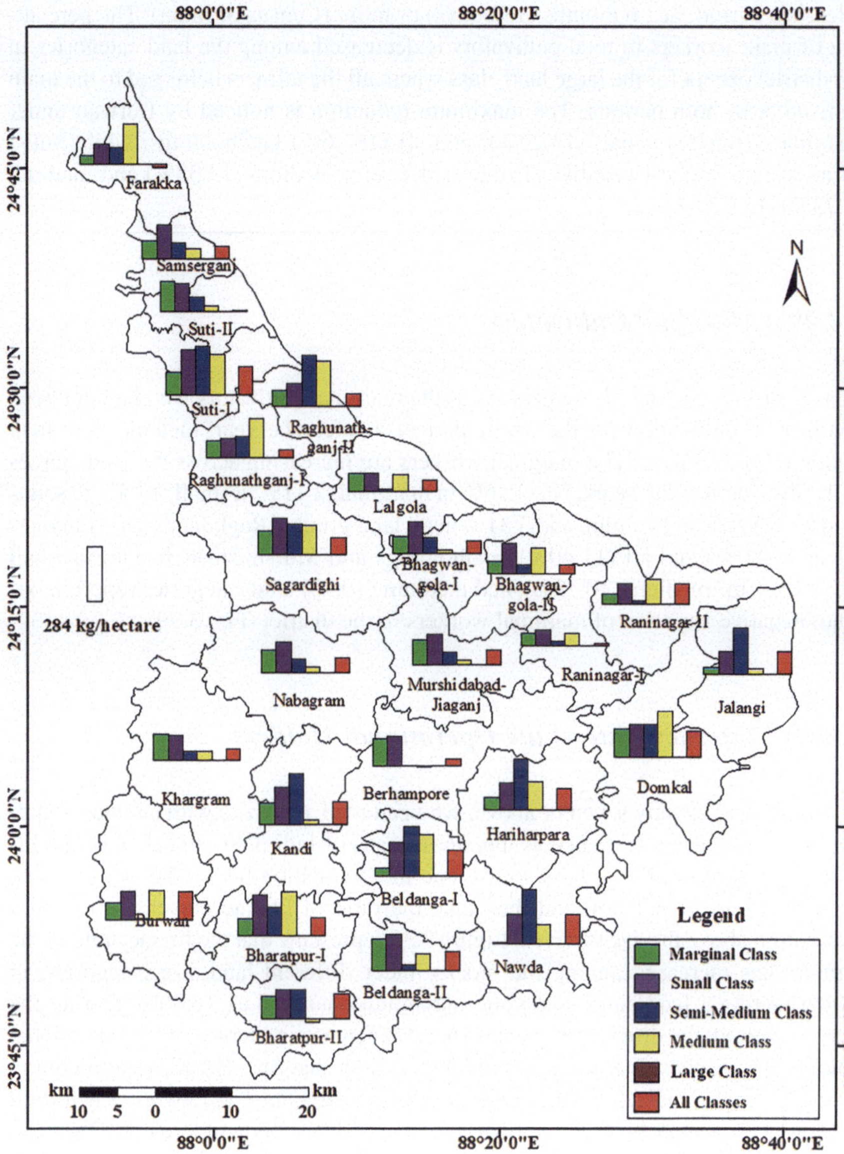

Fig. 3.27 Land size class-wise changes in use of urea, DAP, and superphosphate mixture, 1996–1997 to 2016–2017. (Source: Based on Annexure C.28)

3.4.28 *Main Cultivators*

A person is regarded as the main cultivator engaged in the cultivation of land, including adequate supervision and direction in cultivation for the major part of the reference period, i.e., 6 months (183 days) or more (Census of India). The percentage of main workers to total cultivators is decreased among the land categories in the district except for the large land class where all the farmers belonged to the main cultivators in both periods. The maximum reduction is noticed by Burwan under marginal (16.54%), small (13.27%), and all (16.46%) landholdings, while Suti-I witnesses the highest negative change under semi-medium (13.38%) and medium (9.73%) (Fig. 3.28).

3.4.29 *Marginal Cultivators*

Census of India defines those persons as the marginal cultivators who had not been involved in cultivation for the major part of a reference year, meaning less than 6 months or 183 days. The marginal workers are increasing across the land classes in the district over the years, i.e., 3.16% in marginal, 2.11% in small, 1.05% in semi-medium, 0.20% in medium, and 3.31% in all land groups. Raghunathganj-II in marginal (11.80%) and all (11.40%) landholdings and Murshidabad-Jiaganj in small (8.71%), semi-medium (10.79%), and medium (10.73) land categories report maximum negative variation of marginal workers in the district (Fig. 3.29).

3.4.30 *Literacy Rate of the Operational Holders*

A person aged exactly seven or above, who can read and write with an understanding of any language, is treated as literate (Census of India). So, to calculate the literacy rate, first of all, we have to exclude the population below the age of seven from the entire population and then take the ratio of literate persons to the total population above the age of seven. Figure 3.30 represents that the literacy rate of the farmers has increased among the blocks under different land size categories, in which marginal land class marks the maximum change, i.e., 18.36%. During the 20-year period, the district has gained 16.12% increase in literacy of the operational holders. The highest literacy growing blocks under marginal, small, semi-medium, medium, large, and all landholdings are Raghunathganj-I (27.86%), Sagardighi (21.89%), Beldanga-I (19.75%), Bharatpur-I (25.10%), Berhampore (11.93%), and Domkal (23.49%), respectively.

3.4.31 *Average Age of the Operational Holders*

Woodsong (1994), in her study on 'Old Farmers, Invisible Farmers: Age and Agriculture in Jamaica', argued that the older farmers have lesser consequences on agricultural production than relatively younger cultivators in Jamaica, setting the

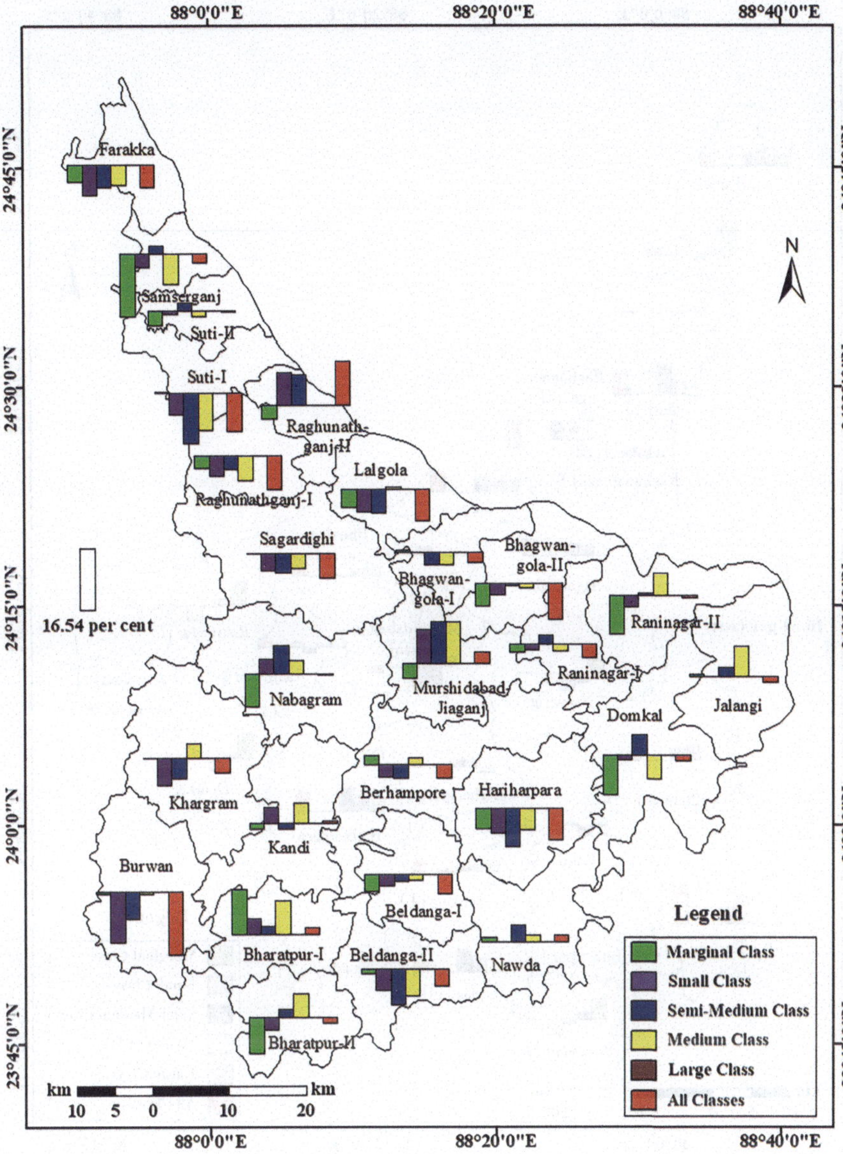

Fig. 3.28 Land size class-wise changes in percentage of main cultivators to total cultivators, 1996–1997 to 2016–2017. (Source: Based on Annexure C.29)

example for developing countries. The average age of operational holders has increased across the land size holdings except for the large class from 1995–1996 to 2015–2016. There is an increase in average age by 2 years in marginal, 3 years in small, 4 years each in semi-medium and medium, and 2 years in all land size categories in the district. The large land group experiences a year reduction in the average age of the farmers (Figs. 3.31 and 3.32).

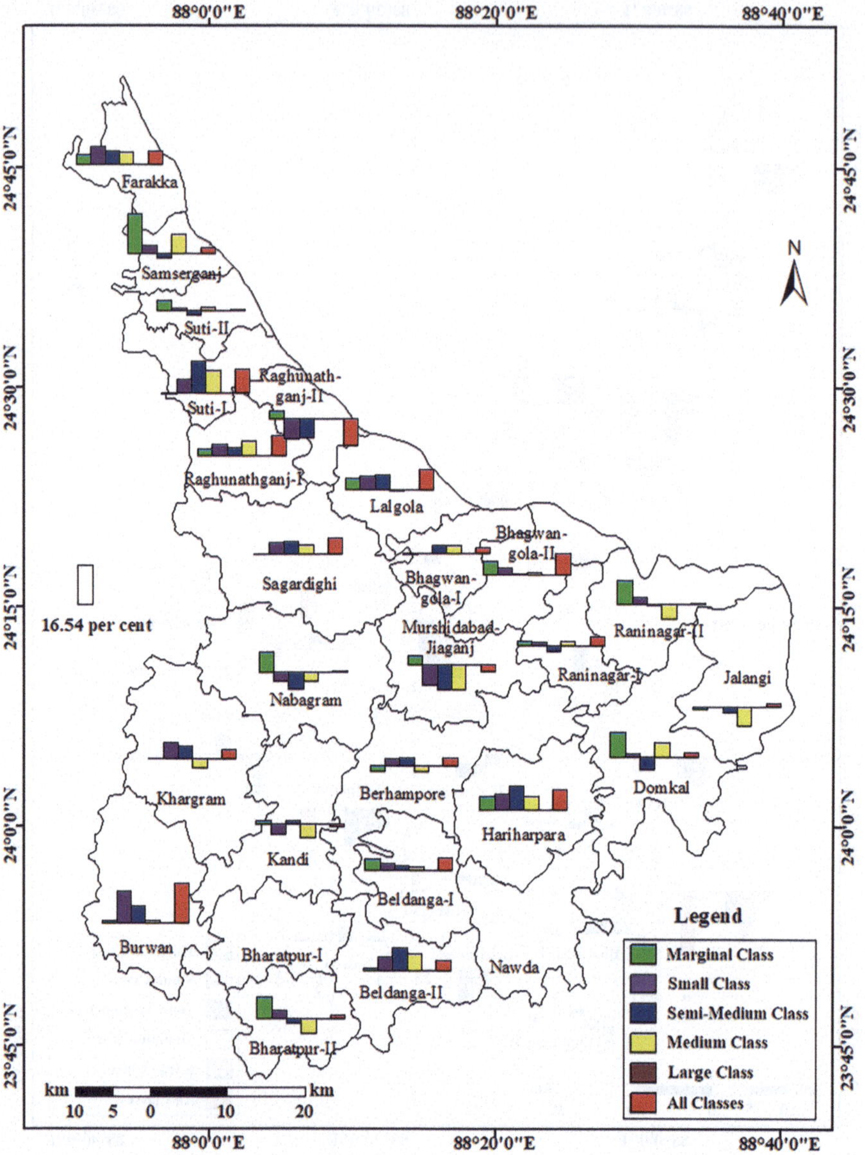

Fig. 3.29 Land size class-wise changes in percentage of marginal cultivators to total cultivators, 1996–1997 to 2016–2017. (Source: Based on Annexure C.30)

3.4.32 *Average Family Size of the Households*

Various works of the literature suggest that a positive relationship exists between the family size of the households and agricultural productivity in a region. Those

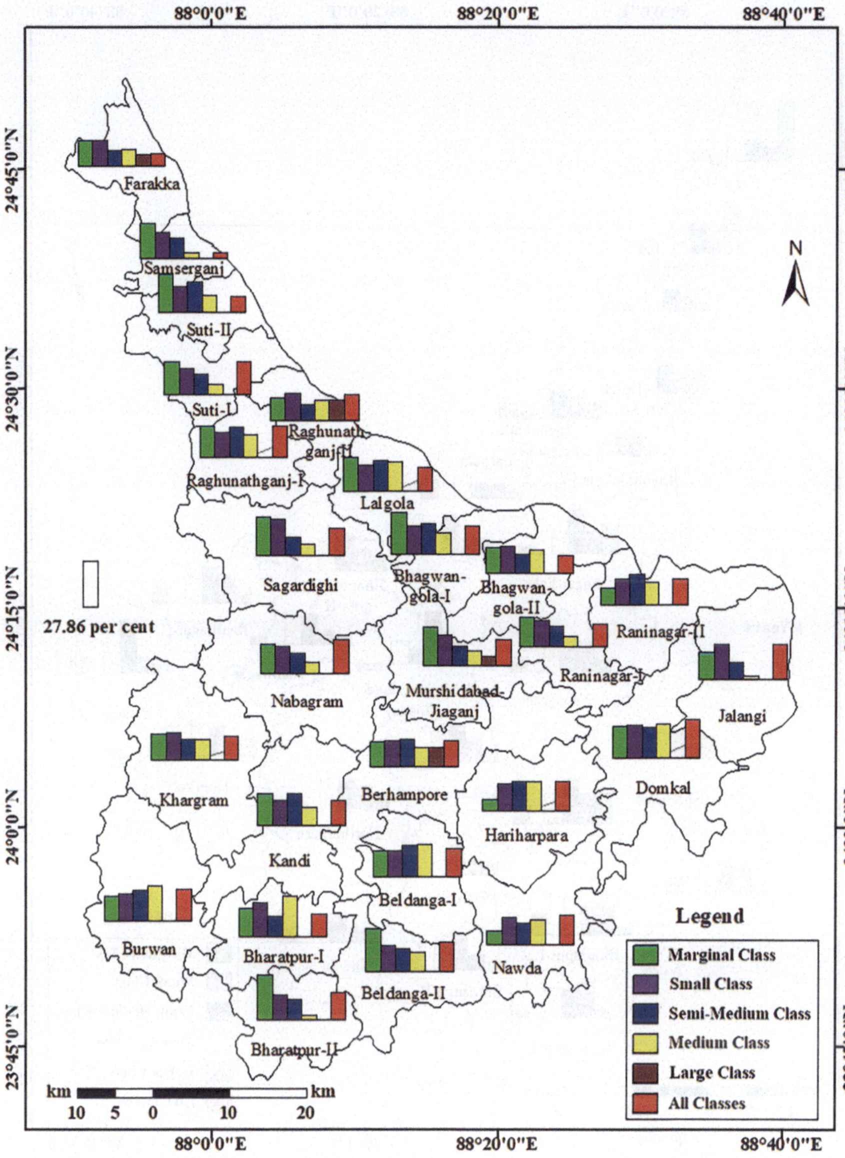

Fig. 3.30 Land size class-wise changes in literacy rate of the operational holders, 1996–1997 to 2016–2017. (Source: Based on Annexure C.31)

farmers with larger household sizes are more productive than those with smaller family sizes because the household members can be used as a source of labour inputs (Nosiru, 2010). The average family size has not changed in the district over the 20 years of the study period. However, some variation is noticed among the blocks under different land size categories. There is an increase of one member in

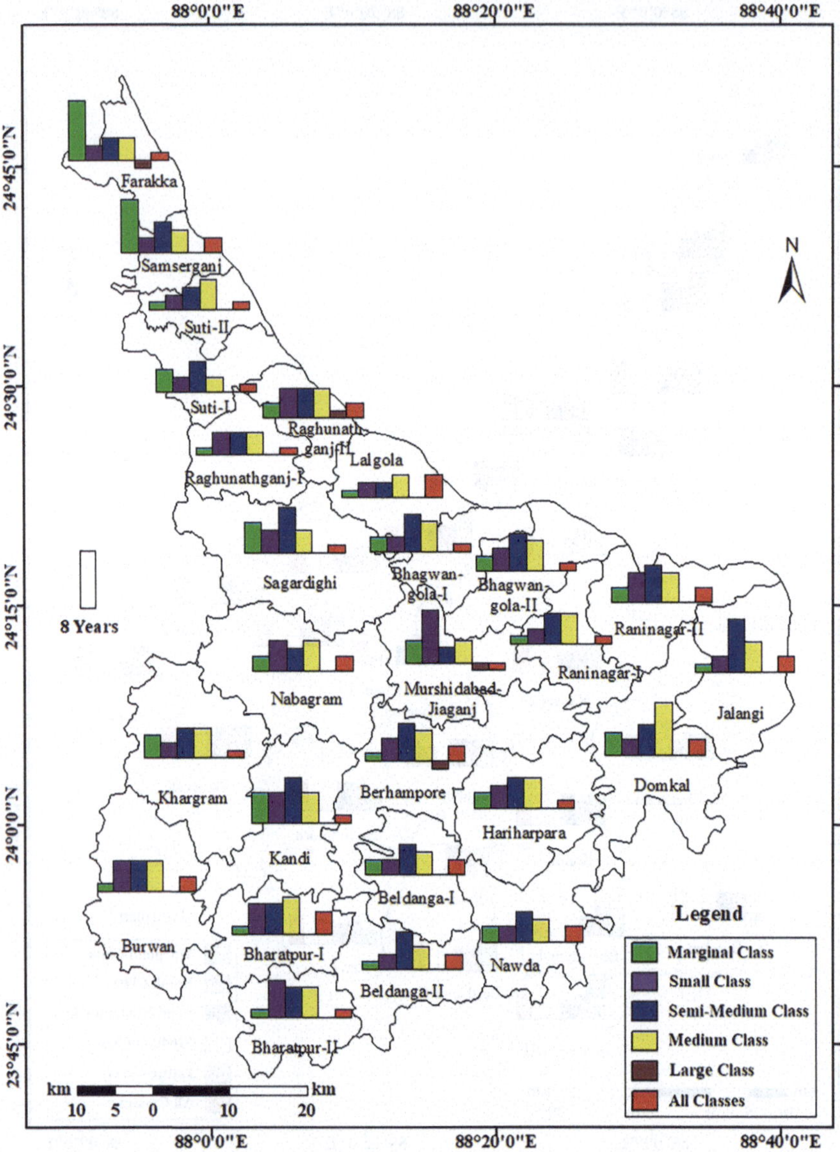

Fig. 3.31 Land size class-wise changes in average age of the operational holders, 1996–1997 to 2016–2017. (Source: Based on Annexure C.32)

the family size in Bharatpur-I in marginal, Beldanga-II and Murshidabad-Jiaganj in small, Sagardighi in semi-medium, Bharatpur-I, Bharatpur-II, and Murshidabad-Jiaganj in medium, and Nawda in all land size classes.

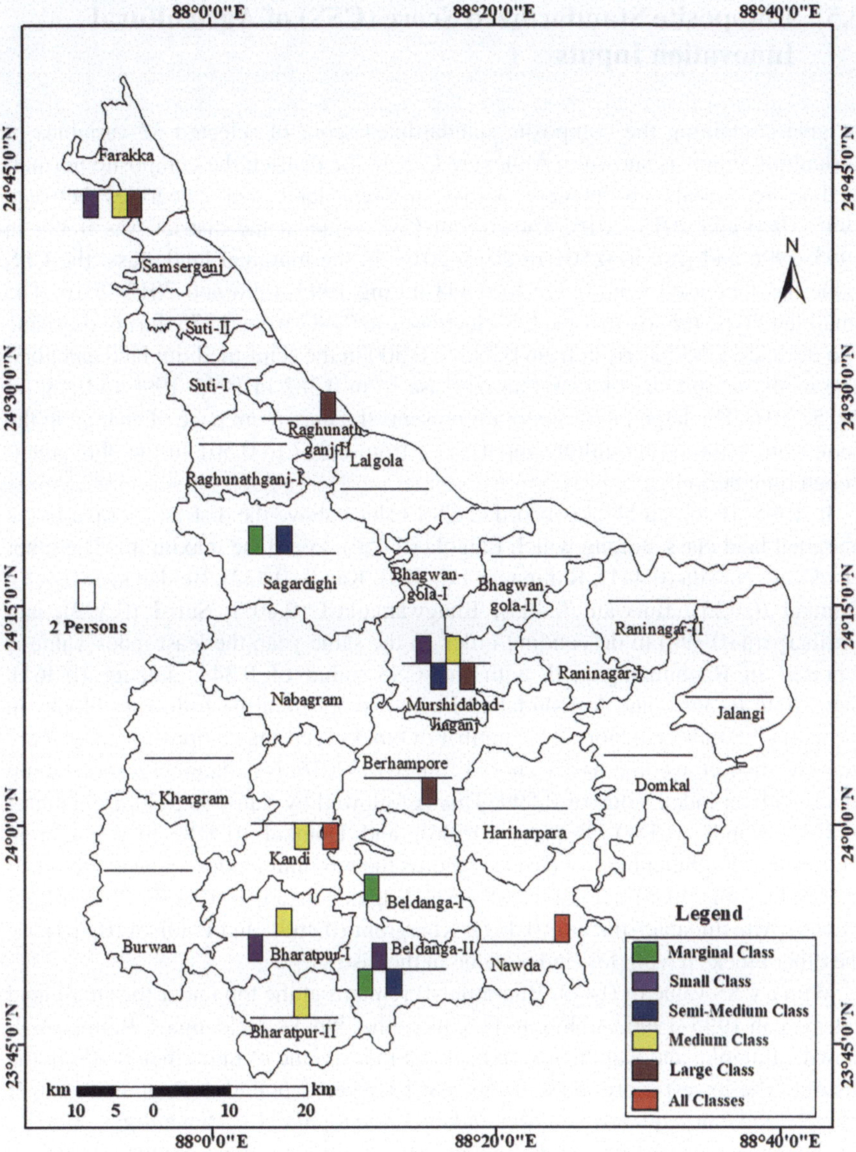

Fig. 3.32 Land size class-wise changes in average family size, 1996–1997 to 2016–2017. (Source: Based on Annexure C.33)

3.5 Composite Standardized Score (CSS) of Agricultural Innovation Inputs

A table containing the composite standardized score of selected 32 variables of agriculture inputs is shown in Annexure C.2. In the district, the composite normalized score reveals an increase across different land size categories between 1995–1996 and 2015–2016. The overall CSS value in the district was 0.399 in 1995–1996 and rose to 0.501 in 2015–2016. In the marginal land class, the CSS value has increased from 0.395 to 0.498 during 1995–1996 and 2015–2016. The small land type reports that the CSS increases to 0.487 from 0.375 during the same two decades. CSS has risen from 0.395 to 0.500 in the semi-medium land category. The medium land class records an increase from 0.382 in 1995–1996 to 0.488 in 2015–2016. The large land category represents the maximum pace of change in the composite value of agriculture inputs, i.e., from 0.317 to 0.501 in the abovementioned time period.

In 1995–1996, ten blocks recorded CSS values above the district average in the marginal land class, among which Lalgola (0.450) scored the maximum. The other blocks are Nawda (0.441), Raninagar-I (0.435), Kandi (0.432), Beldanga-I (0.428), Domkal (0.425), Burwan (0.405), Bhagwangola-I (0.401), Suti-I (0.399), and Hariharpara (0.396) in descending order. In the same year, the least index value is reported in Raghunathganj-II with the CSS value of 0.344. Jalangi (0.361), Sagardighi (0.366), and Murshidabad-Jiaganj (0.369) are the following blocks in the least CSS index category. The number of blocks securing composite scores more than the district average has risen to 12 in 2015–2016. Nawda emerges as the top block with an index value of 0.599. This is followed by Suti-I (0.577), Beldanga-I (0.551), Kandi (0.547), Burwan (0.536), and Domkal (0.518) in succession. Conversely, Raghunathganj-I (0.435) records the minimum index value in the noted year under the marginal land category. Raghunathganj-II (0.436), Sagardighi (0.448), Murshidabad-Jiaganj (0.452), Khargram (0.460), and Farakka (0.464) are the other blocks having low index value in the district.

With a CSS value of 0.444, Bharatpur-II remains at the top under the small land category in 1995–1996. Adding to this, five other blocks, Beldanga-I, Raninagar-I, Nawda, Lalgola, and Kandi, also indicate an index value of more than 0.400 in the district. The lowest score in the same year has been captured in Raghunathganj-II (0.289) and subsequently in Murshidabad-Jiaganj (0.316), Sagardighi (0.327), Raghunathganj-I (0.337), and Beldanga-II (0.345) in the small land group. Raghunathganj-II, with a CSS value of 0.403, once again secures the minimum index value in 2015–2016. In the noted year, the highest index is displayed by Suti-I (0.593). Besides this, eight other blocks have also depicted index values above 0.500 under small land type in 2015–2016. These are Beldanga-I (0.562), Kandi (0.542), Domkal (0.541), Nawda (0.522), Bharatpur-II (0.522), Murshidabad-Jiaganj (0.518), Bhagwangola-I (0.506), and Bharatpur-I (0.501).

In the district, Bharatpur-I having a CSS index of 0.445 sits on the top among the blocks under the semi-medium land category in 1995–1996. In the same list,

Raghunathganj-I (0.437), Suti-I (0.428), Nabagram (0.428), Kandi (0.427), Berhampore (0.424), and Beldanga-I (0.421) are positioned in descending order. Contrary to this, the last index value in 1995–1996 is recorded in Jalangi at 0.324. Beldanga-II (0.346), Raghunathganj-II (0.360), Farakka (0.364), Samserganj (0.368), and Bhagwangola-II (0.375) also represent lower CSS values in the district. Eight blocks in the semi-medium land class score more than 0.500 index values in the district during 2015–2016. With an index value of 0.632, Beldanga-I is the top block in terms of CSS value in 2015–2016 under this land category. The other blocks are Kandi (0.614), Suti-I (0.608), Hariharpara (0.593), Nawda (0.570), Domkal (0.557), Bharatpur-I (0.526), and Raghunathganj-I (0.513). In the same year, Farakka (0.434) stands at the bottom of the CSS index value in the district, followed by Murshidabad-Jiaganj (0.441), Raninagar-II (0.441), and Burwan (0.444).

During 1995–1996, Bharatpur-II, with a 0.456 index value, is ranked first among the blocks under the medium land size category in the district. Nawda (0.426), Beldanga-I (0.419), Suti-I (0.418), Raninagar-II (0.410), and Nabagram (0.408) are the succeeding blocks in the list. The bottom three lowest CSS values in the same land class are found in Raghunathganj-II (0.307), Samserganj (0.311), and Raninagar-I (0.332) in ascending order. The index value is very much improved in Beldanga-I from 1995–1996 to 2015–2016. Due to this, it ranks top with an index value of 0.569 among the blocks in the latter time period. The other blocks that remain up in the same list are Lalgola (0.545), Raghunathganj-I (0.543), Suti-I (0.543), and Domkal (0.528). In contrast, Sagardighi (0.419) reports the minimum CSS value among the blocks in 2015–2016, followed by Berhampore (0.433), Nabagram (0.433), and Murshidabad-Jiaganj (0.444).

There were only eight blocks, namely Berhampore, Beldanga-II, Domkal, Farakka, Murshidabad-Jiaganj, Raghunathganj-I, Raghunathganj-II, and Suti-I, which had farmers belonging to the large land category in the district during 1995–1996. The maximum and minimum index values of these blocks are noted in Beldanga-II (0.354) and Farakka (0.246), respectively. The number of blocks having large farmers was reduced to seven in 2015–2016. Raghunathganj-II (0.537) represents the maximum index value in the district, while the lowest value is in Murshidabad-Jiaganj (0.445).

In 1995–1996, as many as 14 blocks acquired index values higher than the district's average (0.399) under all land size class. Bharatpur-II secures the top position in terms of maximum CSS value with an index value of 0.437. Beldanga-I (0.435), Lalgola (0.433), Raninagar-I (0.427), Nabagram (0.425), Nawda (0.420), and Bharatpur-I (0.419) are the successive blocks in all land size category. In the same time period, Raghunathganj-II (0.343) obtains the lowest index value among the blocks, followed by Farakka (0.350), Jalangi (0.362), Murshidabad-Jiaganj (0.373), Sagardighi (0.376), and Beldanga-II (0.380) in increasing order.

Again in 2015–2016, 14 blocks lie above the district's average (0.501) under all land group. With an index value of 0.565, Nawda and Suti-I combined denote the highest composite score in the district. Beldanga-I (0.557), Kandi (0.540), Bharatpur-I (0.535), Bharatpur-II (0.530), and Lalgola (0.524) are the other blocks

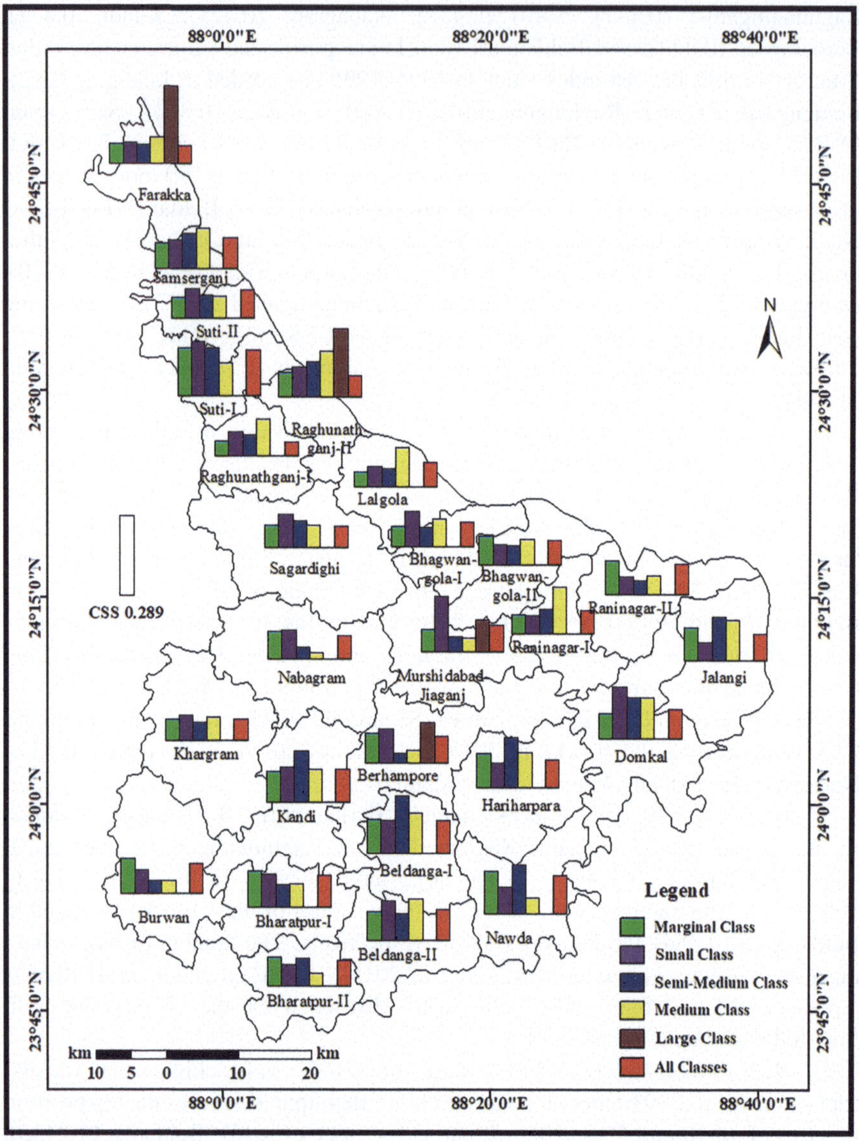

Fig. 3.33 Land size class-wise changes in composite standardized score of agricultural inputs, 1995–1996 to 2015–2016. (Source: Based on Annexure C.34)

that score higher index values in the district. The least CSS value during 2015–2016 in the district is registered in Farakka (0.419), followed by Raghunathganj-II (0.422), Sagardighi (0.452), Raghunathganj-I (0.457), and Jalangi (0.459).

Figure 3.33 represents the land size class-wise changes in CSS index value in the district during 1995–1996 and 2015–2016. The large land category denotes the maximum change of 0.184 index value among the land classes in the district. This is followed by small land category, where the amount of change is 0.111. The lowest change of CSS value is reported in all land size class (0.101). Medium, semi-medium, and marginal land groups record the difference in composite value between two points of time as 0.106, 0.105, and 0.103, respectively.

The block-level data reveals that the nature of the change of CSS value is persistent throughout the blocks under different land categories. Suti-I (0.179) in marginal class witnesses the maximum difference from 1995–1996 to 2015–2016, followed by Nawda (0.159), Bharatpur-I (0.134), Burwan (0.131), Hariharpara (0.129), and Raninagar-II (0.126). Under the same land class, Raghunathganj-I (0.054) shows the minimum index value change. Other blocks such as Lalgola (0.055), Bhagwangola-I (0.076), Farakka (0.077), and Khargram (0.079) have also highlighted the lower pace of development of agricultural inputs. Murshidabad-Jiaganj indicates the maximum growth under the small land group having an index value change of 0.202, while the slightest difference is found in Jalangi (0.064). The highest and lowest CSS changes under the semi-medium land category appear in Beldanga-I (0.211) and Berhampore (0.035), respectively. In the medium land size class, Raninagar-I (0.172) reports the maximum amount of change and Nabagram (0.025) denotes the minimum amount of change in the district. All land category depicts the highest amount of change in Suti-I (0.171), followed by Nawda (0.145), Kandi (0.123), and Beldanga-I (0.111) in the district. In contrast, the least difference in composite value is represented by Raghunathganj-I (0.051). Farakka (0.069), Sagardighi (0.077), Raghunathganj-II (0.079), and Khargram (0.081) are the other blocks where the least amount of change is found.

3.6 Agricultural Innovation: Output

Agricultural innovation output is defined as the total outcomes of a particular holding in a reference year or in a particular cropping season applying the total inputs. 'Final agricultural innovation output measures the value of agricultural products which, free of intra-branch consumption, is produced during the accounting period and is available for export and consumption' (Economic Accounts for Agriculture, OECD). The extensive literatures suggest that productivity, crop diversity, and cropping patterns assert the output dimensions of agriculture. Therefore, these three main dimensions are taken as output variables in the present analysis.

3.6.1 Productivity

Productivity is commonly defined as a ratio of a volume measure of output to a volume measure of input use (OECD, 2001). According to Shafi (1972), 'Productivity is essentially a measure of the efficiency with which inputs are utilized in production'. There are many methods and techniques to measure agricultural productivity. The productivity measurement can be broadly classified as single-factor productivity, where output is computed against a single input measure and multifactor productivity relates to the extent of output to a bundle of inputs. The choice of techniques and methods depend upon the productivity measurement and the nature of data availability. 'Crop yield index' method given by W.Y. Yang (1965) is used for data convenience in this study to calculate agricultural productivity. To calculate the yield index, first of all, we have to take the actual area in hectares under each crop as well as the average yield of those crops in the blocks and district into consideration. Here in this analysis, the top seven crop categories are listed for calculating yield in the blocks. Next, we have to calculate the crop yield in the block as the percentage to the district. After this, the obtained value is multiplied by the actual area occupied by each crop category in the block. Finally, the crop yield index is computed by adding all the values obtained previously and then dividing by the sum of the area devoted to each crop category (Table 3.1).

In 1995–1996, a total of 10 blocks scored the yield index value more than 100 in the district, in which Bharatpur-I reports the maximum, i.e., 129.60, followed by Beldanga-I (114.82), Burwan (114.12), and Khargram. In contrast, the lowest index value is noted in Nawda (85.02), followed by Raghunathganj-II (86.10) and Raninagar-II (86.10). In the same year, the marginal land class locates Bharatpur-I (134.34) and Raghunathganj-II (84.84) as the highest and lowest scorers, respectively, in the district. Raninagar-I (117.07), Bharatpur-I (125.32), and Bharatpur-II (125.11) register the top yield index under small, semi-medium, and medium land categories, respectively. The least score under small (81.99) and medium (83.21) landholdings is notified by Bhagwangola-I, while Samserganj (81.32) shows the minimum value in semi-medium land classes. There are eight blocks in the large

Table 3.1 Method for calculating crop yield index (After W.Y. Yang, 1965)

Name of the crops	The area under crop in the block	Yield in quintal/ha Average yield in the block	Average yield in the district	Crop yield in the block as the percentage to the district	Percentage multiply by area (in ha)
1	2	3	4	5 = Col. 3/ col4 * 100	6 = col.5 * col. 2
Cereals	36,366	2705	2801	96.59	3,512,412.51
Pulses	686	344	601	57.19	39,229.95
Spices	79	2605	2580	100.98	7977.33
Fruits	833	6580	7598	86.61	72,144.24
Vegetables	4404	15,971	17,070	93.56	412,049.93
Oilseeds	2842	910	1087	83.74	237,987.39
Fibres	10,293	1,770,200	2,268,704	78.03	803,131.15
Total	**55,503**				5,084,932.50
Crop yield index for Berhampore block, marginal class, 1995–1996					**5,439,868.62/55,503**
					91.62

Source: Calculated by authors

land type where the yield index values are available, in which Murshidabad-Jiaganj (126.28) shows the maximum and Farakka (80.94) indicates the minimum.

Bharatpur-I, with an index of 129.16 and 126.09, once again records the highest value in the marginal and all land categories during the 2015–2016 census. Burwan, Bharatpur-II, Khargram, and Kandi also report a higher yield index in both land classes. The maximum value under small (127.75), semi-medium (123.01), and medium (135.74) are found in Bharatpur-II. Conversely, Bhagwangola-I reports the lowest value in marginal (81.84), small (83.82), medium (71.35), and all (82.59) landholding, while in semi-medium class, it is shown by Samserganj (84.20). Seven out of the 26 blocks have devoted area under the large land category in 2015–2016, where Berhampore (120.07) and Nawda (80.52) secure the top and bottom positions, respectively.

Figure 3.34 illustrates the block-wise changes in the yield index of crops under different land size categories in the district from 1995–1996 to 2015–2016. The maximum positive change of yield index in marginal land class is denoted in Murshidabad-Jiaganj (7.20). The highest negative difference under the same land group is noted by Bhagwangola-I (6.65). Bharatpur-II depicts the maximum positive variation in index value in small (14.17), semi-medium (9.56), and all land classes. Again, the highest growth of crop yield is noticed by Khargram (15.17) in medium and Berhampore (11.75) in large land categories. However, the top negative variation of index values under small, semi-medium, medium, large and all landholdings is reported in Raninagar-I (13.05), Bharatpur-I (13.64), Bhagwangola-I (11.87), Murshidabad-Jiaganj (9.36), and Raninagar-I (6.44), respectively.

3.6.2 *Crop Diversification*

Broadly, the concept of diversification carries different meanings to different scholars at different levels. A plethora of terms are linked with crop diversification concept (Siddiqui & Rahaman, 2016). For instance, Joshi et al. (2004) defined diversification as 'a movement from low-value crops to high-value crops for the betterment of the agricultural system'. It simply refers to the cultivation of multiple crops in a particular region at a given time. The diversification of crops towards high-value commodities not only ensures better income opportunities and employment generation but also offers better nutrition security and sustainability of natural resources. The present crop diversification enriches the food basket mainly from horticulture and vegetable crops in third world countries like India because the diversification gets impetus from irrigation and fertilizer uses (Rahaman & Ansari, 2024). Apart from the innovation of institutional, infrastructural, and technological facilities, different policies at macro- and micro-level planning pace up the diversified cropping pattern in a region. There are quite a few methods to measure the intensity of crop diversification, such as Gibbs–Martin's technique, Herfindahl Index, Simpson Diversification Index, Entropy Index, Ogive Index, and others.

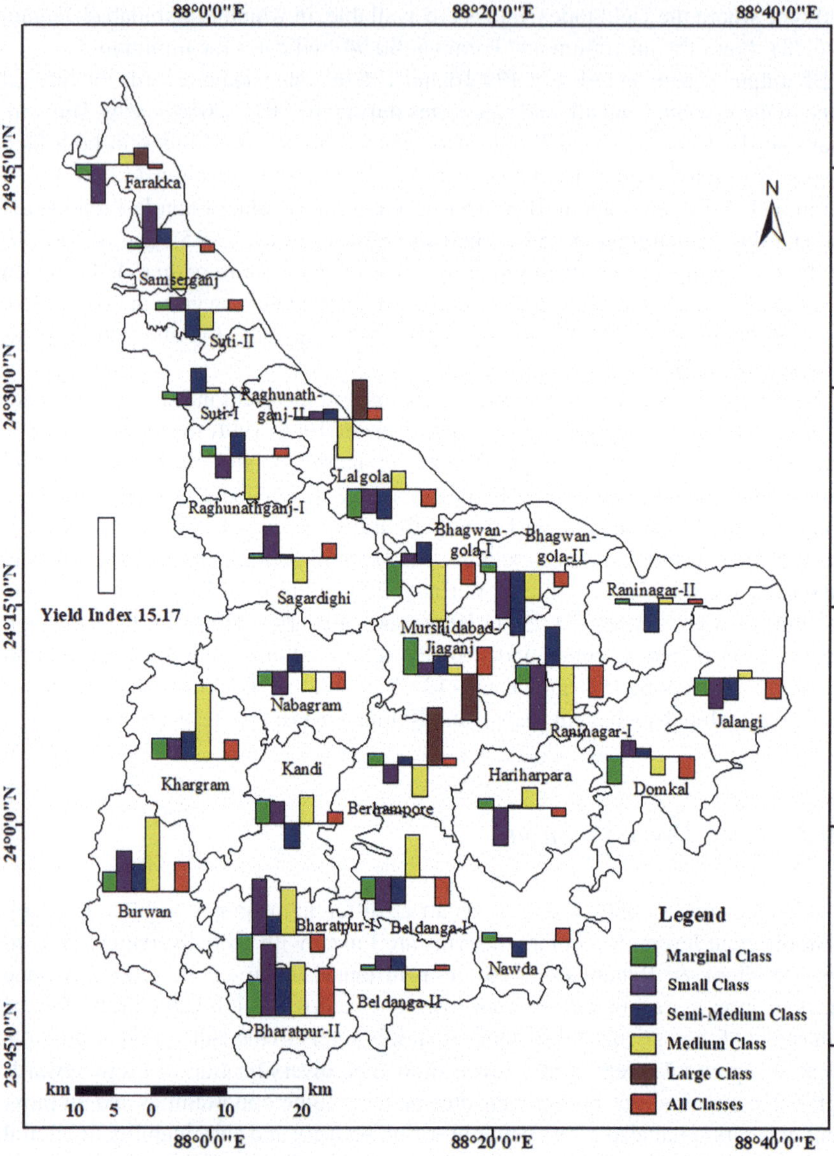

Fig. 3.34 Land size class-wise changes in crop yield index, 1995–1996 to 2015–2016. (Source: Based on Annexure C.35)

Gibbs and Martin's (1962) technique is employed to assess the crop diversification index in the present study because it considers the percentage of the areal extent of all the crops out of the total cropped area. The index is computed as follows:

$$\text{Crop diversification index} = 1 - \frac{\Sigma X^2}{(\Sigma X)^2}$$

Here, X is the percentage of the total cropped area occupied by an individual crop at a point of time. The index value of this technique ranges from 0 to 1. The crop diversification seems to be higher when the index value approaches 1, while the value reaching near 0 reflects a specialization of crop cultivation.

Figure 3.35 reveals an increasing trend of crop diversification across land size categories in the district from 1995–1996 to 2015–2016. The large land class (0.287) reports the highest index gain among the land groups, followed by small (0.124), all (0.099), marginal (0.097), semi-medium (0.073), and medium (0.055) land categories. All the blocks under the marginal landholdings posit a positive change in the diversification index except for Hariharpara (0.108), Beldanga-I (0.062), and Raghunathganj-I (0.034), where reverse change has been noticed due to the dominance of major cereal crops in the existing cropping pattern. Kandi (0.264), Bharatpur-II (0.252), Murshidabad-Jiaganj (0.213), and Bharatpur-I (0.200) are the blocks that gained a significant amount of index value increase in the land class mentioned above. The maximum increases in diversification in small, semi-medium, medium, large, and all landholdings are recorded in Kandi (0.408), Bhagwangola-II (0.239), Burwan (0.336), Murshidabad-Jiaganj (0.363), and Kandi (0.273), respectively. There are only three blocks in small, six blocks in semi-medium, seven blocks in medium, and three blocks in all land groups that register an adverse change in index value in the district. Overall, the highest decrease in diversification index is located by Hariharpara, with an index value of 0.103.

3.6.3 *Cropping Intensity*

The ratio between gross cropped area (GCA) and net sown area (NSA) in a particular region is known as cropping intensity. Thus, the more the percentage share of area sown more than once to the NSA in a region, the higher the land use intensity would be. It is the number of crops planted annually in the same agricultural field (Jain et al., 2013). The intense cropping system of agricultural land depicts the extent to which the productive capacity is utilized through modern technological and infrastructural facilities. It intensifies production by expanding agricultural land use in a reference year. The formula for calculating cropping intensity is stated below:

$$\text{Cropping intensity} = \frac{\text{Gross cropped area}}{\text{Net sown area}} \times 100$$

It is evidenced from Fig. 3.36 that the district has displayed remarkable progress in cropping intensity across the land size categories from 1995–1996 to 2015–2016.

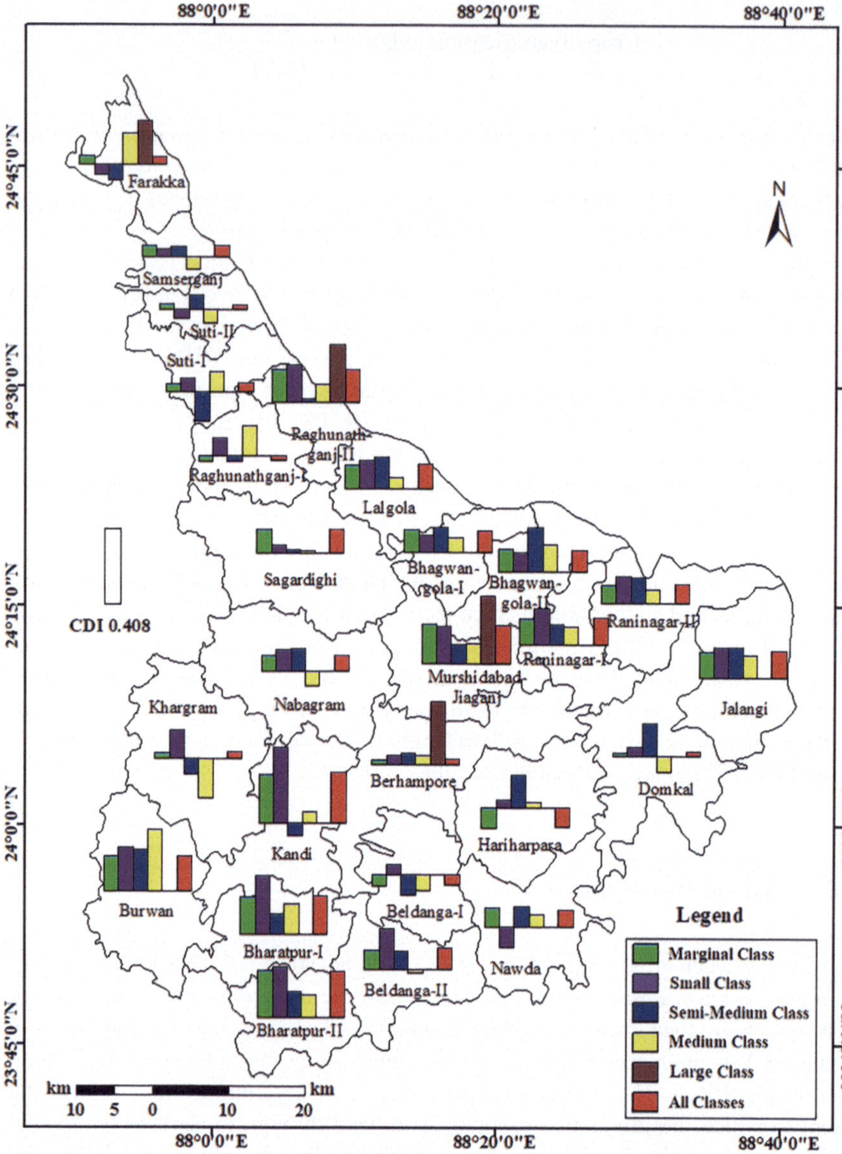

Fig. 3.35 Land size class-wise changes in crop diversification index, 1995–1996 to 2015–2016. (Source: Based on Annexure C.36)

The overall cropping intensity was 181.95% in 1995–1996, which increased to 230.81% in 2015–2016. The marginal land type shows the maximum increase with a 49.35% improvement of cropping intensity in the district. Apart from this, small (46.83%), semi-medium (33.37%), medium (36.09%), large (23.23%), and all (48.87%) landholdings also notified an increase on the positive side. Among the

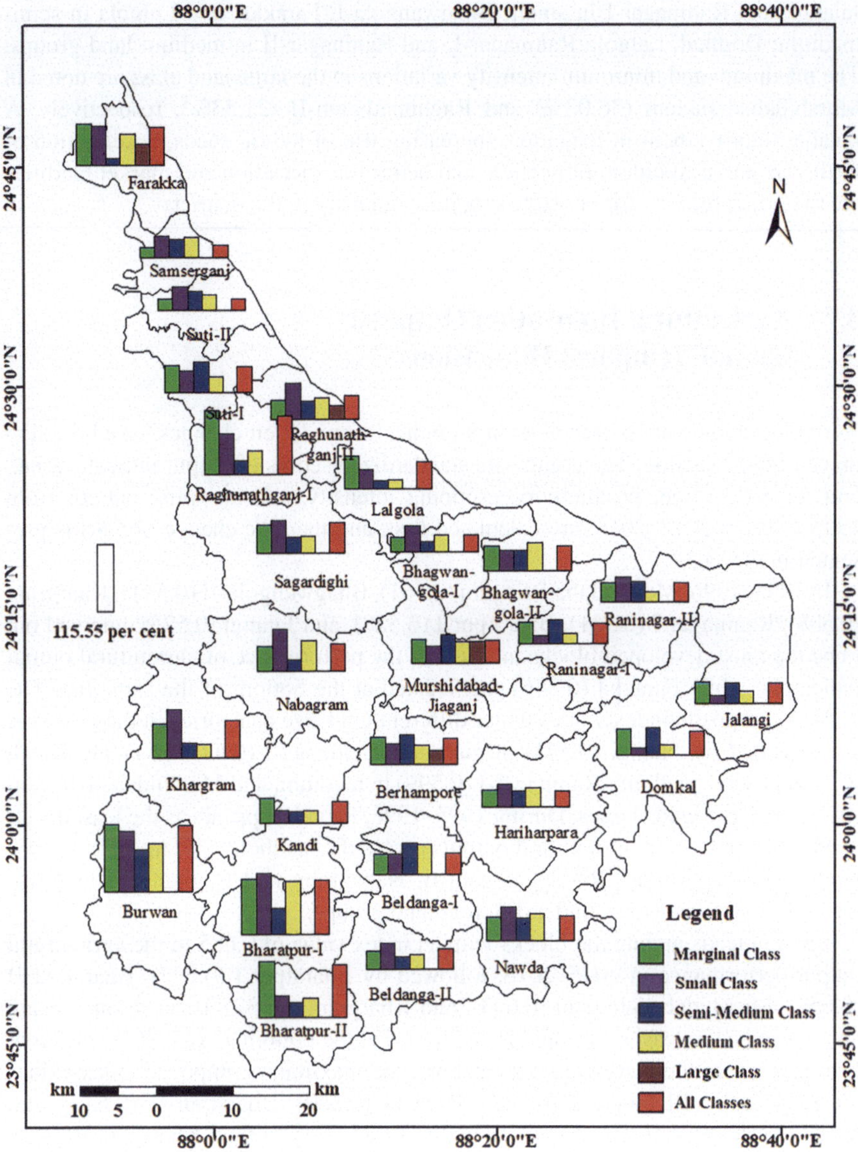

Fig. 3.36 Land size class-wise changes in cropping intensity, 1995–1996 to 2015–2016. (Source: Based on Annexure C.37)

blocks, Burwan reports the highest raising of intensity in marginal (115.55%), semi-medium (70.73%), and all (113.28%) land classes, while in the small (103.09%) and medium (89.67%) land categories, it is indicated by Bharatpur-I. However, some blocks still have not seen a significant improvement in intensification (more than 20%) over the years. These are Samserganj and Suti-II in marginal; Domkal,

Jalangi, and Raninagar-I in small; Bhagwangola-I, Farakka, and Lalgola in semi-medium; Domkal, Lalgola Raninagar-I, and Raninagar-II in medium land groups. The maximum and minimum intensity variations in the large land class are noted in Murshidabad-Jiaganj (36.03%) and Raghunathganj-II (21.53%), respectively. A notable improvement in irrigation, increasing use of hybrid seeds, acceleration of fertilizers and pesticides application, and better transportation and market facilities are the prime reasons for boosting cropping intensity in the district.

3.7 Agricultural Innovation Outputs: Spatial-Temporal Dimensions

Agricultural innovation based on output indicators and their changes have been displayed in this section. The composite standardized scores of output indicators, i.e., crop diversification, productivity, cropping intensity, and cropping pattern from 1995–1996 to 2015–2016, are calculated first, and then the change of CSS is presented in Fig. 3.37.

In 1995–1996, Murshidabad-Jiaganj (0.561), Bhagwangola-II (0.541), Khargram (0.538), Raninagar-I (0.531), Bharatpur-I (0.528), and Jalangi (0.525) emerged out to be the most developed blocks in terms of the performance of agricultural output indicators, while Farakka (0.346) was located at the bottom of the same list. The highest composite index values under different land size categories in the said year are indicated by Jalangi (0.541) in marginal, Beldanga-I (0.637) in small, Kandi (0.559) in semi-medium, Raninagar-I (0.599) in medium, and Murshidabad-Jiaganj (0.585) in large land classes. During 1995–1996, Farakka appears as the least developed block in CSS of agricultural outputs across the landholdings. Apart from this, blocks like Sagardighi, Raghunathganj-II, Nabagram, and Raghunathganj-I also report lower index values under different land groups.

Burwan leads among the blocks with an index value of 0.685 in the agricultural output performance in 2015–2016, followed by Bharatpur-I (0.677), Bharatpur-II (0.652), Murshidabad-Jiaganj (0.633), and Khargram (0.632). Farakka once again appears to be the least scorer and is positioned at the bottom across all the different land classes. In the marginal land category, the maximum composite index values are secured by Bharatpur-I (0.698), Burwan (0.680), Bharatpur-II (0.647), and Murshidabad-Jiaganj (0.645), sequentially. In addition to this, Khargram (0.646), Raninagar-I (0.633), Burwan (0.730), and Berhampore (0.546) are placed at the top in agricultural output capability in small, semi-medium, medium, and large land groups, respectively. These highest index values are induced through higher diversification and intensity of cropping in the blocks mentioned above.

Figure 3.37 represents the land size class-wise changes in the composite score of agricultural output among the blocks, where it has been noticed that the maximum difference is recorded by the marginal land category (0.069). Only the large landholding shows a negative change (0.091) because the number of farmers and area

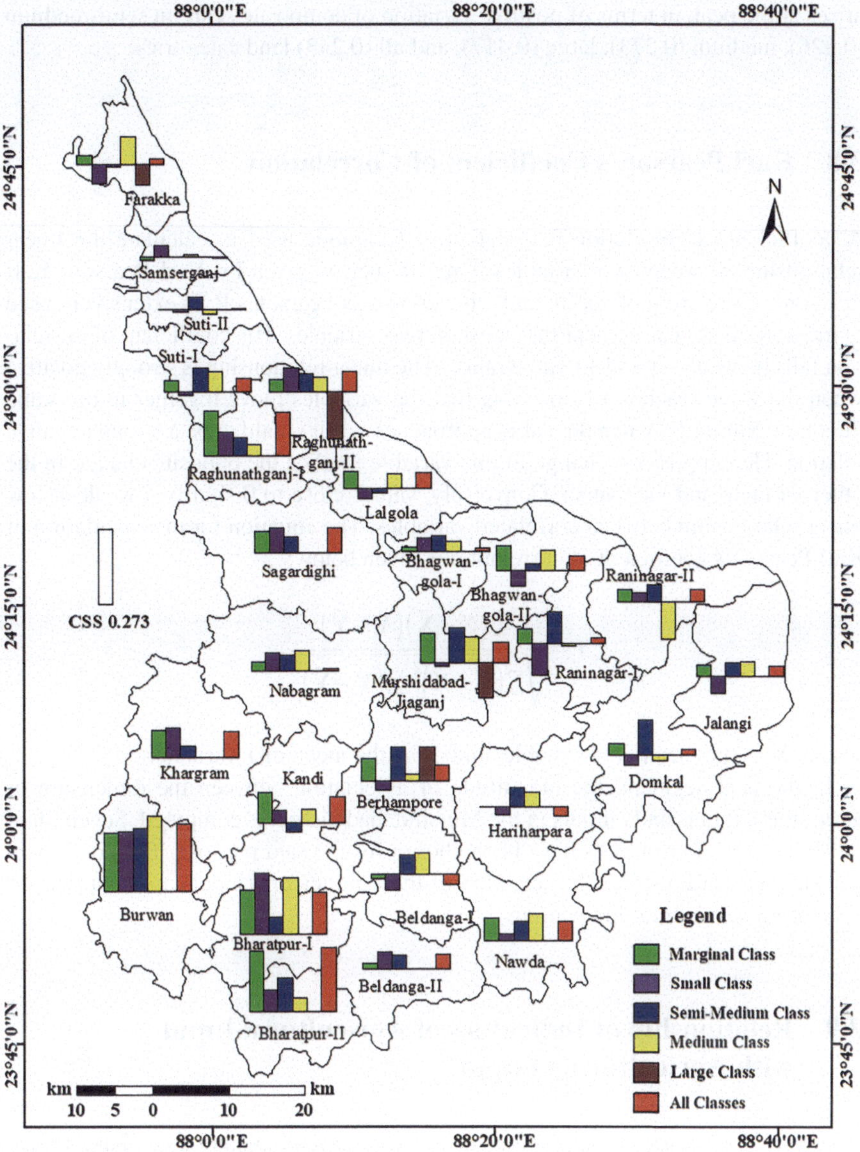

Fig. 3.37 Land size class-wise changes in composite standardized score of agricultural outputs, 1995–1996 to 2015–2016. (Source: Based on Annexure C.38)

devoted to cultivation has reduced over time. All the blocks under marginal land class register positive change except for Beldanga-I (0.017) and Samserganj (0.013). Bharatpur-II (0.220) gains the maximum positive change in the same land type. Under the small land group, the highest positive difference is shown by Bharatpur-I (0.219), while Raninagar-I (0.116) locates the maximum adverse change. Burwan arises at the peak in terms of positive variation of composite score in semi-medium (0.226), medium (0.273), large (0.417), and all (0.243) land categories.

3.8 Karl Pearson's Coefficient of Correlation

A coefficient of correlation is a statistical technique used to calculate the linear relationship between two variables. One of such is given by Karl Pearson. Karl Pearson's correlation of coefficient, also known as Pearson's R, is extensively used to measure the linear association between two variables. His coefficient of correlation falls between −1 and +1 value range. The linear relationship is strongly positive when the value reaches +1, meaning that the variables move together in the same direction. Similarly, when the value approaches −1, it is said to be a strong negative relation. Here a positive change in one variable implies the opposite change in the other variable and vice versa. Conversely, values close to 0 signify a weak or low linear relationship between correlated variables. The equation for the calculation of Karl Pearson's coefficient of correlation is given below:

$$\mathbf{r} = \frac{\Sigma(\mathbf{X}-\bar{\mathbf{X}})(\mathbf{Y}-\bar{\mathbf{Y}})}{\sqrt{\Sigma(\mathbf{X}-\bar{\mathbf{X}})^2\,\Sigma(\mathbf{Y}-\bar{\mathbf{Y}})^2}}$$

where $\bar{X}$ is the mean of X variable, and $\bar{Y}$ is the mean of Y variable.

In the next segment, the magnitude of association between the dimension of agricultural inputs and outputs in the Murshidabad district is computed. Seven input indicators and four outputs have been chosen for the said purpose. The coefficient correlation result justifies the selection of these indicators (inputs and outputs) for measuring agricultural innovation.

3.9 Relationship of Indicators of Agricultural Input with Agricultural Output

Annexure C.39 conveys the relationship between seven indicators, i.e., land, irrigation, seeds, credits, agricultural machinery, fertilizers and pesticides, and socio-economic status with agricultural outputs, crop diversification, productivity, cropping intensity, and cropping pattern under different land size classes for 1995–1996 and 2015–2016. The crop diversification is significantly and positively correlated with irrigation and agricultural machinery across different land categories during 1995–1996, and these associations become more robust in 2015–2016. However, the land indicator is negatively correlated with crop diversification under marginal, semi-medium, large, and all landholdings in 1995–1996. In 2015–2016, the negative relation exhibits only in the marginal land class. The remaining input indicators indicate a partial positive correlation with crop diversification in 1995–1996 (except for socio-economic in semi-medium and seeds in large land types) and 2015–2016 (except for credits in semi-medium and seeds and credits in medium land groups).

The productivity posits strong positive relationships with the land, irrigation, seeds, agricultural machinery, and fertilizers and pesticides in the different land size categories during 1995–1996 and 2015–2016. The credits input indicator positively and significantly correlates with productivity only in the marginal landholding during 1995–1996. The relationship of socio-economic status with this output is found to be positive but not significant (except for the large category in 1995–1996 and the marginal and all land classes in 2015–2016, where it is positive as well as significant).

The cropping intensity has a significant positive relationship with the land, irrigation, and socio-economic under the marginal land category in 1995–1996. The same becomes more robust in 2015–2016. Adding to this, credits and agricultural machinery also significantly positively correlate under the small land class in 1995–1996 but remain the same in 2015–2016. The relationship is strong and positive with land and irrigation under the semi-medium and with seeds, credits, and fertilizers and pesticides under medium landholdings in 1995–1996. While in 2015–2016, it is found with credits in the semi-medium and with irrigation, agricultural machinery, and socio-economic in medium land groups. There is a positive but insignificant association of cropping intensity with the input indicators under the large land category in both time periods (except for irrigation in 1995–1996 and fertilizers and pesticides in 2015–2016).

Overall, the irrigation and agricultural machinery register a strong positive relationship with cropping intensity across land categories during 1995–1996 and 2015–2016. It is seen that the land, credit facility, and socio-economic indicators have no such significant association with cropping intensity under different land size classes in both time periods. In 1995–1996, the significant positive relationships existed between cropping intensity with seeds under the small and medium land groups, and with fertilizers under the medium land type, while the relationship becomes stronger in 2015–2016.

3.10 Spatial-Temporal Pattern of Overall (Inputs–Outputs) Agricultural Innovation

Agricultural innovation is the result of the combined effects of both physical and non-physical factors. It simply denotes the growth and overall changes of either input or output determinants or both positively between two points of time in a particular referred region. Therefore, the extent of agricultural innovation changes strengthens the overall agrarian structure.

Table 3.2 Common breaking points for agricultural innovation

Quartile division	1995–1996	2015–2016
Lower quartile	0.381	0.452
Upper quartile	0.451	0.533

Source: Calculated by authors

The status of agricultural innovation is presented here across the land size categories for 1995–1996 and 2015–2016 at the block level. The quartile technique is used to show the spatial patterns of changes in the agricultural innovation index. The composite value of agricultural inputs and outputs has been calculated first. Then the values are categorized into three classes where lower and upper quartiles are taken as the breaking points of each category. The common breaking points of composite score across the land groups in 1995–1996 are 0.381 for the lower quartile and 0.451 for the upper quartile. The same becomes 0.452 and 0.533, respectively, in 2015–2016. The common breaking point of CSS value is not preserved as equal in both periods as the composite standardized score increased from 1995–1996 to 2015–2016 (Table 3.2).

3.10.1 Agricultural Innovation in Marginal Land Size Category

Figure 3.38 represents the agricultural innovation in all the blocks under marginal land classes in 1995–1996 and 2015–2016 in the district. Eight blocks lay above the upper quartile (0.451) in 1995–1996. These were Raninagar-I, Lalgola, Domkal, Beldanga-I, Bharatpur-I, Kandi, Khargram, and Murshidabad-Jiaganj in descending order. Three blocks, i.e., Raninagar-I, Lalgola, and Beldanga-I, have lost their position, and four more blocks, namely Burwan, Bhagwangola-II, Bharatpur-II, and Nawda, rise above the upper quartile (0.533) in 2015–2016. The noted higher development is supported by better agricultural innovation in the form of ensured and regular irrigation, improved infrastructure, HYV seeds, chemical fertilizers and pesticides, intensive farming, etc. Raninagar-I, with an index of 0.472, was the most developed block in the district during 1995–1996, while from 2015–2016, the top position was held by Bharatpur-I (0.610). Farakka, Raghunathganj-I, and Sagardighi are the least developed blocks placed below the lower quartile in both 1995–1996 and 2015–2016. As many as 15 blocks in 1995–1996 and 14 blocks in 2015–2016 are recorded between the lower and upper quartile in the district.

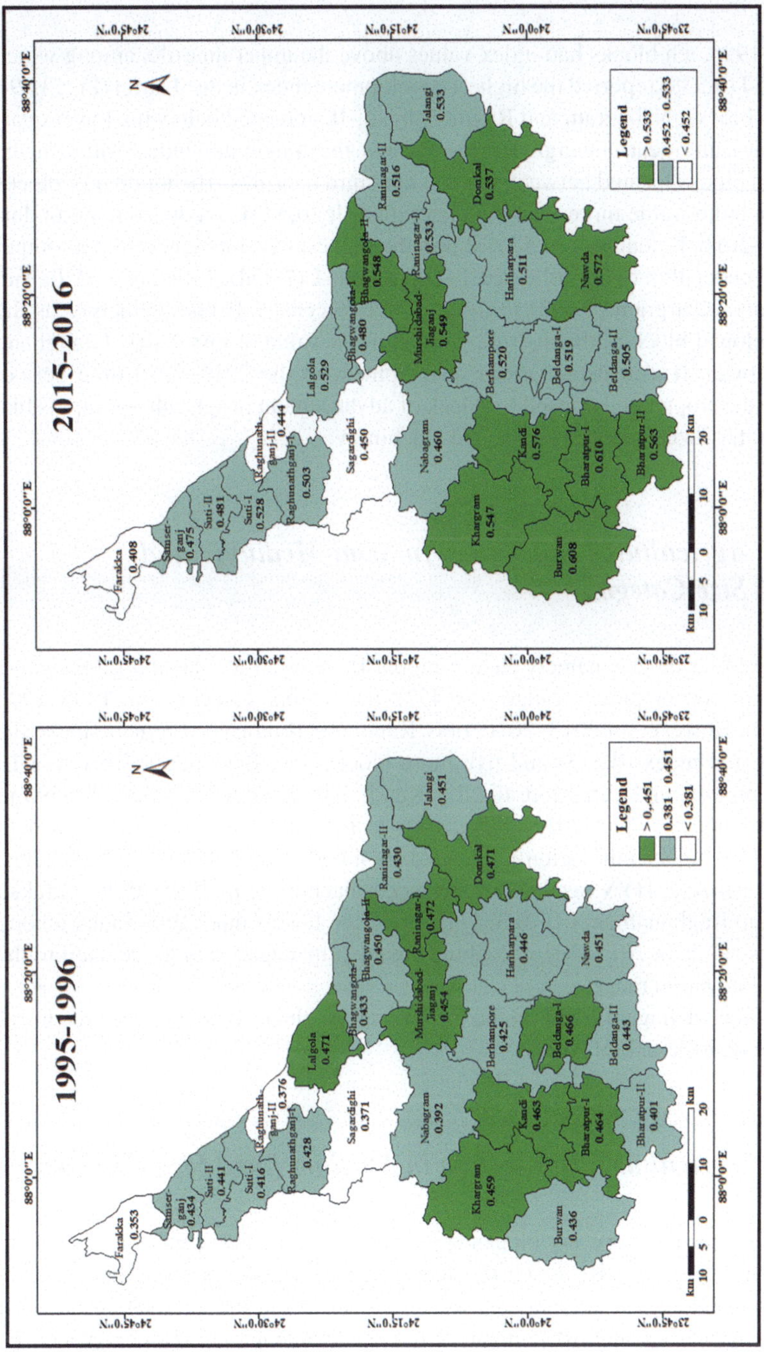

Fig. 3.38 Agricultural innovation (inputs–outputs) under marginal land size category in Murshidabad district, 1995–1996 and 2015. (Source: Based on Annexure C.40)

3.10.2 Agricultural Innovation in Small Land Size Category

In 1995–1996, ten blocks had index values above the upper quartile, among which Beldanga-I (0.538) reported the highest development index in the district (Fig. 3.39). Only two blocks, Nabagram and Raghunathganj-II, appeared below the lower quartile in the same year. Jalangi (0.446) records the maximum index value in the remaining blocks located between the first and third quartiles. The number of blocks having an index value more than the upper quartile (0.533) is reduced to eight during 2015–2016. Beldanga-I (0.570) shows the highest development of input–output variables under this class, followed by Bharatpur-I (0.554), Murshidabad-Jiaganj (0.553), and Khargram (0.543) in the 2015–2016 census. Farakka emerges as the least developed block in the district, with an index value of just 0.401. Jalangi and Raghunathganj-II also depict lesser development in the 2015–2016 time period. Such low development is caused by a lack of advancement in agricultural inputs like irrigation, fertilizers, credits seeds, and machinery.

3.10.3 Agricultural Innovation in Semi-Medium Land Size Category

There were four blocks, namely Bhagwangola-II, Kandi, Lalgola, and Raninagar-I, in the high development range (>0.452) under this category in 1995–1996 (Fig. 3.40). However, in 2015–2016, only Kandi and Raninagar-I remain above the upper quartile index (0.533) and five more blocks, i.e., Beldanga-I, Bharatpur-II, Domkal, Nawda, and Suti-I join the list. Kandi, having an index value of 0.493 in 1995–1996 and 0.571 in 2015–2016, leads to the high innovation index in both periods. This higher innovation is induced by modern agricultural infrastructures, such as fertilizers, HYV seeds, and advanced machinery. In 1995–1996, Farakka (0.348) and Raghunathganj-II (0.355) reported the lowest innovation index among the blocks, while in 2015–2016, Raghunathganj-II managed to upgrade itself to the medium innovation range.

Murshidabad-Jiaganj stands at the top among the blocks lying between the lower and upper quartile in both years.

3.10.4 Agricultural Innovation in Medium Land Size Category

In this land size category, six blocks appeared below the lower quartile (0.381) in 1995–1996. These were Beldanga-II, Bhagwangola-I, Bharatpur-I, Farakka, Jalangi, and Raghunathganj-II. Once again, six blocks sat in the low innovation range in 2015–2016, but Beldanga-II, Bharatpur-I, and Raghunathganj-II are replaced by Nabagram, Raninagar-II, and Sagardighi blocks. From 1995–1996, the higher

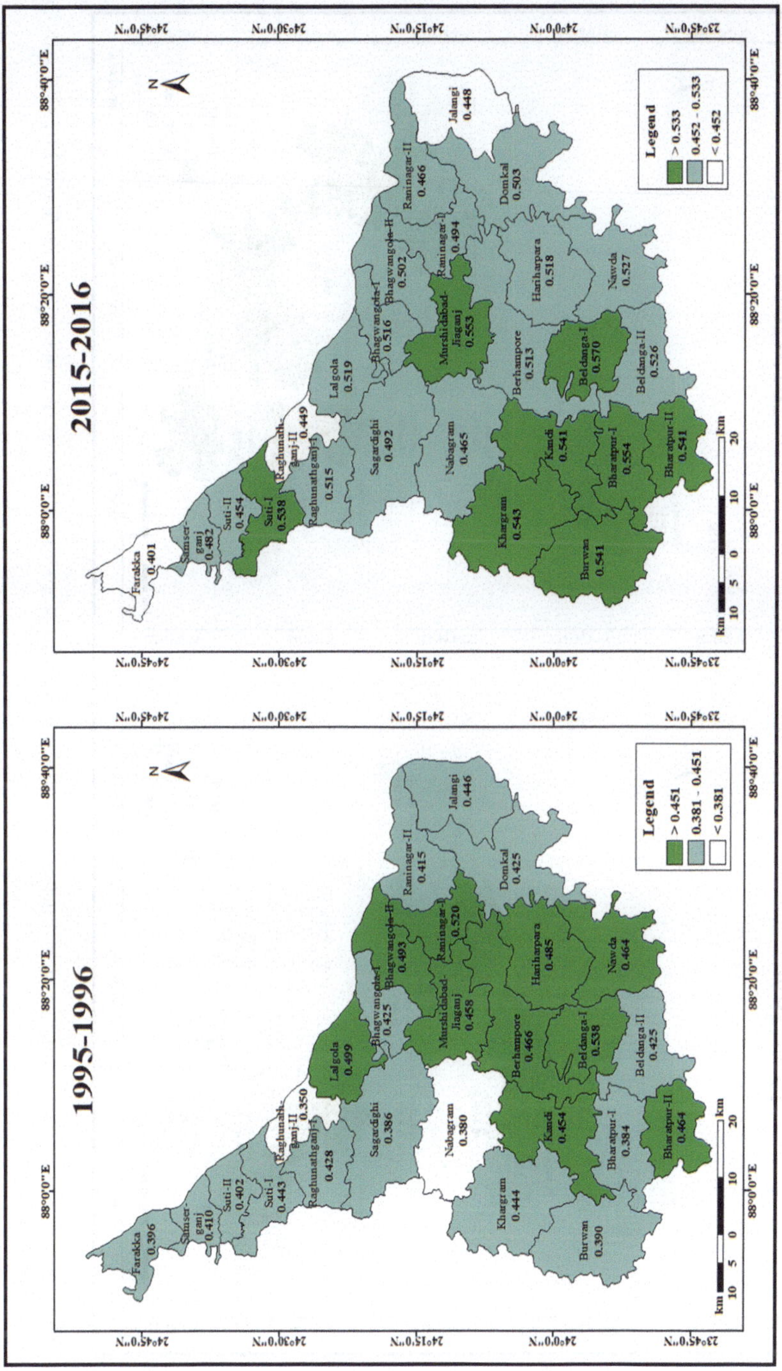

Fig. 3.39 Agricultural innovation (inputs–outputs) under small land size category in Murshidabad district, 1995–1996 and 2015–2016. (Source: Based on Annexure C.40)

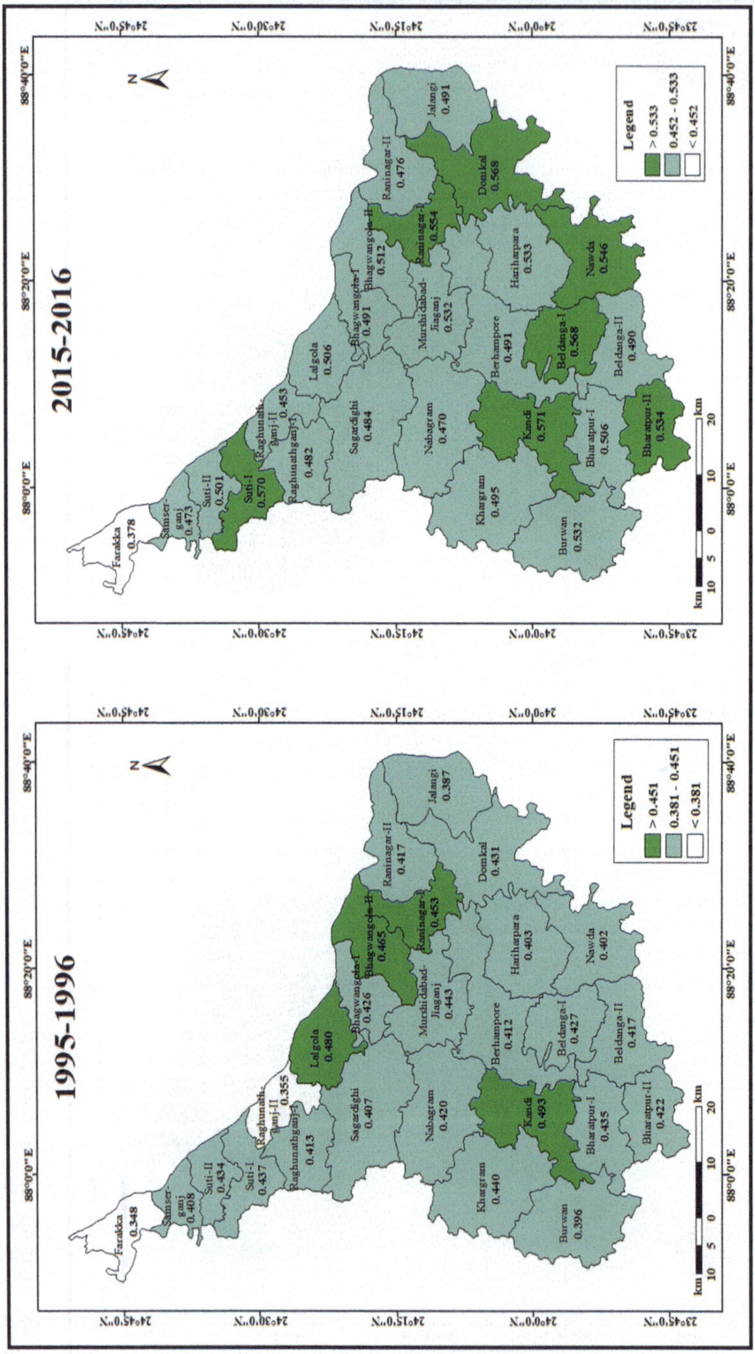

Fig. 3.40 Agricultural innovation (inputs–outputs) under semi-medium land size category in Murshidabad district, 1995–1996 and 2015–2016. (Source: Based on Annexure C.40)

innovation index values are recorded in Kandi (0.486), Bharatpur-II (0.484), Raninagar-I (0.466), Berhampore (0.457), and Raninagar-II (0.452). Except for Kandi and Raninagar-I, the other three blocks are degraded to the medium innovation range in 2015–2016. Moreover, three blocks, namely Burwan, Beldanga-I, and Suti-II, have replaced themselves from medium to high innovation index in the said year. The development in input indicators such as high fertilizer consumption, extended and available irrigation facilities, increasing use of HYV seeds, and intensive farming results in better agricultural outputs, thus higher agricultural innovation.

3.10.5 Agricultural Innovation in Large Land Size Category

The 1995–1996 Agriculture Census revealed that there were only eight blocks in the district where large farmers were found. These were Berhampore, Beldanga-II, Domkal, Farakka, Murshidabad-Jiaganj, Raghunathganj-I, Raghunathganj-II, and Suti-I. Only the Murshidabad-Jiaganj block (0.457) is located above the upper quartile index (0.451). Beldanga-II and Domkal fall under the medium innovation range, and the remaining five blocks display the index value below the lower quartile (0.381). In 2015–2016, seven blocks reported the farmers under the large land size class. It may be noted that Beldanga-II, Domkal, Raghunathganj-I, and Suti-I have lost their status of having large cultivators in the 2015–2016 census. Conversely, Burwan, Bhagwangola-II, and Nawda add farmers with more than 10 ha of landholdings in the same census. All the blocks mentioned above remain in the low innovation range except for Berhampore (0.513) and Burwan (0.459), and none of the block reports development above the upper quartile (0.533) in 2015–2016 (Figs. 3.41 and 3.42).

3.10.6 Agricultural Innovation in All Land Size Category

It is seen from Fig. 3.43 that in 1995–1996, ten blocks, namely Beldanga-I, Bhagwangola-II, Bharatpur-I, Domkal, Hariharpara, Kandi, Khargram, Lalgola, Murshidabad-Jiaganj, And Raninagar, were in high agricultural innovation range (>0.451). Raninagar-I, with an index value of 0.479, ranks first among them, followed by Beldanga-I (0.477), Bharatpur-I (0.474), and Murshidabad-Jiaganj (0.467). In 2015–2016, the number of blocks that remained above the upper quartile (0.533) was reduced to eight. Beldanga-I, Bhagwangola-II, Hariharpara, and Raninagar-I blocks could not pace themselves to stay at the higher innovation range. At the same time, Burwan, Bharatpur-II, and Nawda transformed from the medium innovation range to the high innovation range in 2015–2016. Bharatpur-I (0.606) emerges as the most developed among the blocks in the district. Farakka, Raghunathganj-II, and Sagardighi lay below the lower quartile index (0.452) in 1995–1996. In 2015–2016, Sagardighi block managed to skip out from the low

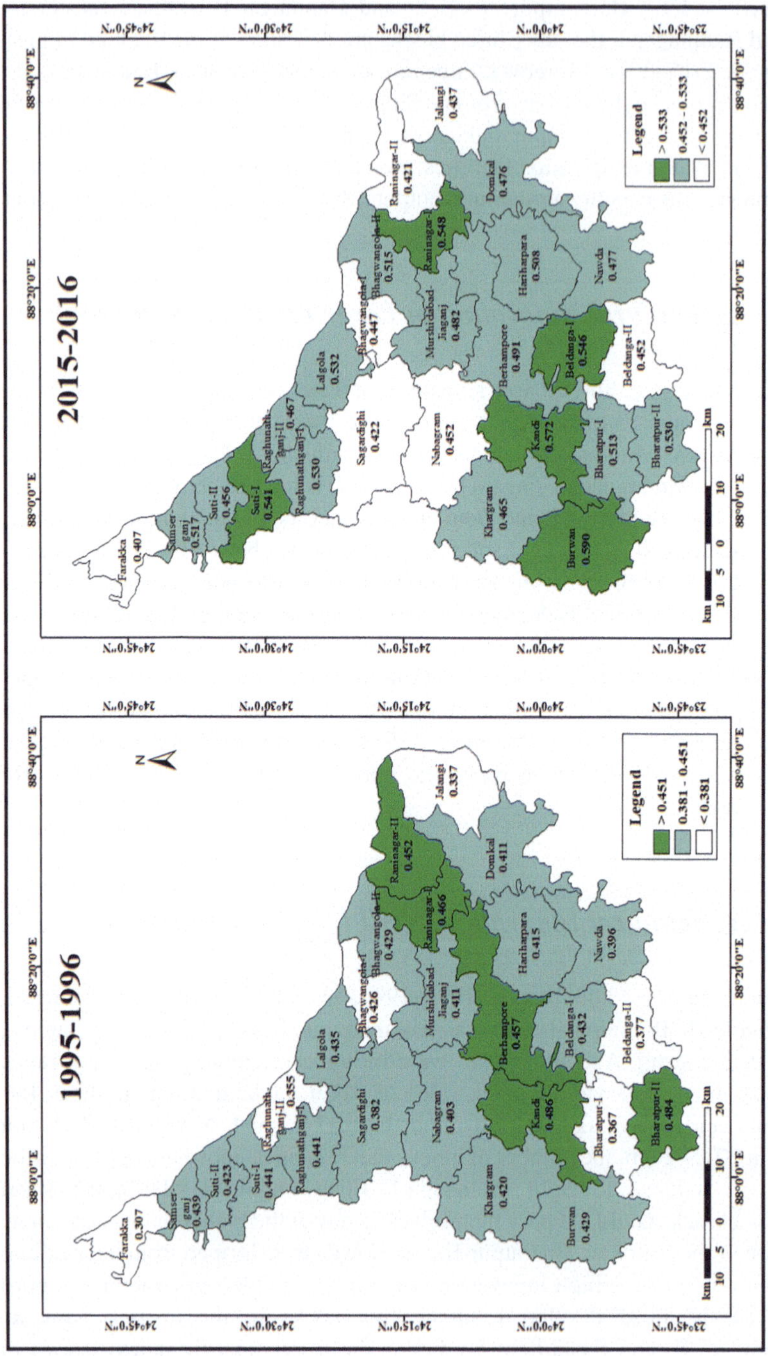

Fig. 3.41 Agricultural innovation (inputs–outputs) under medium land size category in Murshidabad district, 1995–1996 and 2015–2016. (Source: Based on Annexure C.40)

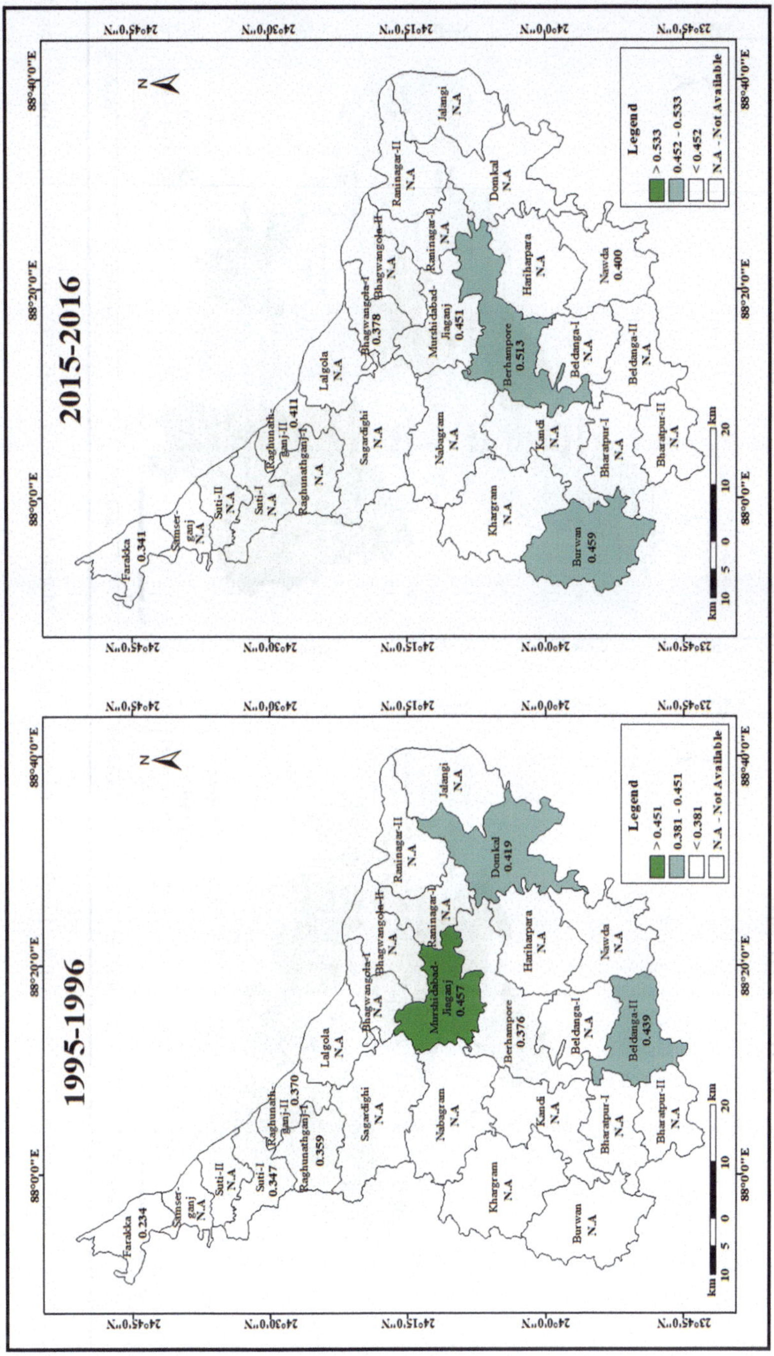

Fig. 3.42 Agricultural innovation (inputs–outputs) under large land size category in Murshidabad district, 1995–1996 and 2015–2016. (Source: Based on Annexure C.40)

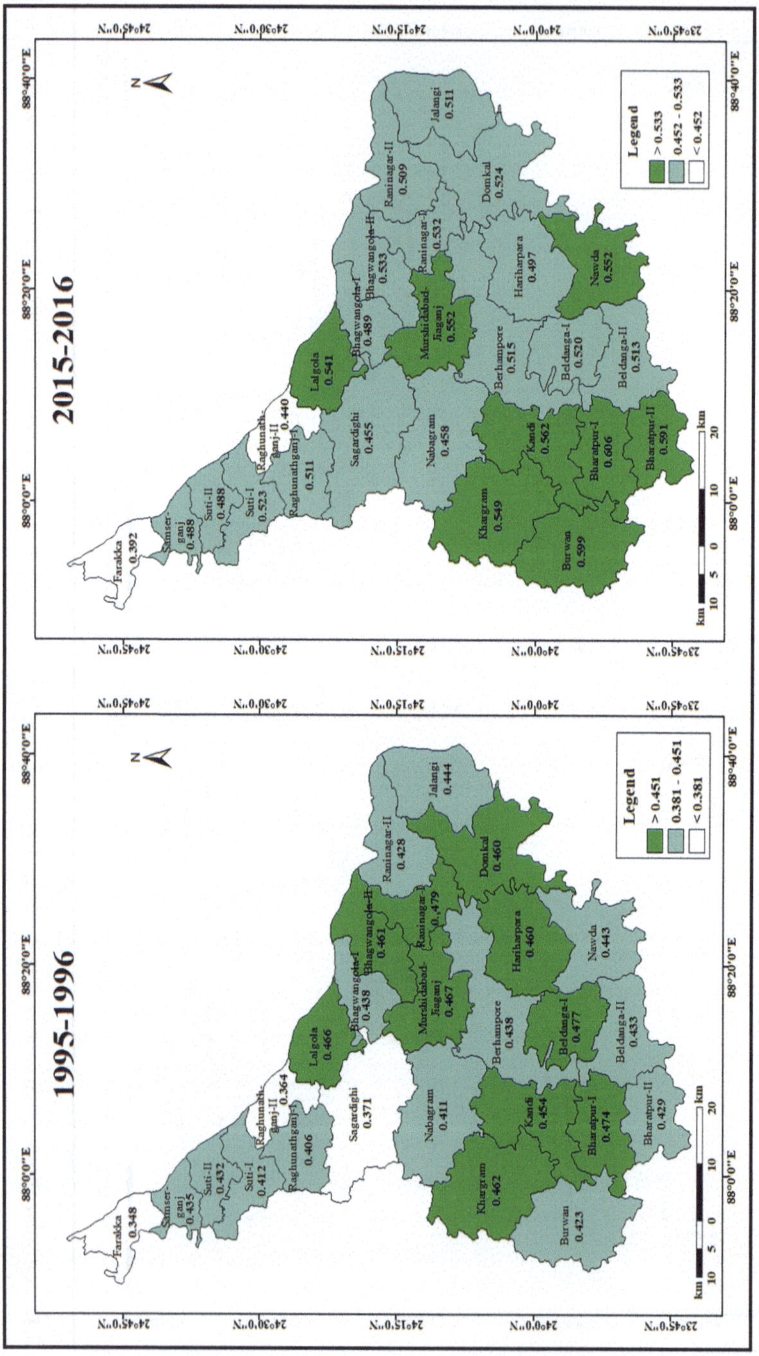

Fig. 3.43 Agricultural innovation (inputs–outputs) under all land size category in Murshidabad district, 1995–1996 and 2015–2016. (Source: Based on Annexure C.40)

innovation range and places itself in the medium range of innovation. In the remaining blocks that lie between the upper and lower quartile, Nawda and Raninagar-I represent the maximum innovation index in 1995–1996 and 2015–2016, respectively.

3.11 Pace of Agricultural Innovation in Murshidabad District

The spatial and temporal analyses of overall (inputs–outputs) agricultural innovation suggest an improvement in the overall agricultural situation in the district. It can be easily pointed out from the above discussion that the pace of agricultural innovation over the years is not the same; rather, it differs significantly. The following characteristics are identified about the nature and pace of agricultural innovation in the district across the land size categories:

1. Burwan (0.172) reports the maximum pace of change under the marginal land class, followed by Bharatpur-II (0.162), Bharatpur-I (0.146), Nawda (0.121), and Kandi (0.113). Conversely, the slowest pace of innovation is indicated in Samserganj-I, which added only a 0.041 index value from 1995–1996 to 2015–2016. Suti-II (0.042), Bhagwangola-I (0.047), Farakka (0.055), and Lalgola (0.058) also display the slowest pace of agricultural innovation.
2. Bharatpur-I gains a 0.171 index value from 1995–1996 to 2015–2016, the highest under small land categories. Four more blocks, Burwan, Beldanga-I, Khargram, and Sagardighi, have added index values above 0.100. During the study period, Raninagar-I fails to gain and loses a 0.025 index value. The pace of innovation in Jalangi and Farakka remains almost stagnant (only a change of 0.002 and 0.004, respectively).
3. The maximum and minimum pace of innovation in semi-medium landholding is noted in Beldanga-I (0.140) and Lalgola (0.026), respectively. Burwan, Domkal, Nawda, Suti-I, and Hariharpara also show a blooming pace of innovation under this category. In contrast, the slower pace of innovation is recorded in Farakka, Bhagwangola-II, Nabagram, and Khargram.
4. With an index value of 0.162, Burwan locates the maximum gain of innovation pace in the medium land group, followed by Bharatpur-I (0.146), Beldanga-I (0.114), and Raghunathganj-II (0.112). Raninagar-II experiences a reduction of index value under this land class during 1995–1996 and 2015–2016.
5. Under the large land type, Berhampore (0.136) denotes the highest tempo of agricultural innovation, followed by Farakka (0.106) and Raghunathganj-II (0.041). In contrast, Murshidabad-Jiaganj (0.007) reports a negative pace of innovation.
6. Burwan, with a change of 0.177 index value, reports the maximum pace of innovation in all land size category. This is followed by Bharatpur-II (0.162), Bharatpur-I (0.132), Suti-I (0.111), and Kandi (0.108). The minimum tempo of

agricultural innovation is represented in Hariharpara (0.037), Beldanga-I (0.043), Farakka (0.044), and Nabagram (0.047) in ascending order.

References

Adams, J., & Bumb, B. (1978). Patterns of agricultural development in India: Determinants of land productivity in Maharashtra's districts, 1971. *The Developing Economies, 16*(1), 74–82.

Alam, S. M. (Ed.). (1974). *Planning atlas of Andhra Pradesh* (p. 3). Pilot Map Production Plan, Survey of India.

Alston, J. M., & Pardey, P. G. (2021). The economics of agricultural innovation. *Handbook of Agricultural Economics, 5*, 3895–3980.

Barnes, D. F., & Samanta, B. B. (1984, January). *Rural electrification and the village economy in India*. International Association of Energy Economists.

Bhadrapur, V. S., & Naregal, S. S. (1992). Levels of agricultural development and its correlates. New dimensions in agricultural geography. *Dynamics of Agricultural Development, 7*, 197.

Gibbs, J., & Martin, W. (1962). Index of diversification. In *Taken from quantitative techniques in geography: An introduction. R. Hammond and PS McGullagh* (p. 21). Clarendon Press.

Jain, M., Mondal, P., DeFries, R. S., Small, C., & Galford, G. L. (2013). Mapping cropping intensity of smallholder farms: A comparison of methods using multiple sensors. *Remote Sensing of Environment, 134*, 210–223.

Jathar, G. B., & Beri, S. G. (1949). *Indian economics; A comprehensive and critical survey*. Oxford University Press.

Johnston, B. F., & Mellor, J. W. (1961). The role of agriculture in economic development. *The American Economic Review, 51*(4), 566–593.

Joshi, P. K., Gulati, A., Birthal, P. S., & Tewari, L. (2004). Agriculture diversification in South Asia: Patterns, determinants and policy implications. *Economic and Political Weekly*, 2457–2467.

Krishna, G. (1992). The concept of agricultural development. *Developmental Dynamics, 7*, 29–36.

Mellor, J. W. (1976). The agriculture of India. *Scientific American, 235*(3), 154–163.

Nath, V. (1969). The growth of Indian agriculture: A regional analysis. *Geographical Review of India, IX*(3), 369.

Nosiru, M. O. (2010). Microcredits and agricultural productivity in Ogun State, Nigeria. *World Journal of Agricultural Sciences, 6*(3), 290–296.

OECD. (2001). *Measuring productivity—OECD manual: Measurement of aggregate and industry-level productivity growth*. OECD Publishing. https://doi.org/10.1787/9789264194519-en

Ohlan, R. (2013). Pattern of regional disparities in socio-economic development in India: District level analysis. *Social Indicators Research, 114*(3), 841–873.

Rahaman, H. (2021). Agricultural development: Inputs-outputs dimension. In *Diversified cropping pattern and agricultural development* (pp. 71–106). Springer.

Rahaman, H., & Ansari, S. (2024). Status of crop diversification by land size classes: A case study from state of West Bengal, India. *Journal of Asian and African Studies, 59*(1), 247–258. https://doi.org/10.1177/00219096221106092

Rahaman, H., Siddiqui, S. H., Asthana, S. P., & Rahman, F. (2019). Status of agricultural development in Malda district: A geographical analysis. *National Geographical Journal of India, 65*(3), 246–256.

Sayer, J., & Cassman, K. G. (2013). Agricultural innovation to protect the environment. *Proceedings of the National Academy of Sciences of the United States of America, 110*(21), 8345–8348.

Seers, D. (1969). *The meaning of development*. p. 3.

Shafi, M. (1972). Measurement of agricultural productivity of the great Indian Plains. *The Geographer, 19*(1), 4–13.

Sharma, P. S. (1971). *Agricultural regionalisation of India*. Economic and Socio-cultural Dimensions of Regionalisation. Registrar General.

Siddiqui, H. S., & Rahaman, H. (2016). Crop diversification in relation to time and space: A study from Malda district. *International Journal of Informative & Futuristic Research, 4*(2), 5133–5142.

Tewari, R. T., & Singh, N. (1985). Development and productivity in Indian agriculture: A cross-section temporal analysis. *Indian Journal of Regional Science, 17*(1), 65–75.

Winters, P., De Janvry, A., Sadoulet, E., & Stamoulis, K. (1998). The role of agriculture in economic development: Visible and invisible surplus transfers. *The Journal of Development Studies, 34*(5), 71–97.

Witt, L. W. (1965). Role of agriculture in economic development—A review. *American Journal of Agricultural Economics, 47*(1), 120–131.

Woodsong, C. (1994). Old farmers, invisible farmers: Age and agriculture in Jamaica. *Journal of Cross-Cultural Gerontology, 9*(3), 277–299.

Yang, W. Y. (1965). *Methods of farm management investigation* (p. 80). FAO. Agric. Dev., Paper.

Chapter 4
Status of Sustainable Land Use Practices and Agricultural Innovation

Contents

In the previous two chapters, the dynamics of sustainable land use and input–output dimensions of agricultural innovations have been discussed. It is now considered essential to find out whether the changing sustainable land use pattern impacts agricultural innovations and, if so, how much and in which direction. Therefore, this chapter is devoted to analysing the relationship between sustainable land use and innovation in terms of changing data from 1995–1996 to 2015–2016. In the first section, the association is measured statistically through multiple regression analysis under different land size categories in the district. Following this, a unit-wise change of those two domains has been explained at the block level across land size classes, where the input and output dimensions of agricultural innovations have been measured with respect to per unit change of sustainable land use.

H. Rahaman, S. Ansari, *Sustainable Land Use Practices and Agricultural Innovations in India*, https://doi.org/10.1007/978-3-031-83246-8_4

4.1 Relationship Between Sustainable Land Use and Agricultural Innovation

Sustainable land use is the concept associated with using land to produce agricultural products for human or animal consumption (Todkari et al., 2010). The cultivable land of a region, including net sown area and current fallow land, comes under the agricultural land. In short, sustainable land use means a kind of cropping pattern which includes the proportion of area under various crops at a given point of time. The proportion of area and the spatial arrangement of crops follow certain pattern based on physical and socio-economic factors. Therefore, it is a dynamic concept that changes over space and time. Land use varies due to the spatial distribution of physical and non-physical elements in any region. It is a well-known fact that the intensity of agricultural land use is changing for many reasons, and profit opportunity is the leading cause. The intensity of change in sustainable land use is closely associated with the sustainable development of the environment (Li & Wang, 2003). In this study, the intensity of sustainable land use is measured with the help of 22 variables (see Annexure B.33) covering land use area devoted to nine-fold classification and gross cropped area under major crops and different crop categories.

Agricultural innovation is a complex concept which mainly compiles the improvement of cropped land, farming practices, farm implements, irrigation systems and irrigated areas, varieties of seeds, utilization of fertilizers and pesticides, the intensity of cropping, and commercialization of agriculture. Thus, agricultural innovation is inevitably a normative concept almost synonymous with improvement (Seers, 2010). It denotes the quality and progress of the agricultural system achieved over a given period of time in a region in terms of crop productivity, diversification, and commercialization.

The pattern of sustainable land use is an essential aspect of geographical studies as it directly engages with the innovation and development of agriculture in any region (Siddiqui & Rahaman, 2019). It can be deduced from the above-discussed statement that the improvisation of the intensity of sustainable land use indeed leads towards agricultural innovation as a whole. In this study, a relationship between sustainable land use and agricultural innovation is therefore established to check the impact of these two dimensions on each other under different land size classes in the Murshidabad district.

4.2 Regression Analyses of Sustainable Land Use and Agricultural Innovation

The impact of various agricultural input and output indicators on sustainable land use under different land size categories in the Murshidabad district is estimated with the help of multiple linear regression (MLR). This statistical technique describes the simultaneous associations of a single continuous dependent or response variable and several independent or explanatory variables (Eberly, 2007). The sustainable land use is taken as the dependent variable and is regressed against the set of independent variables of agricultural innovation (seven indicators of agricultural input, then four indicators of agricultural output, and finally 11 indicators of overall agricultural innovation input–output). The deviation values from 1995–1996 to 2015–2016 have been taken as the dataset for the present analysis. The multiple linear regression model, focusing on the ordinary least square (OLS) method used for the estimation of the model's parameter, is statistically formalized as:

$$y_i = \beta_0 + \beta_1 x_{i1} + \beta_2 x_{i2} + \cdots + \beta_{p-1} x_{i,p-1} + \varepsilon_i$$

where y_i represents the outcome or dependent variable, β_0 is the intercept (the mean of y when all $x_k = 0$), each β_k ($k = 1, 2, \ldots, p - 1$) represents a slope or estimated coefficient with respect to the explanatory variables x_i ($i = 1, 2, \ldots, n$ unit of observations), and ε_i is the random error with a prior expectation of 0 and variance σ^2.

4.2.1 Regression Result Under Marginal Land Size Category

The regression results of sustainable land use with overall (input–output) agricultural innovation under this land size category reveal a high coefficient of determination (R^2) value of 0.823. Thus, it denotes that around 82% variability in the sustainable land use in the district can be predicted by all the indicators of agricultural innovation in combined. The sustainable land use is significantly related with agricultural innovation with a *p*-value less than 0.01. Moreover, the model explains about 66% variance of sustainable land use when it considers the seven indicators of agricultural input separately. Conversely, four variables of agricultural output explain around 61% of variations in sustainable land use. Both times the results are significant at a 1% significance level (Fig. 4.1).

4.2.2 Regression Result Under Small Land Size Category

The changes in sustainable land use are described around 78% by the assembled variables of agricultural innovation through the coefficient of determination ($R^2 = 0.782$). The probability value for the result is 0.005, which implies the significance of the model. The coefficients of seven indicators of agricultural input, namely, land, irrigation, seeds, credits, machinery, fertilizers and pesticides, and socio-economic aspects, have a positive and significant relationship with changing sustainable land use pattern with a coefficient of determination (R^2) of 0.663 at 1% significance level. In contrast, the combination of variables of agricultural output explains approximately 53% of the changes in sustainable land use in the district (Fig. 4.2).

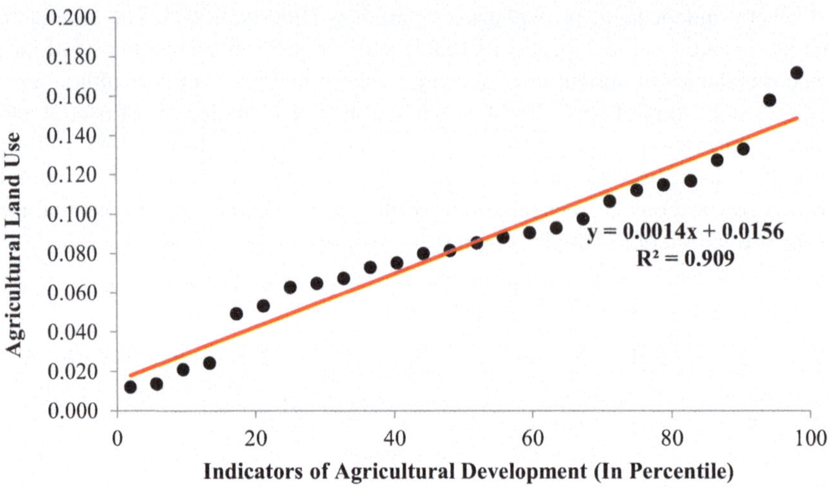

Fig. 4.1 Normal probability plot (marginal land size category). (Source: Based on Table 4.1 and Annexure 4.1 accordingly)

Table 4.1 Regression summary under marginal land size category

Regression statistics	Input dimension	Output dimension	Overall innovation
R square	0.665	0.612	0.823
Regression	0.028	0.026	0.035
Residual	0.014	0.016	0.007
Total	0.042	0.042	0.042
p-value	0.002	0.000	0.001

Source: Computed by authors (based on Annexure D.1)

4.2.3 Regression Result Under Semi-Medium Land Size Category

The regression result in the semi-medium land class shows that agricultural input indicators report nearly 60% (coefficient of determination $R^2 = 0.590$) of the variation in sustainable land use changes in the district. In comparison, the same is explained around 51% (coefficient of determination $R^2 = 0.512$) by the variables of agricultural output. The former explanatory variables are noteworthy at a 5% significant level, while the latter is at a 1% significant level. The coefficient of determination (R^2) of the overall 11 agricultural innovations' indicators is 0.818, with a *p*-value of 0.002. It means that the regressor agricultural innovation explains around 82% variability of the sustainable land use changes (Fig. 4.3).

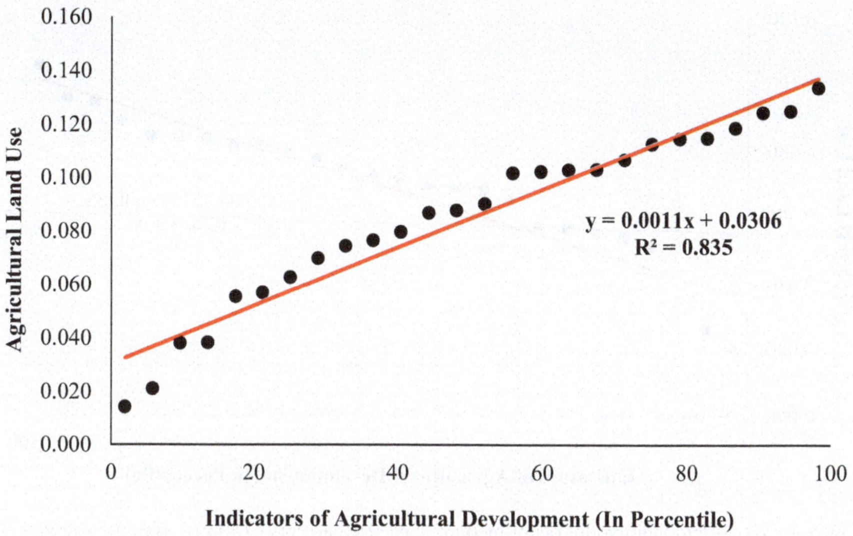

Fig. 4.2 Normal probability plot (small land size category). (Source: Based on Table 4.2 and Annexure D.2 accordingly)

Table 4.2 Regression summary under small land size category

Regression statistics	Input dimension	Output dimension	Overall innovation
R square	0.663	0.526	0.782
Regression	0.018	0.014	0.021
Residual	0.009	0.013	0.006
Total	0.027	0.027	0.027
p-value	0.003	0.003	0.005

Source: Computed by authors (based on Annexure D.2)

4.2.4 Regression Result Under Medium Land Size Category

The result of the regression model shows a high coefficient of determination ($R^2 = 0.799$) between sustainable land use and indicators of agricultural innovation. It simply denotes that nearly 80% variability of sustainable land use is explained by overall agricultural innovation in the district under this land class. The result reveals a *p*-value of 0.003, which is statistically significant. Around 54% variation in sustainable land use is reported when only the seven indicators of agricultural input are taken as the regressed or explanatory variables. Conversely, when the four agricultural output indicators are considered predictors or independent variables, the representation of the variability of sustainable land use changes reaches approximately 60% (Fig. 4.4).

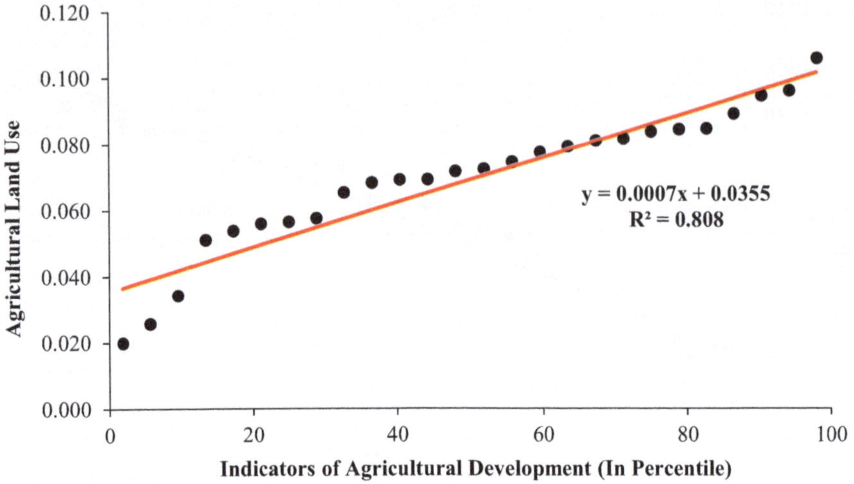

Fig. 4.3 Normal probability plot (semi-medium land size category). (Source: Based on Table 4.3 and Annexure D.3 accordingly)

Table 4.3 Regression summary under semi-medium land size category

Regression statistics	Input dimension	Output dimension	Overall innovation
R square	0.590	0.512	0.818
Regression	0.006	0.006	0.009
Residual	0.004	0.005	0.002
Total	0.011	0.011	0.011
p-value	0.012	0.003	0.002

Source: Computed by authors (based on Annexure D.3)

4.2.5 Regression Result Under Large Land Size Category

In the district, only four blocks have noted large land size class farmer such as Berhampore, Farakka, Murshidabad-Jiaganj, and Raghunathganj-II. The regression statistic mentioned in Table 4.5 is not consistent due to (inaccurate data) limited number of blocks displaying such value.

4.2.6 Regression Result Under All Land Size Category

Table 4.6 represents the regression result of sustainable land use and different indicators of agricultural innovation under all land size class in the district. The result shows a strong positive coefficient of determination (R^2), 0.821 of sustainable land use with agricultural innovation, which is statistically significant at 0.002 *p*-value.

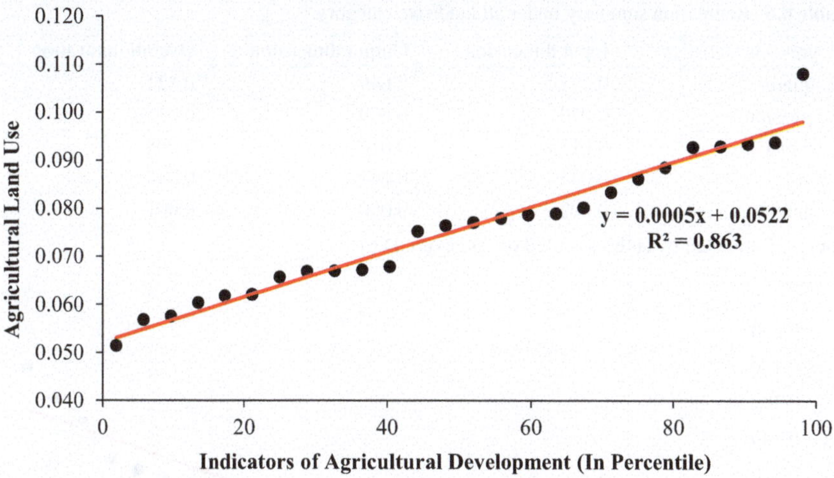

Fig. 4.4 Normal probability plot (medium land size category). (Source: Based on Table 4.4 and Annexure D.4 accordingly)

Table 4.4 Regression summary under medium land size category

Regression statistics	Input dimension	Output dimension	Overall innovation
R square	0.544	0.591	0.799
Regression	0.003	0.003	0.004
Residual	0.002	0.002	0.001
Total	0.005	0.005	0.005
p-value	0.026	0.001	0.003

Source: Computed by authors (based on Annexure D.4)

Table 4.5 Regression summary under large land size category

Regression statistics	Input dimension	Output dimension	Overall innovation
R square	0.805	0.375	0.910
Regression	0.002	0.001	0.002
Residual	0.000	0.001	0.000
Total	0.002	0.002	0.002
p-value	0.564	0.615	0.573

Source: Computed by authors (based on Annexure D.5)

That means agricultural innovation explains about 82% variability of sustainable land use in the district. The regression result of sustainable land use with input indicators indicates a coefficient of determination value of 0.671, which signifies about 67% variability of sustainable land use by the indicators of input variables. Similarly, when only the output variables of agricultural innovation are taken, the model displays a result of a 0.649 coefficient of determination (R^2) (Fig. 4.5).

Table 4.6 Regression summary under all land size category

Regression statistics	Input dimension	Output dimension	Overall innovation
R square	0.671	0.649	0.821
Regression	0.030	0.029	0.036
Residual	0.015	0.016	0.008
Total	0.044	0.044	0.044
p-value	0.002	0.000	0.001

Source: Computed by authors (based on Annexure D.6)

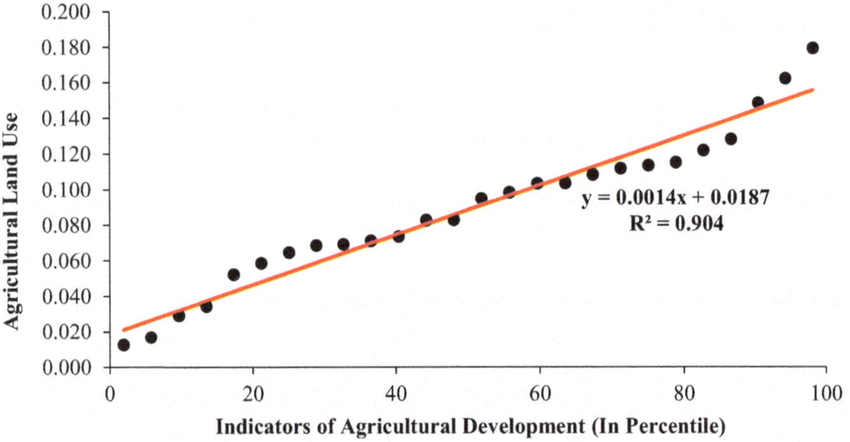

Fig. 4.5 Normal probability plot (all land size class). (Source: Based on Table 4.5 and Annexure D.5 accordingly)

4.3 Unit-Wise Changes in Sustainable Land Use and Innovation

In this section, a calculation of the unit-wise changes in input and output dimensions of agricultural innovation with respect to sustainable land use under different land size categories has been presented. The overall changes in the composite index from 1995–1996 to 2015–2016 of sustainable land use have been compared with the various indicators of input and output dimensions of agricultural innovation to obtain the desired result. What amount of change in agricultural innovation is implied by a one-unit change in sustainable land use? This would be the main concern of this discussion. The unit-wise change analyses are important to know the exact amount and degree of impact of dependent variables to independent variables over space and time (Rahaman, 2021). The nature of changes has been extensively discussed below under all the blocks across land size class categories one by one.

4.3.1 Unit-Wise Changes in Marginal Land Size Category

The one-unit change in sustainable land use from 1995–1996 to 2015–2016 leads to the change of 1.79 units of overall agricultural innovation in the marginal land class, in which the input variables report a change of 2.11 and the output variables record 1.46 per unit of sustainable land use change. The maximum and minimum unit-wise changes in the overall agricultural innovation are noted in Suti-I (9.02) and Farakka (1.00), respectively. In terms of overall innovation, a total of 12 blocks scored higher than the district average (1.79). In the same category, it is found that Bharatpur-I and Beldanga-I are such blocks that depict a similar picture prevalent in the district (Fig. 4.6).

In terms of the input dimension, the maximum change is displayed by Suti-I with a score of 15.42, followed by Khargram (8.18), Bharatpur-II (7.81), and Raghunathganj-II (7.00). However, Raninagar-I scores the slightest change of 0.86. The land indicator change compared to per unit of sustainable land use is highest in Suti-I (1.95) and lowest in Nawda (0.20). Suti-I reports the maximum unit-wise change under all the remaining input indicators, i.e., seeds (16.64), credit (4.43), agricultural machinery (48.20), fertilizers and pesticides (31.01), and socio-economic (5.16) except for irrigation, where Khargram enjoys the highest change of 4.64. Conversely, the least change under irrigation, seeds, credits, agricultural machinery, fertilizers and pesticides, and socio-economic indicators is represented by Hariharpara (0.08), Raghunathganj-I (1.27), Jalangi (0.22), Samserganj (0.47), Lalgola (0.18), and Murshidabad-Jiaganj (0.22), respectively.

Raghunathganj-II (4.55) and Farakka (0.66) are the blocks in the district having the maximum and minimum scores, respectively, in the unit-wise changes of output variables of agricultural innovation. Among the different variables, Raghunathganj-II (5.70) scores the biggest change in the crop diversification index, and the minimum value is seen in Raghunathganj-I (0.27). The changes in crop yield index compared to a one-unit change in sustainable land use are maximum in Khargram (2.04) and lowest in Jalangi (0.07). With a change of 8.66% sustainable land use change, Raghunathganj-II leads in the cropping intensity variable of agricultural output, while Farakka experiences a minimal change of 0.67 units. Moreover, Bharatpur-II (2.24) ranks top in cropping pattern score change per sustainable land use in the district. Raghunathganj-I (0.43) sits at the bottom of the same list.

4.3.2 Unit-Wise Changes in Small Land Size Category

In this land class, a total of 1.96 units of change in overall agricultural innovation have been noticed in the district when the sustainable land use gets changed by one unit. As many as 11 blocks score the change value more than the district average. Suti-I (3.45) registers the maximum among these blocks, followed by Raninagar-I

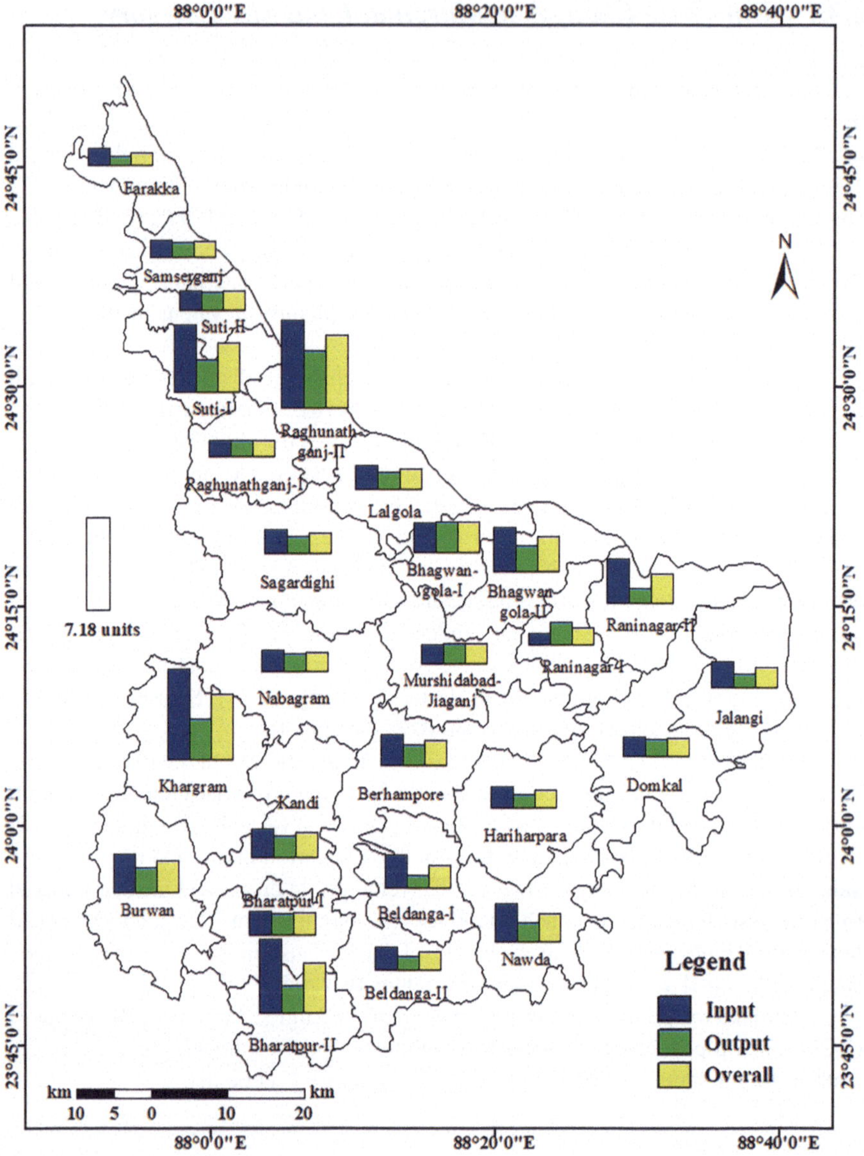

Fig. 4.6 Changes of agricultural innovation per unit of sustainable land use change, marginal land size class (1995–1996 to 2015–2016). (Source: Based on Table 4.6 and Annexure D.6 accordingly)

(3.00), Raghunathganj-II (2.84), Beldanga-I (2.64), and Hariharpara (2.32). In contrast, the lowest change is seen in Raninagar-II (1.41). Sagardighi (1.48), Bhagwangola-I (1.49), Lalgola (1.51), and Bhagwangola-II (1.54) blocks also denote a lower change in the district.

A change of 2.31 units in input variables is recorded in the district during the period of 1995–1996 and 2015–2016. Suti-I remains at the top of the list with a 5.60 units of index value. It is immediately followed by Raghunathganj-II (4.64) and Bharatpur-II (3.14). The least changes, however, are recorded by Kandi (1.90), Raninagar-II (1.97), and Jalangi (1.99) in ascending order. The maximum and minimum changes under the land indicator are reported in Raghunathganj-II (1.29) and Raninagar-II (0.34) in the district. It is Raninagar-I (2.72) and Suti-I (0.33) under irrigation, Suti-I (21.17) and Raninagar-I (3.42) in the case of seeds, Suti-I (3.06) and Beldanga-I (0.72) in credits, Bharatpur-I (8.02) and Bhagwangola-I (2.35) in agricultural machinery, Raghunathganj-I (4.27) and Lalgola (1.59) in fertilizers and pesticides, and finally, Raghunathganj-II (4.70) and Raninagar-I (0.02) in socio-economic indicator of agricultural input dimension. There is much variability of unit-wise changes of input indicators against the sustainable land use seen among the blocks (Fig. 4.7).

The variables of agricultural outputs combined see a change of 1.61 units. The changes in the output dimension of agricultural innovation in per unit of sustainable land use remain lower than the input dimension. This is due to the addition of various agricultural machinery, seeds, credits, and application of fertilizers and pesticides over the past 20 years (1995–1996 to 2015–2016), helping to increase the pace of the input sector. Among the blocks, Raninagar-I (3.36) indicates the most significant change, while the lowest change is experienced by Farakka (0.67). Under the different variables of agricultural output, the top place in unit-wise change is held by Raninagar-I, except for the cropping pattern. Suti-I (2.35) reports the maximum change under this variable. However, the lowest changes in crop diversification, productivity, cropping intensity, and cropping pattern are represented by Bhagwangola-I (0.72), Raghunathganj-II (0.25), Raghunathganj-I (0.24), and Domkal (0.29), respectively.

4.3.3 Unit-Wise Changes in Semi-Medium Land Size Category

The district under the semi-medium land class experiences a change of 2.82 units in overall agricultural innovation when sustainable land use implies one-unit change. Bhagwangola-I (3.98), Lalgola (3.96), and Suti-II (3.62) are the top three blocks in terms of maximum change value in the district. The lowest change is acquired by Raghunathganj-I (2.09). Additionally, Nabagram (2.20), Raghunathganj-II (2.29), and Raninagar-I (2.45) have also been placed at the bottom of the list. Jalangi (2.78) and Domkal (2.77) are the two blocks that resemble nearness to the district regarding unit-wise changes in agricultural innovation against sustainable land use.

The changes in agricultural innovation exclusively in the input dimension are found to be 3.16 units compared to the rise of one unit of sustainable land use in the district. Lalgola, with a value of 5.35 units, emerges as the top block in the district, followed by Bhagwangola-I (5.18), Bharatpur-II (4.26), and Nawda (4.00) in descending order. Meanwhile, Nabagram (2.60) records the lowest change in the

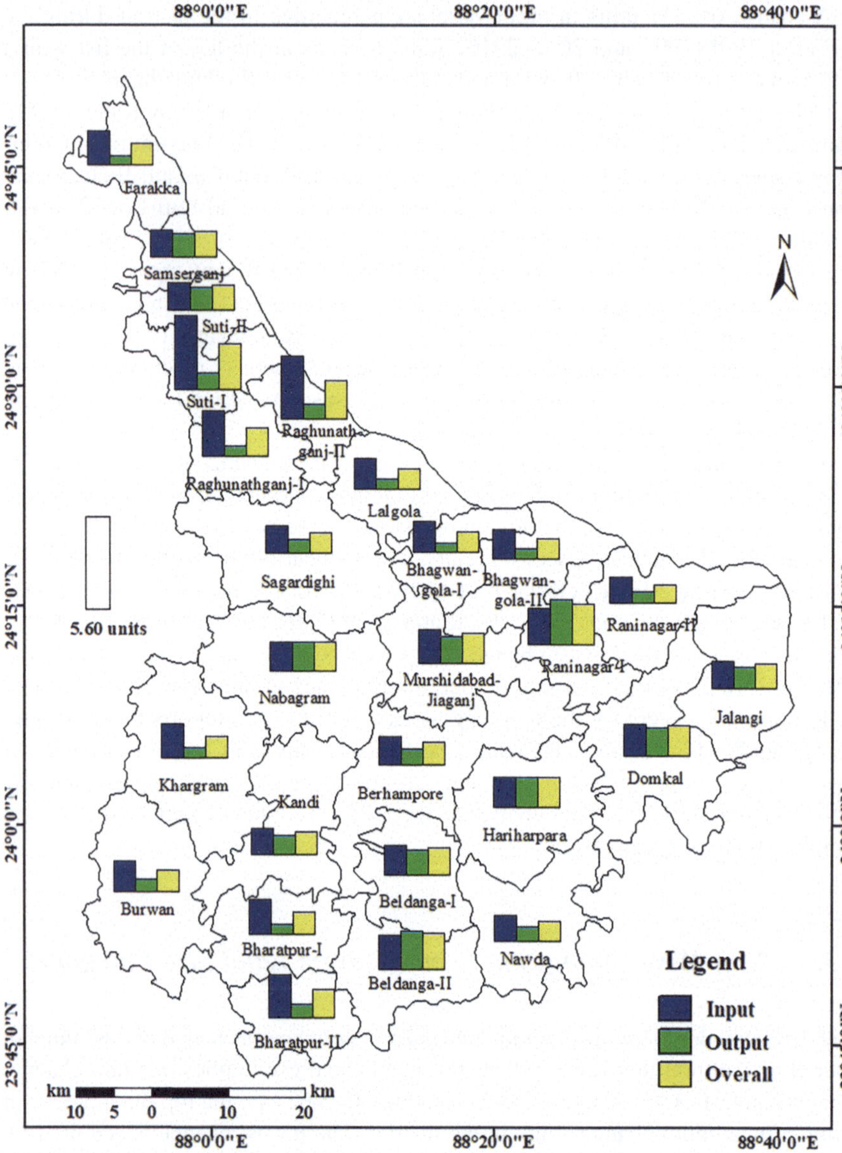

Fig. 4.7 Changes of agricultural innovation per unit of sustainable land use change, small land size class (1995–1996 to 2015–2016). (Source: Based on Annexure D.7)

input dimension along with Berhampore (2.64) and Khargram (2.64). The maximum change under the different indicators of the input sector is noticed in seeds (6.34). The highest and lowest changes in the said indicator are located by Lalgola (17.44) and Murshidabad-Jiaganj (4.88), respectively. Block performance interms of higher and lower achivement varies by category: Bharatpur-II and Nawda in land

(2.07, 0.85); Lalgola and Raghunathganj-II in irrigation (5.18, 2.34); Lalgola and Khargram in credits (3.96, 1.15); Hariharpara and Berhampore in equipment (5.68, 1.90); and Bharatpur-II, Lalgola, Bhagwangola-I and Murshidabad-Jiaganj in socio-economic factors (7.10, 1.70, 6.61, 0.38).

The change in agricultural output concerning sustainable land use is 2.48 unit in the district, in which Suti-II (3.91) reports the maximum changes among the blocks, followed by Murshidabad-Jiaganj (3.36), Samserganj (3.36), and Beldanga-II (3.01). In contrast, Raghunathganj-I, with a score of just 1.29, appears to be the least changed block in terms of agricultural output changes per one unit of sustainable land use change. The maximum changes under crop diversification (2.90) and crop productivity (2.86) variables are noticed in Beldanga-II, while the Lalgola block registers the lowest changes with a value of 0.70 and 0.85 units, respectively. Interestingly, Lalgola (6.67) stands at the top of changes in cropping intensity per one unit of sustainable land use change in the district. Raghunathganj-I (1.68) sits at the bottom of the same list. The biggest and lowest changes under the cropping pattern variable are recorded in Nawda (5.65) and Bhagwangola-II (0.32), respectively (Fig. 4.8).

4.3.4 Unit-Wise Changes in Medium Land Size Category

The unit-wise change data under this land size category informs that the district experiences a change of 2.26 units in overall agricultural innovation when the sustainable land use is increased by one unit. Suti-II (3.42) represents the maximum changes among the blocks. Besides, Jalangi (2.97), Murshidabad-Jiaganj (2.97), Bharatpur-II (2.75), and Bhagwangola-II (2.51) are also situated in the higher amount changes category. However, Burwan records only 1.02 units changes per one unit of sustainable land use change, placing it at the bottom. Adding to this, blocks like Sagardighi, Bhagwangola-I, Khargram, Domkal, etc., also register lower values regarding the unit-wise change in agricultural innovation (Fig. 4.9).

If the output indicators are kept constant, then the input variables jointly expose 2.54-unit changes in agricultural innovation per one-unit change of sustainable land use in the district. Suti-II (3.72), Jalangi (3.13), and Bhagwangola-II (3.11) are the top three blocks in terms of unit-wise changes. Conversely, Burwan (1.18) posits the least changes among the blocks, followed by Sagardighi (1.33), Raninagar-II (1.86), Bharatpur-I (2.02), and Raninagar-I (2.05) in ascending order. The seeds (5.48) as indicator display the maximum changes among different indicators of agricultural input, in which Nabagram (6.56) reveals the top change and the minimum is recorded in Jalangi (4.61). Burwan experiences the lowest change under the remaining indicators of agricultural input (land, irrigation, credits, machinery, fertilizers and pesticides, and socio-economic aspects). However, Beldanga-II (3.87), Suti-II (4.57), Murshidabad-Jiaganj (2.26), Suti-I (5.15), Suti-II (5.28), and Sagardighi (1.34) are the blocks that represent the maximum change of agricultural input per

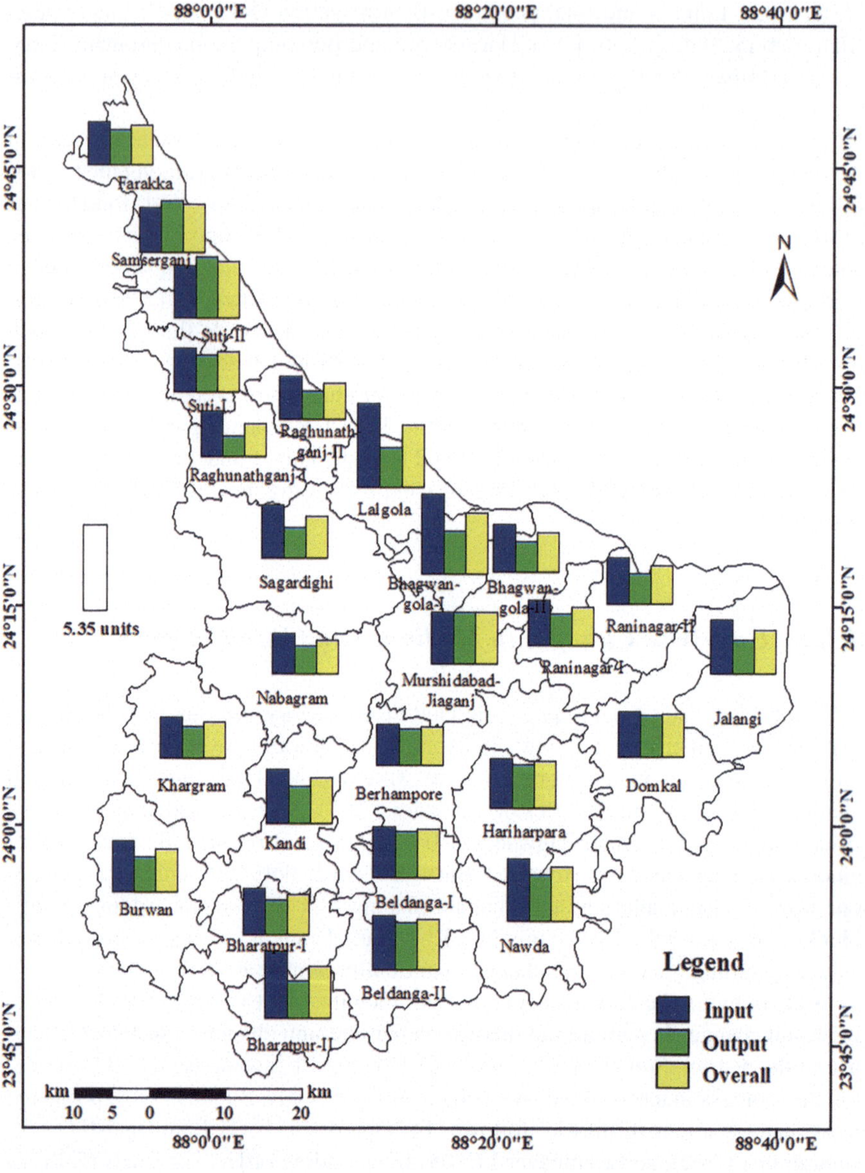

Fig. 4.8 Changes of agricultural innovation per unit of sustainable land use change, semi-medium land size class (1995–1996 to 2015–2016). (Source: Based on Annexure D.7)

unit of sustainable land use under land, irrigation, credits, machinery, fertilizers and pesticides, and socio-economic aspects, respectively.

There is a change of 1.98 units of agricultural innovation per one unit of sustainable land use change in the district when the variables of agricultural output are only considered. Suti-II reflects the highest change of 3.12 units among the blocks,

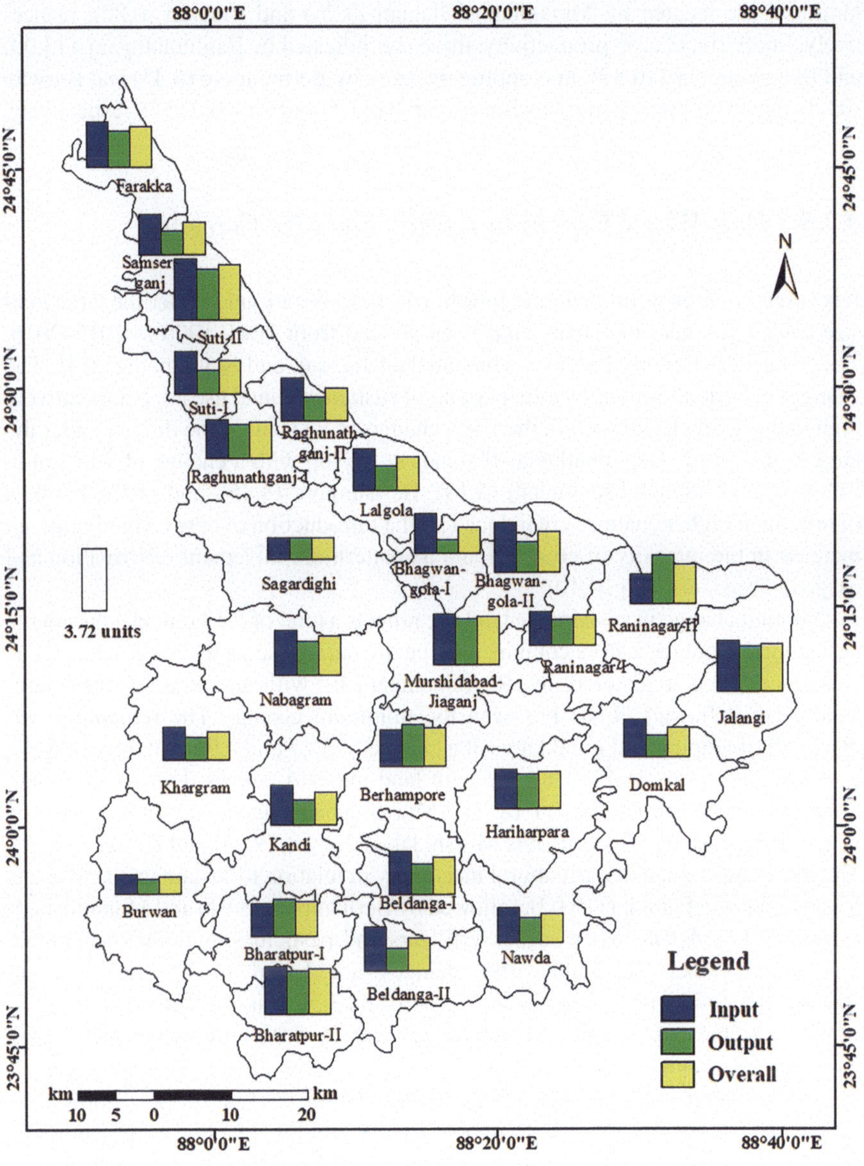

Fig. 4.9 Changes of agricultural innovation per unit of sustainable land use change, medium land size class (1995–1996 to 2015–2016). (Source: Based on Annexure D.7)

followed by Raninagar-II (3.04), Murshidabad-Jiaganj (2.83), Jalangi (2.82), and Bharatpur-II (2.60). However, the district has seen a low change of 0.83 units in the Bhagwangola-II block. Burwan (0.86), Domkal (1.31), Sagardighi (1.33), and Beldanga-II (1.38) have also noticed a lower value in the district. The maximum and minimum unit-wise changes, particularly in crop diversification with sustainable land use, are located by Murshidabad-Jiaganj (3.26) and Burwan (0.82), respectively. Similarly, in crop productivity, these are indicated by Raghunathganj-I (3.00) and Bhagwangola-I (0.37); in cropping intensity by Berhampore (6.45) and Burwan (0.88); and in cropping pattern by Raninagar-II (6.63) and Suti-II (0.52), respectively.

4.3.5 Unit-Wise Changes in Large Land Size Category

It is essential to mention that only four blocks have been found under the large land size category, where the change has been noticed from 1995–1996 to 2015–2016. These are Berhampore, Farakka, Murshidabad-Jiaganj, and Raghunathganj-II. The changes in agricultural innovation per unit of sustainable land use are comparatively high in this land class. Overall, there is a change of 5.11unit in the district under the large landholding. Raghunathganj-II stands at the top with a change of 10.74 unit, followed by Murshidabad-Jiaganj (5.19), Berhampore (4.45), and lastly, Farakka (4.06). Such higher changes are induced by the introduction of new technologies, an increase in the intensity of cropping and its pattern, improvement of irrigation and usage of seeds, etc. (Fig. 4.10).

One-unit change in sustainable land use reflects a total of 5.18 unit changes in the agricultural input indicators combined in the district. As far as the biggest change is concerned, it is registered by Raghunathganj-II, with a value of 6.12 unit. Murshidabad-Jiaganj (4.64) has seen the minimum change. The remaining two blocks, Berhampore and Farakka posit a change of 5.47 and 4.99 unit, respectively. The relation of sustainable land use with land indicator shows the maximum and minimum change in Farakka (1.18) and Murshidabad-Jiaganj (0.90), respectively. While in the case of irrigation, it is Murshidabad-Jiaganj (9.05) and Farakka (6.91). The unit-wise change of agricultural machinery in relation to sustainable land use is found highest in Farakka (10.02), followed by Berhampore (8.91) and Murshidabad-Jiaganj (7.17). Again, when seeds, fertilizers and pesticides, and socio-economic aspects are taken separately, the changes are found to be maximum in Raghunathganj-II with a value of 14.53, 9.49, and 3.57 unit, respectively.

The variables of agricultural output increase by 5.03 units when there is an increase of one unit of sustainable land use change in the district. Raghunathganj-II has seen a change of 15.36, which is the highest among the four blocks, followed by Murshidabad-Jiaganj (5.75), Berhampore (3.42), and Farakka (3.13), sequentially. Raghunathganj-II block also represents the highest changes of the variables of agricultural output, taking individually per unit change of sustainable land use (13.41unit in crop diversification, 9.57 unit in productivity, 17.03 unit in cropping intensity, and 21.42 unit in cropping pattern).

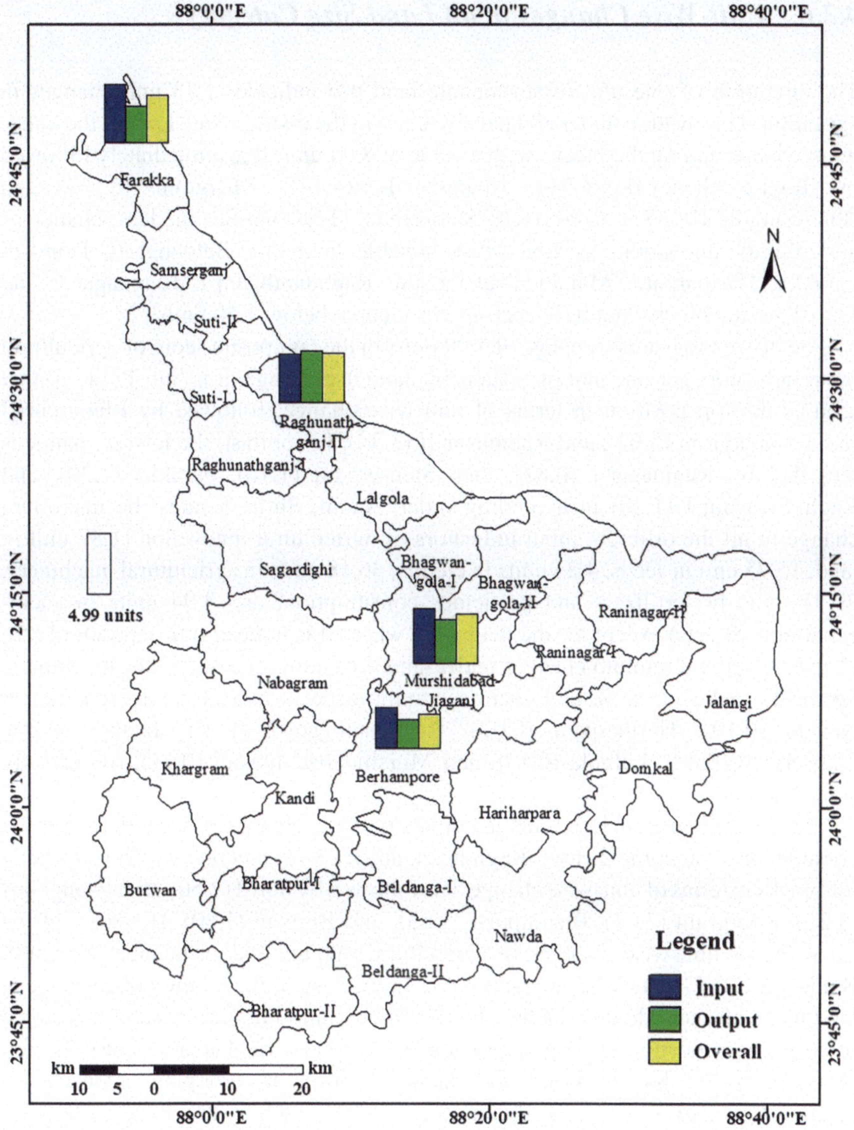

Fig. 4.10 Changes of agricultural innovation per unit of sustainable land use change, large land size class (1995–1996 to 2015–2016). (Source: Based on Annexure D.7)

4.3.6 Unit-Wise Changes in all Land Size Category

The alteration of one unit of sustainable land use indicates 1.73 unit changes in agricultural innovation under all land size class in the district. Suti-I posits the maximum change among the blocks with a value of 8.64 unit. It is immediately followed by Raghunathganj-II (4.71), Bharatpur-II (4.14), Khargram (3.53), and Bhagwangola-II (2.33). Conversely, Samserganj (1.08) notifies the least change of agricultural innovation per unit of sustainable land use. Beldanga-II, Domkal, Farakka, Hariharpara, Murshidabad-Jiaganj, Raghunathganj-I, Raninagar-I, and Suti-II are the blocks that have registered a change below 1.50 units.

The district registers a change of 1.99 units in the composite score of agricultural input indicators per one unit of sustainable land use change. It is Suti-I (14.78) that secures the top position in terms of unit-wise change, followed by Bharatpur-II (6.53), Khargram (5.02), and Raninagar-II (3.34). In contrast, the lowest change is recorded by Raninagar-I (0.82), then Samserganj (1.16), Farakka (1.20), and Raghunathganj-I (1.20) in ascending order. Again, Suti-I locates the maximum change in all the different input indicators of agricultural innovation (1.87 unit in land, 15.95 unit in seeds, 4.24 units in credits, 46.18 units in agricultural machinery, 29.71 units in fertilizers and pesticides consumption, and 4.94 units in socio-economic aspects) except for the irrigation, where it is noticed in Khargram (2.85). Conversely, the minimum changes under land, irrigation, seeds, credits, machinery, fertilizers and pesticides, and socio-economic aspects indicators are reported in Nawda (0.19), Hariharpara (0.07), Raghunathganj-I (1.26), Jalangi (0.20), Samserganj (0.41), Lalgola (0.19), and Murshidabad-Jiaganj (0.21), respectively (Fig. 4.11).

There is a change of 1.46 units in outputs with the change of one unit in the sustainable land use in the district. Raghunathganj-II (3.71) emerges as the top among the blocks in terms of unit-wise change. It is followed by Suti-I (2.50), Bhagwangola-I (2.29), Khargram (2.03), Bharatpur-I (1.83), and Burwan (1.80). However, in the same list of unit-wise changes in agricultural outputs and sustainable land use, Beldanga-I (0.87) block lies at the bottom. In the case of the relation of sustainable land use with individual variables, the maximum and minimum changes are noted in crop diversification by Raghunathganj-II (4.64) and Beldanga-I (0.66); in crop productivity by Suti-I (1.80) and Jalangi (0.06); in cropping intensity by Raghunathganj-II (7.06) and Raninagar-II (1.01); and in cropping pattern by Bhagwangola-I (2.00) and Raghunathganj-I (0.43), respectively.

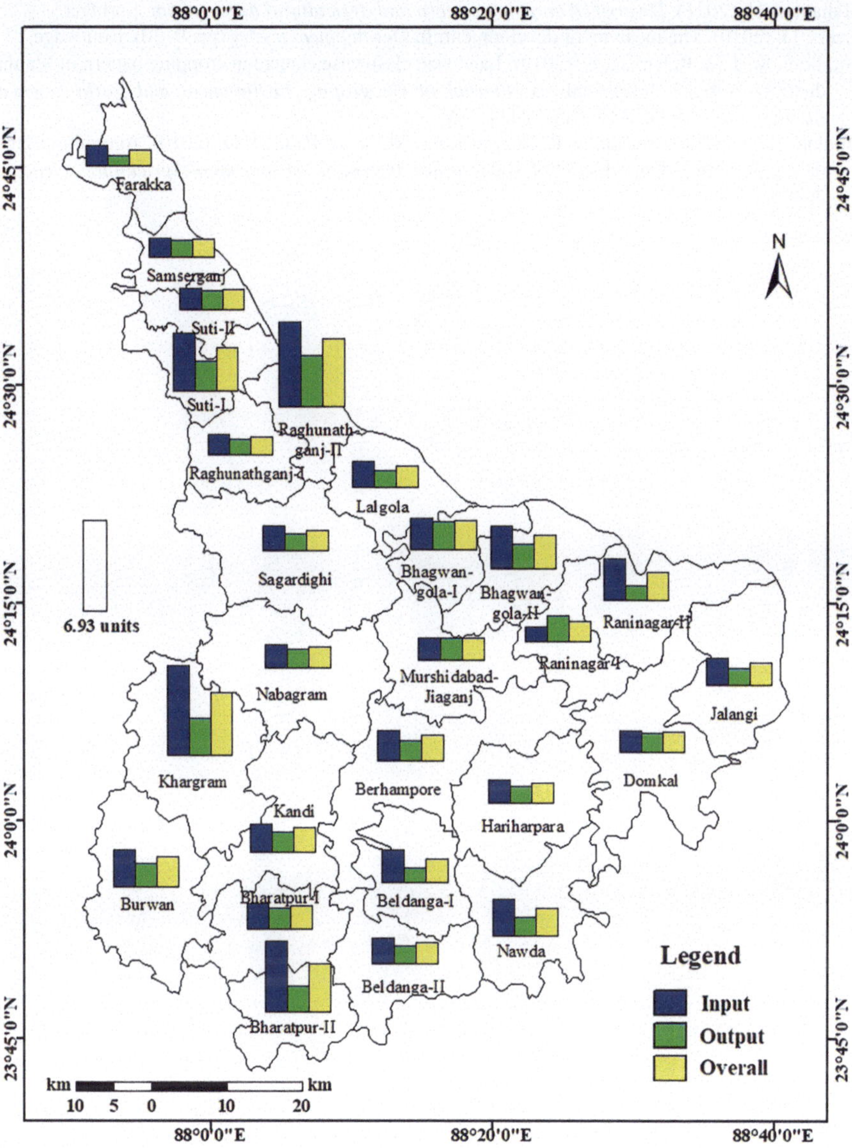

Fig. 4.11 Changes of agricultural innovation per unit of sustainable land use change, all land size class (1995–1996 to 2015–2016). (Source: Based on Annexure D.7)

References

Eberly, L. E. (2007). Multiple linear regression. In *Topics in biostatistics* (pp. 165–187).

Li, X., & Wang, X. (2003). Changes in agricultural land use in China: 1981–2000. *Asian Geographer, 22*(1–2), 27–42.

Rahaman, H. (2021). *Diversified cropping pattern and agricultural development*. Springer.
Seers, D. (2010). The meaning of development. In *Development theory* (pp. 9–30). Routledge.
Siddiqui, S. H., & Rahaman, H. (2019). Land size class wise change in cropping pattern in Malda district—A block level analysis. *Journal of Geography, Environment and Earth Science International, 21*(4), 1–19. ISSN 2454-7352.
Todkari, G. U., Suryawanshi, S. P., Suryawanshi, M. V., & Patil, B. D. (2010). Agriculture landuse pattern in Solapur district of Maharashtra. *International Journal of Agriculture Sciences, 2*(2), 1.

Chapter 5
Agricultural Innovation and Sustainable Land Use in Sampled Villages

Contents

H. Rahaman, S. Ansari, *Sustainable Land Use Practices and Agricultural Innovations in India*, https://doi.org/10.1007/978-3-031-83246-8_5

Chapters 2 to 4 devoted to sustainable land use and levels of agricultural innovation and their relationship are mainly based on secondary data sources. It has been proved there that the performance of the marginal land category is better than other land classes, which invalidates the pre-concept notion that the bigger the land size, the better the innovation. An analysis based on a household survey has been done to check the sustainable land use and innovation relation further at the micro-level. The first section of this chapter describes the physical, socio-economic, and demographic characteristics of the sample villages. This is followed by the soil nutrition index of those selected villages. The agricultural situation of the villages is explained in the next section. Finally, the relationship between agricultural innovation and sustainable land use is analysed through composite index value.

5.1 Brief Description of Sampled Villages

The household survey was conducted in November and December 2021. A total of 26 villages have been selected from each block for this study. A village with the maximum number of cultivators in a particular block in the 2011 Census of India is chosen as a sample village. This research is thoroughly focused on sustainable land use patterns and agricultural innovation; therefore, 'the maximum number of cultivators' is taken as a base for the selection criterion. Moreover, the purposive sampling technique is used to select respondents from each village. Thirty (30) respondents have been selected from each village, of which twenty-four (24) are taken from the marginal landholding, three (3) from small, two (2) from semi-medium, and one (1) from medium land classes. No farmers with larger holdings were found in the sample villages; therefore, the large land category is eliminated from the present analysis. Thus, the total number of respondents accounts for 780.

The researcher has collected information regarding the respondents' socio-economic aspects and essential information related to agricultural innovation and sustainable land use patterns in the sample villages. The following section discusses a brief description of the sampled villages after gathering the required information from secondary sources collected from various government officials and websites, such as the Census of India, Soil Health Card, and District Statistical Handbook.

5.1.1 Chhayghari

Around 10 km east of the block's headquarters, Chhayghari is a village in the Berhampore block situated at 24° 09′ 24″ North latitude and 88° 23′ 43″ East longitude. This village is spread over an area of 1020.25 ha and inhabited by 12,388 people in 3143 households.

The total number of workers in the village is 4741, of which 901 are cultivators. The soil of the village is moderately alkaline (84.88%), and the electric conductivity is 99.93% normal. The soil's organic carbon varies from medium (58.82%) to low (33.09%). The availability of nitrogen, phosphorus, and potassium is medium (75.36%), high (49.78%) to medium (37.28%), and medium (64.52%) to low (22.54%), respectively. In terms of micronutrients of soil, the village has a sufficiency of copper (97.47%), iron (99.64%), manganese (88.24%), and sulfur (65.15%), while it is deficient in zinc (88.66%) and boron (99.64%).

5.1.2 *Jhikarhati*

The village is located (24° 01′ 56″ N latitude and 87° 53′ 35″ E longitude) about 15.6 km northwest of the community development block of Burwan. There are 1933 households spreading over a 958.63 ha area in the village, where 8163 people live. This village represents 781 cultivators out of the 2579 total workers. The soil's acidity ranges from moderately acidic to moderately alkaline, while the electrical conductivity is entirely normal, i.e., 100%. The soil contains medium (58.88%) to low (35.30%) organic carbon. The availability of nitrogen is medium (81.73), phosphorus is high (52.77%), and potassium is medium (67.69%). The micronutrients copper, iron, manganese, and sulfur are found to be sufficient, but the soil lacks adequate zinc and boron (69.80%, and 97.02% deficiency, respectively).

5.1.3 *Mirzapur*

Mirzapur, one of the densest villages, is positioned at 23° 55′ 01″ N latitude and 88° 13′ 20″ E longitude and is the home of 27,364 people with 5380 households. The village's total area is 754.73 ha, situated 2.2 km southwest of the block's headquarters, where 1341 out of 8420 total workers are farmers. The electrical conductivity of soil is 100% normal, and the acidity is 74.23% moderately alkaline. The soil reports medium to low organic carbon, medium level of nitrogen and potassium, and high phosphorus content. Though copper (99.68%), iron (99.84%), manganese (96.95%), and sulfur (98.07%) are sufficient enough, there is a deficiency of zinc (77.81%) and boron (99.52%).

5.1.4 *Andul Baria*

Andul Baria (23° 49′ 54″ N latitude and 88° 18′ 34″ E longitude) is one of the agriculturally important villages in the Beldanga-II block, located 11.9 km away east from the CDB. The population of the village is 13,929, accounting for 3230 households. Out of the 4515 total workers, 1438 people are recognized as cultivators. The village's soil denotes the sufficiency of copper (97.22%), iron (99.81%), manganese (53.77%), and sulfur (91.67%) in the soil. In contrast, boron (99.07%) and zinc (98.15%) nutrients are deficient. The electrical conductivity of soil is purely normal, and 96.30% of it records moderately alkaline in terms of acidity. The organic carbon content of the soil is low (27.78%) to medium (68.52%). Further, the village shows medium availability of nitrogen (70.37%) and potassium (71.30%), and the phosphorus availability is medium (50.00%) to high (46.30%).

5.1.5 *Dakshin Hanumanta Nagar*

Spreading over an area of 2133.09 ha, Dakshin Hanumanta Nagar is marked at 24° 21′ 39″ N latitude and 88° 21′ 10″ E longitude. According to the 2011 Census of India, the village has a total of 19,813 population living in 4345 different houses, out of which 6218 are working people. The number of total cultivators in the village is 704. This village's soil is moderately alkaline (68.15%). The electrical conductivity of soil is neither critical for germination nor harmful or injurious to crop growth. The organic carbon of soil is low to medium range, nitrogen and potassium are in the medium, and phosphorus is in the medium to high category. All the micronutrients of soil under this village are found to be sufficient except for zinc and boron.

5.1.6 *Nasipur*

Nasipur village is situated at 24° 19′ 22″ N latitude and 88° 22′ 36″ E longitude. It covers a 1544.91 ha area with 6047 households containing 25,644 people in the 2011 Census of India. The headquarters of the Bhagwangola-II block is 1.9 km south of this village. The working population accounts for 8009, out of which 1302 are cultivators. The soil pH of the village is moderately alkaline (83.48%), but the electrical conductivity is 100% normal. Organic carbon content is low (63.50%) to medium (29.81%). About 72.51% and 97.62% of nitrogen and potassium, respectively, are in the medium range. Also, phosphorus availability varies from high (74.08%) to medium (23.11%). Micronutrients such as copper, iron, manganese, and sulfur are 99.57%, 100%, 99.78%, and 98.27%, respectively, sufficient in the soil.

5.1.7 *Bharatpur*

The village of Bharatpur located in Bharatpur-I block is expanded over an area of 1021.99 ha. It lies at 23° 53′ 17″ North latitude and 88° 05′ 00″ East longitude and is located within a kilometre of the community development block office. The village has a population of 15,812 from 3445 families. There are 5191 total workers in the village, of which 860 people are engaged in cultivation. The electrical conductivity is normal (100%), and 100% soil is tested as moderately alkaline. The availabilities of organic carbon, nitrogen, and phosphorus are 100% high in this village, while the potassium content shows a medium to high range. Only zinc and boron micronutrients remain deficient in the village (Fig. 5.1).

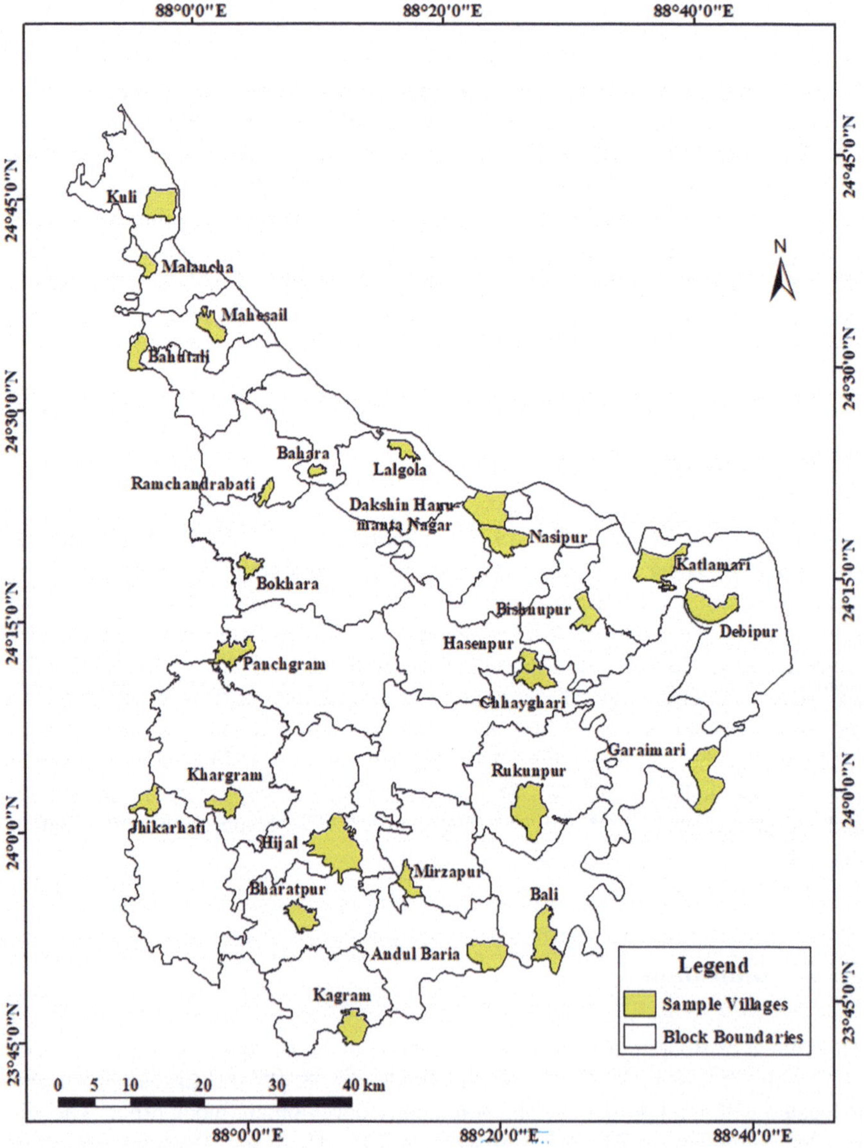

Fig. 5.1 Location of sample village, 2021. (Source: District Census Handbook, Murshidabad District, Part-A, 2011)

5.1.8 Kagram

Kagram, a village of Bharatpur-II block, is located at 23° 45′ 19″ N latitude and 88° 08′ 46″ E longitude covering an area of 1101.12 ha. This village is inhabited by 3481 households having a total population of 14,878. As many as 623 out of the total workers (5673) are involved in agricultural cultivation. It reports a 100% score in moderately alkaline soil and normal condition of electrical conductivity. It means that the soil is not harmful or damaging to the germination and growth of different crops in the village. The availability of macronutrients varies from the medium to the high range. In terms of soil micronutrients, copper, iron, manganese, and sulfur are 100% sufficient in the village.

5.1.9 Garaimari

Garaimari is one of the largest villages of Domkal block, placed at 24° 00′ 47″ N latitude and 88° 37′ 43″ E longitude. This village extends over an area of 2356.70 ha, with a total of 29,260 people living here in 7205 different households. The total number of workers and cultivators in this village are 10,022 and 1748, respectively. According to the Soil Health Card, the acidity of the soil is 100% moderately alkaline here, and the electrical conductivity is also normal (100%). Organic carbon content ranges from low (39.13%) to medium (54.35%). The other macronutrients are also varied from low to high content. The soil is sufficient in copper (100%), iron (100%), and sulfur (98.91%), while manganese (64.29%), zinc (98.37%), and boron (100%) lack sufficiency.

5.1.10 Kuli

The village (24° 43′ 53″ N latitude and 88° 55′ 47″ E longitude) is situated about 7 km north of the headquarters of Farakka block. According to the 2011 Census of India, it comprises 28,092 people settled in 5511 families. This village is stretched over a 1435.42 ha area. The total number of workers in the village are counted as 13,359, out of which 495 are cultivators. The soil pH is mainly alkaline but also extends to moderately acidic in nature. The availability of macronutrients in the soil is reported from the low to medium range. Conversely, micronutrients such as copper (100%), iron (100%), and manganese (90.70%) are sufficient in the soil, but zinc (59.09%), boron (73.86%), and sulfur (59.09%) are insufficient.

5.1.11 Rukunpur

Rukunpur is one of the largest villages (2327.77 ha area) in the Hariharpara block, located at 24° 00′ 07″ North latitude and 88° 24′ 30″ East longitude. The village has a total of 4429 households, in which 18,663 people are living. There are 2149 cultivators in the village, and the total number of workers is 6037. About 97% of the tested soils are moderately alkaline, and the electrical conductivity is absolutely normal (100%). The macronutrients, i.e., organic carbon, nitrogen, phosphorus, and potassium, are low to medium available in the village. Concerning micronutrients, copper, iron, and zinc are detected to be sufficient and manganese, boron, and sulfur to be insufficient.

5.1.12 Hijal

Hijal is the largest village in Murshidabad district with a total area of 3680.22 ha, situated in Kandi block. The geographical location of this village is 23° 57′ 41.7″ N 88° 08′ 12.1″ E, which is almost 12 km from the Kandi block's headquarters. The total households of this village are 2713, with 11,480 population living there. The total workers in this village are 3970, and out of them, 2450 are involved in agricultural cultivation. This village's soil's acidity is mainly of two types: moderately acidic (39.77%) and moderately alkaline (34.77%). The electrical conductivity is normal throughout the village. In terms of organic carbon of the village, it varies from medium to low, at 57.34% and 29.72%, respectively. Nitrogen availability of soil in Hijal village is mostly medium, with 69.58% and only 18.8% high likewise, phosphorus and potassium vary from medium to high. The soil of this village is deficient in zinc (Zn) and boron (B) under micronutrients, while other micronutrients like copper (Cu), iron (Fe) and, manganese (Mn) are sufficient.

5.1.13 Khargram

Khargram village located in the Khargram block has a total of 965.14 ha of land area, located at 24° 01′ 42.7″ N 87° 59′ 13.9″ E. Khargram is very close to the headquarters of the blocks. The total population of this village is 11,209, with 2648 households. The total workers and total cultivators of this village are 3660 and 703, respectively. Three-fourths (74.45%) of the village soil is moderately acidic, and 100% is normal in terms of electric conductivity. More than 90% of the village soil has organic carbon, either medium (48.46%) or low (44.05%). Macronutrients of the villages vary from medium to high for nitrogen and potassium, while macronutrient like phosphorus varies between high (35.68%) and very high (45.37%). Micronutrients of the villages are sufficient for copper (99.56%), iron (100%), manganese (91.03%), and sulfur (90.75%). Conversely, micronutrient is deficient for zinc (80.18%) and boron (99.56%).

5.1.14 *Lalgola*

Lalgola village has 6260 households with a 31,698 population; on average more than five people live per household. The total area of this village is 585.06 ha, situated at 24° 25′ 02.5″ N 88° 15′ 37.4″ E. Out of the total 11,918 workers, only 706 are involved in agricultural cultivation. The acidity of the soil is moderately acidic for 70.04% of the villages, and the electrical conductivity of the soil is normal (100%). If we look at the organic carbon, then it is almost all-around 96% from medium (78.76%) and low (17.57%). In this, 85.33% of the soils has medium nitrogen availability, while macronutrients like potassium and phosphorus were found to be on the higher side with 58.11% and 69.88%. In terms of micronutrients of soil, the village has a sufficiency of copper (100.00%), iron (100.00%), zinc (77.80%) and manganese (84.85%), while it is deficient in sulfur (69.50%) and boron (100.00%).

5.1.15 *Hasenpur*

Hasenpur village of Murshidabad-Jiaganj block is positioned at 24° 10′ 22.9″ N 88° 24′ 16.2″ E, and the village's total area is 549.87 ha. It is the home of 8093 people with 2014 households, situated around 3.5 km from the block's headquarters, where 1121 out of 2898 total workers are farmers. The electrical conductivity of soil is 98.24% normal, and the acidity is 61.31% moderately alkaline. The soil reports medium to low organic carbon, medium level of nitrogen, and high phosphorus and potassium content. Though copper (99.41%), iron (100.00%), manganese (98.82%), and sulfur (94.64%) are sufficient, there is a deficiency of zinc (71.76%) and boron (100.00%).

5.1.16 *Panchgram*

Panchgram village is situated in Nabagram block of Murshidabad district. The total household and population of this village are 4004 and 17,137, respectively. It has a geographical area of 1107.14 ha, located at 24° 12′ 14.8″ N 88° 00′ 38.1″ E. The headquarters of the Nabagram block is 10 km east of this village. The working population accounts for 5611, out of which 767 are cultivators. The soil pH of the village is moderately alkaline (61.31%), but the electrical conductivity is 100% normal. Organic carbon content is medium (62.89%). About 47.42% and 77.65% of nitrogen and potassium, respectively, are in the medium range. In addition, phosphorus availability varies from high (61.18%) to medium (35.29%). All the micronutrients, i.e., copper, iron, manganese, and zinc, are 100% sufficient (except for sulfur, which is 86.60%).

5.1.17 Bali

Bali (22° 41′ 21.1″ N 88° 06′ 37.1″ E) is one of the agriculturally important villages in the Nawda block, with 2009 total cultivators out of 6990 total workers. It is almost 29% of the total workers. The total area of the village is 1741.05 ha. The village's soil denotes a sufficiency of copper (99.10%), iron (100%), manganese (66.16%), and zinc (51.70%) in the soil. In contrast, boron (99.72%) and sulfur (69.50%) nutrients are deficient. The electrical conductivity of soil is purely normal, and 46.03% of it records moderately acidic in terms of acidity. The organic carbon content of the soil is low (23.71%) to medium (62.89%). Further, the village shows medium availability of nitrogen (68.13%) and potassium (54.11%) and phosphorus (51.56%) availability is high.

5.1.18 Ramchandrabati

Ramchandrabati village located in Raghunathganj-I is situated 11 km from the CDB of this block. This village has 1159 households with a 5648 population. The total area of this village is 393.51 ha, located at 24° 23′ 22.6″ N 88° 03′ 57.0″ E. Out of the total 2488 workers, only 314 are involved in agricultural cultivation. The soil's acidity is moderately acidic throughout the village, and the electrical conductivity of the soil is normal (100%). If we look at the organic carbon, it is almost all-around 86% from medium (62.89%) and low (23.71%). Nitrogen availability is low for 100% of the villages, phosphorus is very high for 100% of the villages, and potassium is higher for 54.11% of the village. In terms of micronutrients of soil, the village has a sufficiency of copper, iron, zinc, and manganese (100%), while deficient in boron (100.00%).

5.1.19 Bahara

Around 13.4 km northeast of the block's headquarters, Bahara is a village in the Raghunathganj-II block situated at 24° 24′ 43.8″ North latitude and 88° 07′ 44.5″ East longitude. This village is spread over an area of 301.51 ha and is inhabited by 12,862 people in 2296 households. The total number of workers in the village is 5094, of which 355 are cultivators. The soil of the village is moderately alkaline

(53.66%), and the electric conductivity is 100% normal. The soil's organic carbon varies from medium (70.73%) to high (26.83%). The availability of nitrogen is medium (97.56%) and phosphorus and potassium high (68.29%) and (85.37%), respectively. In terms of micronutrients of soil, the village has a sufficiency of copper (100%), iron (100%), and zinc (100%), while deficient in manganese (68.29%), sulfur (56.10%), and boron (97.56%).

5.1.20 *Bishnupur*

Spreading over an area of 936.61 ha, Bishnupur is marked at 23° 04′ 12.2″ N 87° 18′ 42.1″ E. According to the 2011 Census of India, the village has 8320 population living in 2059 different houses, out of which 2596 are working people. The number of total cultivators in the village is 732. This village's soil is moderately alkaline (78.92%). The electrical conductivity of soil is neither critical for germination nor harmful or dangerous to crop growth. The organic carbon of soil is low to medium range, nitrogen and potassium are in the medium, and phosphorus is in the medium to high category. All the micronutrients of soil under this village are found to be sufficient except for zinc and boron.

5.1.21 *Katlamari*

The village Katlamari of Raninagar-II block expanded over an area of 1610.05 ha. It lies at 24° 16′ 40.5″ N latitude 88° 33′ 57.1″ E latitude and is located within 10 km of the community development block office. The village has a population of 33,290 from 7555 families. There are 10,864 total workers in the village, of which 2632 people are engaged in cultivation. The electrical conductivity is normal (100%), and 99.37% soil is tested as moderately alkaline. The availabilities of organic carbon vary between medium to low; nitrogen and potassium are at a medium level, while phosphorus is 56.45% high in this village. Only zinc and boron micronutrients remain deficient in the village.

5.1.22 *Bokhara*

About 8.3 km west of the headquarters of Sagardighi block, Bokhara village is occupied by 8000 people living in 1789 families. The village spreads over an area of 584.38 ha and lies at 24° 18′ 21.0″ N 88° 02′ 38.0″ E. It reports 3147 total workers, of which 688 people chose cultivation as their main occupation. More than 99% of the soil pH in the mentioned village is moderately alkaline, while the electrical conductivity is in 99.82% normal condition. Here, soil's organic carbon varies from the medium (57.75%) to low (36.28%) range. The nitrogen and potassium availabilities in the village are medium (80.32% and 66.89%, respectively). This village has a high phosphorus content (51.05%) to high (37.10%). Soil quality has achieved sufficiency in all the micronutrients except for zinc and boron.

5.1.23 *Malancha*

Malancha village of the Samserganj block is located at 24° 39′ 56.1″ N 87° 55′ 42.0″ E. The total household and population of this village are 3136 and 16,918, respectively. In terms of geographical coverage, the village has a total of 551.21 area. The headquarters of the Samserganj block is 2.5 km southwest of this village. The working population accounts for 6016, out of which 355 (6%) are cultivators. The soil pH of the village is moderately alkaline (98.44%), but the electrical conductivity is 100% normal. Organic carbon content is low (69.94%) to medium (25.72%). About 65.03% and 79.77% of nitrogen and phosphorus, respectively, are in the medium range. In addition, potassium availability varies from high (70.52%) to very high (12.14%). Micronutrients such as copper, iron, manganese, and sulfur are 100%, 100%, 100%, and 92.49%, respectively, sufficient in the soil.

5.1.24 *Bahutali*

The village (24° 34′ 05.2″ N 87° 54′ 12.4″E) is situated about 19.2 km north of the Suti-I block's headquarters. According to the 2011 Census of India, it comprises 15,963 people settled in 3076 families. This village is stretched over a 1051.46 ha area. The total workers in the village are counted as 6791, out of which 485 are cultivators. The soil pH is mainly moderately alkaline. The availability of macronutrients in the soil is reported from the low to medium range. Conversely, micronutrients such as copper (100%), iron (100%), sulfur (100%), and manganese (67.76%) are sufficient in the soil, but zinc (79.61%) and boron (99.34%) are insufficient.

5.1.25 *Mahesail*

Mahesail village is a village in the Suti-II block of Murshidabad located at 24° 35′ 20.6″ N 88° 00′ 36.3″ E. Total households of this village are 16,178, with 3057 population spreading over an area of 983.29 ha. The headquarters of the Suti-II block is 6.9 km away towards the north of this village. The working population accounts for 7434, out of which 13% are cultivators. The soil pH of the village is moderately alkaline (88.28%), but the electrical conductivity is 99.90% normal. Organic carbon content is medium (54.09%) to low (41.44%). About 83.08% and

51.26% of nitrogen and phosphorus, respectively, are in the medium range. Also, potassium availability varies from high (62.55%) to medium (34.03%). Micronutrients like copper, iron, manganese, and sulfur, with 99.90%, 100%, 99.33%, and 90.46%, respectively, are sufficient in the soil.

5.1.26 Debipur

About 9.5 km north of Jalangi block's headquarters, Debipur village is occupied by 17,877 people living in 4091 families. The village spreads over an area of 1872.26 ha and lies at 24° 12′ 58″ North latitude and 88° 38′ 52″ East longitude. It reports 6221 total workers, of which 2114 people chose cultivation as their primary occupation. More than 95% of the soil pH in the mentioned village is moderately alkaline, while the electrical conductivity is in 100% normal condition. Here, soil's organic carbon varies from the medium (66.95%) to low (17.38%) range. The nitrogen and potassium availabilities in the village are medium (80.91% and 66.38%, respectively). This village has a high phosphorus content (98.51%). Soil quality has achieved sufficiency in all the micronutrients except for boron.

5.2 Soil Nutrition Index in the Sample Villages

The correct proportions of macro- and micro-soil nutrients support healthy growth and production of crops. Sustainable and managed soil can lead to a 58% increase in food production (FAO, 2015). Therefore, an attempt is being made to calculate the composite index of macro, micro, and overall soil nutrients in the sample villages. The same technique (quartile technique) used earlier is applied here to show the spatial variation of the soil nutrient index.

The nutrients that different plants and crops require in a large proportion are known as macronutrients of the soil. These are organic carbon (OC), nitrogen (N), phosphorus (P), and potassium (K). Figure 5.2 reveals that in terms of macronutrients, Bharatpur and Kagram report better soil quality with an index value of 0.927. Five other villages, i.e., Bahara, Dakshin Hanumanta Nagar, Lalgola, Mahesail, and Hijal, have also scored the index value above the upper quartile (0.850). Katlamari village displays the least index in macronutrients of soil among the sample villages, followed by Malancha, Ramchandrabati, Nasipur, and Garaimari in ascending order. Twelve villages lie in between the upper (0.850) and lower (0.791) quartile index, out of which Jhikarhati village (0.838) secures the maximum index value.

The elements of micronutrients include copper (Cu), iron (Fe), manganese (Mn), zinc (Zn), boron (B), and sulfur (S). These nutrients are required in a relatively small quantity than macronutrients in the soil. However, the role of these nutrients in plants' growth cannot be ignored. Seven villages show better soil quality in terms of micronutrients and are placed above the upper quartile index (702), where

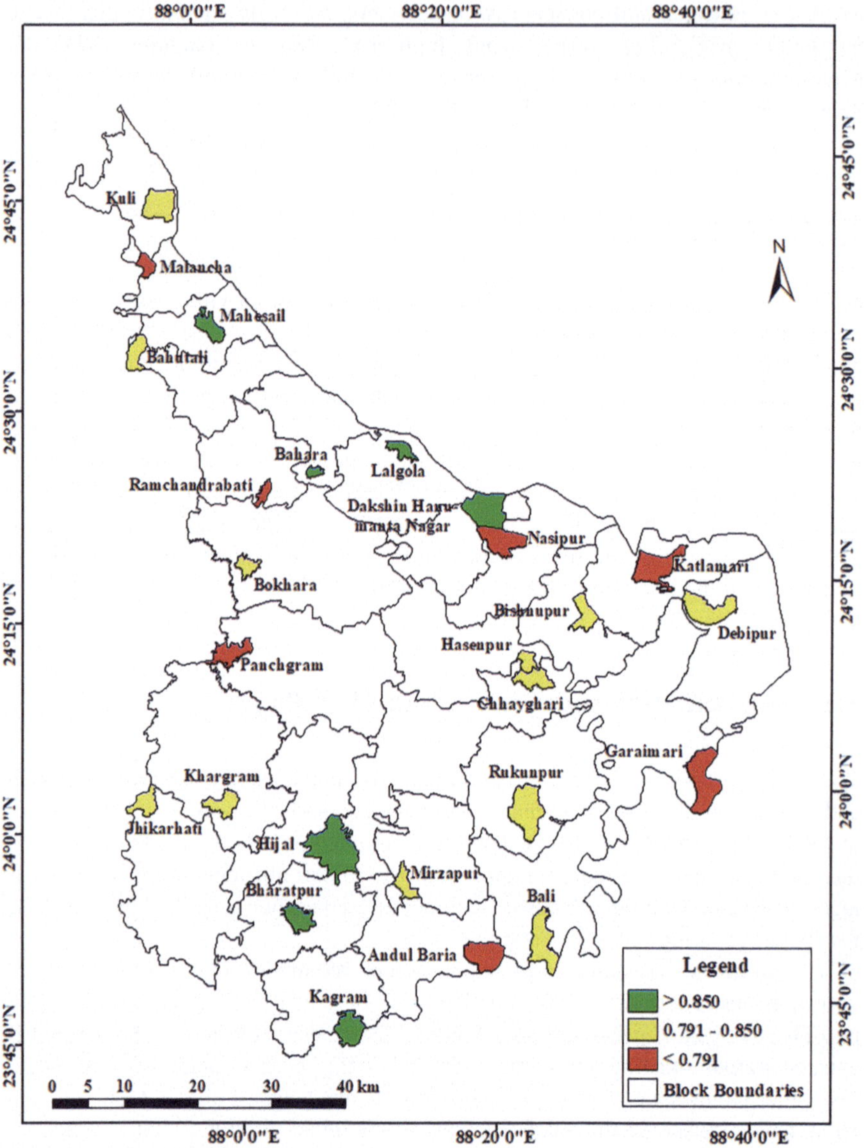

Fig. 5.2 Macro-soil nutrient index in the sample villages, 2021. (Source: Based on Annexure E.5)

Panchgram (0.868) emerges as the healthiest village. The others are Ramchandrabati (0.833), Debipur (0.805), Nasipur (0.738), Bishnupur (0.717), Dakshin Hanumanta Nagar (0.707), and Jhikarhati (0.703). Figure 5.3 represents that the index value below the lower quartile (0.650) is found in seven villages. With a weight of 0.557, the village Bali records the minimum micronutrient index among the villages. Garaimari, Andul Baria, and Rukunpur also report a soil macronutrient index below

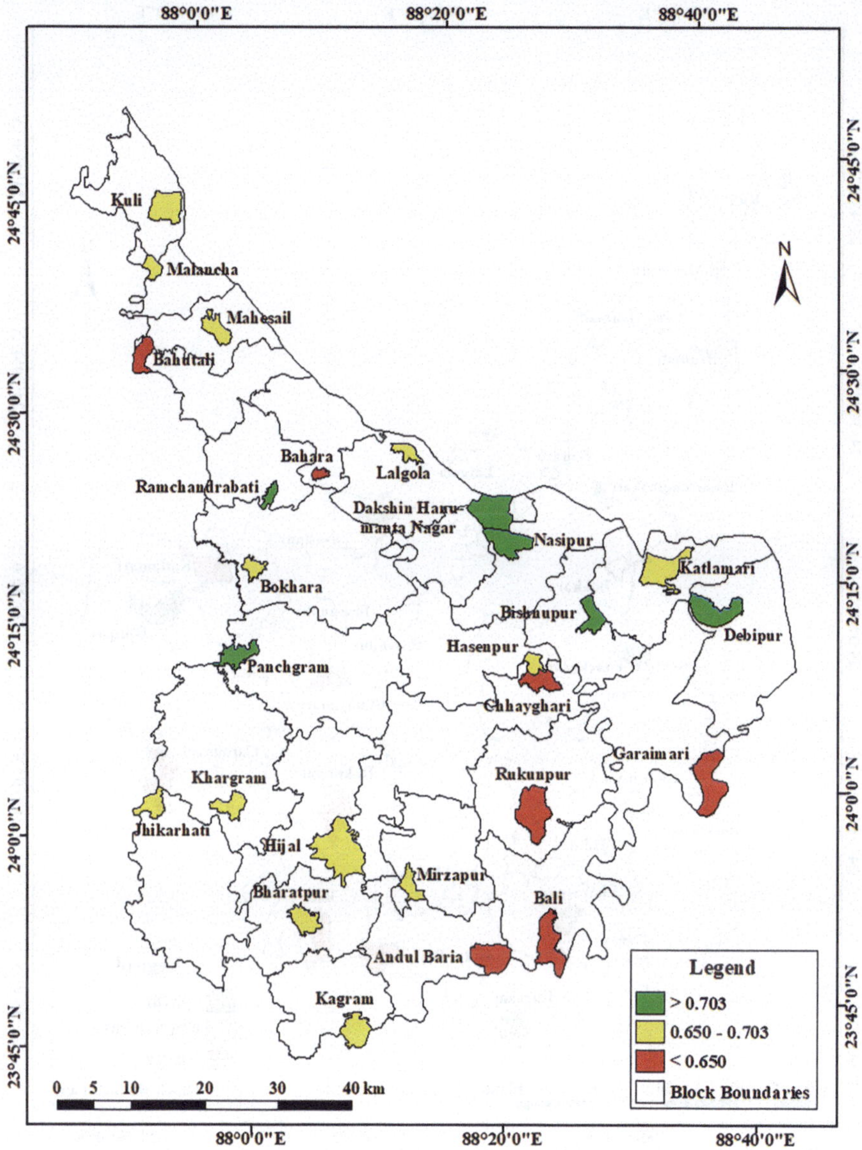

Fig. 5.3 Micro-soil nutrient index in the sample villages, 2021. (Source: Based on Annexure E.5)

0.600 value. The remaining villages noted between the upper and lower quartile signify the moderate micronutrient index.

It has been found from the overall soil nutrient index (Fig. 5.4) that Panchgram is the healthiest village (0.810), which is followed by Bharatpur (0.810), Dakshin Hanumanta Nagar (0.810), and Debipur (0.805). However, Garaimari having an index value of 0.669 sits at the bottom of the overall nutrient index of soil. Figure 5.4

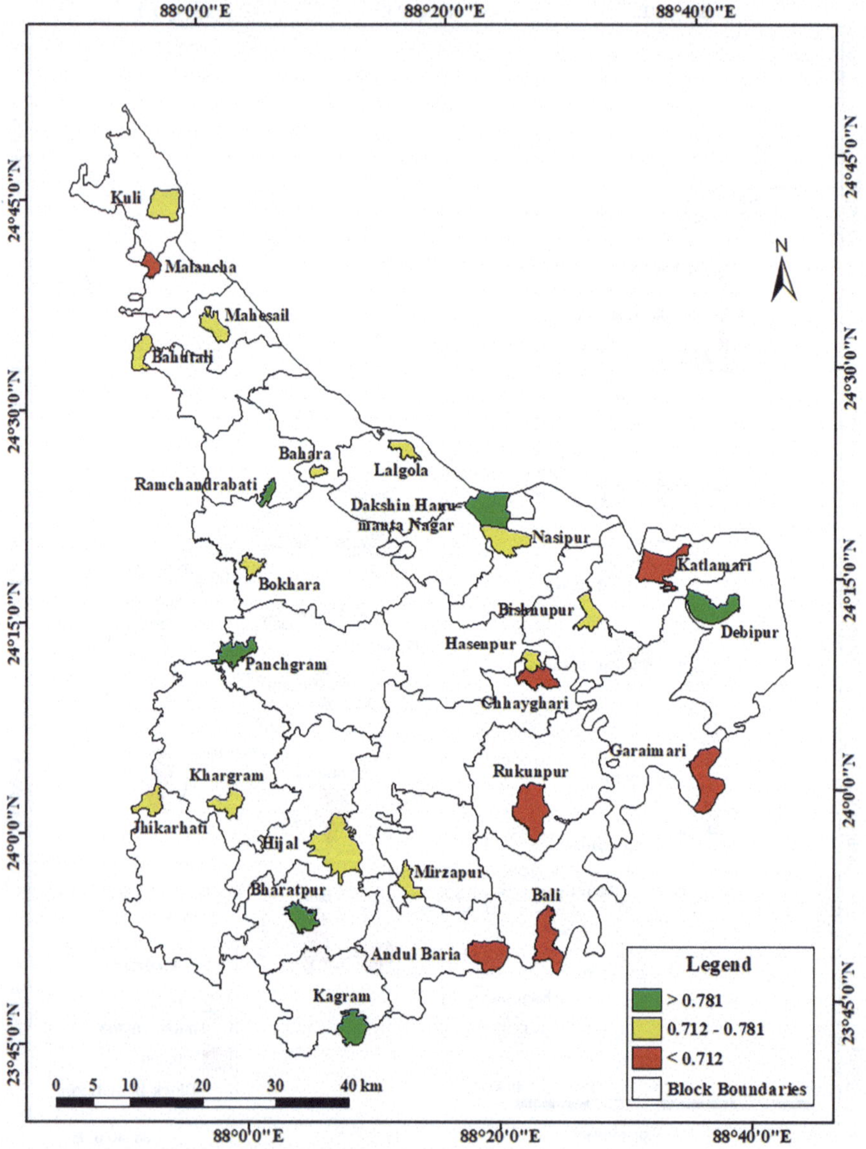

Fig. 5.4 Overall soil nutrient index in the sample villages, 2021. (Source: Based on Annexure E.5)

depicts the spatial patterns of soil nutrients in the sample villages. It denotes that six villages have an index value above the upper quartile (0.781), seven villages report below the lower quartile (0.712), and the remaining 13 villages locate moderate soil nutrients. Finally, it may be concluded that all the sample villages are in good condition in terms of soil nutrition availability.

5.3 Agricultural Situation in Sampled Villages in Murshidabad District

Based on the literature survey, as many as 40 variables have been selected for the present analysis of agricultural innovation in the chosen villages. Moreover, these variables are categorized under eight indicators, i.e., land, irrigation, seeds, income and credits, agricultural machinery, soil and fertilizers, market and cold storage, and socio-economic aspects. The spatial information on each variable of agricultural innovation has been illustrated under land size class category-wise on the total sample of 780, and their spatial analysis has also been explained on the basis of collected field survey data.

5.3.1 Average Size of Landholdings

The average size of landholding of all the sampled villages is 0.88 per household. The average land size of the households under the marginal, small, semi-medium, and medium land categories is 0.47, 1.49, 2.58, and 5.42, respectively (Fig. 5.5). The size of landholdings of the marginal and small classes is a matter of concern here as these two groups together account for more than 95% of the total cultivators in the district. The average land sizes of these categories are low due to the disintegration of landholdings caused by the rapid population growth and splitting of the joint family into a nucleated one.

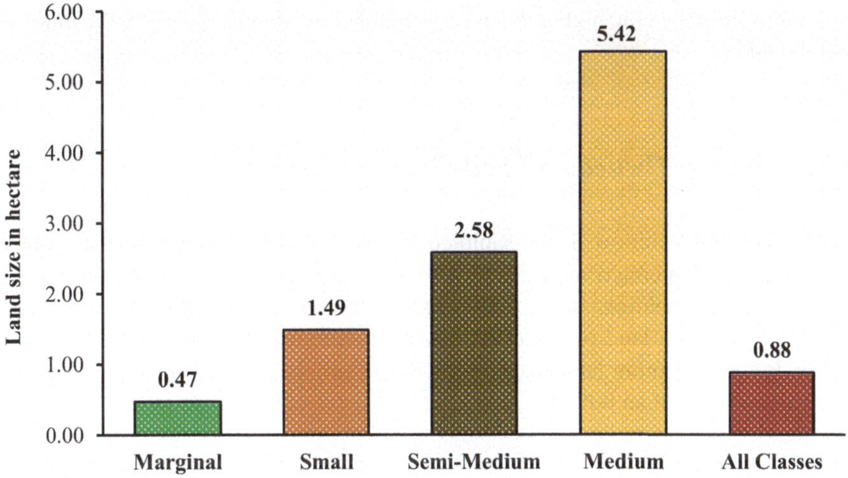

Fig. 5.5 Average size of landholdings, 2021. (Source: Field Survey, 2021 (Annexure E.7))

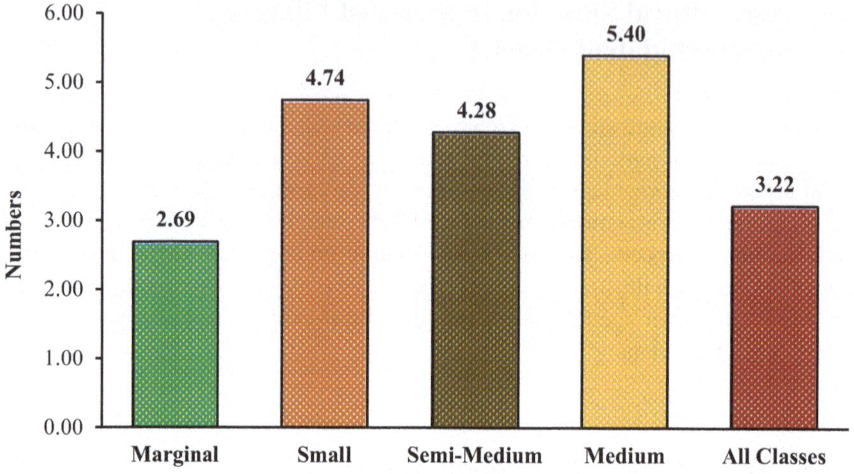

Fig. 5.6 Average number of parcels per holdings, 2021. (Source: Field Survey, 2021 (Annexure E.7))

5.3.2 Average Number of Parcels

A parcel is a piece of land, irrespective of its size, surrounded by other holding or land not forming part of any holding. A landholder may have more than one parcel. From all the surveyed villages, the average number of parcels per household is found to be 3.22. The maximum number of average parcels is recorded in the medium (5.40), followed by small (4.74), semi-medium (4.28), and marginal (2.69) land size classes (Fig. 5.6). The number of parcels is conversely related to agricultural innovation, i.e., the higher the average number of parcels, the lower the innovation would be.

5.3.3 Wholly Owned and Self-Operated Area

Out of the total holdings in the sampled villages, 83.58% belongs to the wholly owned and self-operated area. Figure 5.7 illustrates that the medium land class reports complete holdings under this category among the different land groups. The farmers of medium land type are wealthy enough to maintain their land by themselves. In the remaining land size classes, the wholly owned and self-operated area accounts for more than 80%.

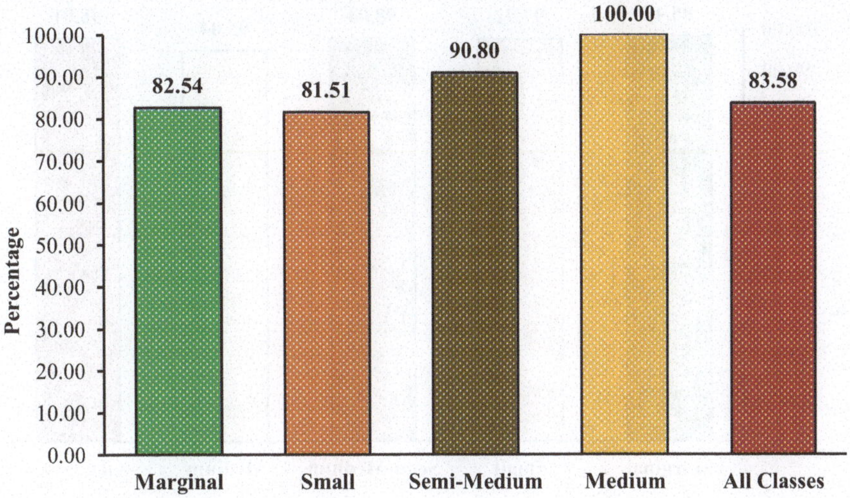

Fig. 5.7 Wholly owned and self-operated area, 2021. (Source: Field Survey, 2021 (Annexure E.7))

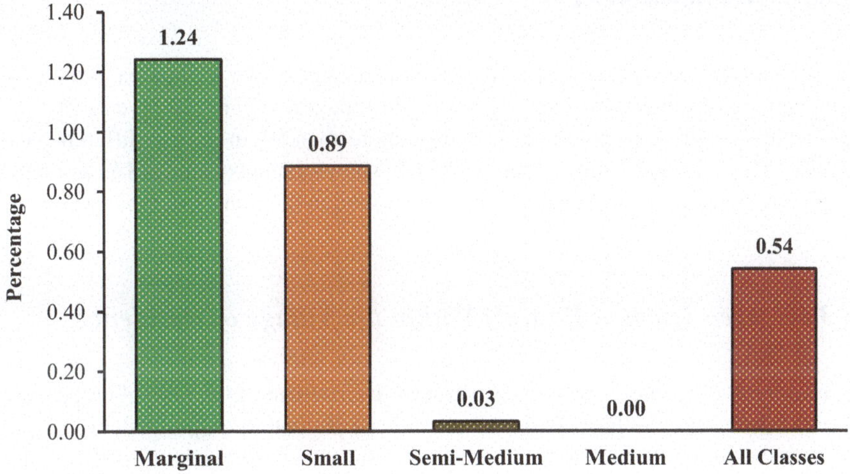

Fig. 5.8 Wholly leased-in area, 2021. (Source: Field Survey, 2021 (Annexure E.7))

5.3.4 Wholly Leased-In Area

About 1.24% of farmers in the marginal land category report the wholly leased-in area in the sampled villages (Fig. 5.8). This is followed by small (0.89%), all (0.,54%), and semi-medium (0.03) land size classes. Cultivators belonging to the medium land group have not registered any area under leased-in. The marginal land-holding farmers take the land on a lease, not for profit but for family consumption.

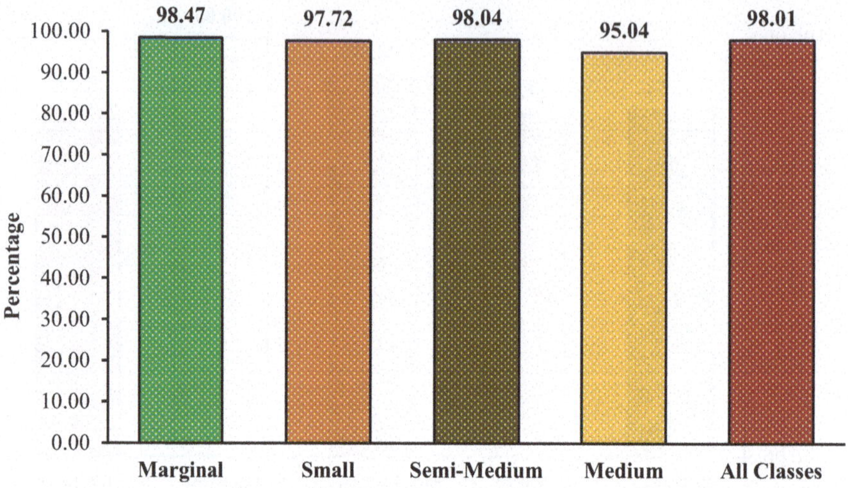

Fig. 5.9 Net sown area to the total area, 2021. (Source: Field Survey, 2021 (Annexure E.7))

5.3.5 Net Sown Area

An attempt has been made to calculate the percentage of net sown area in the total area captured by the farmers of the surveyed villages. Figure 5.9 reveals that the percentage of the net sown area is scored more than 95 in all the different land classes. The marginal land category (98.47%) shows the maximum NSA, while the minimum percentage of the net sown area is found in medium land type (95.04%).

5.3.6 Area Located Entirely Within the Village of Residence

The location of the net area sown is a crucial determinant of the type of crops to be grown. Generally, the farmers prefer to cultivate plantation crops, cash crops, and fruits if the area falls outside their village of residence. Conversely, if the landholdings are near the farmers' houses, they grow cereals, vegetables, oilseeds, floriculture crops, etc. It is seen in Fig. 5.10 that the marginal land class (79.08%) records the maximum percentage of the area within the village of residence, followed by small (67.80%), all (65.57%), semi-medium (60.73%), and medium (54.66%) land size categories in decreasing order.

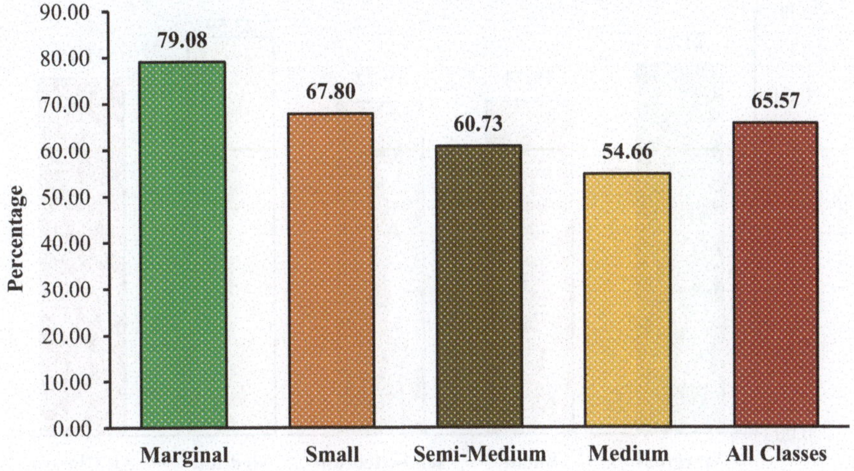

Fig. 5.10 Area located within the village, 2021. (Source: Field Survey, 2021 (Annexure E.7))

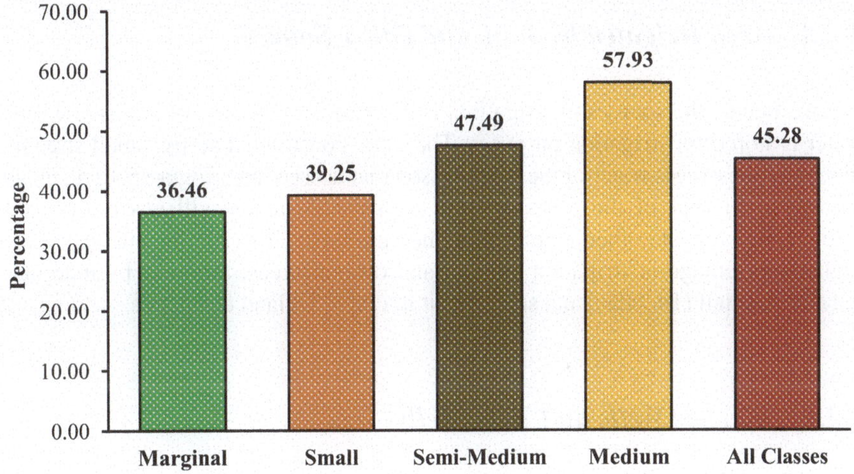

Fig. 5.11 Net irrigated area, 2021. (Source: Field Survey, 2021 (Annexure E.7))

5.3.7 Net Irrigated Area

The percentage of net irrigated area to the total gross cropped area in the villages is comparatively low, i.e., 45.28%. The marginal land class displays the lowest net irrigated area of 36.46% among the land categories. Limited availability of irrigation sources such as pump sets and tube wells is the reason for the lower net irrigated area. Most of the farmers in the district depend on monsoon rainfall, mainly between May and July, for irrigating their fields (Fig. 5.11).

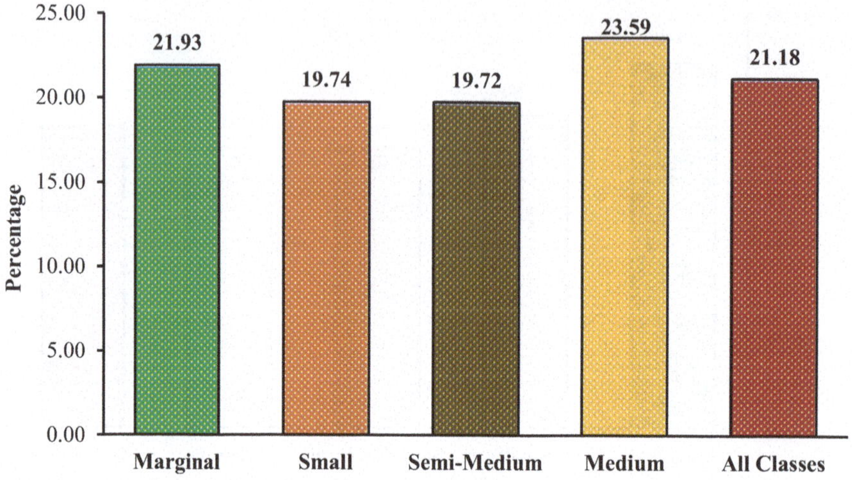

Fig. 5.12 Area irrigated by tank and other sources, 2021. (Source: Field Survey, 2021 (Annexure E.7))

5.3.8 Area Irrigated by Tank and Other Sources

A tank, one of the sources of irrigation, is an artificial reservoir of any size where water is stored for irrigation purposes. These are constructed across small streams to reserve the excess water of the rainy season that is used later. Figure 5.12 displays that about 21.18% of the gross cropped area in the sample villages is irrigated through the tank and other sources. The medium land class (23.59%) indicates the highest percentage of irrigation by tank and other sources, followed by marginal (21.93%), small (19.74%), and semi-medium (19.72%) land categories.

5.3.9 Electric Wells and Tube Wells

Ensuring regular irrigation through different sources is the prime requirement for better intensive subsistence and commercial agriculture in a particular region. Irrigation through wells and tube wells is the most common among the farmers as these are cost and time-efficient. An electric tube well is available per five hectares gross cropped area. However, the lack of regular and timely electricity hinders the irrigation system in the surveyed villages (Fig. 5.13).

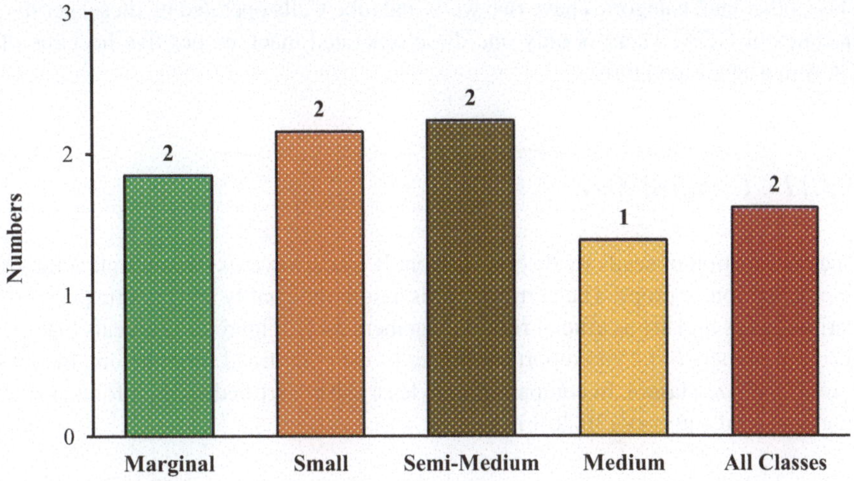

Fig. 5.13 Electric well and tube well per ten hectare GCA, 2021. (Source: Field Survey, 2021 (Annexure E.7))

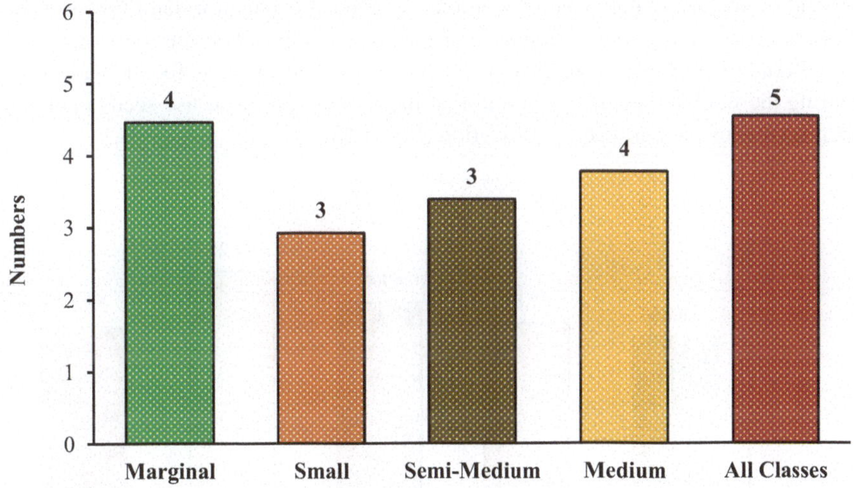

Fig. 5.14 Diesel operated well and tube well per ten hectare GCA, 2021. (Source: Field Survey, 2021 (Annexure E.7))

5.3.10 Diesel-Operated Wells and Tube Wells

The wells and tube wells irrigation are getting electrified over time. Still, there is the dominance of diesel-operated wells and tube wells in Indian agriculture. The cost of electric-generated machines is cheaper than the diesel-operated ones, but the plantation cost of such is very high. Figure 5.14 locates that except for small land size class, other land categories have two wells and tube wells operated by diesel per five hectares of GCA. There is only one diesel-operated machine per five hectares of GCA in a small land type.

5.3.11 *Certified Seeds*

The certification of seeds by the government is a step towards the multiplication of the production of crops. The certified seeds assure the quality and genetic purity of various seeds and act as crucial risk management tools. Figure 5.15 reveals that the medium farmers (44.25%) report the higher usage of certified seeds among the different land size classes. In comparison, the least use of certified seeds is found in the marginal land category (36.26%).

5.3.12 *Hybrid Seeds*

Hybrid seeds ensure better yield with the assurance of irrigation facilities, insecticides and pesticides, infrastructure, and fertilizers. Only a few numbers of operational holders (6.80%) in the sampled villages use hybrid seeds for their farming. During the field survey, it was found that the farmers prefer to use seeds made by themselves for the purpose of cultivation (Fig. 5.16).

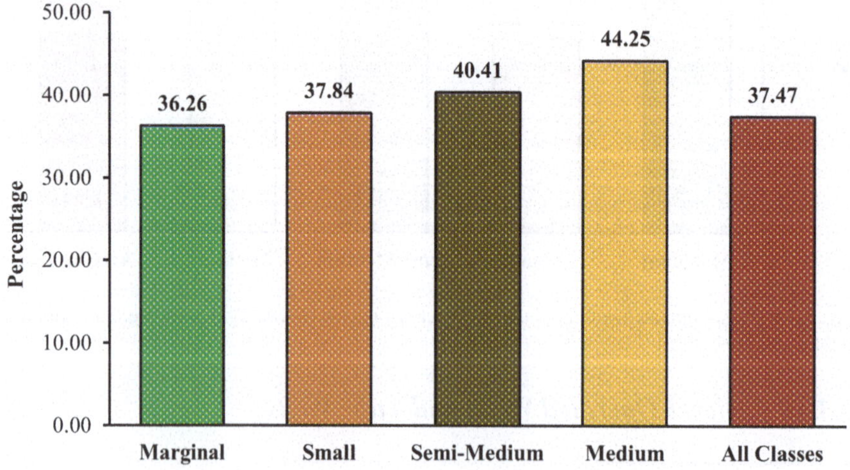

Fig. 5.15 Usage of certified seeds, 2021. (Source: Field Survey, 2021 (Annexure E.7))

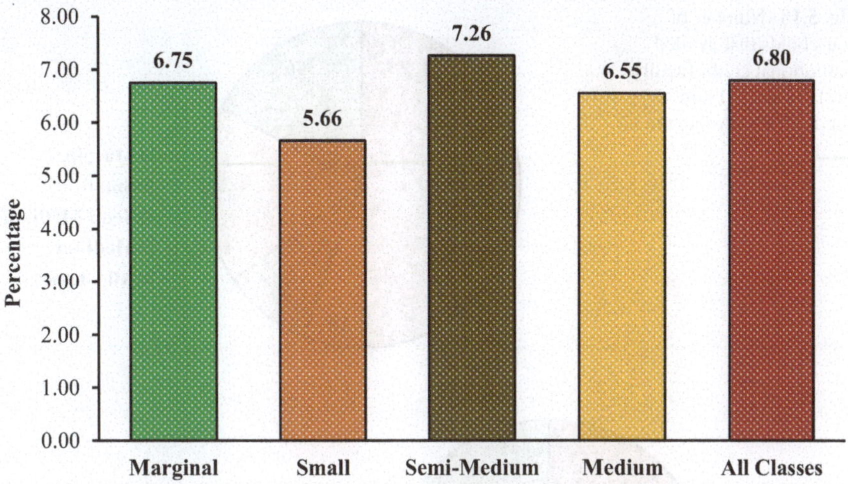

Fig. 5.16 Usage of hybrid seeds, 2021. (Source: Field Survey, 2021 (Annexure E.7))

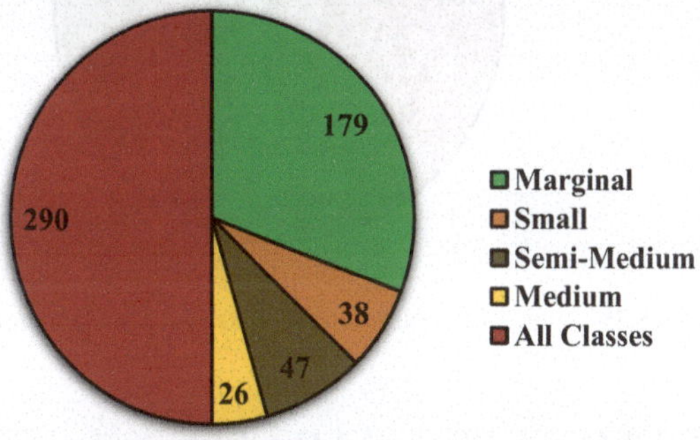

Fig. 5.17 Number of households earn >3000 rupees per month from agriculture, 2021. (Source: Field Survey, 2021 (Annexure E.7))

5.3.13 Monthly Income from Agriculture

Figure 5.17 represents the number of farmers having a monthly income of more than 3000 rupees from agriculture alone. Around 37% of the operational holders earn more than 3000 rupees in a month. It is evident that the semi-medium and medium landholders get a higher return from agriculture as they have a more extensive farm where they can practice intensive cum commercial agriculture. The main concern is with the marginal farmers, where more than 70% of operational holders do not manage to earn 3000 rupees or more. The fall in agricultural prices, loss of perishable products during transportation, volatile market, and natural calamities are the reasons for such low returns from agriculture.

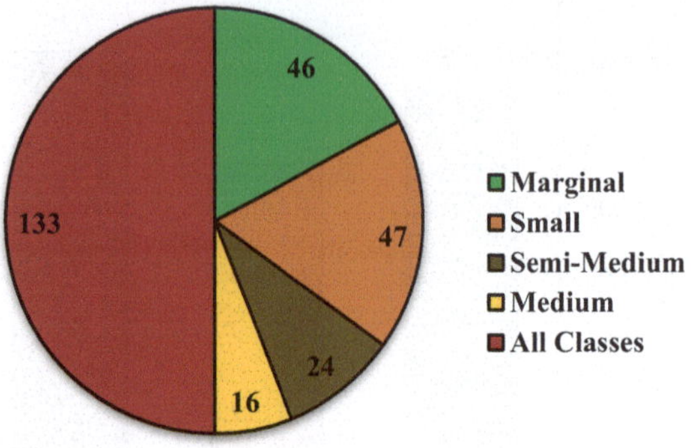

Fig. 5.18 Number of households that availed institutional credit facility, 2021. (Source: Field Survey, 2021 (Annexure E.7))

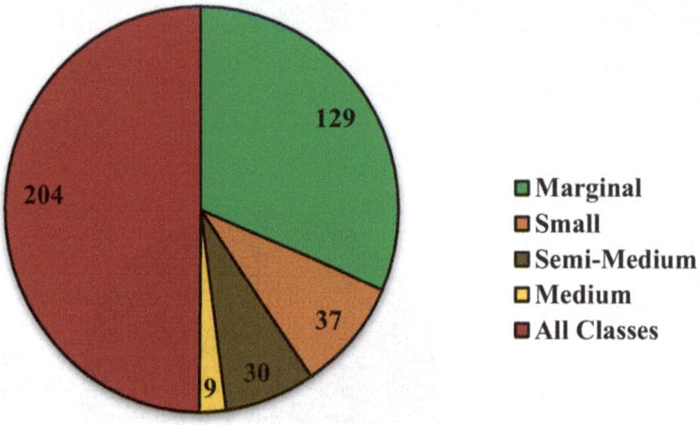

Fig. 5.19 Number of households that took credit from money lender, 2021. (Source: Field Survey, 2021 (Annexure E.7))

5.3.14 Institutional Credits

Out of the total households surveyed, 133 (17.05%) farmers availed institutional credits from different government banks and institutions. These credit facilities are mostly for a short period of time (less than 1 year) and demand the paper of the land owned by the farmers as a mortgage. The small and medium land categories report the maximum percentage of farmers to assist the institutional credit (Fig. 5.18).

5.3.15 Credit from Money Lender

Many operational holders borrow cash from shopkeepers or village money lenders for buying seeds, fertilizers, insecticides and pesticides, and different agricultural implements because of the limited institutional credits. Even though the money lenders charge higher interest, the farmers take loans and credit because of the easy access. Figure 5.19 demonstrates that more than 26% of the surveyed households

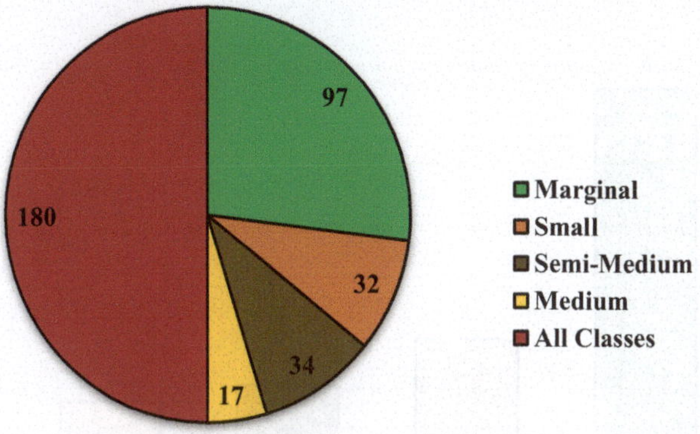

Fig. 5.20 Number of households with Kisan Credit Card, 2021. (Source: Field Survey, 2021 (Annexure E.7))

avail credit from money lenders, in which the share of semi-medium landholders (57.69%) is the highest.

5.3.16 Kishan Credit Card (KCC)

The Kishan Credit Card scheme was launched in 1998 by the Reserve Bank of India (RBI) in collaboration with the National Bank for Agriculture and Rural Development (NABARD) to provide quick, timely, and affordable credits to farmers. It is supplied by all commercial banks, regional rural banks (RRB), and state cooperative banks. About 23% of farmers receive credits from KCC (Fig. 5.20). The semi-medium and medium land size classes have higher farmers befitted through KCC (more than 60%) compared to marginal and small land categories.

5.3.17 Hand- and Animal-Operated Machinery

The use of hand- and animal-operated machinery prevails from the very beginning of agricultural practices. It is seen from Fig. 5.21 that the number of hand- and animal-operated machines is higher in marginal and small land classes compared to the semi-medium and medium land categories due to the smaller and fragmented parcel of land. In addition, marginal and small farmers cannot afford expensive agricultural implements; thus, they rely on traditional tools operated by hand and animal means. On average, three hand- and animal-operated machinery are used per hectare in the surveyed households.

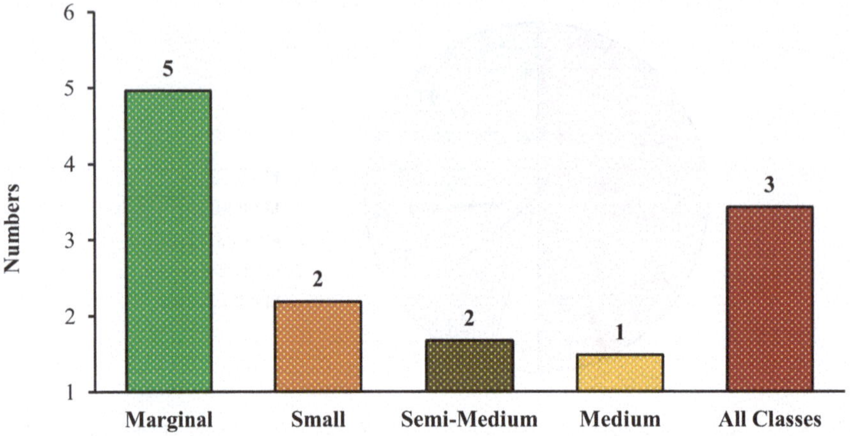

Fig. 5.21 Number of hand- and animal-operated machines per hectare GCA, 2021. (Source: Field Survey, 2021 (Annexure E.7))

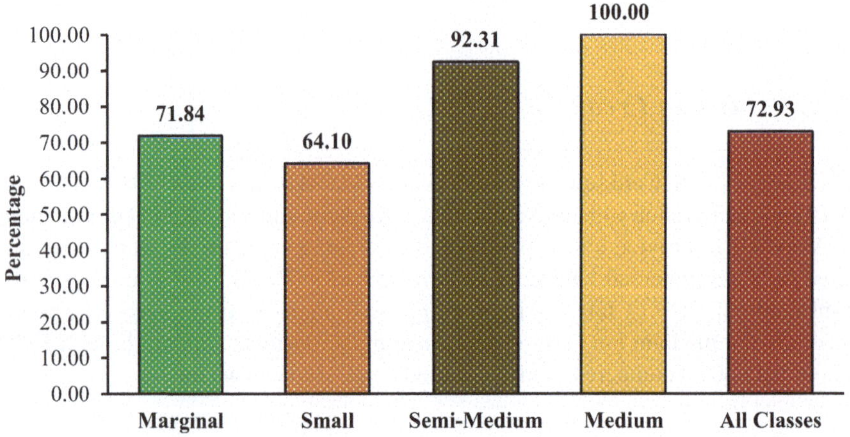

Fig. 5.22 Households use power-operated machineries, 2021. (Source: Field Survey, 2021 (Annexure E.7))

5.3.18 Power-Operated Machinery

All the households under the medium land category respond positively to the use of power-operated machines, including pump sets, power tillers, tractors, power threshers, crushers, etc. (Fig. 5.22). The access of power-operated machines in other land classes is 92.31% in semi-medium, 76.57% in all, 71.84% in marginal, and 64.10% in small land groups. It may be noted that power-operated machines are increasing due to the enhancement in electrification and income of the farmers. Power-operated machines provide ease in agricultural practices and improve the net return from agriculture.

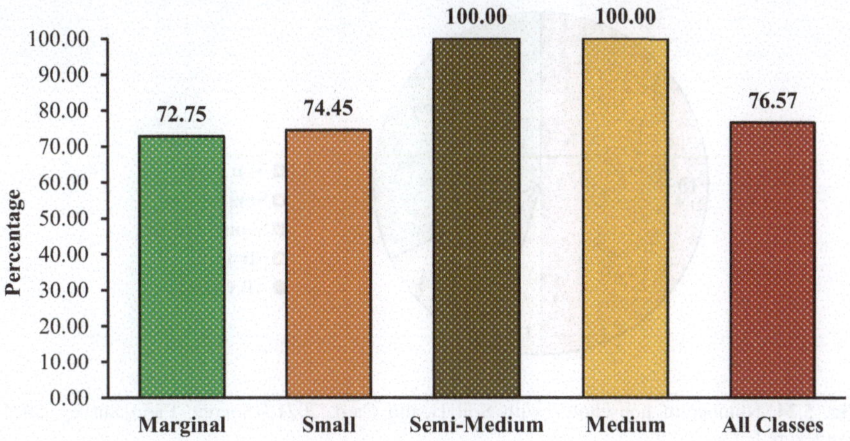

Fig. 5.23 Households use tractors and harvesters, 2021. (Source: Field Survey, 2021 (Annexure E.7))

5.3.19 *Tractor, Harvester, and Sprayer*

The use of tractors, harvesters, and sprayers in the agricultural field is 100% in semi-medium and medium land groups. However, the operational holders of marginal and small land classes represent a low share of modern machinery, as mechanized farming is not cost-efficient when the size of the parcel is too small (Fig. 5.23).

5.3.20 *Soil Health Card*

Soil Health Card (SHC) scheme is a new initiative launched in 2015 by the Government of India and managed by the Department of Agriculture and Co-operation. It provides information regarding the soil's nutrient status and then suggests recommendations on the dosage of key inputs to be utilized for improving its fertility and health. Figure 5.24 depicts that 13.46% of the farmers in the sample villages have Soil Health Cards. The medium land category (38.46%) reports the maximum SHC across the land classes, followed by semi-medium (25.00%), small (17.95%), and marginal (10.90%) in descending order.

5.3.21 *Insecticides and Pesticides*

The diseases of crops are higher in vegetable and fruit crop categories than in cereals and pulses. Therefore, those crops need a proper dose of insecticides and pesticides to tackle insects and pests for better productivity. The field survey displays

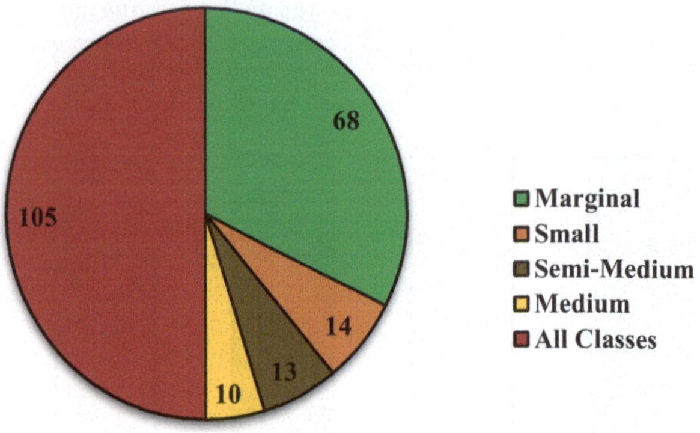

Fig. 5.24 Number of households with Soil Health Card, 2021. (Source: Field Survey, 2021 (Annexure E.7))

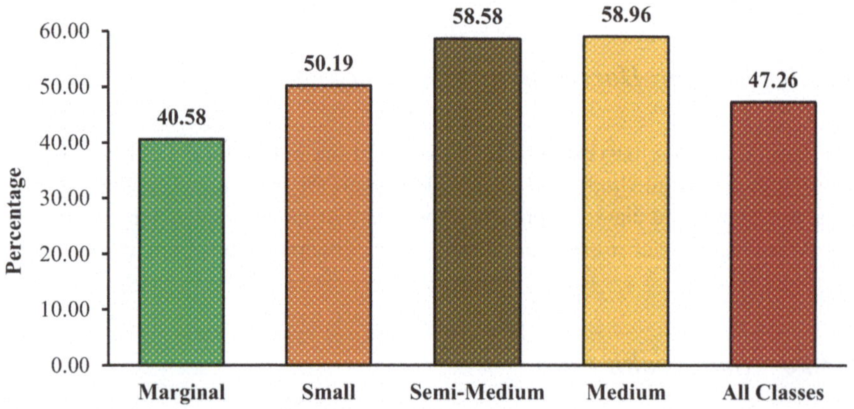

Fig. 5.25 Area treated with insecticides and pesticides, 2021. (Source: Field Survey, 2021 (Annexure E.7))

that less than 50% of the farmers use insecticides and pesticides on their farms to get relief from insects, pests, and different diseases. The medium and semi-medium land classes report more use of insecticides and pesticides than marginal and small land categories (Fig. 5.25).

5.3.22 *Chemical Fertilizers*

Chemical fertilizers are used to enhance plant growth and soil fertility, ultimately providing better crop productivity. There are mainly three types of fertilizers used by the farmers, i.e., nitrogen (N), phosphorus (P), and potassium (K). Figure 5.26 shows that the percentage of the gross cropped area treated with chemical fertilizers is more than 80% across the land size categories.

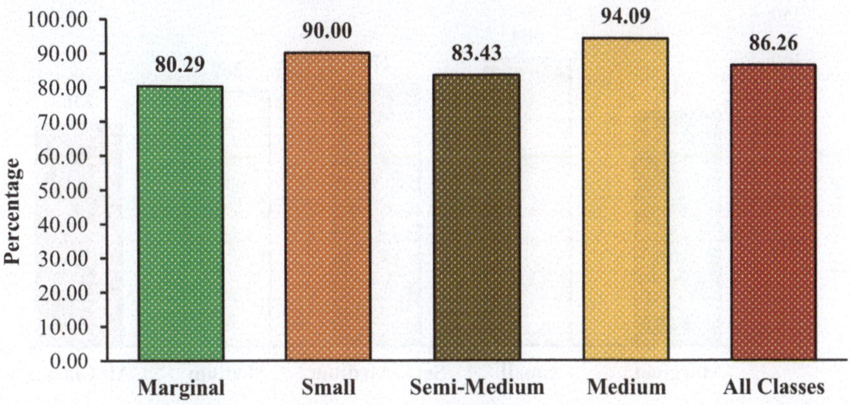

Fig. 5.26 Area treated with chemical fertilizers, 2021. (Source: Field Survey, 2021 (Annexure E.7))

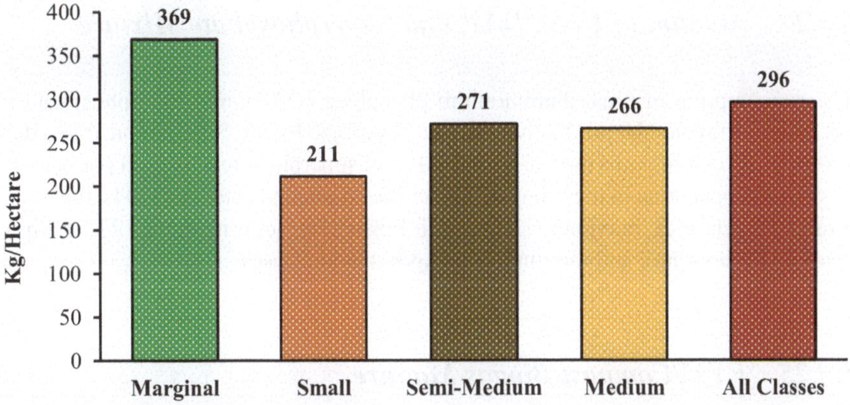

Fig. 5.27 Amount of NPK mixtures, 2021. (Source: Field Survey, 2021 (Annexure E.7))

5.3.23 *Amount of NPK Mixture*

In the surveyed households, the use of the NPK mixture is 296 kg per hectare. The use of NPK mixture is maximum in the marginal land classes (369 kg/ha), followed by semi-medium (271 kg/ha), medium (266 kg/ha), and small (211 kg/ha) landholdings. The use of fertilizers is high in marginal land type because of the intense cultivation of vegetable crops. (It is mentioned earlier that the vegetable and fruit crops are more disease-prone than the rest of the crop categories.) (Fig. 5.27).

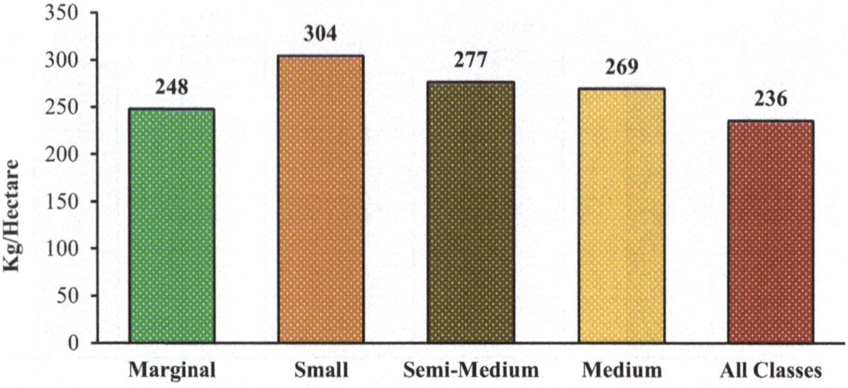

Fig. 5.28 Amount of urea, DAP, and superphosphate mixtures, 2021. (Source: Field Survey, 2021 (Annexure E.7))

5.3.24 Amount of Urea, DAP, and Superphosphate Mixture

The consumption of urea, diammonium phosphate (DAP), and superphosphate is essential for raising the productivity of various crops. Figure 5.28 demonstrates that a total of 236 kg of a mixture of urea, DAP, and superphosphate is used per hectare of gross cropped area in the villages. Under the small land category (304), this mixture per hectare is maximum. This is followed by semi-medium (277 kg/ha), medium (269 kg/ha), and marginal (248 kg/ha) land classes.

5.3.25 FYM/Compost/Biogas Manure

Organic manures are a natural form of fertilizers derived from dead plants and animals, used to enhance soil quality, and increase the yield and health of crops. Compost manures are prepared by mixing farm waste materials, animal wastes, vegetable wastes, sewage, and domestic wastes. These are a great source of humus, a natural form of fertilizers. The usage of compost in Murshidabad district is maximum under medium land class, i.e., 33.23% followed by semi-medium (20.59%), small (16.99%), and marginal (11.88%) land groups.

5.3.26 Oil Cake Manure

The remaining solid portion after the extraction of oil from the oilseed crops is dried as cakes which is used as manure. Both edible and non-edible oil cakes are used as manures, especially for horticulture crops. Only 2.74% of the GCA in Murshidabad

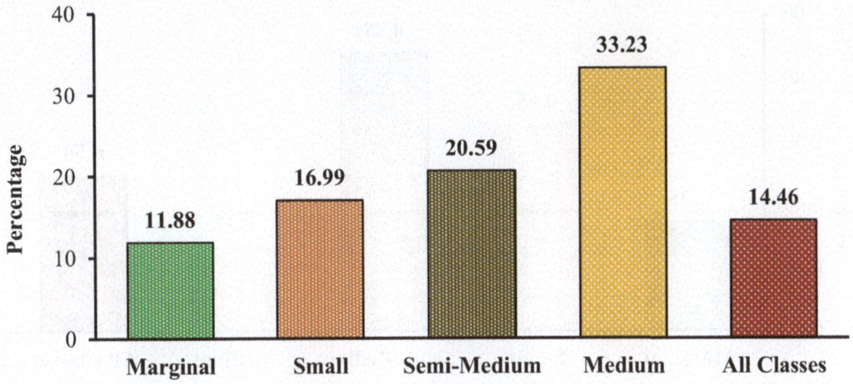

Fig. 5.29 Area treated with FYM/compost/biogas manure, 2021. (Source: Field Survey, 2021 (Annexure E.7))

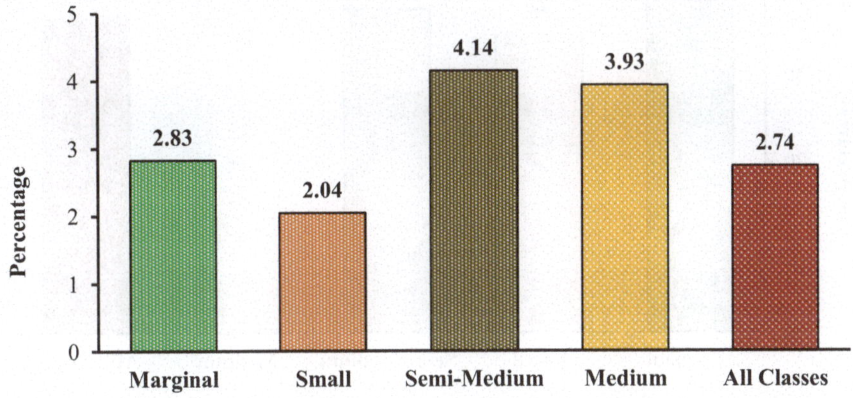

Fig. 5.30 Area treated with oil cake manure, 2021. (Source: Field Survey, 2021 (Annexure E.7))

district use oil cakes as an organic fertilizer. Semi-medium land type reports 4.14% usage of oil cakes, followed by medium (3.93%), marginal (2.83%), and small (2.04%) land classes (Figs. 5.29 and 5.30).

5.3.27 Green Manure

According to the Agriculture Census, 'Green manure refers to cultivation of a specific type of vegetation with the intention of ploughing it back in the soil when the leaves are tender and easily decomposable'. Sesbania (Dhencha), sunhemp (Sanai), indigo, urd, and cowpea (lobia) are some of the popular types of green manures. The area devoted for green manure in the district is less than a percentage across all the land size categories (Fig. 5.31).

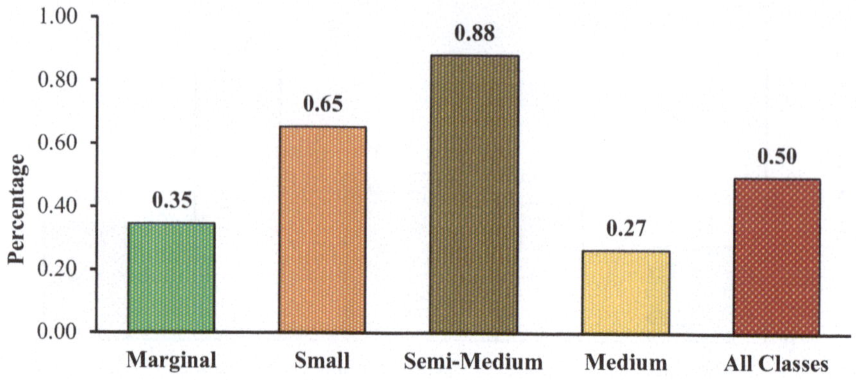

Fig. 5.31 Area treated with green manure, 2021. (Source: Field Survey, 2021 (Annexure E.7))

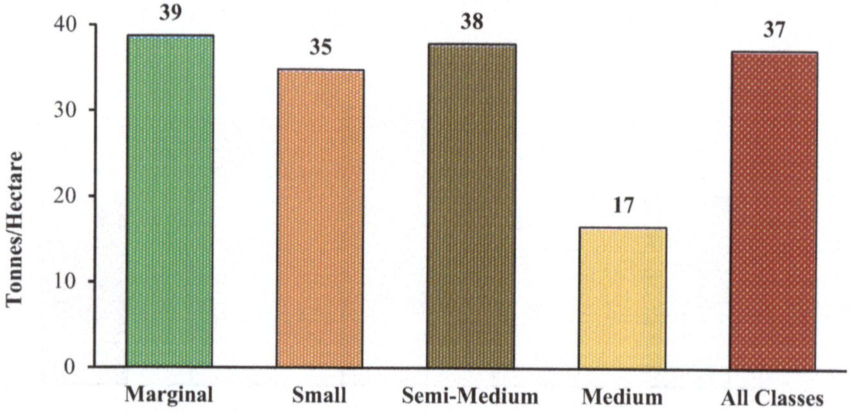

Fig. 5.32 Amount of organic manure, 2021. (Source: Field Survey, 2021 (Annexure E.7))

5.3.28 Amount of Organic Manures

Organic manures are used to replenish the nutrient loss from the soil as a result of cultivation to maintain the soil fertility. The usage of organic manures in the villages per hectare gross cropped area is illustrated here. On an average, 37 tonnes organic manures are used per hectare of GCA in the sample villages. Marginal and medium land classes record highest (39 tonnes) and lowest (17 tonnes) usage of organic manures, respectively (Fig. 5.32).

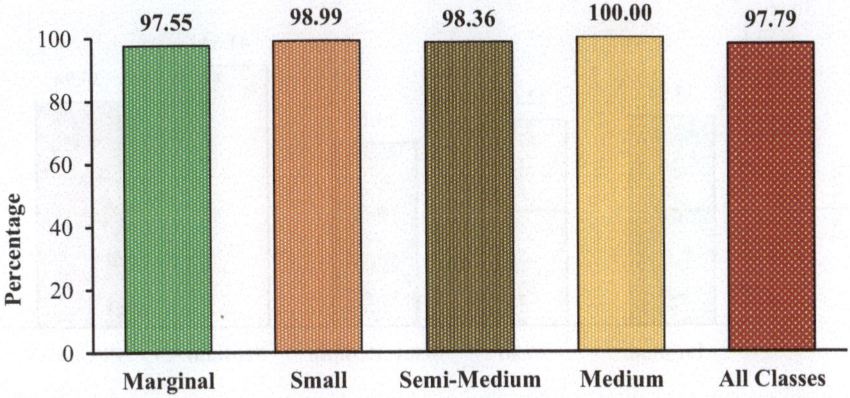

Fig. 5.33 Integrated pest management, 2021. (Source: Field Survey, 2021 (Annexure E.7))

5.3.29 Integrated Pest Management (IPM)

The practices adopted by farmers for the protection of their crops against insects and pests are labelled as integrated pest management. These practices of controlling insects and pests are categorized into four groups by Agriculture Census, viz., agronomic and cultural control, mechanical control, biological control, and chemical control. It is to be mentioned that only a few farmers were found during the field visit who had not used any of the abovementioned preventative methods for crop protection (Fig. 5.33).

5.3.30 Market Opportunities

Around 36% of the total product is sold directly to the market by the farmers. In the marginal and small land categories, the percentages of products directly sold in the market are reported as 33.34% and 32.66%, respectively. Various vegetable crops are mainly sold directly to the market under these two land classes as these crops are perishable in nature. The production sale in the market is highest in medium landholding (41.54%) due to the larger amount of production (Fig. 5.34).

5.3.31 Crop Sold Directly from Home and Field

Figure 5.35 reveals that the percentages of products sold directly from home and field in marginal, small, semi-medium, medium, and all land size classes are 66.66%, 67.34%, 70.81%, 58.46%, and 64.06%, respectively. Most farmers choose to sell produced crops directly from their fields or home due to the inaccessibility of

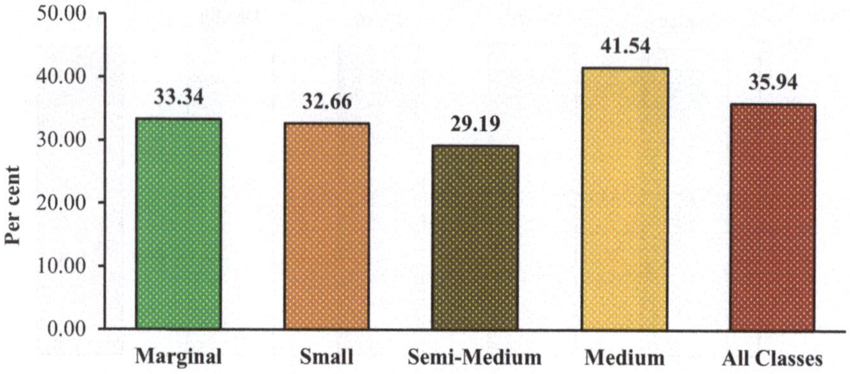

Fig. 5.34 Agro product directly sold in the market, 2021. (Source: Field Survey, 2021 (Annexure E.7))

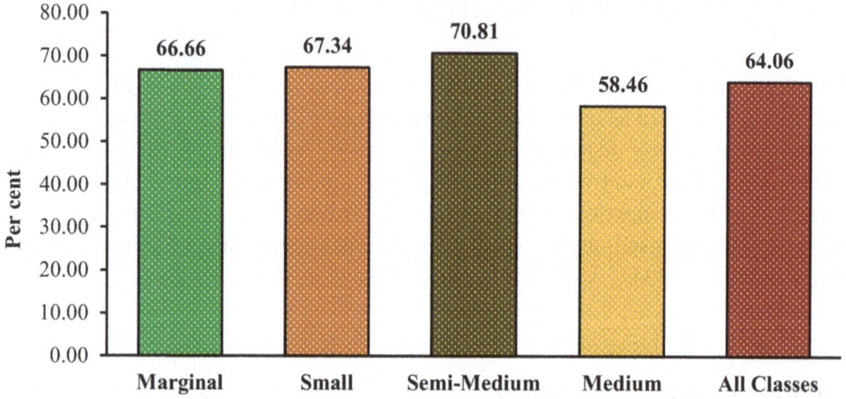

Fig. 5.35 Agro product directly sold from home and field, 2021. (Source: Field Survey, 2021 (Annexure E.7))

the market. Another reason for this is that the vegetables and fruits get spoiled quickly, so the farmers tend to sell those crops as soon as possible.

5.3.32 Cold Storage Facility

Cold storage is a space where crops can be stored for a longer time to fulfil the demands of different crops all over the year as well as to get a better price for those commodities. Proper cold storage facilities for crops, especially vegetables and fruits, help to minimize the loss and increase the farm income. It was found during the field survey that only 44 farmers out of 780 get benefitted from cold storage. Such a number is deficient; therefore, it is essential to increase cold storage facilities in the district.

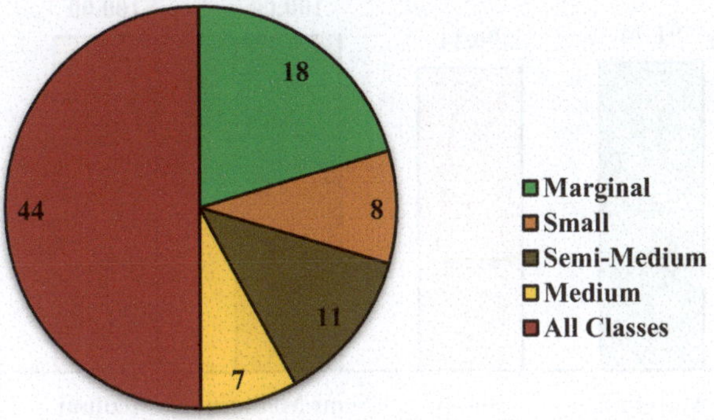

Fig. 5.36 Number of households that avail cold storage facility, 2021. (Source: Field Survey, 2021 (Annexure E.7))

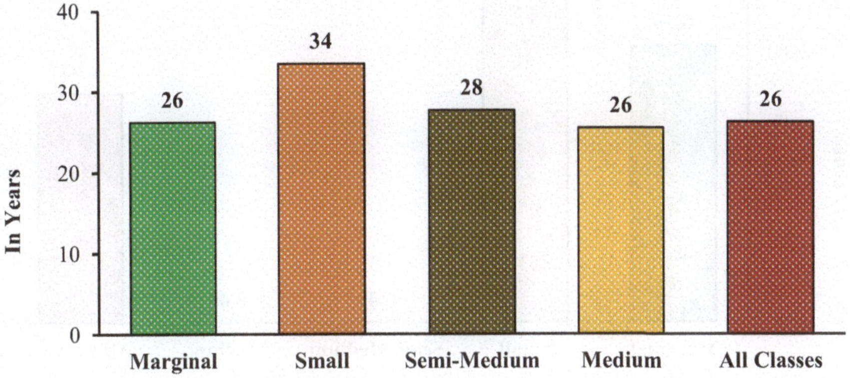

Fig. 5.37 Average farming experience, 2021. (Source: Field Survey, 2021 (Annexure E.7))

5.3.33 Farming Experience

The average farming experience in the sample village is 26 years. Further, the small land class reports a maximum farming experience of 34 years. This result shows that the farmers are mainly of old age. Farming experience is necessary for better cultivation and profitable agriculture (Figs. 5.36 and 5.37).

5.3.34 Main Cultivators

According to the Census of India, a farmer is called the main cultivator when he engages in agricultural activities for 6 months or more in a reference year. The semi-medium and medium land categories record 100% main cultivators in the sample villages. Due to their larger farm size, they involve themselves in agricultural

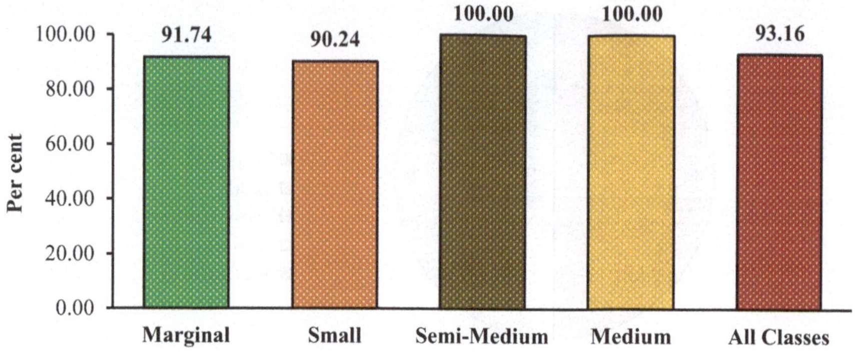

Fig. 5.38 Main cultivators, 2021. (Source: Field Survey, 2021 (Annexure E.7))

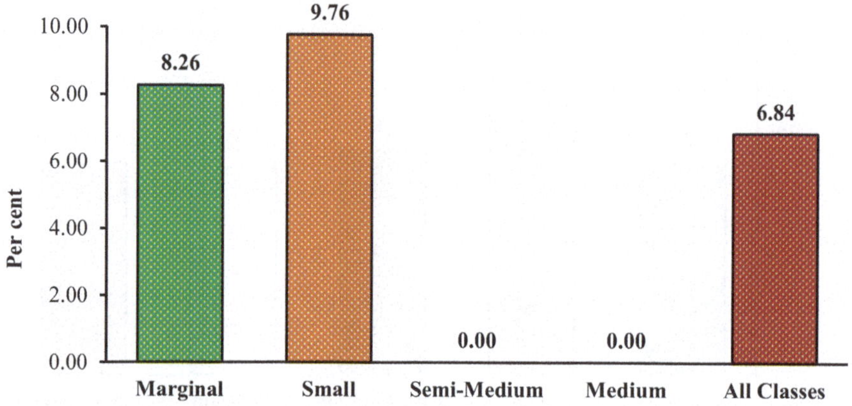

Fig. 5.39 Marginal cultivators, 2021. (Source: Field Survey, 2021 (Annexure E.7))

activities throughout the year and need to engage in any other activities. Conversely, the main cultivators are comparatively low in the marginal and small land classes (Fig. 5.38).

5.3.35 Marginal Cultivators

When a cultivator is involved in agricultural practices for less than 6 months in a reference year, he is called a marginal cultivator (Census of India n.d.). The fewer the marginal cultivators, the higher the agricultural performance of a region will be. Here in the sample villages, the small land category has recorded the maximum percentage of marginal farmers, i.e., 9.76% (Fig. 5.39).

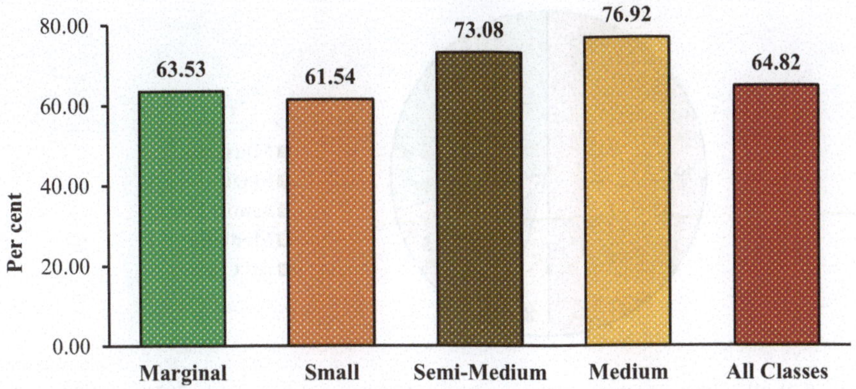

Fig. 5.40 Literacy rate, 2021. (Source: Field Survey, 2021 (Annexure E.7))

5.3.36 Literacy

Figure 5.40 shows that the rate of literacy is low in the marginal (63.53%) and small (61.54%) land classes compared to the semi-medium (73.08%) and medium (76.92) landholdings. Literacy is low among the smaller farmers due to the higher dropout rate of children who engage in child labour and migrate to the bigger cities in search of jobs. Literacy has a direct impact on agricultural activities.

5.3.37 Capacity-Building Program

Capacity-building may be defined as the process of developing and strengthening skills, abilities, instincts, strategies, and resources. Capacity-building program for the farmers helps to update knowledge regarding the best cultivation practices and agricultural technologies through training, techniques, and extension supports. Besides, it informs farmers how to respond to the changing agricultural environment by adopting agricultural innovation. It is found during the field survey that out of 780 respondents, 79 have ever attained any capacity-building program (Fig. 5.41).

5.3.38 Farmer Agitation

The dissatisfaction with governmental policies and programs, such as reduction in various governmental incentives, sudden price falling, reduction in minimum support price (MSP), uneven credit facilities, and negligence in proposed promises, are the leading causes of farmer agitation throughout the country. Figure 5.42 represents that about 14% of the total respondents have participated in farmer agitation

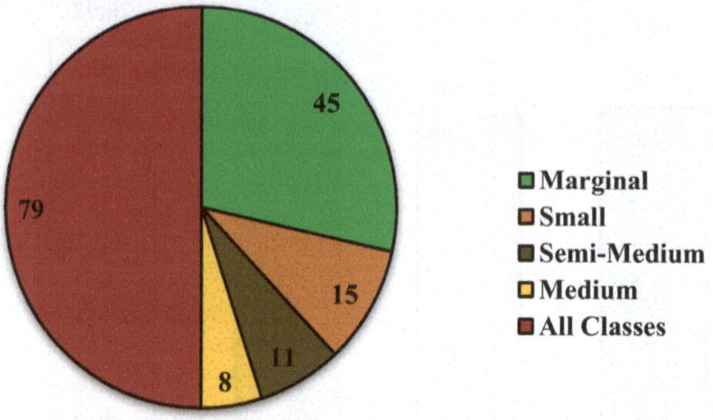

Fig. 5.41 Number of farmers attended capacity building programs ever, 2021. (Source: Field Survey, 2021 (Annexure E.7))

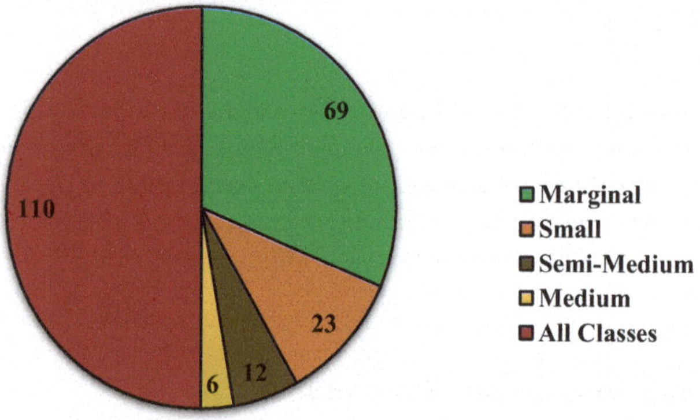

Fig. 5.42 Number of farmers participated in farming agitation, 2021. (Source: Field Survey, 2021 (Annexure E.7))

in their lifetime. Farmers of the small land categories (29.49%) record maximum engagement in farmers' agitation.

5.3.39 Average Size of Households

Figure 5.43 depicts the average family size of the farmers, where all the land classes report the average family size as six persons except for the marginal land category (five). Sometimes the higher number of family members can provide needed labour for agricultural activities, thus reducing the factor cost of production.

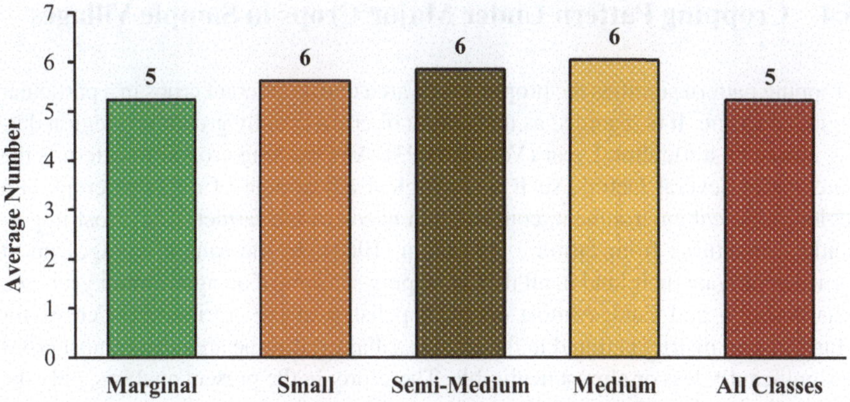

Fig. 5.43 Average size of households, 2021. (Source: Field Survey, 2021 (Annexure E.7))

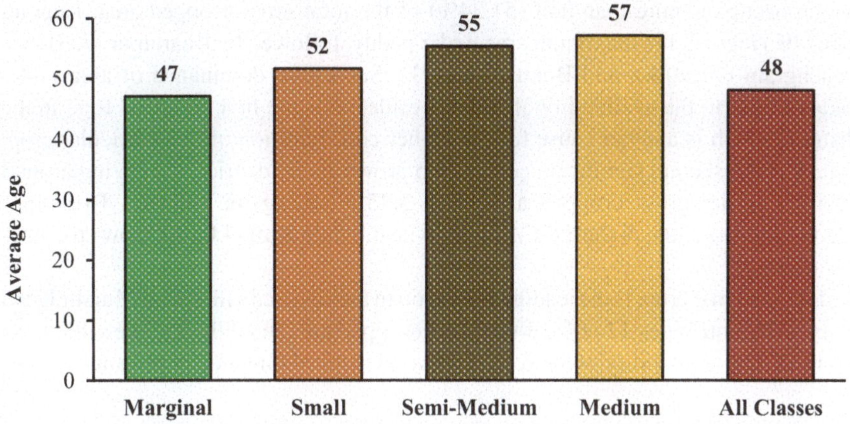

Fig. 5.44 Average age of respondents, 2021. (Source: Field Survey, 2021 (Annexure E.7))

5.3.40 Average Age

The average age of the selected respondents across land size classes is highest in the medium land group (57), followed by semi-medium (55), small (52), and marginal (47) land categories. The participation and engagement of young and adult farmers from the age of 15 to 40 years are essential as they lead to more progressive, productive, and innovative agricultural practices (Fig. 5.44).

5.4 Cropping Pattern Under Major Crops in Sample Villages

Cropping pattern signifies the proportion of area under different crops in a particular region at a time. It is regarded as the quality of crops usually grown in a region during a specific agricultural year (Verma, 1993). Any existing cropping pattern is the outcome of several factors, such as ecological adjustment of different crops in a region, physical environment, consumption needs of the farmers, and most importantly, net returns from farming (Oommen, 1963). In the sample villages, most farmers cultivate their land in all three cropping seasons of an agricultural year, i.e., Kharif, Rabi, and Zaid. Almost every crop listed in the Agriculture Census for Murshidabad district is found in the sample villages. But the area under most crops is significantly less or almost negligible. Therefore, in the present analysis, only the principal crops are discussed. These crops are paddy, wheat, jute, mango, potato, mustard, gram, masur, maize, and other crops.

Figure 5.45 reveals that paddy is the most dominant crop in the sample villages, which occupies more than half (51.34%) of the total gross cropped area. Kagram (76.74%) records the maximum area under paddy, followed by Bharatpur (72.03%), Panchgram (71.90%), and Bokhara (70.91). Such high dominance of area under paddy is due to the availability of water. Besides, it is the first preferred food in the district, which is another cause for the higher concentration of this particular crop. Wheat is the second significant cereal crop grown in the district, occupying around 12.10% of the gross cropped area (Fig. 5.45). Among the villages, Bishnupur (23.09%) reports the highest GCA under wheat, while Kuli (4.08%) shows the lowest area share.

Jute, a Kharif crop, is the leading fibre crop in the selected villages of Murshidabad district. It contributes 12.60% of the gross cropped area in all the sample villages as a whole. The availability of the required humid tropical climate, loamy alluvial soil,

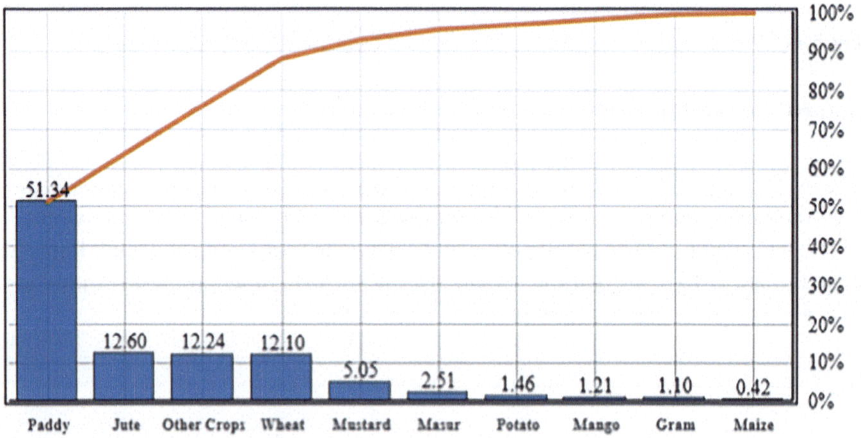

Fig. 5.45 Percentage of GCA under major crops in sample villages of Murshidabad district (2021). (Source: Field Survey, 2021 (Annexure E.8))

and abundant water is the prime reason for the higher concentration of jute in the sample villages. Across the villages, it is found that the maximum area under jute is recorded by Mahesail (32.67%) due to the feeder canal passing through this village. Mango contributes 1.21% of the GCA in the selected villages. The maximum concentration of gross cropped area is noted in Kuli (8.10%), followed by Hasenpur (2.09%), Bahutali (2.06%), and Bali (2.00%). Potato occupies 1.46% of the gross cropped area. The percentage of area share under this crop dominates in Jhikarhati (5.75%) and Bishnupur (2.85%). Conversely, Ramchandrabati (0.34%) denotes the lowest area concentration among the villages.

Mustard is the most important oilseed crop in the villages under consideration, contributing 5.05% of the gross cropped area. Mahesail (13.56%) leads in terms of share of gross cropped area in this particular crop, while Bali (2.00%) is at the bottom of the list. The concentration of area under gram, masur, and maize is 1.10%, 2.51%, and 0.42%, respectively (Fig. 5.45). 12.24% of the gross cropped area is engaged in other crop cultivation in the selected villages. Here, the highest percentage of area is located by Malancha (22.38%) and Bahara (20.78%), thus indicating a diversified cropping system. In contrast, Bokhara (5.69%) and Kagram (6.20%) are reported as the minimum percentage of area in other crops.

5.5 Relationship Between Sustainable Land Use and Agricultural Innovation

The multiple linear regression used in Chap. 4 has revealed a positive and significant relationship between sustainable land use and agricultural innovation in the district. Further, the unit-wise change (also used in Chap. 4) analysis shows that the agricultural inputs like irrigation, seeds, fertilizers and pesticides, and agricultural machinery have strongly related to sustainable land use under the different land size categories in the district. An attempt has been, therefore, made in this present section to scrutinize the spatial relationship between agricultural innovation and sustainable land use across land size classes in the sample villages.

5.5.1 Sustainable Land Use and Agricultural Innovation in Marginal Land Category

Figure 5.46 depicts the relationship between agricultural innovation and sustainable land use under this land size category in the sample villages. Six such villages exist where the composite standardized scores of both agricultural innovation and sustainable land use lie above the upper quartile (>0.516 and >0.497, respectively). These villages are Chhayghari, Mirzapur, Garaimari, Lalgola, Bali, and Bishnupur. Similarly, two villages, Ramchandrabati and Bahara indicate a composite index

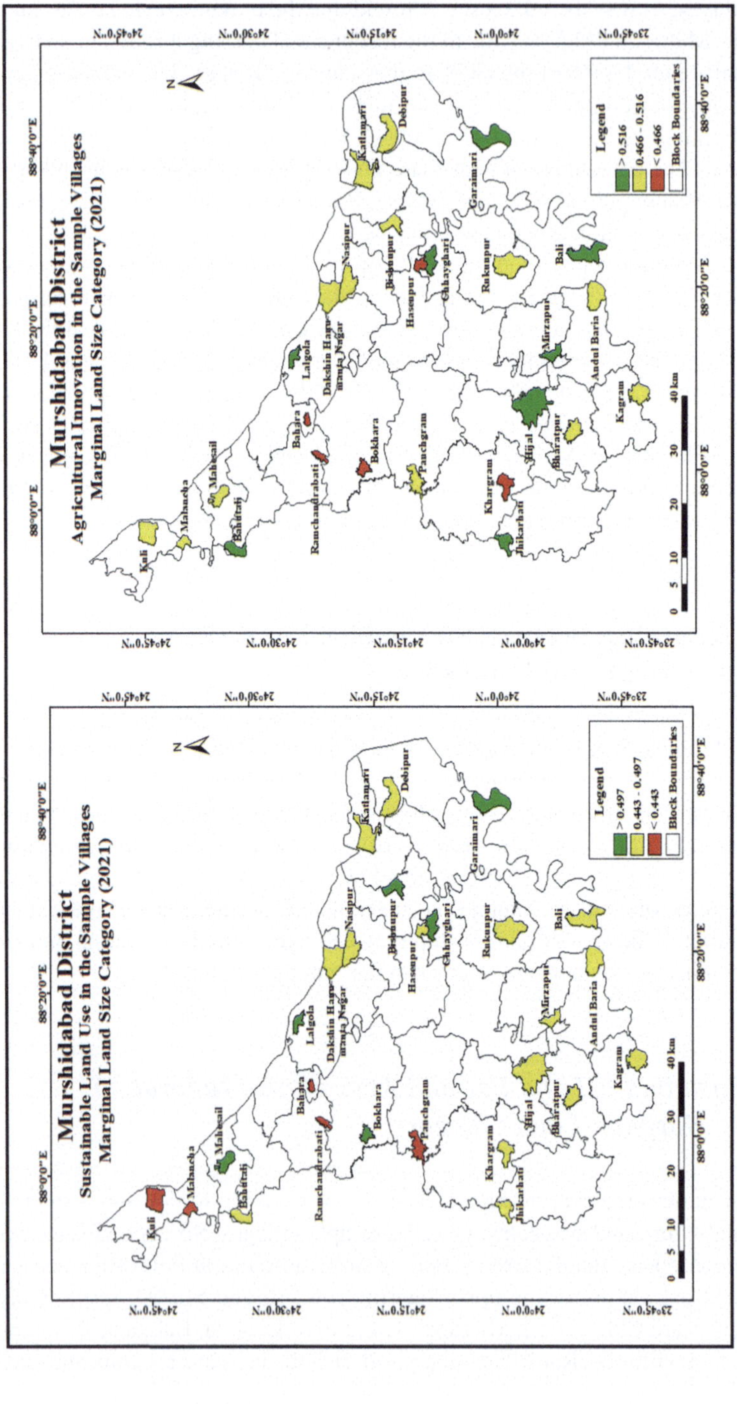

Fig. 5.46 Sustainable land use and agricultural innovation under marginal land size category in sample villages, 2021. (Source: Field Survey, 2021 (Annexure E.9))

value below the lower quartile for sustainable land use (<0.443) as well as agricultural innovation (<0.466). The medium range of sustainable land use and agricultural innovation (lying between the third and first quartile) are reported by nine villages, i.e., Andul Baria, Dakshin Hanumanta Nagar, Nasipur, Bharatpur, Kagram, Rukunpur, Debipur, Khargram, and Katlamari.

However, the Bokhara Village records high sustainable land use index but a lower agricultural innovation index. At the same time, Mahesail reveals medium sustainable land use with higher agricultural innovation. Jhikarhati, Hijal, and Bahutali are categorized in the higher agricultural innovation range but fall in the medium range in sustainable land use. Hasenpur represents higher agricultural innovation and low sustainable land use. Lower sustainable land use with a medium range of agricultural innovation is found in Kuli, Panchgram, and Malancha.

5.5.2 Sustainable Land Use and Agricultural Innovation in Small Land Category

Agricultural innovation and sustainable land use under small land class have been illustrated in Fig. 5.47. It reveals that the four villages, namely Mirzapur, Hijal, Bali, and Bokhara, come under the category of higher agricultural innovation and higher sustainable land use (above the upper quartile) at the same time. Conversely, the lower sustainable land use with lower agricultural innovation (below the first quartile) is found in four villages. These villages are Bharatpur, Kuli, Ramchandrabati, and Bahara. Another six villages, i.e., Dakshin Hanumanta Nagar, Garaimari, Rukunpur, Bishnupur, Katlamari, and Malancha, denote the same medium range category in the case of sustainable land use and agricultural innovation.

Chhayghari, Jhikarhati, Lalgola, and Mahesail are villages with a higher range of sustainable land use, but the agricultural innovation falls under the medium level. When the agricultural innovation is high, but the sustainable land use is in the medium range, such a scenario occurs in Kagram, Debipur, and Bahutali. A medium level of sustainable land use leads to a low range of agricultural innovation in Andul Baria, Nasipur, Khargram, and Hasenpur. Panchgram village indicates an integrated category of low sustainable land use and a medium range of agricultural innovation.

5.5.3 Sustainable Land Use and Agricultural Innovation in Semi-Medium Land Category

Figure 5.48 shows that the higher sustainable land use leads to a higher level of agricultural innovation in Bharatpur and Garaimari villages under this land category. Similarly, the lower level of sustainable land use guides the lower range of agricultural innovation in four villages: Andul Baria, Nasipur, Kuli, and Debipur.

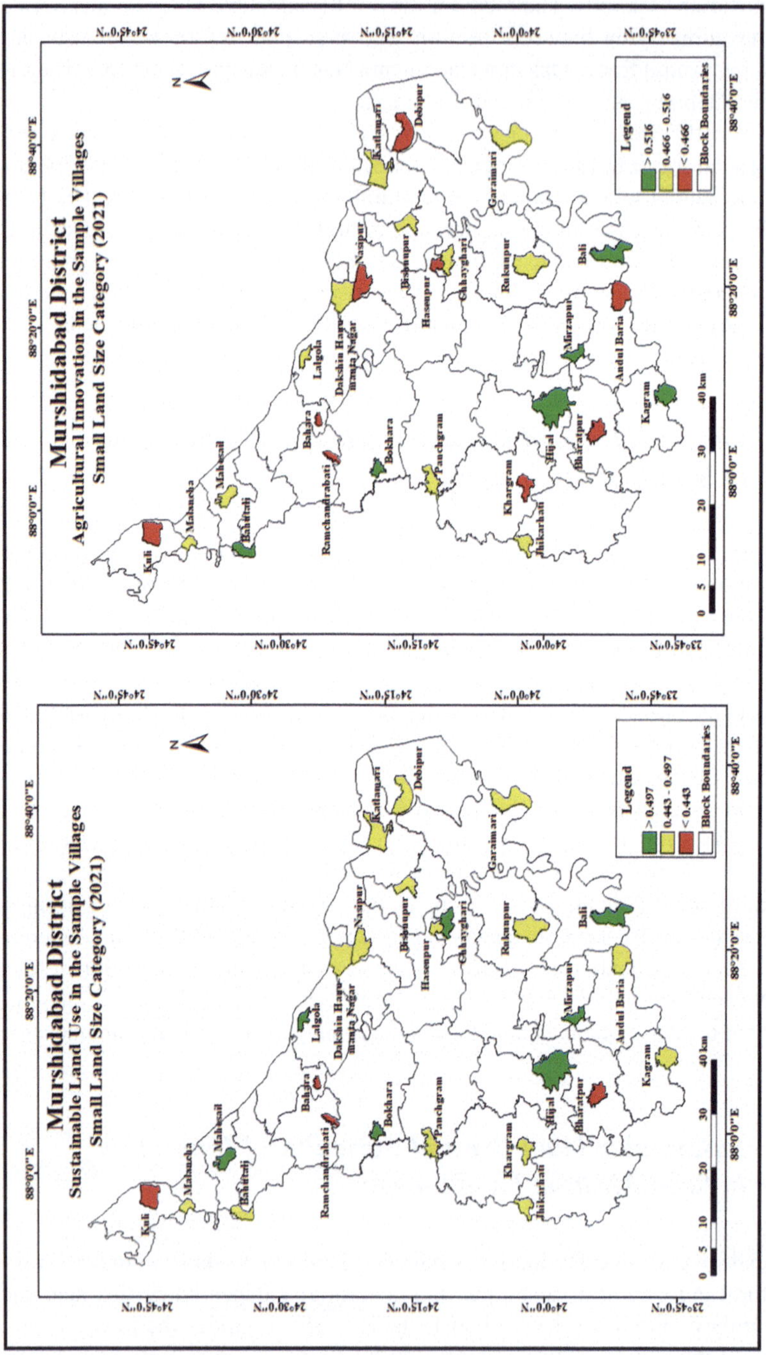

Fig. 5.47 Sustainable land use and agricultural innovation under small land size category in sample villages, 2021. (Source: Field Survey, 2021 (Annexure E.9))

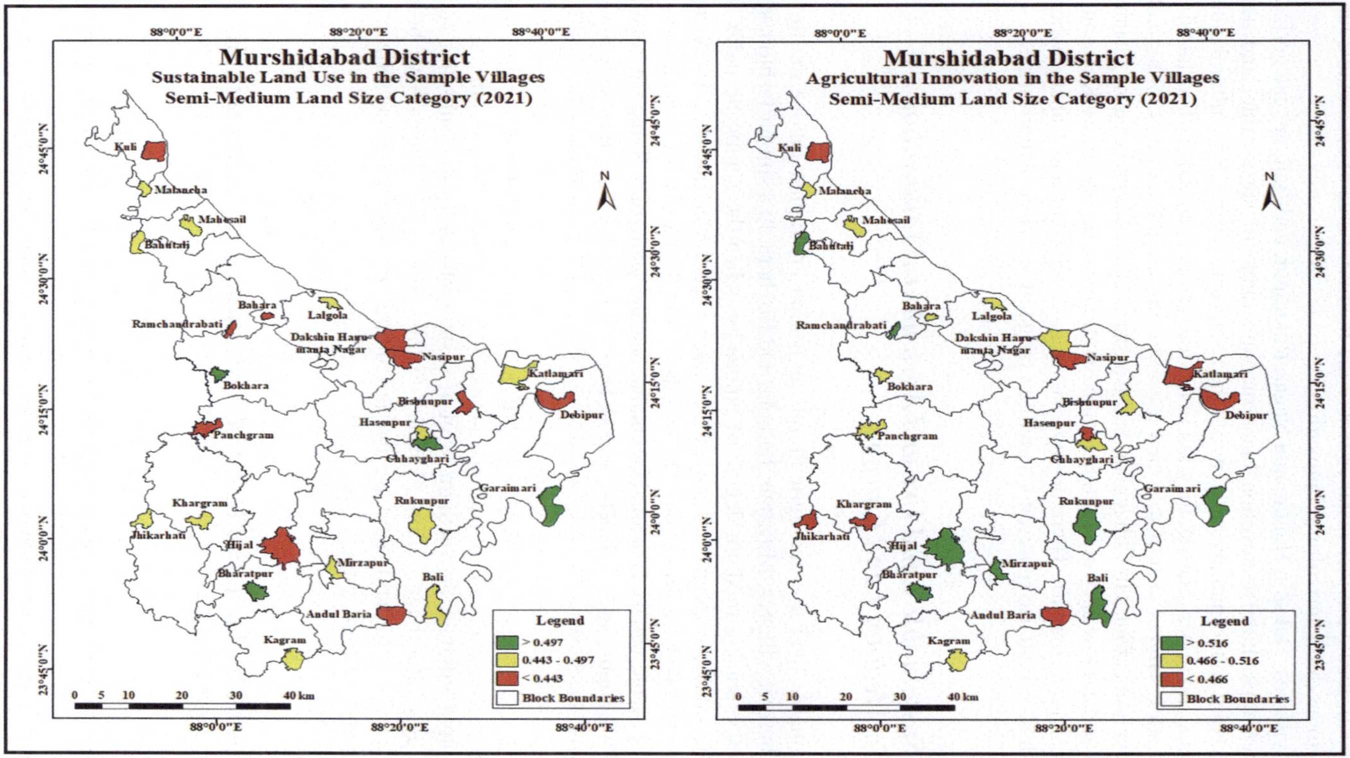

Fig. 5.48 Sustainable land use and agricultural innovation under semi-medium land size category in sample villages, 2021. (Source: Field Survey, 2021 (Annexure E.9))

Seven villages, i.e., Jhikarhati, Mirzapur, Kagram, Lalgola, Bishnupur, Malancha, and Mahesail, locate another instance where the medium range of sustainable land use leads to medium agricultural innovation.

Three villages, Dakshin Hanumanta Nagar, Rukunpur, and Bali, are categorized into higher agricultural innovation but a medium range of sustainable land use. Although Hijal and Ramchandrabati notify the composite score of agricultural innovation above the upper quartile, they fall in the low category in terms of sustainable land use index. Higher sustainable land use reporting a medium range of agricultural innovation is found in Chhayghari and Bokhara. The medium range of sustainable land use and low agricultural innovation parallelly are registered by Khargram, Hasenpur, and Katlamari. There are two villages, Panchgram and Bahara, where low sustainable land use represents a medium range of agriculture innovation.

5.5.4 Sustainable Land Use and Agricultural Innovation in Medium Land Category

The sustainable land use and agricultural innovation are higher (above the third quartile) in three villages: Mirzapur, Lalgola, and Bokhara. In a similar fashion, six villages experience a medium degree of innovation when the composite index of sustainable land use indicates between the upper and lower quartile. These are Jhikarhati, Andul Baria, Rukunpur, Debipur, Bali, and Katlamari. As many as seven villages, i.e., Chhayghari, Dakshin Hanumanta Nagar, Panchgram, Ramchandrabati, Bahara, Bishnupur, and Malancha, report lower sustainable land use with lower innovation (Fig. 5.49).

The high index of sustainable land use leads to the medium innovation of agriculture in the Garaimari and Mahesail villages. A medium index of sustainable land use and higher agricultural innovation is noticed in Kagram and Bahutali. In Nasipur and Hasenpur villages, the medium level of sustainable land use results in a low range of agricultural innovation. Contrary to this, four such villages have been found where the sustainable land use and agricultural innovation are categorized as low range and medium range, respectively.

5.5.5 Sustainable Land Use and Agricultural Innovation in All Land Category

The variation in sustainable agricultural land use and agricultural innovation among the villages under all land size category is displayed in Fig. 5.50. It is seen that three villages, namely Kuli, Ramchandrabati, and Bahara denote low sustainable land use as well as low agricultural innovation in the district. Again, Mirzapur and Lalgola are the two villages where higher sustainable land use leads to a high range of

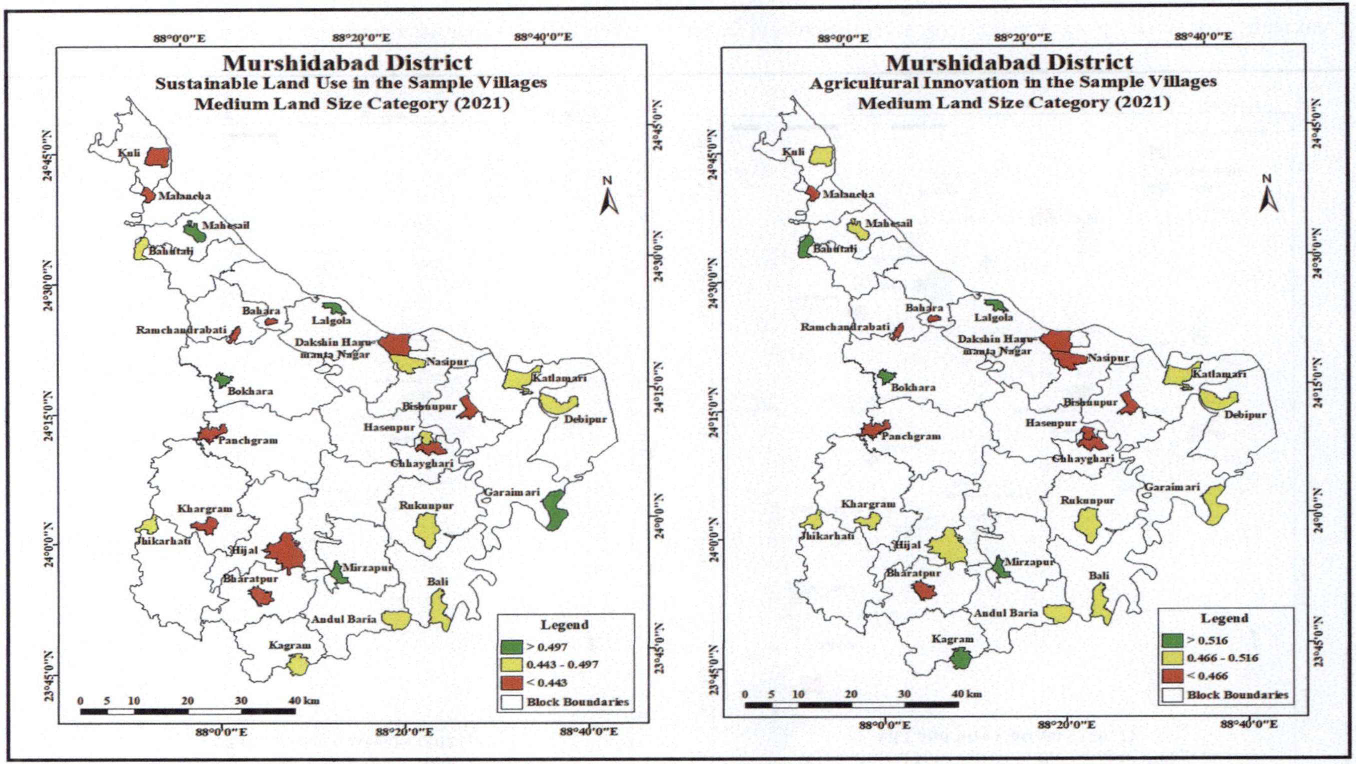

Fig. 5.49 Sustainable land use and agricultural innovation under medium land size category in sample villages, 2021. (Source: Field Survey, 2021 (Annexure E.9))

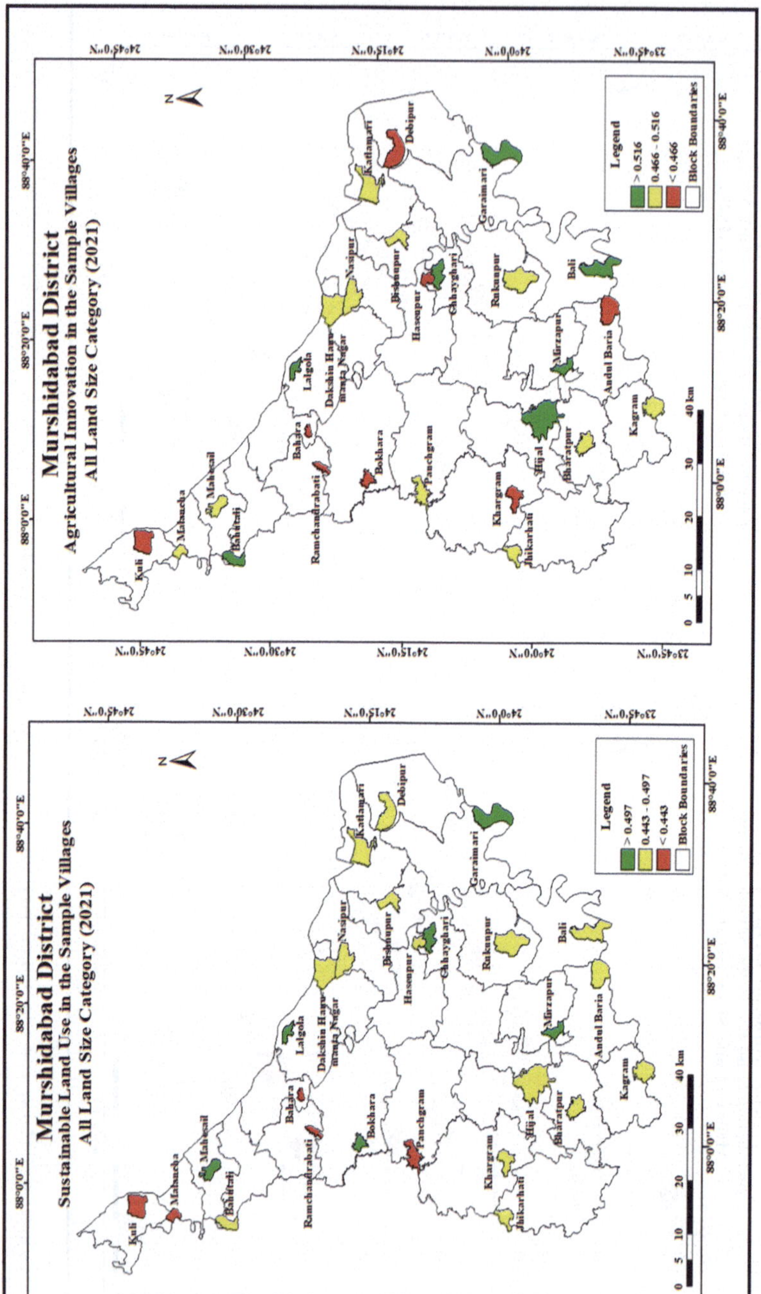

Fig. 5.50 Sustainable land use and agricultural innovation under all land size category in sample villages, 2021. (Source: Field Survey, 2021 (Annexure E.9))

agricultural innovation. The medium sustainable land use concludes in a medium degree of innovation in nine villages: Jhikarhati, Andul Baria, Dakshin Hanumanta Nagar, Nasipur, Bharatpur, Kagram, Rukunpur, Bishnupur, and Katlamari.

Chhayghari, Garaimari, and Mahesail villages have been categorized into higher sustainable land use and a medium range of agricultural innovation. Bokhara is the only village representing a low agricultural innovation despite having a higher sustainable land use. Conversely, Panchgram and Malancha record a lower sustainable land use index but fall in the medium innovation range. The higher agricultural innovation and medium sustainable land use index prevail in Hijal, Bali, and Bahutali villages. Debipur, Khargram, and Hasenpur are the remaining villages with medium sustainable land use and low innovation.

5.6 Principal Component Analysis (PCA)

Principal Component Analysis (PCA) is the most popular multivariate statistical technique used by almost all scientific disciplines. It is also the oldest multivariate technique, and its origin can be traced back to Karl Pearson in 1901. Hotelling (1933) further developed and formalized the modern instantiation of it. In fact, he also coined the term 'principal component' (Abdi & Williams, 2010). Its goal is to pull out the critical information from the datasets and to symbolize it as a set of new orthogonal variables denoted as principal components. It is a method to generate new sets of variables replacing large and small datasets. The strongly correlated variables are determined in the first step, and then the remaining less correlated variables are considered. These less correlated variables are called factors or components, hence the name factor analysis (Helmy & El-Taweel, 2009). The fundamental idea of PCA is to minimize the dimensionality of a large number of interrelated variables while preserving as much variability in the datasets as possible (Jolliffe, 2002).

The calculation of principal component analysis is done in Jamovi 2.3.2 software. In the present model, the mathematical equation follows linear functions chosen for orthogonal analysis later. The final principal components are composite of the original variables in which the first component gets the maximum variance. The second component is orthogonal to the first, derived from the maximum variance of all linear functions in given variables. The varimax rotation technique is used in the present model to get a rotated matrix to interpret factors better and derive reasonable conclusions. The equation applied here is

$$F^{*} = ZAD - \frac{1}{2}$$

where F^* is replaced by F, the standardized unit variance. Here, Z represents standardized equivalents of X variables. A is the matrix of unit-length eigenvectors R, which derives from the matrix ($X'X$) and becomes R as a correlation matrix. And D drives from a diagonal matrix of the eigenvalue of $X'X$.

There are a total of 40 variables of agricultural innovation categorized under eight indicators. These are further reduced into components with the help of principal component analysis. It is already proven in the previous chapter that sustainable land use and agricultural innovation are significantly causally correlated. Thus, it implies that the variables (or factors in the present context) that impact sustainable land use also influence agricultural innovation.

In the marginal land size category, the first three components together explain around 76.70% of the total variance of the selected eight indicators (Annexure). The first component explains 34.60% of the total variance in component loading. Around 23.90% and 18.20% of the total cumulative variance are denoted by the second and third components, respectively. Land, irrigation, and seeds are loaded positively under the first component, while in the second component, income and credits, soil and fertilizers, and market and storage facilities have been loaded positively. Similarly, socio-economic aspects found maximum loading in marginal landholding under the third component.

The first three components under the small land size class have described 65.40% in the variance with the sum square loading of 1.39. These three components, i.e., first, second, and third, explain 28.00%, 20.00%, and 17.40%, respectively. Under the first component, land, seeds, and income and credits loaded positively, which fetched the sum of the square of the component up to 2.24. The second component is found with positive loading of irrigation, income and credits, and soil and fertilizers. The third component reports a positive loading of income and credits, agricultural machinery, market and storage facilities, and socio-economic aspects.

Under three components, the semi-medium land size class category explains 63.10% of cumulative variance. In the first component, land and seeds are found to be loaded positively, and it gets 2.08 as a sum of the square. In the same category, the second component loaded 1.64 as a sum of the square of the component in which income and credits, agricultural machinery, and soil and fertilizers have been loaded positively, contributing 20.50% in total cumulative variance. The third component shows 16.60% in cumulative variance, where the sum of a square of component contributes 1.33, and the indicators like market and storage facilities and socio-economic are loaded positively in it (Annexure E.10).

Annexure E.10 reveals that the medium land size class category explains 70.10% total cumulative variance adding from the first three components. With 2.33 as a sum of a square of components loaded under the first component, land and seeds are loaded positively and explain 29.10% of the total cumulative variance. The second component denotes 22.20% in total variance, with 1.78 as a sum of a square of components contributed positively from income and credits, agricultural machinery, and soil and fertilizers. The third component in the medium land size class category explains 18.80% in total cumulative variance with positive loading from irrigation, market and storage facilities, and socio-economic aspects, which fetched a sum of squares as 1.51.

The first three components in combined explain 65.15% cumulative variance under all land size classes. 32.30% of the total cumulative variance with 2.59 sum of a square of components is explained in the first component. It loads land, irrigation, and seeds positively. The second component records positive loading of irrigation, soils and fertilizers, and market and storage facilities which explain 21.90% of the total cumulative variance (Annexure E.10). The sum of the square of the component from the second and third component loading is 1.75 and 1.43, respectively. With positive loading from irrigation and income and credits, the third component explains a 17.90% variance in total cumulative variance.

The cumulative variance of principal component analysis under different land size categories suggests that agricultural innovation and sustainable land use are best explained under the marginal and medium land classes.

References

Abdi, H., & Williams, L. J. (2010). Principal component analysis. *Wiley Interdisciplinary Reviews: Computational Statistics, 2*(4), 433–459.

Census of India. (n.d.). *Census of India*. Office of the Registrar General & Census Commissioner, Government of India, https://censusindia.gov.in/census.website/

Food and Agriculture Organisation. (2015). *Economic and social development department*. FAO.

Government of India. (2015). *All India report on agriculture census*. Department of Agriculture, Cooperation and Farmers Welfare. Government of India, 2015-16.

Helmy, A. K., & El-Taweel, G. S. (2009). Authentication scheme based on principal component analysis for satellite images. *International Journal of Signal Processing, Image Processing and Pattern Recognition, 2*(3), 1–10.

Hotelling, H. (1933). Analysis of a complex of statistical variables into principal components. *Journal of Educational Psychology, 24*(6), 417.

Jolliffe, I. T. (2002). *Principal component analysis for special types of data* (pp. 338–372). Springer.

Oommen, M. A. (1963). The economics of cropping pattern—a case study of Kerala. *Indian Journal Of Agricultural Economics, 18*(902–2016-67003), 120–132.

Pearson, K. (1901). Principal components analysis. *The London, Edinburgh, and Dublin Philosophical Magazine and Journal of Science, 6*(2), 559.

Verma, N. M. P. (1993). *Irrigation in India: Themes on development, planning, performance and management*. MD Publications.

Chapter 6
Contemporary Issues and Challenges with Sustainable Agricultural Innovation

Contents

In this chapter, an attempt has been made to assess the issues and challenges related to agricultural innovation with respect to existing policies, programs, schemes, and projects introduced by the government for the district. Assessment of existing problems and their solutions is necessary to boost and revitalize sustainable agricultural growth and innovation process of the district. Therefore, the very first section of this chapter starts with the analysis of strengths, weaknesses, opportunities, and threats (SWOT) in order to identify the key factors of agriculture in particular and the innovation process in general. A brief highlight of area-specific (village-level in this case) constraints faced by the farmers using a rating scale is discussed in the next section. In addition to the household responses, interactive discussions and in-depth

H. Rahaman, S. Ansari, *Sustainable Land Use Practices and Agricultural Innovations in India*, https://doi.org/10.1007/978-3-031-83246-8_6

interviews with farmers and Pradhan of respective villages are taken into considerations for the present analyses. Evidence-based policy intervention has been suggested in the last section of this chapter to sort out existing problems.

6.1 SWOT Analysis

The SWOT analysis, one of the most popular strategy tools, helps in framework development for researchers, academicians, business managers, planners, and policy analysts to identify, formulate, and prioritize the strategy for better possible solutions (Benzaghta et al., 2021; Leigh, 2009; Rahaman, 2021; Rozmi, 2018). The origin and consensus among the researchers and academicians regarding the SWOT analyses are complicated and unclear. Puyt et al. (2020) stated in their work entitled *Origin of SWOT Analysis* that the origin of SWOT begun in the early 1950s. They have also claimed that it is Robert Franklin Stewart who should be credited for the responsibility and direction of the research as well as the development of this framework. SWOT sets for strength, weakness, opportunity, and threat. Strengths are internal aspects of an organization that facilitates reaching its goals, while weaknesses are those internal aspects that consistently interfere with organizational success. Opportunities, the external elements that help an organization achieve its goals, are positive environmental aspects and opportunities to address gaps and initiate new activities. Conversely, threats are aspects of the organization's external environment that are barriers or potential barriers to reaching its goals (Aldehayyat & Anchor, 2008; Benzaghta et al., 2021; Fleisher & Bensoussan, 2003; Lee & Lin, 2008; Shrestha et al., 2004).

More than 96% of the farmers belonging to Murshidabad district are marginal and small landholders whose landholdings are below 2 ha (Agriculture Census, 2015–2016). The intensive low subsistence cum commercial agriculture is the only boost for their income and livelihood. Here, the SWOT analysis is used to identify the strategies for agricultural innovation. Identifying strengths and opportunities is necessary to minimize the risk of internal weaknesses and external threats to agriculture in the district. The analysis will help to identify and set a strategic plan to keep the agriculture sector competitive (Rahaman, 2020). This analysis considers social, economic, and environmental factors directly or indirectly linked to agriculture. Annexure B.1 contains the list of the different types of strengths, weaknesses, opportunities, and threats.

6.1.1 Strengths

Strengths are the positive internal factors of any system. The efficiency and effectiveness of any organization or economic portfolio largely depend on it. Therefore, the district planners and policymakers must focus on enhancing and burnishing its internal strengths in order to achieve the desired growth and innovation of the agricultural sector. In the district, the following specific strengths emerge from the field survey.

6.1.1.1 Intensity of Cropping Increased

The cropping intensity has increased almost twofolds from 1990 to 2020, which is a good sign for the overall agricultural innovation impact of a region. Adding to this, the productivity and yields of staple crops have also been raised. This clarifies the more and more engagement of farmers in cultivation process throughout the year.

6.1.1.2 Improvisation of Transport Facilities

Roads and railways are the two essential means of the transport network in the Murshidabad district. The means and medium of transportation networks have been improved over the last decade, especially under Pradhan Mantri Gram Sadak Yojana. The district is served by one national highway, NH12 (previously known as NH34), and three state highways, i.e., SH7, SH11, and SH11A, connecting different parts of the study area. Murshidabad district has a total of 12,172.24 km of road (surfaced and unsurfaced) with a road density of 2.29 km per sq. km (District Statistical Handbook, Murshidabad, 2016). These roads are maintained by Public Works Department (PWD), Zilla Parishad, Gram Panchayats and Panchayat Samiti, and Pradhan Mantri Gram Sadak Yojana. Besides, two main rail routes are found in the study area, running north to south on both banks of the Bhagirathi River. Farakka Bridge makes an essential link between Murshidabad and Malda districts through road and rail.

6.1.1.3 Relatively Low Labour Cost

The agricultural sector demands a huge labour workforce for various activities on and off field processes. Murshidabad district provides cheap and productive labour force. The total cultivators in the district account for 3.8 lakh and are assisted by 8.4 lakh agricultural labourers (Census of India, 2011). Hence, the Murshidabad district reports a cheap labour cost compared to neighbouring districts.

6.1.1.4 Availability of Fertile Soil

The district is situated at the alluvial track of the Ganga River, making the soil fertile and favourable for cultivation. It has already been discussed in the previous chapter that each and all sample villages are sufficient and fall into the excellent category regarding macro- and micro-soil nutrients required for the healthy growth of different crops. Moreover, more than 90% of soils are loamy and clay types suitable for agriculture. Therefore, available fertile soil is one of the most vital strengths in the district.

6.1.1.5 Good Climate for Agriculture

The climate of the district can be categorized as the tropical monsoon type. Mainly four seasons are found here: summer, autumn, winter, and spring. The average annual temperature does not exceed 30 °C in the district, which is ideal for the good health of crops. The rainfall frequently occurs due to the southwest monsoon from June to September. It is to be mentioned that paddy is the main crop grown in the district, which requires regular watering. Apart from this, oilseeds, spices, fibre crops, vegetables, horticultures, and floriculture are increasing in terms of areal and labour forces engagement due to variable climatic characteristics. The good climate and soil help the district to diversify crops towards high-value crops.

6.1.1.6 Increasing Agricultural Productivity

Agricultural productivity denotes the ratio of outputs to any or all associated agricultural inputs (Christensen, 1975). Fertile soil, farm mechanization, more and more area brought under irrigation, utilization of high-yielding varieties of seeds, and consumption of fertilizers, pesticides, and insecticides are the crucial factors leading towards the increase in agricultural productivity as a whole. Increasing productivity is a good symbol of agricultural innovation induced by sustainable land use. The district is also improving on high-value crop productivity in recent past.

6.1.1.7 Increase of Diversification in Cropping System

The sufficient water availability, good climate, fertile alluvial soil, and use of modern technology offer the potential scope for crop diversification at wider range from pulses and horticultural crops in the Murshidabad district. It has been noted in the third chapter that there is an increasing tendency of diversification towards fruit and vegetable crops in the district. Beyond cultivating multiple crops, diversification contributes to socioeconomic resilience (higher incomes, employment) and ecological benefits (resource management), aligning with sustainable land-use objectives.

6.1.1.8 Availability of Land for Agriculture

At present, the net sown area in the Murshidabad district is around 75% of the total reporting area. Therefore, the availability of land for cultivation is not an issue in the district. Instead, it becomes a strength for agricultural innovation. In the last two decades, the barren land, wasteland, and land not suitable for cultivation were brought into agricultural processes, and therefore, the total agricultural land has increased.

6.1.2 *Weaknesses*

Despite the vast strengths prevailing in the agricultural system of Murshidabad district, many obstacles are found which affect the development and innovation processes in the district. All types of deterrent factors of agricultural innovation are considered as weaknesses. Based on primary survey, the following eight major weaknesses have been identified and, as mentioned in Annexure F.1, are discussed below.

6.1.2.1 Huge Concentration of Small Landholdings

One of the severe issues in the district regarding agricultural activity is the small and fragmented landholdings. The concentration of marginal and small farmers is over 96%, with average landholdings below 2 ha (Agriculture Census, 2015–2016).

6.1.2.2 Lacking of Advanced Technology

Most of the cultivators in the district are still dependent on the traditional method of cultivation because of lack of awareness regarding the benefits of modern technology and the low literacy rate. Besides, another important factor for lacking advanced technology is the very small and fragmented landholdings, which are split into many parcels. (According to the Agriculture Census 2015–2016, the average number of parcels accounts for 3.79.)

6.1.2.3 Degradation of Soil Health

The farmers are using a heavy dose of chemical fertilizers, pesticides, and insecticides for the sake of maximizing production of different crops. This adversely affects soil quality in terms of nutrients and fertility, ultimately leading to soil health degradation. Instead of going for a soil test when the soil becomes incapable of producing higher production, the farmers apply more and more fertilizers and pesticides, worsening the soil's situation and health.

6.1.2.4 Low Return

One of the drawbacks of the farmers, especially the marginal and small farmers, is the relatively low return from agriculture. The irregularity in the price of products, poor marketing system, victimization by the middlemen or dalal, high cultivation cost, etc., are some of the major causes for such low returns in cultivation.

6.1.2.5 Lack of Farmer Awareness and Training Program

The lack of farmer awareness and training programs is also a weakness in the district. The awareness of the scientific method of cultivation and its benefit is very much lacking in the study area. Besides, they also lack training programs, capacity-building programs, and focus group discussions regarding selecting suitable crops, seeds, pesticides, and insecticides, applying the correct proportion of NPK mixtures, sowing and harvesting techniques, etc.

6.1.2.6 Limited Infrastructure and Marketing

The market infrastructure such as regulated markets is limited in two to three blocks, fair-price shops, public godowns, and cold stores. This is why farmers do not get the actual benefit of the minimum support price (MSP) fixed by the government every year. The limited market information regarding the selling of the crops leads to the vulnerability of price.

6.1.2.7 Limited Agro-Based Industries

Though the vast area is under cultivation in the Murshidabad district, the agro-based industries are still underdeveloped. Agro-based industries have the potential to provide job opportunities not only in farming but also in off-farm activities. Thus, this plays an essential role in the development of agriculture in the region.

6.1.2.8 Low Institutional Credit Facilities

The farmers need financial supports for various activities, such as ploughing fields, seeds, fertilizers, machinery, and irrigation. Still, there is a lack of adequate and timely credits from the financial institution, especially from government agencies. Moreover, the interest rate has also increased, which makes it difficult for marginal and small farmers to obtain loans for themselves.

6.1.3 *Opportunities*

There are abundant opportunities which can be deployed to achieve the desired growth and development in agriculture of Murshidabad district. Some of these opportunities are discussed here.

6.1.3.1 Organic Farming

Organic farming, one of the eco-friendly agricultures, is a system that avoids using chemical fertilizers, pesticides, insecticides, etc. Organic farming includes crop residues, green manures, animal dunks, and domestic kitchen wastes for the growth of crops. It is also an opportunity to maintain soil quality and agricultural productivity. This is why the government has raised special attention to organic farming in the district.

6.1.3.2 Agro-Based Industries

The establishment of agro-based industries is itself an opportunity to cope with the market and sell related problems of the farmers. These industries not only encourage the farmers to produce maximum but also ensure employment opportunities.

6.1.3.3 Export Facilities

The proper export facility is and always be an opportunity for the growth and development of agriculture in the district because it secures the farmers from the loss of surplus production. As we know, the district is improving in crop diversification which requires proper supply chain process to sell horticultural and vegetable crops to other states and countries.

6.1.3.4 Shifting to Commercial Agriculture

The agriculture of Murshidabad district is mainly subsistence type because most farmers are marginal and small landholders. These farmers produce crops only for consumption by the family. The advance in modernization and technology helps to increase productivity a lot. Therefore, the commercialization of agriculture is also an opportunity to push agriculture towards development.

6.1.3.5 Involvement of Female Participation

The engagement of female participation in the district is very low. They are mainly involved in household activities. Hence, the involvement of female participation can be promoted for the sake of the sustainable innovation of agriculture.

6.1.3.6 Farm Mechanization

Mechanization can be defined as the process of using modern machinery to increase efficiency in agriculture. The mechanized farm ensures an increase in work productivity, improves the quality of products, and reduces crop losses. Thus, the improvement of farm mechanization will impact agriculture in the district positively.

6.1.3.7 Intensification of High-Value Crops

Murshidabad district is still dominated by traditional cereal crops. The paddy crop alone covers more than half of the total gross cropped area. Therefore, there is ample opportunity to expand and intensify the gross cropped area towards high-value crops, such as fruits, vegetables, spices, oilseeds, etc.

6.1.3.8 Introduction of More HYV Seeds

During the field survey, it was revealed that most farmers use seeds which are self-produce. Some do not prefer to buy hybrid seeds, while others have no knowledge. Therefore, government agencies should involve in this matter and encourage and make them aware about the scope of hybrid seeds.

6.1.4 Threats

Threats are external factors that have the potential to cause serious harm and severe damage to the existing agricultural system. It is to be mentioned that threats are not driving problems presently but have the potential to affect adversely in the near future. The recurrent threats faced by the farmers in the district are noted below.

6.1.4.1 Flood and Soil Erosion

Floods are regular and common phenomena in the Murshidabad district, damaging crops. Conversely, soil erosion is another serious threat to the agriculture practice in the district. Excessive doses of chemical fertilizers, unscientific methods of cultivation, and natural causes (landslides) result in soil fertility loss and soil erosion.

6.1.4.2 Increasing Crop Diseases

Crops are getting affected by various diseases seasonally. These diseases are serious threats to crops and hinder the growth and production of different crops. The district has seen frequent attacks of certain insects and pests in various crops, especially vegetables, in the past few years. Unscientific cultivation methods are the leading cause of an increase in several crop diseases.

6.1.4.3 Overdose of Chemical Fertilizers

Overdose of chemical fertilizers, especially among the marginal and small landholders, is growing up as a key concern for the sustainability of soil health. The farmers increase the use without predicting the adverse effect of the chemical fertilizers on soil health.

6.1.4.4 Increasing Fertilizer Cost

The continuous hike of fertilizer price is another threat to the farmers, mainly marginal and small farmers. This scenario ultimately leads to the fall in net return to the cultivators from agricultural activities.

6.1.4.5 Variability of Monsoon Rain

Recently, the monsoon rainfall has become more unpredictable in nature. The vast mass of farmers in the district is still depended on rain for irrigation. Therefore, delays in monsoons become a serious threat. Conversely, uneven distribution of rainfall in the Kharif season resulted in huge crops lost in the district.

6.1.4.6 Degradation of Soil Fertility

Loss of soil fertility is a damaging threat that needs to be concerned as soon as possible. The old-age cultivation method and application of chemical fertilizers in massive amounts are the prominent reasons that degrade the soil fertility in the district. Besides, the cultivation of the same crop continuously in the same field also leads to the loss of soil fertility.

6.1.4.7 Hailstorms and Kalbaishakhi

The hailstorms and Kalbaishakhi, known as nor'westers or short-lived thunderstorms, damage the existing cropping patterns almost every summer (during April and May), resulting in colossal crop loss, especially cereals (paddy), spices, fruits (mango, litchi), vegetables, and fibres (jute) crops. The nature of losses varies as the intensity of hailstorms fluctuates every year.

6.1.4.8 Sudden Price Falls of Agro-Products

Fluctuation in the market and price fall severely affect the cropping system in the district. Farmers often fear the instability and volatility of agro-products prices, as they have experienced many times in the past.

6.2 SWOT Mapping

An attempt has been made here to list out the total number of strengths, weaknesses, opportunities, and threats based on extensive discussions with different land size class farmers in the district. The open-ended SWOT questionnaires yielded an uneven distribution of responses: strengths (22), weaknesses (22), opportunities (21), and threats (20). Each item of SWOT has been listed in alphabetical order in Annexure F.2. Identifying the extent to which it builds on strengths and minimizes weaknesses is a crucial area involving strategies concerning its competitor. The opportunities and threats are influenced by the external environment, and therefore, focusing on the proper evolution of strategic intention is dependable on other aspects.

6.2.1 Strengths Mapping in Murshidabad District

Figure 6.1 reveals that the availability of fertile soil is the number one strength in nine villages, namely Chhayghari, Mirzapur, Nasipur, Rukunpur, Debipur, Bishnupur, Malancha, Bahutali, and Mahesail due to their location on Ganga Plain. In six villages, namely Dakshin Hanumanta Nagar, Kagram, Lalgola, Hasenpur, Panchgram, and Bokhara, crop productivity increase emerged as the best strength. Increasing agricultural productivity stood as the number one strength in Jhikarhati, Andul Baria, Garaimari, and Katlamari. Besides, available organic waste, easy access to groundwater, fodder cultivation, and high-value crop farming are also appeared as the number one strength in some villages.

Conservation of rainwater appears as a strength only in the two villages of Kuli and Bahara. The demand-driven farming appears in five villages, i.e., Kagram, Hijal, Lalgola, Bahara, and Bokhara. Similarly, a good number of dealers is also placed as a strength in five villages. Transportation has significantly improved during the past decades in the district, and as a result, it appears in six villages as an important strength. These villages are Andul Baria, Hasenpur, Ramchandrabati, Katlamari, Malancha, and Mahesail.

Murshidabad District

Strengths Mapping in the Sample Villages (2021)

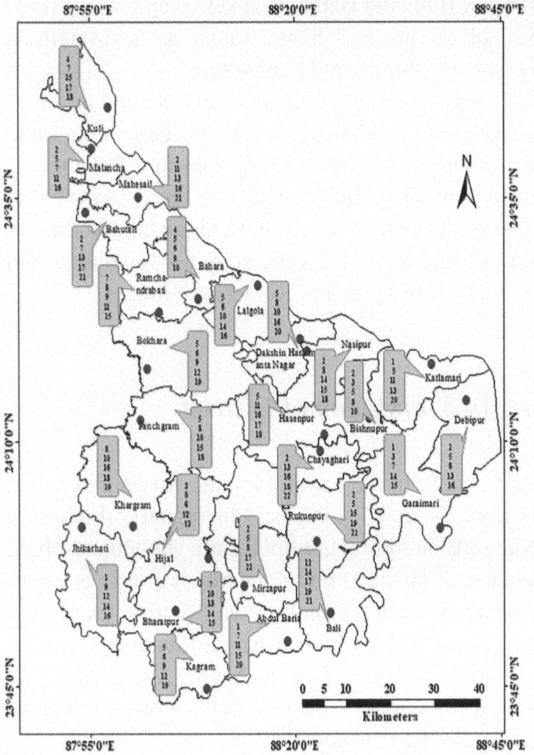

S.No	Strengths
1	Agricultural productivity increased
2	Availability of fertile soil
3	Available organic waste
4	Conservation of rainwater
5	Crop productivity increased
6	Demand driven farming
7	Easy access to ground water
8	Fodder cultivation
9	Good number dealers
10	Good number of cattle
11	Good road network
12	High-value crops farming
13	Intensity of cropping increased
14	Outsourcing increasing
15	Plenty of labour
16	Presence of market
17	Presence of water bodies
18	Self-help group (SHG)
19	Small agro-industry
20	Use of organic manure
21	Water retention is high
22	Women participation

Fig. 6.1 Strengths mapping in the sample villages (2021). (Source: Based on Annexure F.3)

It is to be noted that high-value crop farming is found as a strength only in five villages, namely Jhikarhati, Kagram, Hijal, Bali, and Bokhara. The district is lacking in cultivating high-value crops due to the dominance of cereal crops in the gross cropped area. The cropping intensity, one of the critical aspects of agricultural output, has increased significantly. Due to this reason, it appears as a strength in seven villages, i.e., Chhayghari, Bharatpur, Debipur, Hijal, Katlamari, Bahutali, and Mahesail.

Because of the migration of young working labour forces and the opening up of small and medium shops in the district, the outsourcing of the farmers from Jhikarhati, Nasipur, Bharatpur, Garaimari, Lalgola, and Bali has been increasing continuously. Plenty and cheap labour are an advantage in eight villages: Andul Baria, Nasipur, Bharatpur, Garaimari, Kuli, Rukunpur, Panchgram, and Ramchandrabati. The presence of weekly and monthly markets is another strength in Chhayghari, Jhikarhati, Dakshin Hanumanta Nagar, Debipur, Khargram, Lalgola, Hasenpur, Malancha, and Mahesail.

The district receives a good amount of rainfall in the monsoon season, due to which different ponds and shallow wetlands are got filled. Therefore, it identifies as a strength in Mirzapur, Kuli, Hasenpur, Bali, and Bahutali. A self-help group (SHG) is actively working as a receiver of credits and loans, found as a strength in Chhayghari, Nasipur, Kuli, Khargram, Hasenpur, and Panchgram.

Small agro-based industries such as rice and jute mill are found in good numbers in Kagram, Rukunpur, Khargram, Bali, and Bokhara. The use of organic manure to retain soil fertility is considered a strength in Andul Baria, Dakshin Hanumanta Nagar, and Katlamari. High water retention capability helping in the growth of certain crops such as paddy, jute, and mesta, is sufficient in Bali, Bahutali, and Mahesail. The participation of women in agricultural activities, especially for vegetables and spice cultivation, is high in Chhayghari, Mirzapur, and Rukunpur villages.

6.2.2 Weaknesses Mapping in Murshidabad District

The most prominent weakness that occurs in the sample villages of Murshidabad district is land fragmentation. It appears in 12 villages: Chhayghari, Jhikarhati, Mirzapur, Dakshin Hanumanta Nagar, Bharatpur, Hasenpur, Bali, Ramchandrabati, Bahara, Bokhara, Bahutali, and Mahesail. The land fragmentation is a serious weakness as it hinders farmers from using modern technology and getting maximum profit from small land. The farmers believe that the high cost of cultivation is another critical weakness of agricultural innovation. Those farmers having less than 0.5 ha of landholdings are the most prone to this noted weakness. This stops them from cultivation and gives them a lower return than expected.

Farmers require access to affordable credit for essential agricultural operations including sowing, plowing, harvesting, machinery purchase, irrigation, and fertilizers. However, timely low-interest loans remain largely inaccessible through government institutional channels. That is why many villages imply it as one of the weaknesses of agricultural innovation. A heavy dose of chemical fertilizers and pesticides has a short-term benefit but degrades the soil quality in the long run. Besides, it also adds up the cultivation cost. The marginal and small farmers have failed to adopt new technological innovations due to their poor socio-economical background and small parcels of land size.

There is a lack of agro-based industries in the district, which can boost up the agricultural economy of the district. These industries can provide ample job opportunities to the working-age population engaged in agriculture and thus reduce the family-dependent ratio. The proper and ensured goods and cold storages are the prime requirements of farmers, especially those cultivating fruit and vegetable crops. But it has been indicated by the farmers that very few storage centres are actually there in the different blocks. Another weakness reported by the farmers is the low yield of various crops compared to the neighbouring districts. All these factors together ultimately resulted in low returns from the agriculture sectors.

The uncontrolled exploitation of groundwater is a serious problem in the district. As a result, villages like Chhayghari, Rukunpur, Hijal, Hasenpur, Bali, and

Katlamari reported lowering water tables as a weakness. The engagement of child labour (below the age of 14) in agricultural activities is very common in the district. More than 70% of child labour is attributed to the agricultural sector. Therefore, it is also regarded as a weakness of the agricultural innovation system. The continuous lower return from agriculture discourages the farmers from the cultivation process. That is why the lack of interest in new farming is considered a weakness in five villages.

Lack of convenient transport facilities, especially for the final product, is another problem of farmers, as responded by Dakshin Hanumanta Nagar, Debipur, Hijal, and Bishnupur. Recently, the government has withdrawn incentives provided for farmers' benefits. The unscientific cultivation method often reduces the agricultural output and thus arises as a weakness in Andul Baria, Malancha, and Mahesail.

Traditional crop farming is a growing concern, as reported by Chhayghari and Bahutali villages. Farmers do not quickly shift themselves to commercial or high-value crops; instead, they cultivate traditional cereal crops over the years. They are also slowing in adopting new cultivation methods. The workforce migration from the productive age group hinders the agricultural growth of any region (Fig. 6.2).

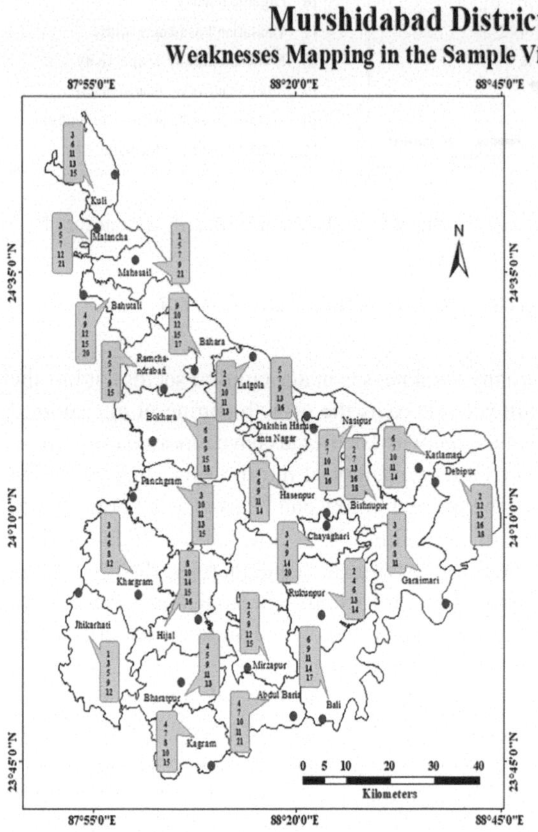

S.No	Weaknesses
1	Deteriorating soil quality
2	Engagement of Child labour
3	Heavy doses of pesticides
4	High cost of cultivation
5	Lack of new technology
6	Lack of industries
7	Lack of goods and cold storage
8	Lack in interest to new farming
9	Land fragmentation
10	Large number of small landholdings
11	Low institutional credit facilities
12	Low returned
13	Low yield
14	Lowering water Tables
15	Overdoses of fertilizer
16	Poor transport
17	Rain fed agriculture
18	Reduction of government incentives
19	Slow adaptation of new change
20	Traditional crop farming
21	Unscientific method of cultivation
22	Migration of workforce

Fig. 6.2 Weaknesses mapping in the sample villages (2021). (Source: Based on Annexure F.3)

Murshidabad District
Opportunities Mapping in the Sample Villages (2021)

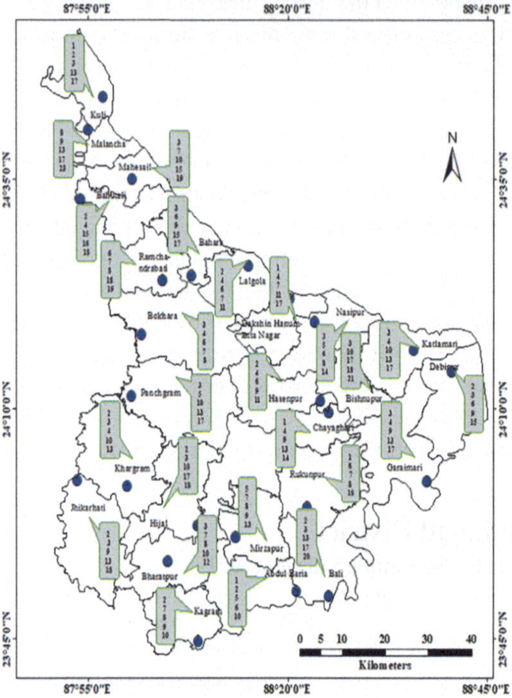

S.No	Opportunities
1	Agro-based industries
2	Bringing fallow land to the cultivation
3	Crop diversification
4	Demand based production
5	Efficient irrigation systems
6	Export oriented cultivation
7	Extension of fodder cultivation
8	Extension of jute factory
9	Fair and regulated market system
10	Increase productivity
11	Intensification of high value crop
12	Intensive and mixed cultivation
13	Introduction of more HYV seeds
14	Involvements of NGOs
15	Nearness to urban market
16	Organic farming
17	Plantation based agriculture
18	Road and railway connectivity
19	Soil test nutrition management
20	Vertical diversification of agriculture
21	Wetland based cultivation

Fig. 6.3 Opportunities mapping in the sample villages (2021). (Source: Based on Annexure F.3)

6.2.3 Opportunities Mapping in Murshidabad District

Though the district has recorded many weaknesses in agricultural sectors, it has the potential of multi-facet opportunities to serve as the growth engine of agricultural innovation. Figure 6.3 displays that as many as 16 villages have reported crop diversification as a vital opportunity for better agricultural return in the district. Along with this, farmers also believe that the unused fallow can be used as an opportunity to uplift the development process of agriculture. The certified and high-yielding varieties (HYV) of seeds are essential for faster growth and increase the production of different crops. Therefore, Chhayghari, Jhikarhati, Mirzapur, Garaimari, Kuli, Khargram, Panchgram, Bali, Katlamari, and Malancha suggest introducing more and more HYV seeds to the sustainable agricultural system.

The farmers need to cultivate crops according to the demand in the market. This will help them earn more profit and reduce the risk of loss of finished products. The farmers also need to focus on export-oriented crops as an option to boost up sustainable agricultural innovation reported by Andul Baria, Nasipur, Rukunpur, Debipur, Lalgola, Hasenpur, Ramchandrabati, Bahara, and Bokhara. A fair and regulated market can also be seen as an opportunity. The government needs to boost awareness regarding the various issues and challenges the farmers face. Hence, it will help increase productivity, which is an option for the betterment of agriculture. One of the options is plantation-based agriculture to cope up with soil erosion and fertility.

The opportunity also lies in the extension of fodder cultivation as agreed by Mirzapur, Bharatpur, Kagram, Rukunpur, Lalgola, Ramchandrabati Bokhara, and Mahesail. The fodder cultivation, along with livestock farming, will definitely lift up the economy. The jute cultivation is regular and prominent in the district. But the number of jute factories is falling down day by day. Therefore, the promotion and extension of jute industries are essential to boost the income of the farmers. It is seen as an option to Mirzapur, Nasipur, Bharatpur, Kagram, Rukunpur, Ramchandrabati, Bokhara, and Malancha.

The district receives a good amount of rainfall in the monsoon season. But for the rest of the year, irrigation gets limited. It also lacks different sources of irrigation, such as sprinkles and water tanks, which can help to reduce water wastage. Therefore, an efficient irrigation system could be an opportunity by introducing low water-wastage techniques. The good road and railway connectivity can also be an opportunity for the farmers because it helps them save time and cost through fast and cheap transportation.

The farmers of Debipur, Bahara, Bahutali, and Mahesail urge that the nearer the cultivated land to the market, the more they benefit from it. The nearness to urban market is a fruitful option for farmers, especially those engaged in fruits and vegetables cultivation. Similarly, the cultivation of high-value crops brings higher income and promotes a demand-based production system for export. The involvement of NGOs can sort out many problems and act as a helpful organization for the farmers, as suggested by Chhayghari and Nasipur villages.

Soil test based nutrition management can help to reduce unwanted and excess usage of chemical fertilizers noted as an option by Ramchandrabati and Mahesail villages. The intensive and mixed cultivation will undoubtedly be an option to boost up the district's agriculture. The organic farming replacing chemical fertilizers and pesticides not only helps to retain soil fertility but is also an eco-friendly option. The multiple wetlands, including small lakes, ponds, and shallow deeps, can be used as an option for pisciculture and wetland-based farming in the district. The horizontal expansion of agriculture is almost impossible due to increased population pressure. Therefore, the only way to boost up production is through vertical diversification of agriculture. It can be achieved through multiple cropping, shifting towards more remunerative and commercial crops, adopting advanced machinery and innovation, modern technology, etc.

6.2.4 Threats Mapping in Murshidabad District

Several problems threaten the agriculture of the district. During the survey, as many as 20 threats have been noted. The most problematic threat in the district is dependent on rainfall-based farming. Twelve villages report it as the number one threat.

These are Jhikarhati, Mirzapur, Nasipur, Bharatpur, Rukunpur, Debipur, Hijal, Lalgola, Panchgram, Bali, Katlamari, and Mahesail. The fluctuation of minimum supporting price of certain crops in the local and national markets emerges as severe threat in the district. The inconsistent rainfall and its variability affect agriculture negatively and result in huge crop loss, as reported by farmers in the district.

The soil gets degraded by excessive use of chemical fertilizers, unscientific method of cultivation, repetitive cultivation of the same crops, etc. Therefore, the degradation of soil fertility is a serious threat to the agricultural system, as considered by Jhikarhati, Andul Baria, Nasipur, Kuli, Rukunpur, Hijal, Khargram, Panchgram, Malancha, and Bahutali villages. Several unwanted natural calamities, such as hailstorms and Kalbaishakhi, threaten the agriculture of the district. The withdrawal of the adult working-age population from the agriculture sector is a critical threat to the overall growth and development processes. The adult groups are the most productive and hardworking people who always try to maximize profit. But there is a trend of migration of this group to the other state in search of job opportunities. As a result, it becomes a threat to the agriculture system indicated by the farmers.

Sometimes the prices of certain crops fall down suddenly as the demand fluctuates. This happens mainly in fruit and vegetable crops. Therefore, the farmers experience a massive loss on these crops as these crops are the most perishable due to the lack of storage facilities. Jhikarhati, Dakshin Hanumanta Nagar, Debipur, Hijal, Bali, Bahara, Malancha, Bahutali, and Mahesail villages have pointed out the sudden fall in crop prices as a threatening problem to the farmers. The inorganic way of farming causes a serious threat too. Due to the climate change phenomenon, the temperature is increasing day by day as it felt by the farmers. It adversely affects the growth of plants.

The farmers also realize that excessive use of fertilizers can be dangerous on long-term for the soil health. It destroys the various macro- and micronutrients of the soil. Therefore, soil erosion becomes another threat to agriculture in the district, as reported by Mirzapur, Andul Baria, Kagram, Hijal, Lalgola, Bali, and Bokhara villages. The occurrence of frequent floods is a problem that causes losses of crops and agricultural land. The land encroachment, mainly of those productive agro lands, for the purpose of settlements and industries, is a common threat all over the district. Adding to this, the insecurity of land is also a drawback for agriculture, especially for those villages situated near the international border of Bangladesh (Fig. 6.4).

One of the threats is crop diseases which affect the overall production system. It also degrades the quality of the crops and thus results in low income. The reduction of wetlands is a threat in the district, which needs to be paid careful attention by the government. The water shortage is a major concern, especially for vegetable and pulse crops. It is to be mentioned that the monsoon has become irregular during the past few years, which disburses unexpected rain in the early summer, causing damage to fruit and vegetable crops. Farmers face persistent problems from waterlogged fields resulting from inefficient drainage infrastructure.

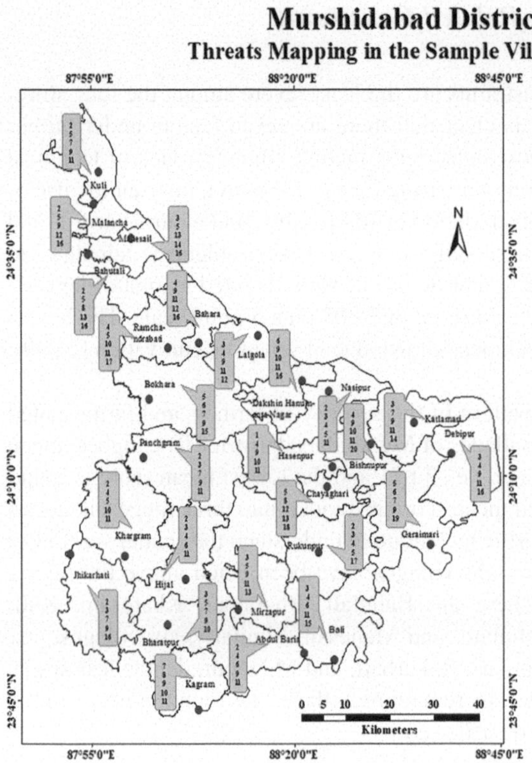

S.No	Threats
1	Decreasing wet land
2	Degradation of soil fertility
3	Depended rainfall-based farming
4	Erratic rainfall
5	Fluctuation in price
6	Frequent occurrence of flood
7	Hailstorm and Kalbaishakhi
8	Increasing crop diseases
9	Increasing migration
10	Inorganic farming
11	Increasing temperature
12	Land encroachment
13	Land insecurity
14	Overdose of chemical fertilizers
15	Soil erosion
16	Sudden fall of price crops
17	Unexpected rain
18	Unpredictable weather
19	Water shortage
20	Waterlogged

Fig. 6.4 Threats mapping in the sample villages (2021). (Source: Based on Annexure F.3)

6.3 Identification of Major Problems Faced by the Farmers

Murshidabad district, one of the most important agricultural districts of West Bengal, is developing at a faster rate. Here, farmers grow various crops on their land throughout the year. During the cultivation of land, the farmers generally face many problems. Therefore, an attempt has been made to identify the major and minor critical issues faced by the farmers in the district. The researcher asked the respondents about their agriculture-related problems to identify these problems. The researcher has also personally observed and determined a few issues after discussing with Pradhan (head of the village). These critical problems are divided into four major constraints: farmers and farm-related constraints, input of cultivation-related constraints, institutional and infrastructural-related constraints, and production-related constraints. The respondents were asked to rate each particular problem on a 5-point rating scale to measure the severity of these problems. The higher the rating score, the more severe is the problem.

6.3.1 Farmers and Farm-Related Constraints

The farmer and farm-related constraints are the least severe among the four aforementioned problematic areas. It is clear that there are seven factors under farmer and farm-related constraints. These constraints include illiteracy, lack of technical knowledge of scientific cultivation, increasing cost of irrigation, unscientific use of water, low farm mechanization, degradation of soil fertility, and natural hazards and soil erosion. The district's average severity score for these problems is calculated at 3.242. Among these problems, unscientific use of water is rated as first-order constraints as it secures the highest mean score of 3.805 on a 5-point scale. The lowest mean score is found for lack of technical knowledge of scientific cultivation (2.549), followed by illiteracy (2.953).

Figure 6.5 shows the spatial pattern of the severity of farmers and farm-related problems in the selected sample villages of Murshidabad district. The highest mean score is secured by Bahutali village, i.e., 3.871 and the lowest mean score is found in Lalgola (2.964). The calculated mean score is divided into three range categories based on the previously used quartile technique to understand the spatial variability at the village level. As many as eight villages have been listed above the higher severity range, i.e. (>3.384). These are Bahutali, Bharatpur, Khargram, Kuli, Hasenpur, Bishnupur, Ramchandrabati, and Mahesail. Similarly, six villages, i.e., Lalgola, Hijal, Panchgram, Garaimari, Jhikarhati, and Mirzapur, denote less severity of these particular constraints. The remaining villages are found in between the first (>3.109) and third (>3.384) quartile scores.

6.3.2 Input of Cultivation-Related Constraints

The input of cultivation-related constraints includes low use of advanced technology, lack of supply of HYV seeds, lack of supply of fertilizers, high cost of fertilizers, unscientific use of pesticides, hiring of inputs from others, and unskilled labours. The average severity score in terms of these constraints together is 3.569 in the district. Among these seven problems, the high cost of fertilizers (3.949) emerges as an intense problem, whereas the less problematic constraint is unskilled labour (3.234).

Eleven villages indicate a severity score above the third quartile (>3.598), in which Bahara (4.235) secures the maximum score value. This is followed by Bahutali (4.231), Kuli (3.990), and Rukunpur (3.897). Conversely, five such villages are found where the score value lies below the first quartile (< 3.182). These villages are Bokhara (3.105), Mirzapur (3.124), Ramchandrabati (3.143), Panchgram (3.167), and Bali (3.167). Eight villages have displayed a medium range of severity in the input-related constraints (Fig. 6.6).

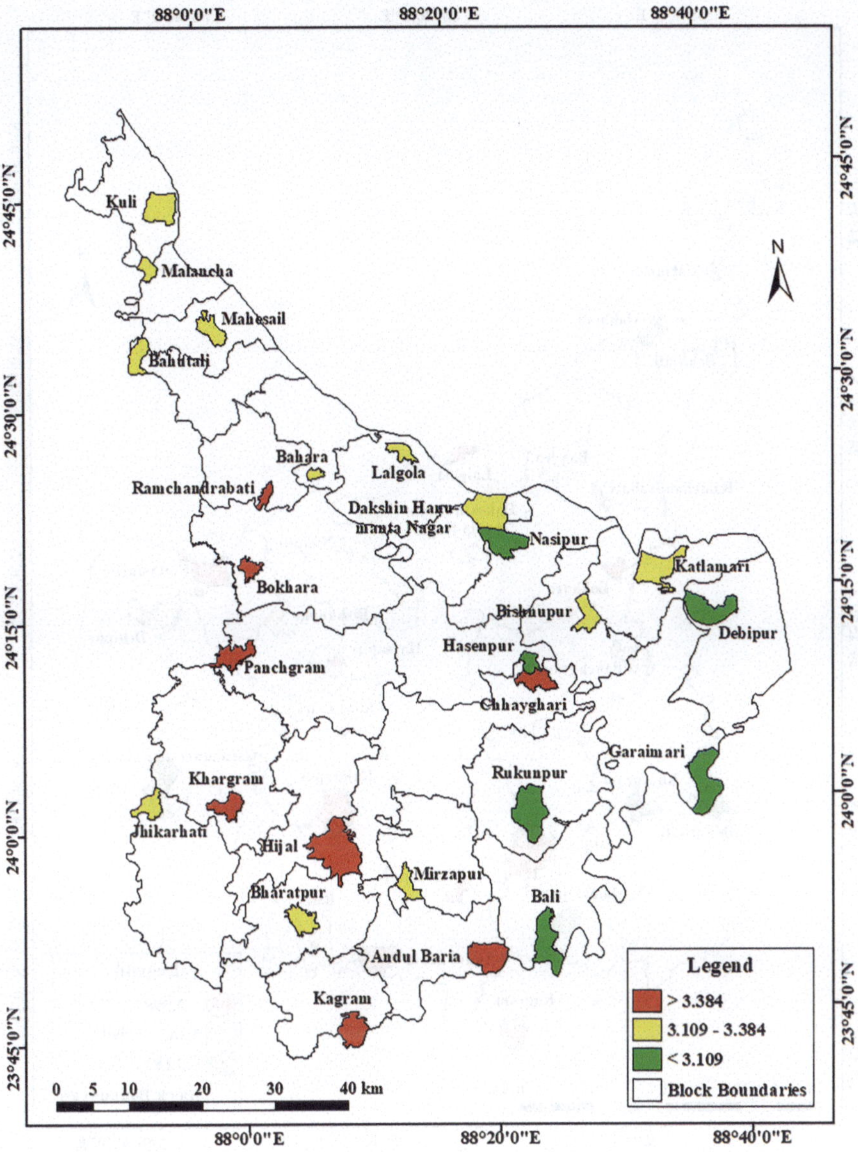

Fig. 6.5 Farmers and farm-related constraints, sample villages (2021). (Source: Based on Annexure)

6.3.3 Institutional and Infrastructural-Related Constraints

The institutional and infrastructural-related constraints are the most severe among the four major constraints of agricultural practices, with a mean score of 3.968 in the Murshidabad district. The problems such as lack of marketing facilities, lack of

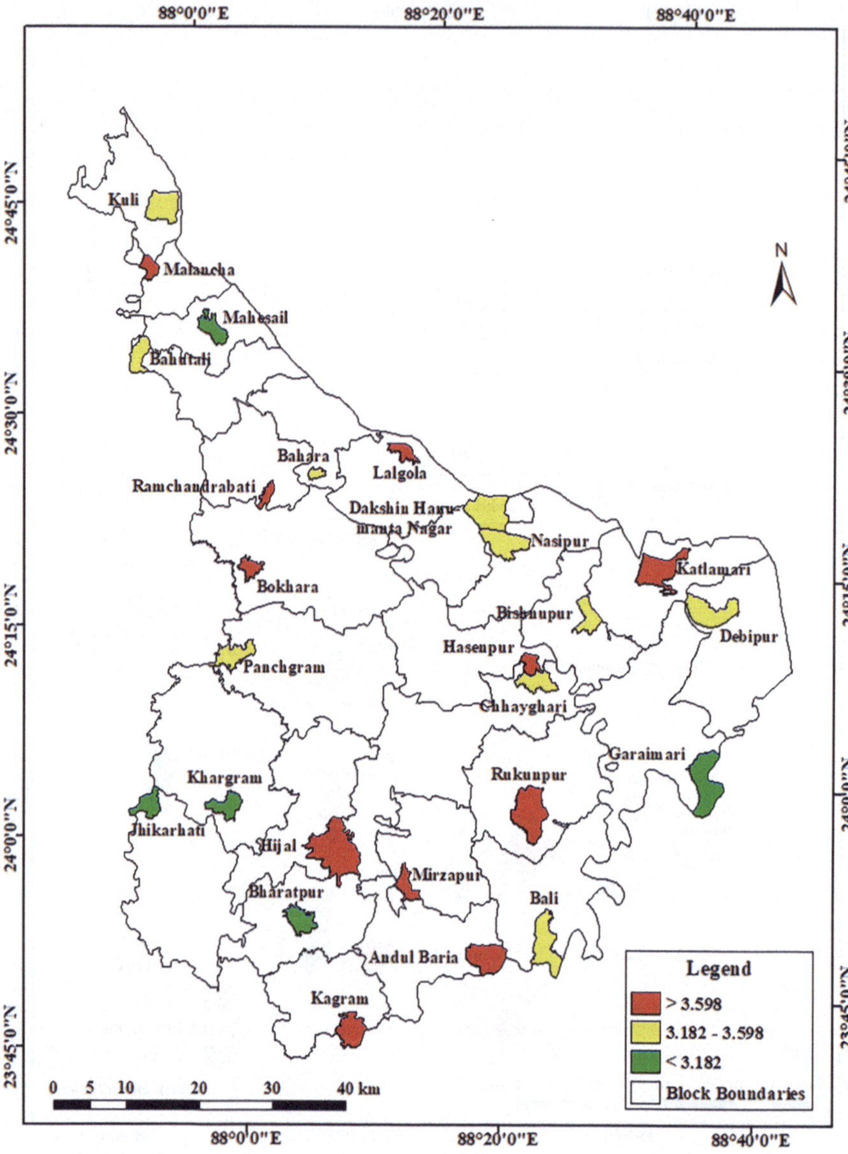

Fig. 6.6 Input of cultivation-related constraints, sample villages (2021). (Source: Based on Annexure)

cold storage and warehousing facility, poor transportation facilities, lack of credit and loan facilities, lack of seed store facilities, lack of fair price systems, and high transportation costs are classified under the institutional and infrastructure-related constraints. Among these seven problems under this constraint-related area, the inadequacy of credit and loan facilities is found to be the highest-rated problem. The

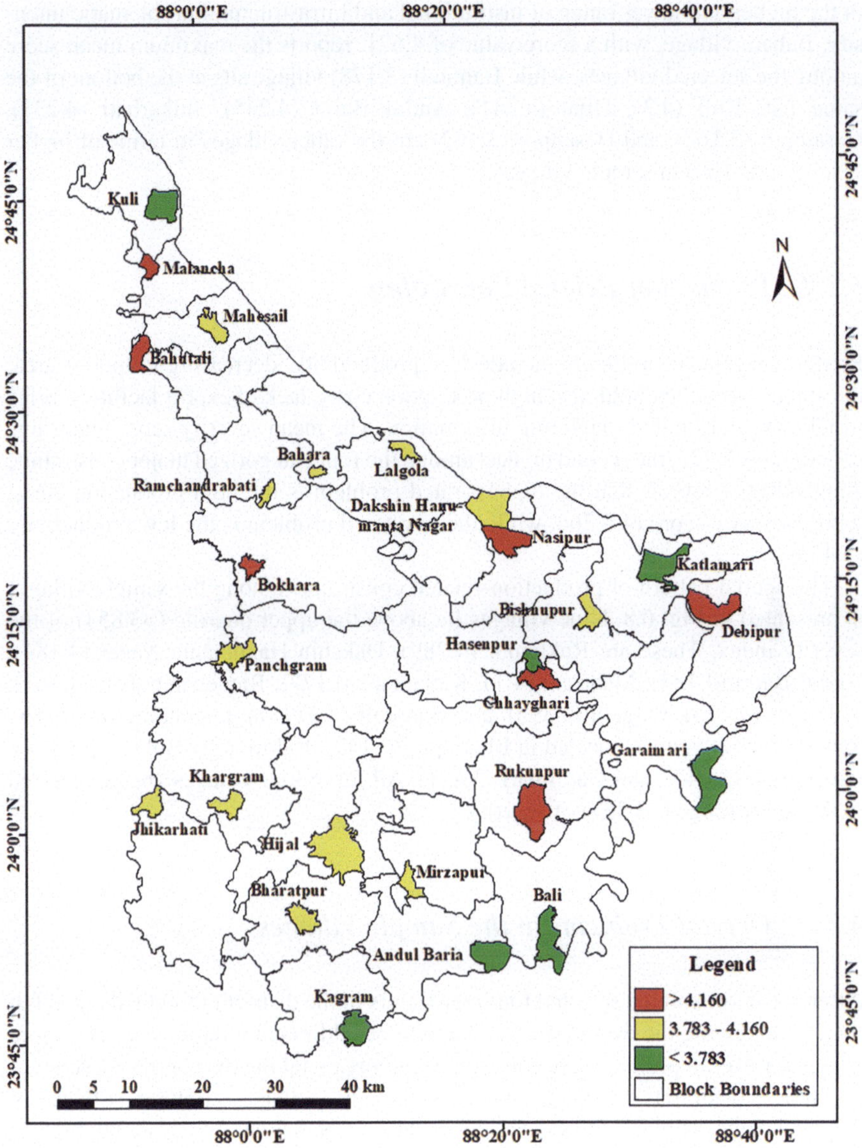

Fig. 6.7 Institutional and infrastructural-related constraints, sample villages (2021). (Source: Based on Annexure)

mean score of this particular problem in the district is calculated as 4.182. A high mean score is also found for the problems such as lack of cold storage facilities, poor transport facilities, and lack of fair price systems.

Figure 6.7 demonstrates the village-wise distribution of severity scores under three categories based on upper and lower quartiles. Seven villages have been found in the higher and lower range of institutional and infrastructural problematic intensity. Bahara village, with a score value of 4.671, reports the maximum mean score among the selected villages, while Bahutali (3.178) village sits at the bottom of the same list. Kuli (4.3), Hijal (4.341), Andul Baria (4.245), Jhikarhati (4.231), Garaimari (3.162), and Hasenpur (3.162) are the other villages in terms of higher severity (>4.160) in sample villages.

6.3.4 Production-Related Constraints

Production-related problems include low productivity, decreasing cropping area, the small size of landholdings, high production costs, lack of export facilities, price instability, and limited marketing information. The mean severity score under this constraint is 3.775, the second highest among the four categorized major constraints. Annexure 6.4 reveals that the highest-rated problem is the high production costs, with a severity score of 4.364, while the low-rated problem is the low productivity (3.475).

The spatial pattern of production-related constraints among the sample villages is presented in Fig. 6.8. Nine villages lie above the upper quartile (>3.854) of the severity index. These are Rukunpur (4.289), Dakshin Hanumanta Nagar (4.184), Andul Baria (4.183), Mahesail (4.1), Katlamari (4.172), Ramchandrabati (4.163), Mirzapur (4.124), Lalgola (3.895), and Bahutali (3.879) in descending order. The lowest severity value is located in Bharatpur (3.142), followed by Bokhara (3.149), Panchgram (3.342), and Garaimari (3.481). All these four villages are categorized in the lower range (<3.557) of severity.

6.3.5 Overall Problems in the Sample Villages

After discussing all the selected four constraints in the district individually, it is felt that, at this point, the overall problems are to be addressed village-wise. The score values for overall problems are constructed simply by taking the composite score of all four constraints. Out of the selected 26 villages, Bahara village of Raghunathganj-II block and Panchgram village of Nabagram block are the highest and lowest problematic villages, with a combined score of 3.899 and 3.272, respectively, in terms of overall problems. The spatial pattern of overall problems is divided into three categories based on the quartile technique on a 5-point rating scale for a better understanding.

Figure 6.9 reveals that six villages denote a severity score below the lower quartile (3.503), and the villages are Panchgram, Bokhara, Chhayghari, Malancha, Lalgola, and Garaimari. In contrast to this, as many as nine villages secure the

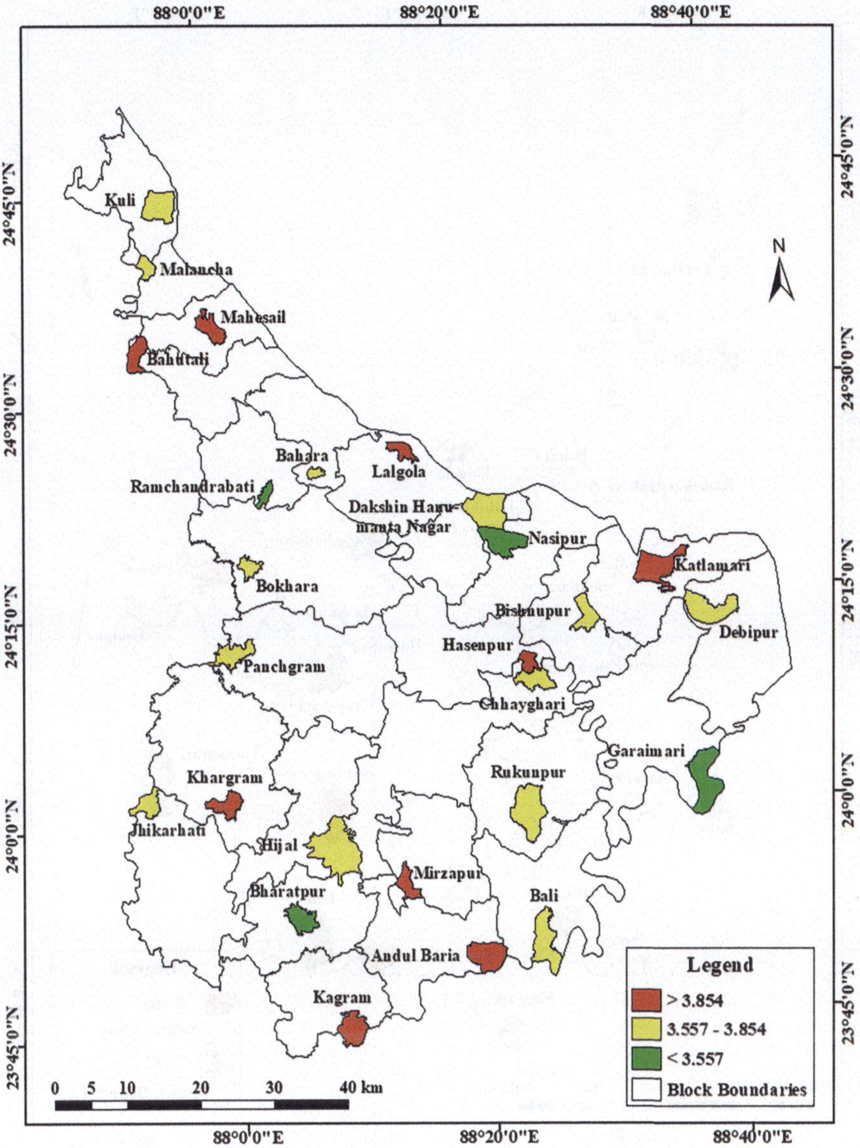

Fig. 6.8 Production-related constraints, sample villages (2021). (Source: Based on Annexure)

severity value above the upper quartile (3.709). These villages are Bahara, Kuli, Mahesail, Bahutali, Rukunpur, Dakshin Hanumanta Nagar, Hasenpur, Andul Baria, and Katlamari, in descending order. The remaining 11 villages are located in the medium range of severity score.

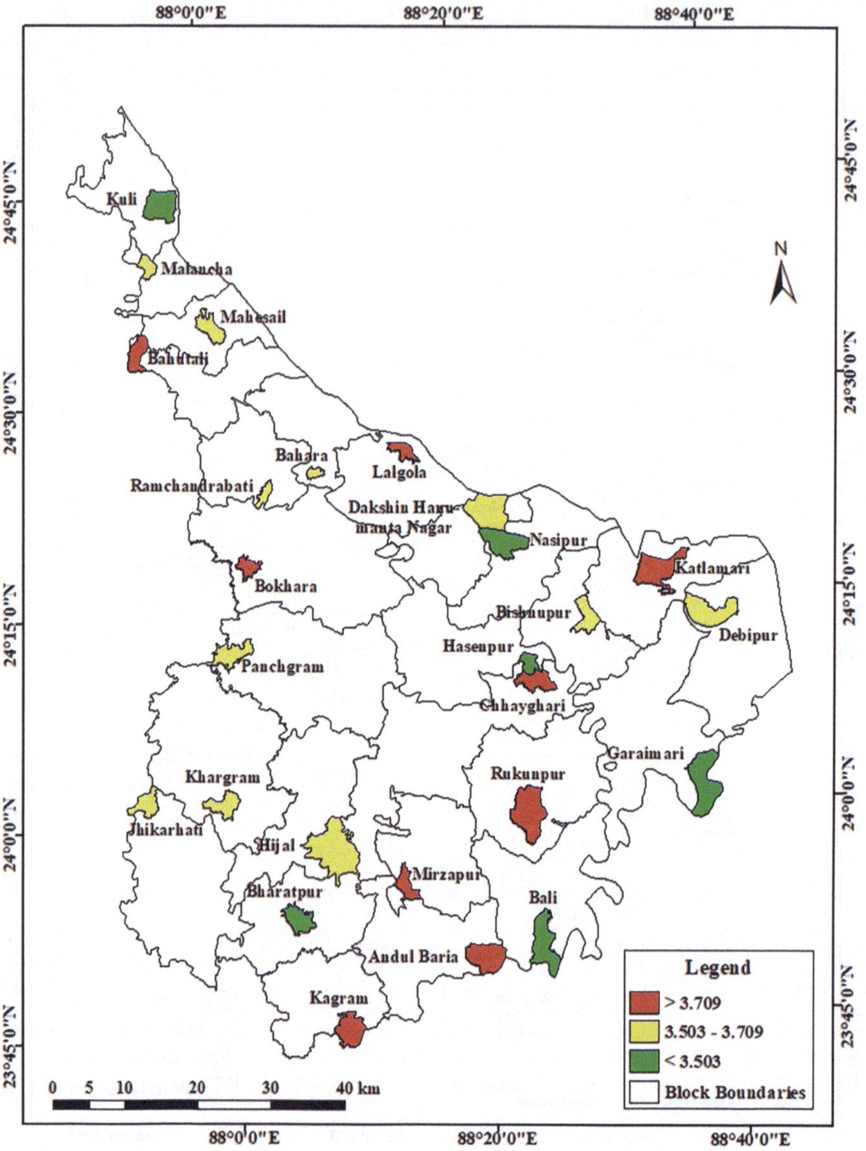

Fig. 6.9 Overall constraints in sample villages (2021). (Source: Based on Annexure)

6.4 Ongoing Programs, Policies, and Schemes in Murshidabad District

It comes out from the above discussion that institutional and infrastructure-related problems are the most severe among the significant issues and their severity, according to the farmers, in the district. The problems such as lack of cold storage and warehousing facilities, lack of credit and loan facilities, lack of fair price system, high production cost, and limited marketing information emerge as the most critical challenges faced by the farmers of the sampled villages. After realising these issues and problems, it is crucial to investigate the ongoing programs, policies, and schemes implemented by the state and central governments (Annexure F.6). In order to minimize the problems mentioned above, a strong regulatory body needs to be formed. The regulatory bodies should work at the ground level to monitor the fair price system in the market and keep the farmers safe from the intermediaries. There are several schemes and programs going on in the district for the benefit of farmers. One major problem is that the farmers lack awareness and knowledge regarding getting those benefits. Therefore, it should be advisable to the government that various awareness programs, capacity-building programs, and focus group discussions (FGDs) should be conducted regularly so that the farmers are able to know the benefit of different schemes, programs, and policies.

6.5 Hypotheses Testing

The hypotheses have been tested through different statistical tools and techniques. All null hypotheses are failed to accept at a different significance level.

1. **There is no significant difference in the amount of change in sustainable agricultural land use across different land size categories.**

Table 6.1 reveals that the level of difference in the amount of change in agricultural land use by land size class categories in between groups and within groups stands as significant with a *p*-value of 0.000 and an *F*-statistic of 25.06637. Hence, the formulated null hypothesis is failed to accept at a 99% confidence level.

2. **Agricultural input negatively correlates with agricultural output under different land size classes.**

The output of the second hypothesis is presented in Table 6.2. It explains that the degree of Karl Pearson's correlation coefficient between input and output dimensions of agricultural innovation is positive and significant in 1995–1996 and 2015–2016 under all land size classes except for the large land category where the degree of relationship is not significant due to a fewer number of blocks having large landholders. Moreover, the association of relationship across land size classes becomes more robust in 2015–2016. It is confirmed that the

Table 6.1 ANOVA summary

Source of variation	SS	df	MS	*F*	*P*-value
Between groups	0.118596	5	0.023719	25.06637	0.000
Within groups	0.141938	150	0.000946		
Total	0.260534	155			
The null hypothesis is failed to accept					

Table 6.2 Karl Pearson's correlation of coefficient

Land size category	1995–1996	2015–2016	df	*P*-value
Marginal class	0.479[a]	0.603[b]	11	0.001
Small class	0.434[a]	0.526[b]		
Semi-medium class	0.396[a]	0.456[a]		
Medium class	0.531[b]	0.594[b]		
Large class	0.314	0.377		
All classes	0.464[a]	0.589[b]		
The null hypothesis is failed to accept				

[a]Correlation is significant at 0.05 significance level (2-tailed)
[b]Correlation is significant at 0.01 significance level (2-tailed)

Table 6.3 *T*-test summary

t-Test: Two-sample assuming unequal variances				
	1995–1996		2015–2016	
Land size category	$P(T <= t)$ two-tail	*t*-critical two-tail	$P(T <= t)$ two-tail	*t*-critical two-tail
Marginal class	0.030	2.01174	0.027	2.00958
Small class	0.007	2.00856	0.002	2.00874
Semi-medium class	0.002	2.01290	0.000	2.00856
Medium class	0.007	2.00856	0.010	2.00972
Large class	0.260	2.14479	0.227	2.07881
All classes	0.014	2.01174	0.009	2.00963
The null hypothesis is failed to accept				

correlation coefficients of agricultural input and agricultural output under different land classes are significant at 99% confidence level (0.001 significant level). Therefore, the selected null hypothesis is failed to accept that the agricultural input is negatively correlated with agricultural output under different land size classes. It further indicates the agricultural policy implication at medium and long term in the study area.

3. **Sustainable agricultural land use is not significantly related to overall agricultural innovation across land size categories.**

The outcome of the third hypothesis is summarized in Table 6.3. Since we have two independent groups (sustainable land use and agricultural innovation), a *t*-test of two samples assuming unequal variances has been applied to the third and final hypothesis, which assumes that sustainable land use has no significant relation with agricultural innovation. The two-tailed t-critical values across land size class categories in 1995–1996 and 2015–2016 confirm indifference relationships which are statistically significant at different confidence levels (except for the large land category where the result is not significant in both the time period due to the lack of strength of blocks having large landholders). Hence, the third null hypothesis is again failed to accept.

References

Aldehayyat, J. S., & Anchor, J. R. (2008). Strategic planning tools and techniques in Jordan: Awareness and use. *Strategic Change, 17*(7–8), 281–293.

Benzaghta, M. A., Elwalda, A., Mousa, M. M., Erkan, I., & Rahman, M. (2021). SWOT analysis applications: An integrative literature review. *Journal of Global Business Insights, 6*(1), 55–73.

Census of India. (2011). *Census of India, Murshidabad district.*

Christensen, L. R. (1975). Concepts and measurement of agricultural productivity. *American Journal of Agricultural Economics, 57*(5), 910–915.

District Statistical Handbook. (2016). *District Statistical Handbook, Murshidabad.* Bureau of Applied Economics & Statistics, Govt. of West Bengal.

Fleisher, C. S., & Bensoussan, B. E. (2003). *Strategic and competitive analysis: Methods and techniques for analyzing business competition.* Pearson.

Lee, K. L., & Lin, S. C. (2008). A fuzzy quantified SWOT procedure for environmental evaluation of an international distribution center. *Information Sciences, 178*(2), 531–549.

Leigh, D. (2009). SWOT analysis. In *Handbook of improving performance in the workplace* (Vol. 1–3, pp. 115–140).

Puyt, R., Lie, F. B., De Graaf, F. J., & Wilderom, C. P. (2020). Origins of SWOT analysis. In *Academy of management proceedings* (p. 17416). Academy of Management.

Rahaman, H. (2020). *Diversified cropping pattern and agricultural development: A case study from Malda District* (p. xvi). Springer Nature.

Rahaman, H. (2021). Challenges and opportunities of agricultural development in the district. In *Diversified cropping pattern and agricultural development* (pp. 189–220).

Rozmi, A. N. A. (2018). The perception of ICT adoption in small medium enterprise: A SWOT analysis. *International Journal of Innovation and Business, 9*(1), 69–79.

Shrestha, R. K., Alavalapati, J. R., & Kalmbacher, R. S. (2004). Exploring the potential for silvo-pasture adoption in south-Central Florida: An application of SWOT–AHP method. *Agricultural Systems, 81*(3), 185–199.

Annexures

Annexure A: Chapter 1

H. Rahaman, S. Ansari, *Sustainable Land Use Practices and Agricultural Innovations in India*, https://doi.org/10.1007/978-3-031-83246-8

Annexure A.1 Administrative units of Murshidabad district, 2016

Sub-division	Block and municipality	Police stations	Area (km^2)	Households	Rural				Urban		
					Villages (inhabited)	Panchayat Samiti	Gram panchayat	Gram Sansad	Municipality		Census town
									Number	Wards	
Sadar	Berhampore	Berhampore (P)	314.19	105,958	124	1	17	317	–	–	9
		Daulatabad									
	Beldanga-I	Beldanga (P)	168.75	68,499	55	1	13	204	–	–	1
	Beldanga-II	Rejinagar	207.93	55,268	61	1	11	171	–	–	–
		Shaktipur									
	Nawda	Nawda	231.39	53,962	28	1	10	162	–	–	–
	Hariharpara	Hariharpara	253.14	62,609	55	1	10	181	–	–	–
	Berhampore (M)	Berhampore (P)	16.19	43,075	–	–	–	–	1	28	–
		Woman P.S.									
	Beldanga (M)	Beldanga (P)	3.98	6580	–	–	–	–	1	14	–
Sub-division total	**5 and 2**	**8**	**1195.57**	**395,951**	**323**	**5**	**61**	**1035**	**2**	**42**	**10**
Kandi	Kandi	Kandi (P)	227.48	48,963	84	1	10	151	–	–	–
	Khargram	Khargram	318.45	65,580	138	1	12	186	–	–	–
	Burwan	Burwan	299.66	60,732	155	1	13	185	–	–	–
	Bharatpur I	Bharatpur	183.72	40,102	82	1	8	121	–	–	–
	Bharatpur II	Salar	158.50	39,832	51	1	7	123	–	–	1
	Kandi (M)	Kandi (P)	12.95	12,237	0	–	0	–	1	17	1
Sub-division total	**5 and 1**	**5**	**1200.76**	**267,446**	**510**	**5**	**50**	**766**	**1**	**17**	**2**

Jangipur	Farakka	Farakka	132.74	54,005	56	1	9	147	–	–	8
	Samserganj	Samserganj (P)	84.21	51,412	24	1	9	154	–	–	11
	Suti-I	Suti	143.68	36,628	46	1	6	106	–	–	4
	Suti-II		111.13	53,619	24	1	10	157	–	–	13
	Raghunathganj-I	Raghunathganj (P)	140.91	39,021	56	1	6	117	–	–	6
	Raghunathganj-II		121.60	50,129	36	1	10	156	–	–	10
	Sagardighi	Sagardighi	345.42	67,889	178	1	11	199	–	–	–
	Jangipur (M)	Jangipur Women P.S.	7.86	17,418	0	–	0	–	1	20	1
	Dhuliyan (M)	Samserganj (P)	10.27	16,545	0	–	0	–	1	19	1
Sub-division total	**7 and 2**	**6**	**1097.82**	**386,666**	**420**	**7**	**61**	**1036**	**2**	**39**	**54**
Lalbagh	Lalgola	Lalgola	184.37	70,216	84	1	12	205	-	-	-
	Bhagwangola-I	Bhagwangola	136.10	45,336	54	1	8	127	-	-	-
	Bhagwangola-II	Ranitala	175.26	37,445	56	1	6	96	-	-	-
	Murshidabad-Jiaganj	Jiaganj (P) Murshidabad (P)	192.13	54,029	127	1	8	153	–	–	–
	Nabagram	Nabagram	306.63	52,954	109	1	10	155	–	–	–
	Murshidabad (M)	Murshidabad (P)	12.95	9829	0	–	–	–	1	16	1
	Jiaganj-Azimganj (M)	Jiaganj (P)	11.66	11,787	0	–	–	–	1	17	1
Sub-division total	**5 and 2**	**6**	**1019.10**	**281,596**	**430**	**5**	**44**	**736**	**2**	**33**	**2**

(continued)

Annexure A.1 (continued)

Sub-division	Block and municipality	Police stations	Area (km^2)	Households	Rural				Urban		
					Villages (inhabited)	Panchayat Samiti	Gram panchayat	Gram Sansad	Municipality		Census town
									Number	Wards	
Domkal	Domkal	Domkal	305.19	88,520	77	1	13	249	–	–	–
	Jalangi	Jalangi	210.63	60,928	39	1	10	171	–	–	–
	Raninagar-I	Islampore	146.93	45,147	57	1	6	125	–	–	2
	Raninagar-II	Raninagar	175.13	44,505	30	1	9	127	–	–	–
Sub-division total	**4 and 0**	**4**	**837.88**	**23,9100**	**203**	**4**	**38**	**672**	–	–	**2**
District total	**26 and 7**	**29**	**5324.00**	**1,570,759**	**1886**	**26**	**254**	**4245**	**7**	**128**	**72**

Source: District Statistical Handbook of Murshidabad District, 2016

Annexure A.2 Information of soils of the Murshidabad district, 2015–2016 (area in hectares)

	Sandy soil		Sandy loam		Clay loam		Red soil	
Blocks	Area	Percent	Area	Percent	Area	Percent	Area	Percent
Berhampore	0	0.00	13,298	55.00	10,882	45.00	0	0.00
Burwan	0	0.00	182	1.00	16,380	99.00	0	0.00
Beldanga-I	0	0.00	10,732	80.00	2683	19.00	0	0.00
Beldanga-II	0	0.00	10,975	79.00	2744	20.00	0	0.00
Bhagwangola-I	0	0.00	5500	55.00	4500	45.00	0	0.00
Bhagwangola-II	0	0.00	6600	55.00	5400	43.00	0	0.00
Bharatpur-I	66	0.40	2383	18.00	10,790	81.00	0	0.00
Bharatpur-II	0	0.00	2139	18.00	9749	82.00	0	0.00
Domkal	0	0.00	22,050	98.00	450	2.00	0	0.00
Farakka	0	0.00	654	15.00	3702	84.90	4	0.00
Hariharpara	0	0.00	13,667	75.00	4556	24.00	0	0.00
Jalangi	0	0.00	11,319	67.70	485	30.00	0	0.00
Kandi	0	0.00	4451	25.00	13,350	75.00	0	0.00
Khargram	0	0.00	14,640	61.00	9360	39.00	0	0.00
Lalgola	0	0.00	3640	30.00	8494	69.00	0	0.00
Murshidabad-Jiaganj	0	0.00	8002	55.00	6547	45.00	0	0.00
Nabagram	0	0.00	2350	10.00	21,150	90.00	0	0.00
Nawda	0	0.00	3000	15.00	17,000	85.00	0	0.00
Raghunathganj-I	0	0.00	450	5.00	8550	95.00	0	0.00
Raghunathganj-II	0	0.00	3400	85.00	600	15.00	0	0.00
Raninagar-I	105	1.00	5775	55.00	4620	44.00	0	0.00
Raninagar-II	114	1.00	6308	54.00	5046	45.00	0	0.00
Sagardighi	690	3.00	9660	42.00	12,650	55.00	0	0.00
Samserganj	0	0.00	1600	40.00	2400	55.00	0	0.00
Suti-I	0	0.00	3855	54.00	3154	44.00	0	0.00
Suti-II	0	0.00	3373	54.00	2760	44.00	0	0.00
Total	**975**	**5.40**	**170,003**	**47.36**	**188,002**	**52.37**	**4**	**0.00**

Source: Deputy Director of Agriculture, Agriculture Department, Murshidabad

Annexure A.3 Soil nutrient sufficiencies in Murshidabad district, 2016–2017

Subdistrict/mandal	Macro-soil nutrients				Micro-soil nutrients					
	N	OC	P	K	Zn	Fe	Cu	Mn	B	S
Berhampore	80.42	66.04	100.00	76.88	11.34	99.64	97.47	88.24	0.36	65.15
Burwan	91.12	64.41	100.00	79.76	30.20	99.78	99.93	89.03	2.98	99.49
Beldanga-I	84.08	60.45	99.84	90.03	22.19	99.84	99.68	96.95	0.48	98.07
Beldanga-II	70.37	72.22	100.00	73.15	1.85	100.00	97.22	53.77	0.93	91.67
Bhagwangola-I	98.67	69.58	100.00	96.96	45.14	99.78	99.57	99.78	0.22	98.27
Bhagwangola-II	75.97	36.50	99.78	98.27	25.10	99.81	100.00	99.24	0.00	100.00
Bharatpur-I	97.66	81.24	100.00	92.44	35.69	100.00	98.57	97.65	0.00	95.68
Bharatpur-II	96.77	82.26	100.00	91.94	32.26	100.00	98.39	98.39	0.00	95.16
Domkal	69.01	66.14	97.12	78.74	29.52	99.80	100.00	66.06	17.27	88.96
Farakka	73.68	55.59	96.17	92.50	45.59	100.00	100.00	93.34	11.18	62.94
Hariharpara	83.02	75.19	99.92	63.98	31.24	99.50	99.09	82.03	0.49	69.07
Jalangi	78.81	78.12	98.51	77.64	49.08	99.77	99.89	79.11	31.31	87.39
Kandi	88.64	69.67	99.47	84.00	24.48	99.91	100.00	84.44	3.76	98.78
Khargram	91.71	49.25	99.82	83.71	30.21	99.97	99.91	98.41	0.81	98.26
Lalgola	88.61	81.85	100.00	92.09	77.80	99.61	100.00	84.85	0.00	30.50
Murshidabad-Jiaganj	82.26	61.91	99.69	76.49	30.81	99.80	99.86	99.38	3.33	81.34
Nabagram	60.25	80.37	99.74	75.12	97.62	99.75	100.00	99.75	41.87	83.00
Nawda	69.12	71.67	100.00	84.71	51.70	99.58	99.10	66.16	0.28	17.14
Raghunathganj-I	86.53	57.56	99.42	57.17	28.49	99.90	100.00	98.45	2.81	87.11
Raghunathganj-II	87.79	83.97	100.00	96.18	80.92	98.47	100.00	74.05	1.53	54.20

Raninagar-I	88.78	44.04	99.83	96.31	21.16	99.94	99.66	99.39	1.45	97.09
Raninagar-II	89.03	50.85	99.96	44.88	22.21	99.91	100.00	97.63	1.66	90.85
Sagardighi	84.80	61.28	98.90	74.84	29.79	99.97	100.00	96.80	5.73	79.44
Samserganj	66.47	26.59	96.53	96.53	24.57	100.00	100.00	100.00	0.00	92.49
Suti-I	84.30	61.10	99.93	89.60	26.16	99.93	100.00	64.05	13.28	93.49
Suti-II	91.25	57.89	99.42	97.72	25.76	100.00	99.90	99.33	1.71	90.46
Total	83.04	64.07	99.39	83.14	35.80	99.79	99.55	88.70	5.52	82.54

Source: Soil Health Card, Department of Agriculture, Cooperation, and Welfare, Govt of India, 2016–2017

Annexure A.4 Area, population distribution, density, literacy, and sex ratio in Murshidabad district, 2011

Blocks	Area		Population		Density	Literacy			Sex ratio
	Actual (km^2)	% to the district	Actual (in numbers)	% to the district		Male	Female	Total	
Berhampore	314.19	5.90	642,110	9.04	1422	76.52	70.34	73.51	954
Burwan	299.66	5.63	257,466	3.62	859	74.97	62.6	68.96	944
Beldanga-I	168.75	3.17	348,527	4.91	1892	71.92	68.08	70.06	945
Beldanga-II	207.93	3.91	250,458	3.53	1205	71.1	64.4	67.86	939
Bhagwangola-I	136.1	2.56	202,071	2.84	1485	67.64	65.89	66.79	954
Bhagwangola-II	175.26	3.29	158,024	2.22	902	63.06	62.58	62.82	958
Bharatpur-I	183.72	3.45	172,702	2.43	940	67.94	57.57	62.93	939
Bharatpur-II	158.5	2.98	176,368	2.48	1113	70.96	60.97	66.07	959
Domkal	305.19	5.73	363,976	5.12	1193	64.47	63.31	63.9	955
Farakka	132.74	2.49	274,111	3.86	2065	65.56	53.77	59.75	969
Hariharpara	253.14	4.75	257,571	3.63	1018	69.97	68.39	69.2	957
Jalangi	210.63	3.96	252,477	3.55	1199	69.36	65.24	67.35	951
Kandi	227.48	4.27	275,777	3.88	968	71.66	58.25	65.13	952
Khargram	318.45	5.98	273,332	3.85	858	69.69	57.16	63.56	959
Lalgola	184.37	3.46	335,831	4.73	1822	65.81	62.76	64.32	964
Murshidabad-Jiaganj	192.13	3.61	330,374	4.65	1221	72.82	65.16	69.12	936
Nabagram	306.63	5.76	227,586	3.20	742	76.44	64.98	70.83	960
Nawda	231.39	4.35	226,859	3.19	980	66.91	65.24	66.09	950
Raghunathganj-I	140.91	2.65	283,792	3.99	1388	68.99	59.74	64.49	951
Raghunathganj-II	121.6	2.28	265,336	3.74	2182	64.71	57.45	61.17	955

Raninagar-I	146.93	2.76	189,105	2.66	1287	68.42	66.03	67.25	965
Raninagar-II	175.13	3.29	190,885	2.69	1090	64.78	62.36	63.6	961
Sagardighi	345.42	6.49	310,461	4.37	899	68.34	62.05	65.26	957
Samserganj	84.21	1.58	379,778	5.35	3373	60.42	49.57	54.98	1000
Suti-I	143.68	2.70	179,908	2.53	1252	62.84	53.08	58.06	958
Suti-II	111.13	2.09	278,922	3.93	2510	60.44	49.98	55.23	992
Total	**5324**	100.00	**7,103,807**	100.00	**1334**	69.95	**63.09**	**66.59**	**958**

Source: District Statistical Handbook, Murshidabad 2016

Annexure B: Chapter 2

Annexure B.1 Trends and pattern of land use in Murshidabad district, 1995–1996 to 2015–2016 (area in hectares)

Types of land use	1995–1996	2000–2001	2005–2006	2010–2011	2015–2016
Total area	441,779	439,087	438,966	438,682	442,677
Net sown area	430,879	433,021	431,332	433,487	433,503
Area under current fallow	2807	2831	2571	1440	3766
Net cultivated area	433,686	435,852	433,903	434,926	437,269
Other uncultivated land excluding fallow land	1351	341	412	541	475
Fallow land other than current fallow	541	523	406	15	343
Culturable waste land	861	116	767	83	898
Total uncultivated land	2753	980	1585	640	1716
Land not available for cultivation	5340	2255	3478	3116	3694

Source: Agriculture Census, Government of India; 1995–1996 to 2015–2016

Annexure B.2 Land size category-wise land use pattern in Murshidabad district, 1995–1996 and 2015–2016 (area in hectares)

Classes	Total area	Net sown area	Area under current fallow	Net cultivated area	Other uncultivated land excluding fallow land	Fallow land other than current fallow	Culturable waste land	Total uncultivated land	Land not available for cultivation
Land use by size category-wise in 1995–1996									
Marginal	194,812	191,436	790	192,226	275	226	279	780	1806
Small	145,388	141,639	955	142,594	410	198	286	894	1900
Semi-medium	85,226	82,475	669	83,144	480	38	259	777	1305
Medium	15,580	14,630	358	14,988	186	74	37	297	295
Large	773	699	35	734	0	5	0	5	34
All classes	**441,779**	**430,879**	**2807**	**433,686**	**1351**	**541**	**861**	**2753**	**5340**
Land use by size category-wise in 2015–2016									
Marginal	238,482	234,742	1638	236,380	173	171	344	688	1414
Small	148,089	143,913	2028	145,941	239	127	361	727	1421
Semi-medium	52,478	51,385	77	51,461	58	27	148	233	784
Medium	3098	2931	24	2955	5	18	45	69	74
Large	532	532	0	532	0	0	0	0	0
All classes	**442,677**	**433,503**	**3766**	**437,269**	**475**	**343**	**898**	**1716**	**3694**

Source: Agriculture Census, Government of India; 1995–1996 and 2015–2016

Annexure B.3 Land size class category-wise total area in Murshidabad district, 1995–1996 and 2015–2016 (area in hectares)

	1995–1996						2015–2016					
Blocks	Marginal	Small	Semi-medium	Medium	Large	All classes	Marginal	Small	Semi-medium	Medium	Large	All classes
Berhampore	10,118	6704	3619	554	296	21,291	14,161	6801	1809	285	175	23,231
Burwan	9436	11,459	4172	552	0	25,619	12,277	6885	2274	122	53	21,612
Beldanga-I	9675	3234	1576	509	0	14,994	9434	4864	1739	82	0	16,119
Beldanga-II	8076	4684	3161	224	51	16,196	11,704	6076	2161	34	0	19,975
Bhagwangola-I	10,191	6190	4412	309	0	21,102	7335	5053	1874	87	70	14,419
Bhagwangola-II	4455	2772	1513	129	0	8869	6428	5387	2490	381	0	14,686
Bharatpur-I	8891	6869	2813	513	0	19,086	9291	5944	1961	171	0	17,366
Bharatpur-II	5367	6271	3376	127	0	15,141	7058	5975	1998	58	0	15,089
Domkal	9370	5328	2513	962	60	18,233	16,264	6294	2133	132	0	24,822
Farakka	2027	1572	676	126	50	4451	3092	4747	1764	31	55	9690
Hariharpara	10,828	6312	4254	1446	0	22,840	10,807	6020	1883	175	0	18,886
Jalangi	5947	4043	1414	269	0	11,673	9908	5720	1918	8	0	17,554
Kandi	10,421	7346	2232	149	0	20,148	8893	3866	1840	16	0	14,614
Khargram	17,172	5968	6242	996	0	30,378	10,873	6993	2568	166	0	20,600
Lalgola	6944	5814	4815	640	0	18,213	10,646	6108	1703	6	0	18,463
Murshidabad-Jiaganj	8471	6115	3612	743	142	19,083	12,484	6882	2061	13	67	21,507
Nabagram	11,363	9697	7917	1707	0	30,684	13,048	8064	2524	226	0	23,861
Nawda	8266	8179	4228	1040	0	21,713	13,217	8057	2264	456	60	24,054
Raghunathganj-I	3457	2996	1883	302	70	8708	1628	1490	1900	123	0	5139
Raghunathganj-II	2445	1214	878	637	50	5224	3851	1938	1813	73	52	7726

Raninagar-I	5722	5255	2691	173	0	13,841	5273	5181	1944	143	0	12,542
Raninagar-II	6018	3721	2213	929	0	12,881	10,267	5539	1948	45	0	17,799
Sagardighi	9609	14,894	8768	1555	0	34,826	18,314	9542	2699	174	0	30,729
Samserganj	3424	2678	1418	83	0	7603	2525	4750	1681	8	0	8964
Suti-I	4313	4420	3866	710	54	13,363	5403	4917	1812	75	0	12,207
Suti-II	2806	1653	964	196	0	5619	4302	4996	1717	8	0	11,023
Total	**194,812**	**145,388**	**85,226**	**15,580**	**773**	**441,779**	**238,482**	**148,089**	**52,478**	**3098**	**532**	**442,677**

Source: Agriculture Census, Government of India, 1995–1996 and 2015–2016

Annexure B.4 Land size class category-wise net sown area in Murshidabad district, 1995–1996 and 2015–2016 (area in hectares)

	1995–1996						2015–2016					
Blocks	Marginal	Small	Semi-medium	Medium	Large	All classes	Marginal	Small	Semi-medium	Medium	Large	All classes
Berhampore	10,014	6613	3575	548	272	21,022	14,078	6777	1786	272	175	23,087
Burwan	9364	11,319	4147	546	0	25,376	12,179	6784	2237	117	53	21,370
Beldanga-I	9646	3211	1571	506	0	14,934	9396	4864	1734	80	0	16,073
Beldanga-II	7848	4540	3074	220	45	15,727	11,684	6073	2153	33	0	19,944
Bhagwangola-I	10,064	6141	4318	266	0	20,789	7267	5053	1874	87	70	14,351
Bhagwangola-II	4297	2598	1454	127	0	8476	5829	5017	1973	370	0	13,189
Bharatpur-I	8744	6707	2813	513	0	18,777	9020	5708	1918	171	0	16,817
Bharatpur-II	5367	6271	3376	127	0	15,141	7003	5966	1998	58	0	15,025
Domkal	9217	5122	2299	855	57	17,550	16,202	6274	2132	121	0	24,729
Farakka	1997	1557	651	123	47	4375	3092	4747	1764	29	55	9688
Hariharpara	10,631	6029	4032	1297	0	21,989	10,765	5865	1882	175	0	18,688
Jalangi	5760	3881	1270	247	0	11,158	9662	5119	1917	8	0	16,707
Kandi	10,409	7334	2232	143	0	20,118	8884	3858	1829	8	0	14,579
Khargram	17,044	5854	6125	925	0	29,948	10,833	6960	2557	166	0	20,516
Lalgola	6240	5008	4202	562	0	16,012	10,213	6034	1666	6	0	17,920
Murshidabad-Jiaganj	8442	5996	3603	588	136	18,765	11,795	6196	2014	8	67	20,079
Nabagram	10,962	9438	7743	1678	0	29,821	12,807	7929	2386	184	0	23,307
Nawda	8122	7900	4021	932	0	20,975	13,095	7829	2262	456	60	23,701
Raghunathganj-I	3433	2971	1800	279	67	8550	1614	1451	18,60	101	0	5025
Raghunathganj-II	2436	1197	817	624	26	5100	3794	1880	1813	73	52	7612

Raninagar-I	5625	5146	2620	173	0	13,564	5197	5106	1941	143	0	12,388
Raninagar-II	5920	3618	2067	900	0	12,505	10,068	5296	1872	0	0	17,236
Sagardighi	9562	14,766	8624	1514	0	34,466	18,184	9367	2645	174	0	30,370
Samserganj	3424	2677	1417	82	0	7600	2508	3915	1680	8	0	8111
Suti-I	4247	4385	3793	698	49	13,172	5318	4858	1774	75	0	12,025
Suti-II	2621	1360	831	157	0	4969	4256	4984	1716	8	0	10,964
Total	**191,436**	**141,639**	**82,475**	**14,630**	**699**	**430,879**	**234,742**	**143,913**	**51,385**	**2931**	**532**	**433,503**

Source: Agriculture Census, Government of India, 1995–1996 and 2015–2016

Annexure B.5 Land size class category-wise area under current fallow in Murshidabad district, 1995–1996 and 2015–2016 (area in hectares)

	1995–1996						2015–2016					
Blocks	Marginal	Small	Semi-medium	Medium	Large	All classes	Marginal	Small	Semi-medium	Medium	Large	All classes
Berhampore	28	51	28	2	0	109	8	0	0	0	0	8
Burwan	11	37	6	5	0	59	16	0	0	0	0	16
Beldanga-I	12	13	5	1	0	31	1	0	0	0	0	1
Beldanga-II	16	0	1	1	0	18	14	0	0	0	0	14
Bhagwangola-I	65	29	93	41	0	228	66	0	0	0	0	66
Bhagwangola-II	4	0	0	1	0	5	120	0	2	0	0	121
Bharatpur-I	114	156	0	0	0	270	129	97	38	0	0	264
Bharatpur-II	0	0	0	0	0	0	55	9	0	0	0	64
Domkal	62	65	46	32	1	206	58	0	0	0	0	58
Farakka	6	11	1	2	2	22	0	0	0	0	0	0
Hariharpara	26	17	8	15	0	66	34	152	0	0	0	186
Jalangi	35	17	3	16	0	71	240	601	0	0	0	841
Kandi	0	0	0	3	0	3	3	0	0	0	0	3
Khargram	53	64	84	11	0	212	27	0	0	0	0	27
Lalgola	51	94	64	8	0	217	140	0	0	0	0	140
Murshidabad-Jiaganj	25	105	8	153	4	295	177	0	0	0	0	177
Nabagram	30	48	21	16	0	115	186	56	37	24	0	303
Nawda	10	14	35	3	0	62	49	127	0	0	0	176
Raghunathganj-I	4	0	2	0	2	8	14	0	0	0	0	14
Raghunathganj-II	9	17	61	13	24	124	55	57	0	0	0	112
Raninagar-I	33	54	0	0	0	87	37	62	0	0	0	99

Raninagar-II	34	38	46	15	0	133	49	59	0	0	0	108
Sagardighi	15	5	37	1	0	58	128	79	0	0	0	207
Samserganj	0	1	1	1	0	3	17	729	0	0	0	747
Suti-I	59	35	72	11	2	179	0	0	0	0	0	0
Suti-II	88	84	47	7	0	226	16	0	0	0	0	16
Total	**790**	**955**	**669**	**358**	**35**	**2807**	**1638**	**2028**	**77**	**24**	**0**	**3766**

Source: Agriculture Census, Government of India, 1995–1996 and 2015–2016

Annexure B.6 Land size class category-wise net cultivated area in Murshidabad district, 1995–1996 and 2015–2016 (area in hectares)

Blocks	1995–1996						2015–2016					
	Marginal	Small	Semi-medium	Medium	Large	All classes	Marginal	Small	Semi-medium	Medium	Large	All classes
Berhampore	10,042	6664	3603	550	272	21,131	14,086	6777	1786	272	175	23,096
Burwan	9375	11,356	4153	551	0	25,435	12,195	6784	2237	117	53	21,386
Beldanga-I	9658	3224	1576	507	0	14,965	9396	4864	1734	80	0	16,074
Beldanga-II	7864	4540	3075	221	45	15,745	11,698	6073	2153	33	0	19,958
Bhagwangola-I	10,129	6170	4411	307	0	21,017	7333	5053	1874	87	70	14,417
Bhagwangola-II	4301	2598	1454	128	0	8481	5949	5017	1974	370	0	13,310
Bharatpur-I	8858	6863	2813	513	0	19,047	9149	5805	1956	171	0	17,081
Bharatpur-II	5367	6271	3376	127	0	15,141	7058	5975	1998	58	0	15,089
Domkal	9279	5187	2345	887	58	17,756	16,260	6274	2132	121	0	24,787
Farakka	2003	1568	652	125	49	4397	3092	4747	1764	29	55	9688
Hariharpara	10,657	6046	4040	1312	0	22,055	10,799	6017	1882	175	0	18,874
Jalangi	5795	3898	1273	263	0	11,229	9903	5720	1917	8	0	17,548
Kandi	10,409	7334	2232	146	0	20,121	8887	3858	1829	8	0	14,582
Khargram	17,097	5918	6209	936	0	30,160	10,859	6960	2557	166	0	20,542
Lalgola	6291	5102	4266	570	0	16,229	10,353	6034	1666	6	0	18,060
Murshidabad-Jiaganj	8467	6101	3611	741	140	19,060	11,972	6196	2014	8	67	20,257
Nabagram	10,992	9486	7764	1694	0	29,936	12,993	7985	2423	208	0	23,609
Nawda	8132	7914	4056	935	0	21,037	13,143	7956	2262	456	60	23,877
Raghunathganj-I	3437	2971	1802	279	69	8558	1628	1451	1860	101	0	5039
Raghunathganj-II	2445	1214	878	637	50	5224	3849	1937	1813	73	52	7723
Raninagar-I	5658	5200	2620	173	0	13,651	5234	5168	1941	143	0	12,486

Raninagar-II	5954	3656	2113	915	0	12,638	10,117	5355	1872	0	0	17,345
Sagardighi	9577	14,771	8661	1515	0	34,524	18,312	9446	2645	174	0	30,578
Samserganj	3424	2678	1418	83	0	7603	2525	4645	1680	8	0	8858
Suti-I	4306	4420	3865	709	51	13,351	5318	4858	1774	75	0	12,025
Suti-II	2709	1444	878	164	0	5195	4272	4984	1716	8	0	10,980
Total	**192,226**	**142,594**	**83,144**	**14,988**	**734**	**433,686**	**236,380**	**145,941**	**51,461**	**2955**	**532**	**437,269**

Source: Agriculture Census, Government of India, 1995–1996 and 2015–2016

Annexure B.7 Land size class category-wise other uncultivated land excluding fallow land in Murshidabad district, 1995–1996 and 2015–2016 (area in hectares)

	1995–1996						2015–2016					
Blocks	Marginal	Small	Semi-medium	Medium	Large	All classes	Marginal	Small	Semi-medium	Medium	Large	All classes
Berhampore	6	11	13	0	0	30	8	1	1	0	0	10
Burwan	3	0	4	0	0	7	0	0	0	0	0	0
Beldanga-I	5	3	0	0	0	8	2	0	0	2	0	5
Beldanga-II	0	0	0	0	0	0	0	0	0	0	0	1
Bhagwangola-I	4	0	0	0	0	4	2	0	0	0	0	2
Bhagwangola-II	39	8	13	0	0	60	26	0	0	0	0	26
Bharatpur-I	0	0	0	0	0	0	13	5	0	0	0	19
Bharatpur-II	0	0	0	0	0	0	0	0	0	0	0	0
Domkal	11	16	17	3	0	47	1	18	0	0	0	19
Farakka	22	4	24	0	0	50	0	0	0	0	0	0
Hariharpara	19	56	72	89	0	236	0	0	0	0	0	0
Jalangi	0	1	0	0	0	1	0	0	0	0	0	0
Kandi	12	12	0	0	0	24	0	0	2	0	0	2
Khargram	10	6	3	0	0	19	2	0	9	0	0	10
Lalgola	0	5	32	0	0	37	0	0	6	0	0	6
Murshidabad-Jiaganj	4	14	0	0	0	18	88	132	19	1	0	240
Nabagram	1	4	0	0	0	5	3	12	16	1	0	32
Nawda	99	231	150	65	0	545	4	7	0	0	0	11
Raghunathganj-I	7	2	0	0	0	9	0	9	6	0	0	15
Raghunathganj-II	0	0	0	0	0	0	0	0	0	0	0	0

Raninagar-I	6	0	0	0	0	6	4	2	0	0	0	7
Raninagar-II	14	15	48	2	0	79	0	0	0	0	0	0
Sagardighi	5	0	76	27	0	108	0	53	0	0	0	53
Samserganj	0	0	0	0	0	0	0	0	0	0	0	0
Suti-I	0	0	0	0	0	0	0	0	0	0	0	0
Suti-II	8	22	28	0	0	58	20	0	0	0	0	20
Total	**275**	**410**	**480**	**186**	**0**	**1351**	**173**	**239**	**58**	**5**	**0**	**475**

Source: Agriculture Census, Government of India, 1995–1996 and 2015–2016

Annexure B.8 Land size class category-wise fallow land other than current fallow in Murshidabad district, 1995–1996 and 2015–2016 (area in hectares)

Blocks	1995–1996						2015–2016					
	Marginal	Small	Semi-medium	Medium	Large	All classes	Marginal	Small	Semi-medium	Medium	Large	All classes
Berhampore	4	0	0	0	0	4	0	0	3	0	0	3
Burwan	0	3	0	0	0	3	11	11	10	0	0	32
Beldanga-I	5	5	0	0	0	10	5	0	0	0	0	5
Beldanga-II	8	78	5	1	5	97	0	0	0	0	0	1
Bhagwangola-I	25	0	0	0	0	25	0	0	0	0	0	0
Bhagwangola-II	76	6	5	0	0	87	31	0	0	0	0	31
Bharatpur-I	3	0	0	0	0	3	0	0	2	0	0	2
Bharatpur-II	0	0	0	0	0	0	0	0	0	0	0	0
Domkal	0	0	0	0	0	0	2	2	0	11	0	15
Farakka	0	0	0	0	0	0	0	0	0	0	0	0
Hariharpara	1	5	9	7	0	22	0	0	0	0	0	0
Jalangi	1	0	0	0	0	1	0	0	0	0	0	0
Kandi	0	0	0	3	0	3	0	0	0	8	0	8
Khargram	61	33	3	21	0	118	0	0	0	0	0	0
Lalgola	17	22	0	0	0	39	0	0	2	0	0	2
Murshidabad-Jiaganj	0	0	0	0	0	0	31	16	0	0	0	46
Nabagram	1	2	0	0	0	3	4	9	0	0	0	14
Nawda	5	0	4	8	0	17	4	8	0	0	0	12
Raghunathganj-I	0	0	0	0	0	0	0	0	0	0	0	0
Raghunathganj-II	0	0	0	0	0	0	0	0	0	0	0	0
Raninagar-I	4	7	0	0	0	11	21	1	1	0	0	22

Raninagar-II	0	4	12	2	0	18	1	0	0	0	0	1
Sagardighi	1	0	0	0	0	1	0	21	0	0	0	21
Samserganj	0	0	0	0	0	0	0	0	0	0	0	0
Suti-I	0	0	0	0	0	0	59	59	9	0	0	127
Suti-II	14	33	0	32	0	79	0	0	0	0	0	0
Total	**226**	**198**	**38**	**74**	**5**	**541**	**171**	**127**	**27**	**18**	**0**	**343**

Source: Agriculture Census, Government of India, 1995–1996 and 2015–2016

Annexure B.9 Land size class category-wise culturable waste land in Murshidabad district, 1995–1996 and 2015–2016 (area in hectares)

Blocks	1995–1996						2015–2016					
	Marginal	Small	Semi-medium	Medium	Large	All classes	Marginal	Small	Semi-medium	Medium	Large	All classes
Berhampore	0	0	0	0	0	0	5	0	0	0	0	5
Burwan	0	0	0	0	0	0	0	0	0	0	0	0
Beldanga-I	0	2	0	0	0	2	30	0	5	0	0	35
Beldanga-II	0	0	0	0	0	0	0	0	0	0	0	0
Bhagwangola-I	12	0	0	0	0	12	0	0	0	0	0	0
Bhagwangola-II	3	0	41	0	0	44	40	84	0	0	0	124
Bharatpur-I	2	0	0	0	0	2	42	53	1	0	0	97
Bharatpur-II	0	0	0	0	0	0	0	0	0	0	0	0
Domkal	37	37	24	5	0	103	0	0	0	0	0	0
Farakka	0	0	0	0	0	0	0	0	0	0	0	0
Hariharpara	0	0	0	0	0	0	0	0	0	0	0	0
Jalangi	140	121	135	0	0	396	0	0	0	0	0	0
Kandi	0	0	0	0	0	0	4	0	2	0	0	6
Khargram	1	0	0	0	0	1	1	0	0	0	0	1
Lalgola	6	0	0	0	0	6	14	11	3	0	0	27
Murshidabad-Jiaganj	0	0	0	0	0	0	43	9	4	0	0	57
Nabagram	6	4	0	0	0	10	15	0	5	0	0	20
Nawda	0	0	0	27	0	27	2	2	0	0	0	4
Raghunathganj-I	12	0	0	5	0	17	0	3	0	0	0	3
Raghunathganj-II	0	0	0	0	0	0	0	0	0	0	0	0
Raninagar-I	0	0	0	0	0	0	1	0	0	0	0	1

Raninagar-II	0	0	0	0	0	0	146	177	73	45	0	441
Sagardighi	1	0	1	0	0	2	2	21	54	0	0	77
Samserganj	0	0	0	0	0	0	0	0	0	0	0	0
Suti-I	5	0	0	0	0	5	0	0	0	0	0	0
Suti-II	54	122	58	0	0	234	0	0	0	0	0	0
Total	**279**	**286**	**259**	**37**	**0**	**861**	**344**	**361**	**148**	**45**	**0**	**898**

Source: Agriculture Census, Government of India, 1995–1996 and 2015–2016

Annexure B.10 Land size class category-wise total uncultivated land in Murshidabad district, 1995–1996 and 2015–2016 (area in hectares)

Blocks	1995–1996						2015–2016					
	Marginal	Small	Semi-medium	Medium	Large	All classes	Marginal	Small	Semi-medium	Medium	Large	All classes
Berhampore	10	11	13	0	0	34	13	1	4	0	0	18
Burwan	3	3	4	0	0	10	11	11	10	0	0	33
Beldanga-I	10	10	0	0	0	20	38	0	5	2	0	45
Beldanga-II	8	78	5	1	5	97	0	1	0	1	0	1
Bhagwangola-I	41	0	0	0	0	41	2	0	0	0	0	2
Bhagwangola-II	118	14	59	0	0	191	97	84	0	0	0	181
Bharatpur-I	5	0	0	0	0	5	56	58	4	0	0	117
Bharatpur-II	0	0	0	0	0	0	0	0	0	0	0	0
Domkal	48	53	41	8	0	150	3	20	0	11	0	33
Farakka	22	4	24	0	0	50	0	0	0	0	0	0
Hariharpara	20	61	81	96	0	258	0	0	0	0	0	0
Jalangi	141	122	135	0	0	398	0	0	0	0	0	0
Kandi	12	12	0	3	0	27	4	0	4	8	0	16
Khargram	72	39	6	21	0	138	2	0	9	0	0	11
Lalgola	23	27	32	0	0	82	14	11	10	0	0	35
Murshidabad-Jiaganj	4	14	0	0	0	18	162	157	23	1	0	343
Nabagram	8	10	0	0	0	18	22	21	21	1	0	65
Nawda	104	231	154	100	0	589	10	17	0	0	0	27
Raghunathganj-I	19	2	0	5	0	26	0	13	6	0	0	18
Raghunathganj-II	0	0	0	0	0	0	0	0	0	0	0	0
Raninagar-I	10	7	0	0	0	17	26	3	1	0	0	30

Raninagar-II	14	19	60	4	0	97	148	177	73	45	0	442
Sagardighi	7	0	77	27	0	111	2	96	54	0	0	151
Samserganj	0	0	0	0	0	0	0	0	0	0	0	0
Suti-I	5	0	0	0	0	5	59	59	9	0	0	127
Suti-II	76	177	86	32	0	371	20	0	0	0	0	20
Total	**780**	**894**	**777**	**297**	**5**	**2753**	**688**	**727**	**233**	**69**	**0**	**1716**

Source: Agriculture Census, Government of India, 1995–1996 and 2015–2016

Annexure B.11 Land size class category-wise land not available for cultivation in Murshidabad district, 1995–1996 and 2015–2016 (area in hectares)

	1995–1996						2015–2016					
Blocks	Marginal	Small	Semi-medium	Medium	Large	All classes	Marginal	Small	Semi-medium	Medium	Large	All classes
Berhampore	66	29	3	4	24	126	62	23	19	13	0	117
Burwan	58	100	15	1	0	174	71	90	27	5	0	193
Beldanga-I	7	0	0	2	0	9	0	0	0	0	0	0
Beldanga-II	204	66	81	2	1	354	6	2	8	0	0	16
Bhagwangola-I	21	20	1	2	0	44	0	0	0	0	0	0
Bhagwangola-II	36	160	0	1	0	197	382	286	516	11	0	1195
Bharatpur-I	28	6	0	0	0	34	86	81	1	0	0	168
Bharatpur-II	0	0	0	0	0	0	0	0	0	0	0	0
Domkal	43	88	127	67	2	327	1	0	1	0	0	2
Farakka	2	0	0	1	1	4	0	0	0	2	0	2
Hariharpara	151	205	133	38	0	527	8	3	1	0	0	12
Jalangi	11	23	6	6	0	46	5	0	1	0	0	6
Kandi	0	0	0	0	0	0	2	8	7	0	0	16
Khargram	3	11	27	39	0	80	12	33	2	0	0	47
Lalgola	630	685	517	70	0	1902	279	63	27	0	0	368
Murshidabad-Jiaganj	0	0	1	2	2	5	350	529	24	4	0	907
Nabagram	363	201	153	13	0	730	33	58	80	17	0	187
Nawda	30	34	18	5	0	87	64	84	2	0	0	150
Raghunathganj-I	1	23	81	18	1	124	0	26	34	22	0	82
Raghunathganj-II	0	0	0	0	0	0	2	1	0	0	0	3
Raninagar-I	54	48	71	0	0	173	13	10	2	0	0	26

Raninagar-II	50	46	40	10	0	146	2	7	3	0	0	12
Sagardighi	25	123	30	13	0	191	0	0	0	0	0	0
Samserganj	0	0	0	0	0	0	0	105	1	0	0	106
Suti-I	2	0	1	1	3	7	26	0	29	0	0	55
Suti-II	21	32	0	0	0	53	10	12	1	0	0	23
Total	**1806**	**1900**	**1305**	**295**	**34**	**5340**	**1414**	**1421**	**784**	**74**	**0**	**3694**

Source: Agriculture Census, Government of India, 1995–1996 and 2015–2016

Annexure B.12 Percentage of gross cropped area in Murshidabad district, 1995–1996 and 2015–2016

	1995–1996						2015–2016					
Blocks	Marginal	Small	Semi-medium	Medium	Large	All classes	Marginal	Small	Semi-medium	Medium	Large	All classes
Berhampore	46.32	31.38	19.11	2.46	0.73	100.00	63.09	29.16	6.47	0.97	0.32	100.00
Burwan	37.45	44.67	15.82	2.06	0.00	100.00	59.21	31.42	8.81	0.48	0.09	100.00
Beldanga-I	67.05	20.95	9.24	2.75	0.00	100.00	66.12	26.84	6.74	0.31	0.00	100.00
Beldanga-II	53.97	27.42	17.44	1.01	0.16	100.00	61.13	29.78	9.02	0.07	0.00	100.00
Bhagwangola-I	49.71	29.28	19.85	1.16	0.00	100.00	56.60	32.02	10.51	0.61	0.26	100.00
Bhagwangola-II	51.95	30.73	16.14	1.18	0.00	100.00	50.12	33.83	14.51	1.54	0.00	100.00
Bharatpur-I	47.01	35.23	15.34	2.43	0.00	100.00	58.97	33.77	6.37	0.89	0.00	100.00
Bharatpur-II	35.74	41.52	21.94	0.80	0.00	100.00	55.40	35.72	8.70	0.17	0.00	100.00
Domkal	54.91	29.18	11.68	4.07	0.16	100.00	66.66	27.83	5.26	0.25	0.00	100.00
Farakka	42.31	36.30	17.71	2.77	0.91	100.00	32.82	54.78	12.01	0.13	0.25	100.00
Hariharpara	53.10	26.38	15.88	4.65	0.00	100.00	61.85	31.57	6.02	0.56	0.00	100.00
Jalangi	53.32	34.13	10.83	1.72	0.00	100.00	61.36	30.29	8.33	0.02	0.00	100.00
Kandi	51.31	36.35	11.56	0.78	0.00	100.00	64.88	27.78	7.31	0.03	0.00	100.00
Khargram	58.42	19.55	19.36	2.67	0.00	100.00	54.33	36.86	8.44	0.38	0.00	100.00
Lalgola	37.23	31.80	27.30	3.67	0.00	100.00	56.62	33.32	10.05	0.02	0.00	100.00
Murshidabad-Jiaganj	46.56	31.57	18.42	3.13	0.31	100.00	61.97	30.59	7.27	0.02	0.15	100.00
Nabagram	36.25	30.91	27.50	5.34	0.00	100.00	53.65	37.80	7.90	0.65	0.00	100.00
Nawda	42.02	37.87	16.75	3.36	0.00	100.00	57.91	34.45	6.22	1.25	0.16	100.00
Raghunathganj-I	39.70	37.14	19.38	3.26	0.53	100.00	40.33	34.95	23.80	0.94	0.00	100.00
Raghunathganj-II	50.57	22.78	15.08	10.95	0.62	100.00	53.98	32.70	12.47	0.50	0.35	100.00
Raninagar-I	39.80	36.19	22.68	1.30	0.04	100.00	44.99	39.55	14.56	0.90	0.00	100.00

Raninagar-II	48.87	29.81	14.96	6.36	0.00	100.00	61.97	27.46	10.57	0.00	0.00	100.00
Sagardighi	28.22	43.17	24.34	4.28	0.00	100.00	61.02	32.35	6.27	0.36	0.00	100.00
Samserganj	44.07	36.05	18.76	1.12	0.00	100.00	34.91	46.76	18.27	0.06	0.00	100.00
Suti-I	32.74	34.17	27.78	5.10	0.21	100.00	51.88	37.09	10.59	0.45	0.00	100.00
Suti-II	53.70	27.16	16.27	2.87	0.00	100.00	45.25	40.87	13.83	0.04	0.00	100.00
Total	**46.22**	**32.19**	**18.45**	**3.04**	**0.10**	**100.00**	**57.49**	**33.20**	**8.81**	**0.44**	**0.06**	**100.00**

Source: Agriculture Census, Government of India, 1995–1996 and 2015–2016

Annexure B.13 Percentage of gross cropped area under cereals in Murshidabad district, 1995–1996 and 2015–2016

	1995–1996						2015–2016					
Blocks	Marginal	Small	Semi-medium	Medium	Large	All classes	Marginal	Small	Semi-medium	Medium	Large	All classes
Berhampore	60.85	62.03	61.54	62.66	94.01	61.64	63.39	66.26	76.01	100.00	100.00	65.51
Burwan	79.61	82.31	81.67	81.90	0.00	81.19	67.83	65.86	76.75	48.55	100.00	67.93
Beldanga-I	53.70	58.63	59.81	60.69	0.00	55.50	75.58	84.44	100.00	100.00	0.00	79.68
Beldanga-II	59.29	59.78	66.88	60.54	79.63	60.79	60.29	59.78	64.95	100.00	0.00	60.59
Bhagwangola-I	48.39	48.95	47.13	43.17	0.00	48.24	46.83	59.46	56.66	48.50	0.00	51.79
Bhagwangola-II	44.69	37.14	44.39	40.59	0.00	42.27	48.62	57.94	48.06	0.00	0.00	50.95
Bharatpur-I	88.03	86.75	86.60	90.86	0.00	87.42	67.80	71.44	93.38	81.07	0.00	70.77
Bharatpur-II	91.76	93.74	94.05	94.44	0.00	93.10	66.78	74.81	89.40	100.00	0.00	71.68
Domkal	45.63	45.14	48.83	46.86	100.00	46.00	63.83	66.67	82.89	100.00	0.00	65.71
Farakka	56.51	62.00	42.11	95.83	100.00	57.44	62.37	72.45	79.18	72.41	0.00	69.76
Hariharpara	48.07	55.29	57.55	69.37	0.00	52.47	77.26	79.40	100.00	100.00	0.00	79.44
Jalangi	49.70	50.86	49.32	78.95	0.00	50.56	47.49	54.69	54.55	0.00	0.00	50.25
Kandi	80.29	80.33	78.69	68.61	0.00	80.03	73.06	72.77	99.73	100.00	0.00	74.93
Khargram	76.25	75.14	76.12	72.81	0.00	75.92	82.45	78.77	98.97	97.94	0.00	82.54
Lalgola	68.52	58.46	54.40	50.62	0.00	60.81	42.06	38.74	43.35	0.00	0.00	41.08
Murshidabad-Jiaganj	66.62	62.33	63.56	67.06	100.00	64.82	47.08	48.84	30.34	0.00	0.00	46.32
Nabagram	76.08	77.22	76.95	81.32	0.00	76.95	84.72	81.16	94.65	90.60	0.00	84.19
Nawda	46.79	49.40	56.75	55.99	0.00	49.76	80.75	81.62	100.00	96.49	100.00	82.48
Raghunathganj-I	63.20	63.57	73.88	72.20	100.00	65.89	75.64	78.50	92.68	100.00	0.00	80.92
Raghunathganj-II	64.35	66.68	60.68	67.65	55.77	64.63	42.76	58.81	1.42	0.00	0.00	42.48
Raninagar-I	53.32	54.02	51.55	46.45	100.00	53.10	39.62	34.40	34.32	50.44	0.00	36.88

Raninagar-II	47.92	45.49	44.87	44.67	0.00	46.53	43.53	46.01	50.85	0.00	0.00	44.99
Sagardighi	75.74	79.19	81.40	80.67	0.00	78.82	79.77	86.27	92.91	73.21	0.00	82.67
Samserganj	54.76	53.73	60.39	69.03	0.00	55.60	23.68	7.71	0.00	0.00	0.00	11.87
Suti-I	64.62	60.25	62.18	62.35	100.00	62.41	85.82	84.82	99.89	100.00	0.00	87.01
Suti-II	51.84	59.72	46.60	44.60	0.00	52.92	17.41	11.76	11.09	0.00	0.00	14.23
Total	**62.39**	**64.93**	**65.74**	**65.94**	**93.38**	**63.96**	**63.50**	**64.72**	**67.57**	**74.23**	**54.14**	**64.31**

Source: Agriculture Census, Government of India, 1995–1996 and 2015–2016

Annexure B.14 Percentage of gross cropped area under pulses in Murshidabad district, 1995–1996 and 2015–2016

	1995–1996						2015–2016					
Blocks	Marginal	Small	Semi-medium	Medium	Large	All classes	Marginal	Small	Semi-medium	Medium	Large	All classes
Berhampore	1.98	2.86	2.37	5.49	0.00	2.41	1.82	0.30	0.00	0.00	0.00	1.24
Burwan	7.63	5.58	4.92	9.64	0.00	6.33	5.61	8.10	3.60	3.26	0.00	6.19
Beldanga-I	5.34	4.77	5.87	3.07	0.00	5.21	0.33	0.46	0.00	0.00	0.00	0.35
Beldanga-II	2.75	3.16	2.38	9.34	5.56	2.87	0.89	2.39	0.60	0.00	0.00	1.31
Bhagwangola-I	4.08	2.56	3.60	0.99	0.00	3.50	2.75	3.71	5.61	16.17	0.00	3.44
Bhagwangola-II	9.47	9.23	8.72	10.89	0.00	9.29	3.70	6.94	1.29	0.00	0.00	4.39
Bharatpur-I	2.25	2.54	1.79	0.00	0.00	2.22	9.43	12.43	0.75	3.67	0.00	9.84
Bharatpur-II	0.02	0.00	0.00	0.00	0.00	0.01	12.10	2.97	0.00	0.00	0.00	7.76
Domkal	7.87	7.48	9.06	10.08	0.00	7.97	1.45	0.60	0.00	0.00	0.00	1.13
Farakka	7.79	5.52	27.42	0.00	0.00	10.15	5.56	1.03	0.00	0.00	0.00	2.39
Hariharpara	7.74	10.27	12.02	17.46	0.00	9.54	0.33	0.38	0.00	0.00	0.00	0.32
Jalangi	24.71	17.17	14.90	0.00	0.00	20.65	4.58	2.67	2.80	0.00	0.00	3.85
Kandi	3.92	3.42	3.60	1.29	0.00	3.68	1.00	0.00	0.00	0.00	0.00	0.65
Khargram	2.92	3.31	2.14	6.75	0.00	2.95	2.29	6.41	0.00	1.55	0.00	3.61
Lalgola	1.75	3.04	0.16	2.72	0.00	1.76	0.45	0.00	6.66	0.00	0.00	0.93
Murshidabad-Jiaganj	2.14	1.42	2.16	0.67	0.00	1.86	9.09	19.46	6.69	0.00	0.00	12.07
Nabagram	2.49	2.26	2.84	0.83	0.00	2.43	2.93	9.27	0.00	0.00	0.00	5.08
Nawda	14.21	14.44	13.42	11.83	0.00	14.09	0.28	0.02	0.00	2.41	0.00	0.20
Raghunathganj-I	13.01	17.99	12.00	15.25	0.00	14.67	9.03	3.39	2.61	0.00	0.00	5.45
Raghunathganj-II	19.99	16.92	18.59	12.85	32.69	18.37	9.38	23.79	94.69	87.67	0.00	25.08
Raninagar-I	3.08	3.08	10.31	4.14	0.00	4.74	0.69	0.00	0.00	0.00	0.00	0.31
Raninagar-II	9.58	11.63	12.74	9.92	0.00	10.68	5.15	7.05	0.00	0.00	0.00	5.13

Sagardighi	4.50	2.72	4.14	4.66	0.00	3.65	5.97	1.17	0.00	0.00	0.00	4.02
Samserganj	27.15	28.75	29.54	15.48	0.00	28.04	3.50	24.26	25.20	0.00	0.00	17.18
Suti-I	20.24	23.48	22.40	25.26	0.00	22.16	1.60	0.00	0.11	0.00	0.00	0.84
Suti-II	19.75	17.80	22.92	2.88	0.00	19.25	0.25	0.00	0.00	0.00	0.00	0.11
Total	**6.78**	**6.70**	**6.74**	**7.75**	**2.50**	**6.77**	**3.79**	**4.98**	**4.35**	**3.22**	**0.00**	**4.23**

Source: Agriculture Census, Government of India, 1995–1996 and 2015–2016

Annexure B.15 Percentage of gross cropped area under spices in Murshidabad district, 1995–1996 and 2015–2016

	1995–1996						2015–2016					
Blocks	Marginal	Small	Semi-medium	Medium	Large	All classes	Marginal	Small	Semi-medium	Medium	Large	All classes
Berhampore	0.11	0.00	1.79	0.00	0.00	0.39	0.23	0.00	0.00	0.00	0.00	0.14
Burwan	0.02	0.01	0.00	0.00	0.00	0.01	0.00	0.00	0.00	0.00	0.00	0.00
Beldanga-I	2.12	2.13	2.24	3.76	0.00	2.18	0.01	0.00	0.00	0.00	0.00	0.01
Beldanga-II	0.59	0.21	0.19	0.00	0.00	0.41	0.17	0.00	0.00	0.00	0.00	0.10
Bhagwangola-I	0.69	0.73	0.05	0.79	0.00	0.58	0.49	0.61	0.00	0.00	0.00	0.47
Bhagwangola-II	0.35	0.15	0.00	0.00	0.00	0.23	1.09	0.66	0.00	0.00	0.00	0.77
Bharatpur-I	0.00	0.00	0.00	0.00	0.00	0.00	0.09	0.00	0.00	0.00	0.00	0.06
Bharatpur-II	0.06	0.03	0.00	0.00	0.00	0.03	0.00	0.00	0.00	0.00	0.00	0.00
Domkal	0.37	0.29	0.09	0.39	0.00	0.32	0.14	0.00	0.00	0.00	0.00	0.09
Farakka	0.00	0.00	0.00	0.00	0.00	0.00	0.06	0.00	0.00	0.00	0.00	0.02
Hariharpara	0.65	0.47	0.47	0.88	0.00	0.58	0.00	0.00	0.00	0.00	0.00	0.00
Jalangi	0.03	0.06	0.00	0.00	0.00	0.04	0.27	0.09	0.00	0.00	0.00	0.19
Kandi	0.01	0.00	0.00	0.00	0.00	0.01	0.00	0.00	0.00	0.00	0.00	0.00
Khargram	0.00	0.00	0.00	0.00	0.00	0.00	0.00	0.00	0.00	0.00	0.00	0.00
Lalgola	0.34	0.17	0.00	0.00	0.00	0.18	0.24	1.32	0.00	0.00	0.00	0.58
Murshidabad-Jiaganj	1.57	1.24	1.48	0.59	0.00	1.41	0.61	0.41	0.70	0.00	0.00	0.55
Nabagram	0.09	0.12	0.12	0.00	0.00	0.10	0.00	0.00	0.00	0.00	0.00	0.00
Nawda	0.01	0.02	0.15	0.00	0.00	0.04	0.00	0.00	0.00	0.00	0.00	0.00
Raghunathganj-I	0.00	0.00	0.00	0.00	0.00	0.00	0.00	0.00	0.00	0.00	0.00	0.00
Raghunathganj-II	0.00	0.10	0.00	0.00	0.00	0.02	0.29	0.46	0.00	0.00	0.00	0.31
Raninagar-I	0.33	0.30	0.20	0.00	0.00	0.28	0.69	1.04	0.84	0.00	0.00	0.85
Raninagar-II	0.85	1.30	1.60	1.57	0.00	1.14	0.79	0.57	1.69	0.00	0.00	0.82

Sagardighi	0.16	0.04	0.44	0.00	0.00	0.17	0.02	0.00	0.00	0.00	0.00	0.01
Samserganj	3.24	0.80	0.97	0.00	0.00	1.90	0.00	0.00	0.00	0.00	0.00	0.00
Suti-I	0.05	0.00	0.06	0.00	0.00	0.03	0.00	0.00	0.00	0.00	0.00	0.00
Suti-II	0.00	0.00	0.00	0.00	0.00	0.00	1.08	0.00	0.00	0.00	0.00	0.49
Total	**0.48**	**0.30**	**0.39**	**0.41**	**0.00**	**0.40**	**0.21**	**0.17**	**0.15**	**0.00**	**0.00**	**0.19**

Source: Agriculture Census, Government of India, 1995–1996 and 2015–2016

Annexure B.16 Percentage of gross cropped area under fruits in Murshidabad district, 1995–1996 and 2015–2016

	1995–1996						2015–2016					
Blocks	Marginal	Small	Semi-medium	Medium	Large	All classes	Marginal	Small	Semi-medium	Medium	Large	All classes
Berhampore	0.06	0	0	1.77	0	0.07	1.33	2.27	0	0	0	1.5
Burwan	0.05	0	0	0	0	0.02	2.96	2.51	0	0	0	2.54
Beldanga-I	0.01	0	0.44	0	0	0.05	3.52	4.21	0	0	0	3.46
Beldanga-II	0.04	0	0	0.3	0	0.02	1.86	1.51	0	0	0	1.59
Bhagwangola-I	1.07	0.73	0.19	0	0	0.78	2.02	0	0	0	0	1.14
Bhagwangola-II	0.53	1.77	3.38	0	0	1.37	2.75	3.23	0	0	0	2.47
Bharatpur-I	0	0	0	0	0	0	1.81	2.07	0	0	0	1.76
Bharatpur-II	0	0	0	0	0	0	1.42	1.56	0	0	0	1.34
Domkal	0.11	0.06	0.39	0.52	0	0.15	3.06	3.56	0	0	0	3.03
Farakka	24.13	17.36	21.44	0	0	20.31	0.21	0	0	0	0	0.07
Hariharpara	0.04	0.17	0.11	0	0	0.08	3.65	4.54	0	0	0	3.69
Jalangi	0.04	0.29	1.07	0	0	0.24	6.17	2.61	0.06	0	0	4.58
Kandi	0	0	0	0	0	0	2.91	4.41	0	0	0	3.11
Khargram	0	0	0	0.19	0	0.01	1.39	1.95	0	0	0	1.48
Lalgola	0.57	1.36	0	0	0	0.64	1.02	4.51	0	0	0	2.08
Murshidabad-Jiaganj	1.18	1.94	5.47	3.11	0	2.27	4.31	1.24	6.87	0	0	3.55
Nabagram	0.02	0.01	0	0.08	0	0.02	1.84	1.86	0	0	0	1.69
Nawda	0.5	1.32	2.61	6.16	0	1.35	3.1	3.29	0	0	0	2.93
Raghunathganj-I	4.99	1.08	1.17	1.79	0	2.67	1.4	1.3	0	0	0	1.02
Raghunathganj-II	0.59	0.79	1.11	0.44	0	0.69	1.72	0	0	0	0	0.93
Raninagar-I	0.03	0.32	1.47	0	0	0.46	2.17	1.73	1.06	0	0	1.82
Raninagar-II	0.42	0.66	0.74	3.08	0	0.71	3.5	5.82	11.84	0	0	5.02

Sagardighi	0.05	0	0	0.09	0	0.02	1.05	0	0	0	0	0.64
Samserganj	0.18	0	0	0	0	0.08	0.14	0	0	0	0	0.05
Suti-I	0	0	0.01	0	0	0	1.83	8.57	0	0	0	4.12
Suti-II	0.17	0.91	1.84	11.51	0	0.97	3.56	3.06	0	0	0	2.86
Total	**0.44**	**0.52**	**0.78**	**1.06**	**0**	**0.54**	**2.47**	**2.35**	**0.89**	**0**	**0**	**2.28**

Source: Agriculture Census, Government of India, 1995–1996 and 2015–2016

Annexure B.17 Percentage of gross cropped area under vegetables in Murshidabad district, 1995–1996 and 2015–2016

	1995–1996						2015–2016					
Blocks	Marginal	Small	Semi-medium	Medium	Large	All classes	Marginal	Small	Semi-medium	Medium	Large	All classes
Berhampore	9.65	9.19	12.98	10.34	5.99	10.13	7.76	10.42	0	0	0	7.93
Burwan	5.85	4.69	5.02	2.94	0	5.14	15.04	16.35	9.33	24.28	0	14.98
Beldanga-I	13.18	12.3	15.26	6.73	0	13.01	20.3	9.93	0	0	0	16.08
Beldanga-II	10.19	8.91	9.52	7.83	0	9.68	13.18	13.6	6.03	0	0	12.65
Bhagwangola-I	8.98	10.2	10.08	13.66	0	9.61	13.37	8.58	9.39	0	0	11.3
Bhagwangola-II	8.6	10.17	9.95	6.93	0	9.28	14.77	3.55	5.83	0	0	9.45
Bharatpur-I	2.67	2.04	1.84	1.33	0	2.28	13.5	8.59	1.74	9.04	0	11.05
Bharatpur-II	2.45	1.54	1.9	0.51	0	1.94	7.5	7.36	0.17	0	0	6.8
Domkal	9.73	10.36	9.06	8.44	0	9.77	15.94	15.27	0	0	0	14.87
Farakka	0.64	0	0.22	1.39	0	0.35	6.18	0.96	5.67	0	0	3.24
Hariharpara	8.96	7.33	6.13	2.44	0	7.78	18.69	15.19	0	0	0	16.35
Jalangi	6.48	6.67	8.55	2.76	0	6.7	15.44	14.4	2.44	0	0	14.04
Kandi	11.79	11.98	12.54	13.27	0	11.96	19.4	18.72	0.22	0	0	17.8
Khargram	11.49	12.12	13.69	12.47	0	12.06	13.12	12.13	0.14	0.52	0	11.61
Lalgola	9.44	8.52	9.51	9.98	0	9.18	3.96	9.67	0	0	0	5.47
Murshidabad-Jiaganj	4.51	4.52	2.28	2.66	0	4.03	11.89	4.21	4.31	0	0	8.97
Nabagram	12.31	10.46	11.67	11.92	0	11.54	8.4	6.42	0.16	3.13	0	6.97
Nawda	8.51	8.51	7.74	8.79	0	8.39	15.86	15.06	0	0	0	14.38
Raghunathganj-I	10.03	8.53	5.96	4.26	0	8.44	8.89	10.76	1.56	0	0	7.72
Raghunathganj-II	1.11	1.52	1.82	4.58	5.77	1.72	10.59	3.74	0	0	0	6.94
Raninagar-I	3.17	2.82	1.96	3.85	0	2.78	12.61	18.74	6.09	1.75	0	13.98
Raninagar-II	7.9	10.79	10.55	13.23	0	9.5	16.2	15.29	0.63	0	0	14.3

Sagardighi	4.12	2.72	2.49	2.57	0	3.05	7.66	7.22	0	0	0	7.01
Samserganj	0.71	1.65	2.39	2.58	0	1.38	14.45	17.05	0	0	0	13.02
Suti-I	8.14	8.13	6.02	8.98	0	7.57	10.75	6.6	0	0	0	8.03
Suti-II	2.23	6.01	2.98	0	0	3.32	16.69	4.97	11.09	0	0	11.12
Total	**8.35**	**7.33**	**7.75**	**7.34**	**2.75**	**7.88**	**12.73**	**10.62**	**2.84**	**2.88**	**0**	**11.1**

Source: Agriculture Census, Government of India, 1995–1996 and 2015–2016

Annexure B.18 Percentage of gross cropped area under oilseeds in Murshidabad district, 1995–1996 and 2015–2016

	1995–1996						2015–2016					
Blocks	Marginal	Small	Semi-medium	Medium	Large	All classes	Marginal	Small	Semi-medium	Medium	Large	All classes
Berhampore	3.6	5.12	4.9	2.79	0	4.28	6.44	2.56	4.79	0	0	5.12
Burwan	6.25	6.78	7.95	3.64	0	6.7	7.9	6.89	10.27	23.55	0	7.86
Beldanga-I	5.69	4.79	3.39	9.7	0	5.4	0.14	0.94	0	0	0	0.34
Beldanga-II	5.83	6.91	5.91	11.45	3.7	6.19	7.68	7.57	8.94	0	0	7.76
Bhagwangola-I	11.15	10.6	11.28	10.89	0	11.01	12.24	11.35	10.97	16.17	0	11.82
Bhagwangola-II	0.95	2.1	2.98	0.5	0	1.62	9.63	11.66	19.61	0	0	11.62
Bharatpur-I	6.27	7.26	8.46	6.31	0	6.96	5.81	3.93	2.34	5.37	0	4.95
Bharatpur-II	5.55	4.52	3.7	5.05	0	4.71	6.16	4.72	6.29	0	0	5.65
Domkal	6.27	8.83	8.4	7.13	0	7.29	3.24	2.44	4.35	0	0	3.07
Farakka	0.5	0	1.85	0	0	0.54	11.88	11.95	3.78	0	0	10.9
Hariharpara	10.2	9.78	11.3	9.35	0	10.22	0	0.34	0	0	0	0.11
Jalangi	0.63	0.81	2.34	0	0	0.87	6.32	5.03	2.02	0	0	5.57
Kandi	3.56	3.2	3.56	16.83	0	3.53	2.99	3.35	0	0	0	2.87
Khargram	8.61	8.32	6.98	4.69	0	8.14	0.75	0.75	0.9	0.52	0	0.76
Lalgola	5.69	7.19	1.88	0	0	4.92	11.63	0.71	0	0	0	6.82
Murshidabad-Jiaganj	4.3	5.57	2.24	2.44	0	4.25	8.19	4.69	12.07	0	0	7.39
Nabagram	5.6	6.06	5.02	3.61	0	5.48	1.78	1.3	5.2	6.27	0	1.9
Nawda	5.24	6.32	3.52	3.18	0	5.29	0	0	0	0	0	0
Raghunathganj-I	4.04	4.93	3.78	5.16	0	4.33	4.78	6.04	3.12	0	0	4.78
Raghunathganj-II	3.73	4.24	2.77	0.76	0	3.35	8.38	6.66	0	0	0	6.69
Raninagar-I	7.56	6.44	6.29	5.62	0	6.84	14.85	10.22	20.76	16.67	0	13.9
Raninagar-II	6.46	7.98	5.91	7.29	0	6.88	7.44	4.27	9.44	0	0	6.78

Sagardighi	8.6	9.13	7.71	6.76	0	8.53	4.49	4.49	7.09	19.2	0	4.71
Samserganj	3.37	4.42	1	0	0	3.27	8.98	1.95	9.45	0	0	5.77
Suti-I	4.12	4.95	5.88	2.94	0	4.82	0	0	0	0	0	0
Suti-II	7.08	5.1	6.73	11.51	0	6.61	19.11	22.93	25.05	0	0	21.48
Total	**6.1**	**6.47**	**5.79**	**5.26**	**0.25**	**6.13**	**5.83**	**4.87**	**7.16**	**5.44**	**0**	**5.62**

Source: Agriculture Census, Government of India, 1995–1996 and 2015–2016

Annexure B.19 Percentage of gross cropped area under fibres in Murshidabad district, 1995–1996 and 2015–2016

	1995–1996						2015–2016					
Blocks	Marginal	Small	Semi-medium	Medium	Large	All classes	Marginal	Small	Semi-medium	Medium	Large	All classes
Berhampore	23.1	19.54	16.01	16.57	0	20.3	19.02	18.18	19.2	0	0	18.54
Burwan	0	0.04	0	0.71	0	0.03	0.42	0.14	0	0.36	0	0.29
Beldanga-I	18.52	16.79	11.8	14.16	0	17.42	0.12	0	0	0	0	0.08
Beldanga-II	18.53	18.87	13.01	8.73	11.11	17.55	14.53	14.65	17.5	0	0	14.82
Bhagwangola-I	25.2	25.93	26.9	30.5	0	25.81	17.83	16.04	17.38	19.76	0	17.43
Bhagwangola-II	34.91	38.55	29.13	41.09	0	35.17	18.37	16.02	24.55	100	0	19.73
Bharatpur-I	0.1	0.46	0	0	0	0.21	1.46	1.32	1.74	0.85	0	1.42
Bharatpur-II	0	0	0	0	0	0	5.84	8.57	4.01	0	0	6.65
Domkal	29.83	27.55	23.13	25.92	0	28.18	12.32	11.46	12.79	0	0	12.07
Farakka	8.52	15.13	5.55	2.78	0	10.15	13.16	13.61	11.37	27.59	0	13.44
Hariharpara	23.9	16.41	12.07	0.51	0	18.96	0.07	0.16	0	0	0	0.09
Jalangi	18.33	24.14	23.17	18.05	0	20.83	19.55	20.48	37.42	100	0	21.34
Kandi	0.07	0.43	1.37	0	0	0.35	0.65	0.76	0	0	0	0.64
Khargram	0.58	0.82	0.9	2.19	0	0.73	0	0	0	0	0	0
Lalgola	13.64	20.85	34.05	36.69	0	22.35	40.59	45.03	49.98	100	0	43.02
Murshidabad-Jiaganj	18.97	20.73	20.81	23.46	0	19.94	16.5	17.32	28.29	100	0	17.75
Nabagram	3.11	3.36	2.92	2.14	0	3.08	0.29	0	0	0	0	0.16
Nawda	24.28	19.68	15.72	14.05	0	20.76	0	0	0	1.1	0	0.01
Raghunathganj-I	4.66	3.74	3.21	1.35	0	3.91	0.28	0	0	0	0	0.11
Raghunathganj-II	8.28	8.54	10.21	9.15	0	8.67	26.86	6.55	3.94	12.33	0	17.55
Raninagar-I	32.41	32.04	27.05	39.94	0	31.15	29.38	33.87	36.93	31.14	0	32.27
Raninagar-II	26.34	20.83	21.53	17.66	0	23.42	22.72	20.68	25.56	0	0	22.46

Sagardighi	6.63	5.9	3.46	5.05	0	5.48	1.04	0.85	0	8.04	0	0.94
Samserganj	5.85	7.35	4.56	12.9	0	6.23	44.42	49.03	65.34	100	0	50.42
Suti-I	2.67	2.94	3.15	0.48	0	2.78	0	0	0	0	0	0
Suti-II	17.37	10.35	18.29	29.14	0	15.95	41.82	57.27	52.74	100	0	49.67
Total	**14.8**	**13.05**	**12.09**	**11.56**	**0.75**	**13.63**	**10.98**	**12.05**	**16.42**	**14.19**	**0**	**11.85**

Source: Agriculture Census, Government of India, 1995–1996 and 2015–2016

Annexure B.20 Percentage of gross cropped area under total food crops in Murshidabad district, 1995–1996 and 2015–2016

	1995–1996						2015–2016					
Blocks	Marginal	Small	Semi-medium	Medium	Large	All classes	Marginal	Small	Semi-medium	Medium	Large	All classes
Berhampore	73.27	75.18	78.92	80.45	100.00	75.32	74.52	79.25	76.01	100.00	100.00	76.32
Burwan	93.75	93.18	92.05	95.65	0.00	93.26	91.69	92.93	89.73	76.09	100.00	91.84
Beldanga-I	74.55	77.99	83.95	74.65	0.00	76.14	99.74	99.06	100.00	100.00	0.00	99.58
Beldanga-II	73.79	72.66	80.56	79.82	85.19	74.74	77.51	77.63	72.29	100.00	0.00	77.09
Bhagwangola-I	63.51	63.47	61.82	58.61	0.00	63.11	65.49	72.37	71.65	64.67	0.00	68.16
Bhagwangola-II	63.64	58.46	67.05	58.42	0.00	62.53	70.94	72.33	55.18	0.00	0.00	68.04
Bharatpur-I	93.62	92.28	91.54	93.69	0.00	92.83	92.73	94.76	95.92	93.79	0.00	93.62
Bharatpur-II	94.45	95.48	96.30	94.95	0.00	95.29	87.94	86.71	89.71	100.00	0.00	87.67
Domkal	63.89	63.62	68.45	66.95	100.00	64.53	84.42	86.10	82.89	100.00	0.00	84.84
Farakka	90.98	84.87	92.60	97.22	100.00	89.31	74.96	74.44	84.85	72.41	0.00	75.67
Hariharpara	65.52	73.59	76.46	90.14	0.00	70.53	99.93	99.50	100.00	100.00	0.00	99.80
Jalangi	80.96	75.05	74.13	81.70	0.00	78.22	73.96	74.46	59.84	0.00	0.00	72.92
Kandi	96.36	96.38	95.07	83.17	0.00	96.12	96.36	95.89	100.00	100.00	0.00	96.50
Khargram	90.80	90.86	92.11	93.12	0.00	91.13	99.25	99.25	99.10	99.48	0.00	99.24
Lalgola	80.67	71.67	64.07	63.31	0.00	72.64	47.78	54.25	49.98	0.00	0.00	50.16
Murshidabad-Jiaganj	76.72	73.70	76.95	74.09	100.00	75.80	72.98	74.15	48.92	0.00	0.00	71.47
Nabagram	91.23	90.31	91.87	94.25	0.00	91.28	97.90	98.70	94.80	93.73	0.00	97.93
Nawda	70.02	73.74	80.68	82.77	0.00	73.65	100.00	100.00	100.00	98.90	100.00	99.99
Raghunathganj-I	91.23	91.17	93.02	93.50	100.00	91.67	94.94	93.96	96.85	100.00	0.00	95.10
Raghunathganj-II	87.47	86.85	86.31	90.09	100.00	87.52	64.76	86.79	96.06	87.67	0.00	75.76
Raninagar-I	60.03	60.79	66.66	54.44	100.00	61.75	55.77	55.91	42.31	52.19	0.00	53.83
Raninagar-II	67.14	70.81	71.64	74.55	0.00	69.38	69.19	74.72	64.98	0.00	0.00	70.27

Sagardighi	84.68	84.79	88.66	87.98	0.00	85.84	94.46	94.66	92.91	73.21	0.00	94.35
Samserganj	90.71	88.23	94.44	87.10	0.00	90.48	46.60	49.03	25.20	0.00	0.00	43.80
Suti-I	93.21	92.11	90.97	96.58	100.00	92.40	100.00	100.00	100.00	100.00	0.00	100.00
Suti-II	75.01	84.48	74.35	59.35	0.00	77.03	38.99	19.80	22.22	0.00	0.00	28.81
Total	**78.83**	**80.27**	**81.97**	**83.05**	**99.00**	**80.02**	**82.85**	**82.87**	**75.85**	**80.35**	**100.00**	**82.21**

Source: Agriculture Census, Government of India, 1995–1996 and 2015–2016

Annexure B.21 Percentage of gross cropped area under total non-food crops in Murshidabad district, 1995–1996 and 2015–2016

	1995–1996						2015–2016					
Blocks	Marginal	Small	Semi-medium	Medium	Large	All classes	Marginal	Small	Semi-medium	Medium	Large	All classes
Berhampore	26.73	24.82	21.08	19.55	0.00	24.68	25.48	20.75	23.99	0.00	0.00	23.68
Burwan	6.25	6.82	7.95	4.35	0.00	6.74	8.31	7.07	10.27	23.91	0.00	8.16
Beldanga-I	25.45	21.99	16.05	25.35	0.00	23.85	0.26	0.94	0.00	0.00	0.00	0.42
Beldanga-II	26.21	27.34	19.44	20.18	14.81	25.26	22.49	22.37	27.71	0.00	0.00	22.91
Bhagwangola-I	36.49	36.53	38.18	41.39	0.00	36.89	34.51	27.63	28.35	35.93	0.00	31.84
Bhagwangola-II	36.36	41.54	32.95	41.58	0.00	37.47	29.06	27.68	44.79	100.00	0.00	31.97
Bharatpur-I	6.38	7.72	8.46	6.31	0.00	7.17	7.27	5.25	4.08	6.50	0.00	6.38
Bharatpur-II	5.55	4.52	3.70	5.05	0.00	4.71	12.06	13.29	10.29	0.00	0.00	12.33
Domkal	36.11	36.38	31.53	33.05	0.00	35.47	15.58	13.90	17.11	0.00	0.00	15.16
Farakka	9.02	15.13	7.40	2.78	0.00	10.69	25.04	25.56	15.15	27.59	0.00	24.33
Hariharpara	34.48	26.41	23.54	9.86	0.00	29.47	0.07	0.51	0.00	0.00	0.00	0.20
Jalangi	19.04	24.95	25.87	18.05	0.00	21.78	26.04	25.54	40.16	100.00	0.00	27.08
Kandi	3.64	3.62	4.93	16.83	0.00	3.88	3.64	4.11	0.00	0.00	0.00	3.50
Khargram	9.20	9.14	7.89	6.88	0.00	8.87	0.75	0.75	0.90	0.52	0.00	0.76
Lalgola	19.33	28.33	35.93	36.69	0.00	27.36	52.22	45.75	49.98	100.00	0.00	49.85
Murshidabad-Jiaganj	23.28	26.30	23.05	25.91	0.00	24.20	27.02	25.85	51.08	100.00	0.00	28.53
Nabagram	8.77	9.69	8.13	5.75	0.00	8.72	2.10	1.30	5.20	6.27	0.00	2.07
Nawda	29.98	26.26	19.32	17.23	0.00	26.35	0.00	0.00	0.00	1.10	0.00	0.01
Raghunathganj-I	8.77	8.83	6.98	6.50	0.00	8.33	5.06	6.04	3.12	0.00	0.00	4.89
Raghunathganj-II	12.53	13.15	13.69	9.91	0.00	12.48	35.24	13.21	3.94	12.33	0.00	24.24
Raninagar-I	39.97	39.21	33.34	45.56	0.00	38.25	44.23	44.09	57.69	47.81	0.00	46.17
Raninagar-II	32.85	29.19	28.34	25.45	0.00	30.61	30.81	25.28	35.00	0.00	0.00	29.73

Sagardighi	15.32	15.21	11.34	12.02	0.00	14.16	5.54	5.34	7.09	26.79	0.00	5.65
Samserganj	9.29	11.77	5.56	12.90	0.00	9.52	53.40	50.97	74.80	100.00	0.00	56.19
Suti-I	6.79	7.89	9.03	3.42	0.00	7.60	0.00	0.00	0.00	0.00	0.00	0.00
Suti-II	24.99	15.52	25.59	40.65	0.00	22.96	61.00	80.20	77.78	100.00	0.00	71.19
Total	**21.17**	**19.73**	**18.03**	**16.94**	**1.00**	**19.98**	**17.15**	**17.13**	**24.15**	**19.62**	**0.00**	**17.79**

Source: Agriculture Census, Government of India, 1995–1996 and 2015–2016

Annexure B.22 Land size category-wise percentage of GCA under major crops in Murshidabad district, 1995–1996 and 2015–2016

1995–1996						
Crops	Marginal	Small	Semi-medium	Medium	Large	All classes
Paddy	48.75	52.29	53.78	56.17	90.64	51.09
Wheat	12.97	12.01	11.42	9.1	2.62	12.25
Jute	14.65	12.72	11.96	11.41	0.75	13.42
Mango	0.28	0.41	0.53	0.72	0	0.38
Potato	1.42	1.44	1.47	1.28	0	1.43
Mustard	5.01	5.18	4.59	4.34	0.25	4.96
Gram	1.88	1.91	2.03	2.93	0.87	1.95
Masur	3.66	3.05	2.98	2.9	1.12	3.31
Maize	0.19	0.17	0.07	0.2	0	0.16
Other crops	11.19	10.81	11.17	10.95	3.75	11.05
Total	100	100	100	100	100	100
2015–2016						
Crops	**Marginal**	**Small**	**Semi-medium**	**Medium**	**Large**	**All classes**
Paddy	50.15	51.12	61.63	71.98	100	51.58
Wheat	12.6	12.81	4.95	2.25	0	11.95
Jute	10.91	11.93	0	14.19	0	11.77
Mango	2.17	2.25	0	0	0	2.04
Potato	1.44	1.35	0	2.27	0	1.48
Mustard	5.38	4.4	0.06	2.99	0	5.13
Gram	0.23	0.3	0	0.18	0	0.25
Masur	1.26	2.64	0	1.07	0	1.7
Maize	0.58	0.78	0.99	0	0	0.68
Other crops	15.28	12.43	32.37	5.08	0	13.43
Total	100	100	100	100	100	100

Source: Agriculture Census, Government of India, 1995–1996 and 2015–2016

Annexure B.23 Percentage of gross cropped area under paddy crop in Murshidabad district, 1995–1996 and 2015–2016

Blocks	1995–1996						2015–2016					
	Marginal	Small	Semi-medium	Medium	Large	All classes	Marginal	Small	Semi-medium	Medium	Large	All classes
Berhampore	48.38	48.38	48.82	49.63	93.38	48.82	56.09	58.75	76.01	100.00	100.00	58.71
Burwan	72.81	76.76	77.07	77.09	N.A	75.34	55.12	51.86	74.09	46.38	100.00	55.77
Beldanga-I	40.93	45.87	47.77	52.67	N.A	42.92	57.18	72.92	100.00	100.00	N.A	64.42
Beldanga-II	50.03	51.59	58.54	57.83	74.07	52.06	48.11	42.00	57.04	100.00	N.A	47.13
Bhagwangola-I	34.86	33.61	35.58	35.84	N.A	34.65	32.22	47.37	56.59	31.74	100.00	39.55
Bhagwangola-II	25.89	15.98	35.45	18.81	N.A	24.31	33.45	39.97	43.73	0.00	N.A	36.63
Bharatpur-I	82.35	81.96	82.03	86.71	N.A	82.27	58.30	61.54	92.91	79.94	N.A	61.79
Bharatpur-II	86.68	90.18	89.33	92.93	N.A	88.76	60.03	65.79	89.40	100.00	N.A	64.71
Domkal	25.37	24.66	29.09	32.53	88.33	25.99	46.89	48.39	82.89	100.00	N.A	49.34
Farakka	53.83	49.84	42.11	95.83	100.00	51.89	56.71	68.90	79.18	72.41	100.00	65.96
Hariharpara	31.79	40.17	41.74	57.85	N.A	36.79	58.60	62.16	100.00	100.00	N.A	62.45
Jalangi	33.57	30.39	29.01	77.69	N.A	32.75	29.48	31.63	40.76	0.00	N.A	31.07
Kandi	73.08	76.08	71.95	67.64	N.A	74.00	57.29	52.32	99.73	100.00	N.A	59.02
Khargram	66.90	67.12	66.54	69.47	N.A	66.94	70.45	65.59	98.67	97.94	N.A	71.15
Lalgola	46.03	46.04	46.05	44.11	N.A	45.97	16.57	15.05	5.01	0.00	N.A	14.90
Murshidabad-Jiaganj	47.90	42.83	41.70	44.78	100.00	45.23	32.75	37.14	10.36	0.00	100.00	32.41
Nabagram	67.48	69.18	69.64	75.98	N.A	69.05	74.48	70.95	93.27	90.60	N.A	74.74
Nawda	27.46	29.37	35.81	42.28	N.A	30.08	64.75	62.70	100.00	96.49	100.00	66.69
Raghunathganj-I	55.84	57.07	67.80	63.68	100.00	59.10	65.60	69.35	89.06	100.00	N.A	72.81
Raghunathganj-II	48.61	54.69	54.83	55.66	50.00	51.71	26.06	43.61	0.60	0.00	100.00	28.41
Raninagar-I	26.36	24.78	24.13	14.20	60.00	25.14	19.56	14.36	20.16	28.07	N.A	17.67
Raninagar-II	27.62	30.09	29.82	29.09	N.A	28.77	22.81	25.43	31.37	N.A	N.A	24.44

(continued)

Annexure B.23 (continued)

Blocks	1995–1996						2015–2016					
	Marginal	Small	Semi-medium	Medium	Large	All classes	Marginal	Small	Semi-medium	Medium	Large	All classes
Sagardighi	61.89	67.62	68.53	72.67	N.A	66.44	74.24	75.20	87.35	72.77	N.A	75.37
Samserganj	39.15	40.71	43.01	58.71	N.A	40.65	18.30	7.71	0.00	0.00	N.A	9.99
Suti-I	57.43	53.09	56.97	59.02	94.34	55.98	75.82	78.22	99.89	100.00	N.A	79.37
Suti-II	31.99	40.85	27.75	29.86	N.A	33.64	7.92	2.43	11.09	0.00	N.A	6.11
Total	**48.75**	**52.29**	**53.78**	**56.17**	**90.64**	**51.09**	**50.15**	**51.12**	**61.63**	**71.98**	**100.00**	**51.58**

Source: Agriculture Census, Government of India, 1995–1996 and 2015–2016

N.A not available

Annexure B.24 Percentage of gross cropped area under wheat crop in Murshidabad district, 1995–1996 and 2015–2016

	1995–1996						2015–2016					
Blocks	Marginal	Small	Semi-medium	Medium	Large	All classes	Marginal	Small	Semi-medium	Medium	Large	All classes
Berhampore	12.28	12.83	12.72	13.04	0.63	12.47	7.08	7.51	0.00	0.00	0.00	6.66
Burwan	6.69	5.54	4.59	4.82	N.A	5.80	12.69	13.88	2.66	2.17	0.00	12.12
Beldanga-I	12.14	12.58	11.42	8.02	N.A	12.05	18.40	11.52	0.00	0.00	N.A	15.26
Beldanga-II	8.87	7.43	7.47	2.71	5.56	8.17	12.16	17.77	7.94	0.00	N.A	13.44
Bhagwangola-I	13.14	14.88	10.62	7.33	N.A	13.08	12.92	8.53	0.07	16.17	0.00	10.15
Bhagwangola-II	15.92	15.98	8.28	17.82	N.A	14.73	14.48	17.98	4.34	0.00	N.A	13.97
Bharatpur-I	5.61	4.78	4.57	4.15	N.A	5.12	9.35	9.84	0.52	1.13	N.A	8.88
Bharatpur-II	4.80	3.56	4.72	1.52	N.A	4.24	6.74	9.03	0.00	0.00	N.A	6.96
Domkal	20.09	20.28	19.67	14.20	11.67	19.84	16.47	18.29	0.00	0.00	N.A	16.06
Farakka	1.73	12.15	0.00	0.00	0.00	5.14	4.36	2.88	0.00	0.00	0.00	3.01
Hariharpara	16.17	14.90	15.62	9.81	N.A	15.45	18.66	17.25	0.00	0.00	N.A	16.98
Jalangi	15.99	20.30	20.31	1.25	N.A	17.67	17.75	22.07	13.76	0.00	N.A	18.72
Kandi	7.07	4.15	6.75	0.97	N.A	5.92	15.69	20.45	0.00	0.00	N.A	15.86
Khargram	9.30	8.02	9.58	3.34	N.A	8.94	11.99	13.18	0.30	0.00	N.A	11.40
Lalgola	21.33	11.56	8.09	6.51	N.A	14.06	15.11	14.97	32.31	0.00	N.A	16.79
Murshidabad-Jiaganj	18.18	19.23	20.78	22.28	0.00	19.06	12.28	7.64	6.14	0.00	0.00	10.39
Nabagram	8.15	8.03	7.21	5.34	N.A	7.71	10.01	10.02	1.38	0.00	N.A	9.27
Nawda	18.93	19.71	20.25	13.22	N.A	19.25	16.00	18.93	0.00	0.00	0.00	15.78
Raghunathganj-I	6.67	6.09	4.91	8.52	0.00	6.14	9.97	9.04	3.62	0.00	N.A	8.04
Raghunathganj-II	11.87	8.28	4.27	5.01	5.77	9.12	12.21	11.77	0.33	0.00	0.00	10.47
Raninagar-I	26.37	28.51	27.22	32.25	30.00	27.41	19.07	20.04	14.16	22.37	N.A	18.77

(continued)

Annexure B.24 (continued)

Blocks	1995–1996						2015–2016					
	Marginal	Small	Semi-medium	Medium	Large	All classes	Marginal	Small	Semi-medium	Medium	Large	All classes
Raninagar-II	18.67	13.75	13.81	14.13	N.A	16.19	20.68	20.39	19.48	N.A	N.A	20.48
Sagardighi	13.80	11.38	12.78	8.00	N.A	12.26	5.53	10.81	2.65	0.00	N.A	7.03
Samserganj	11.16	9.20	10.27	8.39	N.A	10.26	5.27	0.00	0.00	0.00	N.A	1.84
Suti-I	6.10	5.93	4.10	1.59	5.66	5.25	10.00	6.60	0.00	0.00	N.A	7.64
Suti-II	10.33	9.81	14.29	14.39	N.A	10.95	9.51	9.33	0.00	0.00	N.A	8.12
Total	**12.97**	**12.01**	**11.42**	**9.10**	**2.62**	**12.25**	**12.60**	**12.81**	**4.95**	**2.25**	**0.00**	**11.95**

Source: Agriculture Census, Government of India, 1995–1996 and 2015–2016
N.A not available

Annexure B.25 Percentage of gross cropped area under jute crop in Murshidabad district, 1995–1996 and 2015–2016

Blocks	1995–1996						2015–2016					
	Marginal	Small	Semi-medium	Medium	Large	All classes	Marginal	Small	Semi-medium	Medium	Large	All classes
Berhampore	22.82	19.44	15.31	14.62	0.00	19.96	18.95	18.18	19.20	0.00	0.00	18.50
Burwan	0.00	0.04	0.00	0.71	N.A	0.03	0.42	0.14	0.00	0.36	0.00	0.29
Beldanga-I	18.41	16.65	11.80	14.16	N.A	17.32	0.12	0.00	0.00	0.00	N.A	0.08
Beldanga-II	18.38	18.87	13.01	8.73	11.11	17.47	14.51	14.65	17.50	0.00	N.A	14.81
Bhagwangola-I	25.19	25.82	26.31	30.50	N.A	25.66	17.67	13.63	17.38	19.76	0.00	16.57
Bhagwangola-II	34.77	36.37	29.13	41.09	N.A	34.43	18.22	16.02	24.55	100.00	N.A	19.65
Bharatpur-I	0.10	0.46	0.00	0.00	N.A	0.21	1.46	1.32	1.74	0.85	N.A	1.42
Bharatpur-II	0.00	0.00	0.00	0.00	N.A	0.00	5.84	8.57	4.01	0.00	N.A	6.65
Domkal	29.79	27.32	23.13	25.85	0.00	28.08	12.15	10.68	12.79	0.00	N.A	11.74
Farakka	7.97	2.97	5.55	2.78	0.00	5.51	13.05	13.61	11.37	27.59	0.00	13.40
Hariharpara	23.87	16.29	12.07	0.51	N.A	18.91	0.07	0.16	0.00	0.00	N.A	0.09
Jalangi	18.26	23.66	23.17	18.05	N.A	20.63	19.33	20.05	37.42	100.00	N.A	21.07
Kandi	0.07	0.34	1.37	0.00	N.A	0.32	0.60	0.76	0.00	0.00	N.A	0.60
Khargram	0.48	0.50	0.59	2.19	N.A	0.55	0.00	0.00	0.00	0.00	N.A	0.00
Lalgola	13.64	19.86	34.05	35.53	N.A	22.00	40.59	45.03	49.98	100.00	N.A	43.02
Murshidabad-Jiaganj	18.91	20.73	20.81	23.46	0.00	19.92	16.14	17.32	28.29	100.00	0.00	17.53
Nabagram	3.09	3.36	2.92	2.14	N.A	3.08	0.29	0.00	0.00	0.00	N.A	0.16
Nawda	24.09	19.46	15.72	14.05	N.A	20.60	0.00	0.00	0.00	1.10	0.00	0.01
Raghunathganj-I	4.66	3.51	3.21	1.35	0.00	3.82	0.28	0.00	0.00	0.00	N.A	0.11
Raghunathganj-II	8.28	8.54	10.21	9.15	0.00	8.67	26.76	6.55	3.94	12.33	0.00	17.50
Raninagar-I	31.94	31.66	27.05	39.94	0.00	30.82	29.00	33.87	36.74	31.14	N.A	32.08
Raninagar-II	24.67	19.93	20.39	17.66	N.A	22.17	22.66	20.68	25.56	N.A	N.A	22.42

(continued)

Annexure B.25 (continued)

Blocks	1995–1996						2015–2016					
	Marginal	Small	Semi-medium	Medium	Large	All classes	Marginal	Small	Semi-medium	Medium	Large	All classes
Sagardighi	6.55	5.68	3.46	5.05	N.A	5.36	1.04	0.85	0.00	8.04	N.A	0.94
Samserganj	5.47	6.61	4.56	12.90	N.A	5.79	44.42	49.03	65.34	100.00	N.A	50.42
Suti-I	2.67	2.94	3.15	0.48	0.00	2.78	0.00	0.00	0.00	0.00	N.A	0.00
Suti-II	17.08	10.19	17.78	29.14	N.A	15.67	41.82	57.27	52.74	100.00	N.A	49.67
Total	14.65	12.72	11.96	11.41	0.75	13.42	10.91	11.93	16.41	14.19	0.00	11.77

Source: Agriculture Census, Government of India, 1995–1996 and 2015–2016
N.A not available

Annexure B.26 Percentage of gross cropped area under mango crop in Murshidabad district, 1995–1996 and 2015–2016

Blocks	1995–1996						2015–2016					
	Marginal	Small	Semi-medium	Medium	Large	All classes	Marginal	Small	Semi-medium	Medium	Large	All classes
Berhampore	0.00	0.00	0.00	1.77	0.00	0.04	1.26	2.27	0.00	0.00	0.00	1.46
Burwan	0.00	0.00	0.00	0.00	N.A	0.00	2.96	2.51	0.00	0.00	0.00	2.54
Beldanga-I	0.00	0.00	0.00	0.00	N.A	0.00	3.52	4.21	0.00	0.00	N.A	3.46
Beldanga-II	0.01	0.00	0.00	0.00	0.00	0.00	1.86	1.51	0.00	0.00	N.A	1.59
Bhagwangola-I	0.90	0.51	0.19	0.00	N.A	0.63	2.02	0.00	0.00	0.00	0.00	1.14
Bhagwangola-II	0.10	1.03	3.38	0.00	N.A	0.91	2.29	2.77	0.00	0.00	N.A	2.09
Bharatpur-I	0.00	0.00	0.00	0.00	N.A	0.00	1.76	2.07	0.00	0.00	N.A	1.74
Bharatpur-II	0.00	0.00	0.00	0.00	N.A	0.00	1.42	1.56	0.00	0.00	N.A	1.34
Domkal	0.00	0.00	0.25	0.00	0.00	0.03	2.91	3.56	0.00	0.00	N.A	2.93
Farakka	16.85	17.36	15.23	0.00	0.00	16.13	0.21	0.00	0.00	0.00	0.00	0.07
Hariharpara	0.03	0.09	0.11	0.00	N.A	0.06	3.65	4.54	0.00	0.00	N.A	3.69
Jalangi	0.00	0.23	0.00	0.00	N.A	0.08	2.23	2.32	0.00	0.00	N.A	2.07
Kandi	0.00	0.00	0.00	0.00	N.A	0.00	2.91	4.41	0.00	0.00	N.A	3.11
Khargram	0.00	0.00	0.00	0.00	N.A	0.00	1.39	1.95	0.00	0.00	N.A	1.48
Lalgola	0.24	1.25	0.00	0.00	N.A	0.49	0.96	4.51	0.00	0.00	N.A	2.04
Murshidabad-Jiaganj	0.41	1.40	3.02	2.29	0.00	1.26	3.46	0.88	6.87	0.00	0.00	2.91
Nabagram	0.00	0.01	0.00	0.08	N.A	0.01	1.84	1.86	0.00	0.00	N.A	1.69
Nawda	0.30	1.01	2.20	5.33	N.A	1.06	3.10	3.29	0.00	0.00	0.00	2.93
Raghunathganj-I	4.99	1.08	1.17	1.79	0.00	2.67	1.40	1.30	0.00	0.00	N.A	1.02
Raghunathganj-II	0.59	0.68	1.11	0.44	0.00	0.67	1.72	0.00	0.00	0.00	0.00	0.93
Raninagar-I	0.00	0.06	1.47	0.00	0.00	0.36	1.94	1.46	0.00	0.00	N.A	1.45

(continued)

Annexure B.26 (continued)

Blocks	1995–1996						2015–2016					
	Marginal	Small	Semi-medium	Medium	Large	All classes	Marginal	Small	Semi-medium	Medium	Large	All classes
Raninagar-II	0.12	0.25	0.10	0.45	N.A	0.18	2.71	4.17	5.02	N.A	N.A	3.35
Sagardighi	0.00	0.00	0.00	0.00	N.A	0.00	1.05	0.00	0.00	0.00	N.A	0.64
Samserganj	0.00	0.00	0.00	0.00	N.A	0.00	0.14	0.00	0.00	0.00	N.A	0.05
Suti-I	0.00	0.00	0.00	0.00	0.00	0.00	1.83	8.57	0.00	0.00	N.A	4.12
Suti-II	0.17	0.91	0.57	11.51	N.A	0.76	3.55	3.06	0.00	0.00	N.A	2.86
Total	**0.28**	**0.41**	**0.53**	**0.72**	**0.00**	**0.38**	**2.17**	**2.25**	**0.52**	**0.00**	**0.00**	**2.04**

Source: Agriculture Census, Government of India, 1995–1996 and 2015–2016
N.A not available

Annexure B.27 Percentage of gross cropped area under potato crop in Murshidabad district, 1995–1996 and 2015–2016

	1995–1996						2015–2016					
Blocks	Marginal	Small	Semi-medium	Medium	Large	All classes	Marginal	Small	Semi-medium	Medium	Large	All classes
Berhampore	0.79	0.82	3.27	0.28	0.00	1.25	0.96	0.00	0.00	0.00	0.00	0.60
Burwan	4.97	4.16	4.66	2.00	N.A	4.50	6.17	7.81	8.79	24.28	0.00	7.00
Beldanga-I	3.06	3.60	4.19	3.76	N.A	3.29	1.01	1.87	0.00	0.00	N.A	1.17
Beldanga-II	1.13	0.39	1.30	0.30	0.00	0.95	3.16	2.49	4.82	0.00	N.A	3.11
Bhagwangola-I	0.81	0.82	1.04	0.59	N.A	0.86	1.98	1.61	9.39	0.00	0.00	2.63
Bhagwangola-II	0.59	1.24	1.24	0.00	N.A	0.89	1.91	0.70	3.24	0.00	N.A	1.66
Bharatpur-I	1.77	1.44	1.31	1.33	N.A	1.57	3.31	2.02	0.75	3.39	N.A	2.72
Bharatpur-II	2.19	1.33	1.41	0.51	N.A	1.65	0.27	0.04	0.17	0.00	N.A	0.18
Domkal	1.47	1.74	2.56	2.03	0.00	1.70	0.32	0.15	0.00	0.00	N.A	0.26
Farakka	0.00	0.00	0.00	0.00	0.00	0.00	3.71	0.96	0.00	0.00	0.00	1.75
Hariharpara	0.80	1.11	0.88	0.51	N.A	0.88	0.00	0.16	0.00	0.00	N.A	0.05
Jalangi	0.31	0.61	0.48	0.00	N.A	0.42	1.24	0.32	0.66	0.00	N.A	0.91
Kandi	2.31	1.66	1.45	0.32	N.A	1.96	2.96	0.52	0.22	0.00	N.A	2.08
Khargram	1.21	1.13	0.68	1.41	N.A	1.10	0.09	0.18	0.14	0.52	N.A	0.13
Lalgola	0.27	0.09	0.29	0.00	N.A	0.21	1.16	0.00	0.00	0.00	N.A	0.66
Murshidabad-Jiaganj	2.06	1.95	0.89	2.29	0.00	1.81	0.72	0.26	0.86	0.00	0.00	0.59
Nabagram	1.51	1.59	1.15	1.09	N.A	1.42	0.08	0.62	0.16	1.88	N.A	0.31
Nawda	0.64	1.25	1.47	2.56	N.A	1.08	0.00	0.00	0.00	0.00	0.00	0.00
Raghunathganj-I	0.37	0.14	0.15	0.22	0.00	0.23	0.21	0.00	1.56	0.00	N.A	0.45
Raghunathganj-II	0.14	0.94	1.03	1.53	0.00	0.61	0.52	0.00	0.00	0.00	0.00	0.28
Raninagar-I	0.98	0.89	1.59	0.89	0.00	1.09	1.83	7.27	6.09	1.75	N.A	4.60
Raninagar-II	1.12	2.23	2.79	2.63	N.A	1.80	0.77	0.38	0.00	N.A	N.A	0.58

(continued)

Annexure B.27 (continued)

Blocks	1995–1996						2015–2016					
	Marginal	Small	Semi-medium	Medium	Large	All classes	Marginal	Small	Semi-medium	Medium	Large	All classes
Sagardighi	1.65	1.21	1.35	0.86	N.A	1.35	0.02	0.02	0.00	0.00	N.A	0.02
Samserganj	0.20	1.13	0.73	0.00	N.A	0.63	1.02	2.71	0.00	0.00	N.A	1.62
Suti-I	0.78	0.71	0.82	0.32	0.00	0.74	0.00	0.00	0.00	0.00	N.A	0.00
Suti-II	0.40	1.22	0.51	0.00	N.A	0.63	4.33	3.12	11.09	0.00	N.A	4.77
Total	**1.42**	**1.44**	**1.47**	**1.28**	**0.00**	**1.43**	**1.44**	**1.35**	**2.15**	**2.27**	**0.00**	**1.48**

Source: Agriculture Census, Government of India, 1995–1996 and 2015–2016
N.A not available

Annexure B.28 Percentage of gross cropped area under mustard and rapeseed crop in Murshidabad district, 1995–1996 and 2015–2016

	1995–1996						2015–2016					
Blocks	Marginal	Small	Semi-medium	Medium	Large	All classes	Marginal	Small	Semi-medium	Medium	Large	All classes
Berhampore	3.08	4.29	2.73	2.05	0.00	3.34	6.37	2.56	4.79	0.00	0.00	5.07
Burwan	3.24	3.56	4.81	2.23	N.A	3.61	6.84	5.33	8.52	2.90	0.00	6.49
Beldanga-I	5.43	4.55	3.30	9.01	N.A	5.15	0.14	0.94	0.00	0.00	N.A	0.34
Beldanga-II	5.56	6.47	5.35	11.45	3.70	5.83	7.67	7.57	8.94	0.00	N.A	7.75
Bhagwangola-I	7.26	5.88	4.89	4.95	N.A	6.36	10.81	9.43	10.97	16.17	0.00	10.39
Bhagwangola-II	0.78	1.39	2.11	0.00	N.A	1.17	7.95	9.84	16.37	0.00	N.A	9.69
Bharatpur-I	5.79	6.64	8.20	6.31	N.A	6.48	4.89	2.34	2.10	5.37	N.A	3.86
Bharatpur-II	4.72	3.79	3.13	3.03	N.A	3.97	4.23	3.53	5.35	0.00	N.A	4.08
Domkal	5.52	7.46	7.42	6.87	0.00	6.35	2.74	1.84	4.35	0.00	N.A	2.57
Farakka	0.50	0.00	1.85	0.00	0.00	0.54	11.82	11.95	3.78	0.00	0.00	10.88
Hariharpara	9.57	8.94	9.60	8.25	N.A	9.35	0.00	0.34	0.00	0.00	N.A	0.11
Jalangi	0.53	0.81	2.34	0.00	N.A	0.81	6.10	4.50	2.02	0.00	N.A	5.27
Kandi	2.75	2.48	2.91	16.83	N.A	2.78	2.95	3.35	0.00	0.00	N.A	2.84
Khargram	6.35	6.48	6.03	3.66	N.A	6.24	0.66	0.59	0.53	0.52	N.A	0.62
Lalgola	5.60	6.92	1.88	0.00	N.A	4.80	10.97	0.71	0.00	0.00	N.A	6.45
Murshidabad-Jiaganj	3.35	3.72	1.91	1.11	0.00	3.12	7.49	3.90	12.07	0.00	0.00	6.71
Nabagram	4.38	4.37	3.73	3.20	N.A	4.14	1.66	1.23	4.83	2.19	N.A	1.75
Nawda	4.20	4.46	2.95	1.66	N.A	4.00	0.00	0.00	0.00	0.00	0.00	0.00
Raghunathganj-I	3.28	4.85	3.51	5.16	0.00	3.95	4.78	6.04	3.12	0.00	N.A	4.78
Raghunathganj-II	3.61	4.14	2.77	0.76	0.00	3.27	8.03	6.07	0.00	0.00	0.00	6.32
Raninagar-I	6.56	5.38	6.29	2.37	0.00	6.02	14.12	10.08	16.82	5.70	N.A	12.84
Raninagar-II	4.80	5.47	4.00	4.60	N.A	4.87	6.59	3.72	5.56	N.A	N.A	5.69

(continued)

Annexure B.28 (continued)

Blocks	1995–1996						2015–2016					
	Marginal	Small	Semi-medium	Medium	Large	All classes	Marginal	Small	Semi-medium	Medium	Large	All classes
Sagardighi	7.81	8.55	7.24	6.59	N.A	7.94	4.44	4.23	7.09	19.20	N.A	4.59
Samserganj	2.74	4.42	1.00	0.00	N.A	2.99	8.98	1.95	9.45	0.00	N.A	5.77
Suti-I	3.46	4.86	4.93	2.78	0.00	4.31	0.00	0.00	0.00	0.00	N.A	0.00
Suti-II	6.44	4.07	5.02	8.63	N.A	5.63	19.11	22.93	25.05	0.00	N.A	21.48
Total	**5.01**	**5.18**	**4.59**	**4.34**	**0.25**	**4.96**	**5.38**	**4.40**	**6.45**	**2.99**	**0.00**	**5.13**

Source: Agriculture Census, Government of India, 1995–1996 and 2015–2016
N.A not available

Annexure B.29 Percentage of gross cropped area under gram crop in Murshidabad district, 1995–1996 and 2015–2016

	1995–1996						2015–2016					
Blocks	Marginal	Small	Semi-medium	Medium	Large	All classes	Marginal	Small	Semi-medium	Medium	Large	All classes
Berhampore	0.44	0.90	0.46	1.77	0.00	0.61	0.10	0.00	0.00	0.00	0.00	0.00
Burwan	4.86	3.72	3.00	6.11	N.A	4.09	0.92	0.90	1.91	1.09	0.00	0.21
Beldanga-I	2.21	1.92	2.66	3.07	N.A	2.22	0.00	0.00	0.00	0.00	N.A	0.06
Beldanga-II	0.67	0.71	0.16	1.81	0.00	0.60	0.14	0.33	0.28	0.00	N.A	0.15
Bhagwangola-I	0.41	1.02	0.82	0.00	N.A	0.66	0.17	0.18	0.00	0.00	0.00	1.14
Bhagwangola-II	2.45	4.10	0.54	5.94	N.A	2.69	1.15	1.51	0.00	0.00	N.A	0.00
Bharatpur-I	1.87	1.65	1.52	0.00	N.A	1.69	0.00	0.00	0.36	1.13	N.A	0.00
Bharatpur-II	0.00	0.00	0.00	0.00	N.A	0.00	0.00	0.00	0.00	0.00	N.A	1.00
Domkal	1.11	1.21	1.69	3.99	0.00	1.32	0.02	0.00	0.00	0.00	N.A	0.01
Farakka	0.73	0.00	0.00	0.00	0.00	0.31	0.45	1.03	0.00	0.00	0.00	0.71
Hariharpara	1.41	1.77	2.74	3.32	N.A	1.81	0.00	0.00	0.00	0.00	N.A	0.00
Jalangi	1.42	0.97	2.70	0.00	N.A	1.38	0.00	0.00	0.00	0.00	N.A	0.00
Kandi	2.33	2.06	2.26	0.65	N.A	2.21	0.00	0.00	0.00	0.00	N.A	0.00
Khargram	2.40	2.42	1.84	6.23	N.A	2.40	0.07	0.22	0.00	0.52	N.A	0.12
Lalgola	0.72	1.87	0.00	0.91	N.A	0.90	0.00	0.00	0.00	0.00	N.A	0.00
Murshidabad-Jiaganj	0.90	0.43	1.06	0.30	0.00	0.76	0.76	1.32	0.86	0.00	0.00	0.94
Nabagram	1.61	1.47	1.83	0.68	N.A	1.58	0.20	0.09	0.00	0.00	N.A	0.14
Nawda	2.10	3.05	4.68	3.88	N.A	2.95	0.00	0.00	0.00	0.00	0.00	0.00
Raghunathganj-I	8.74	6.01	5.93	4.48	0.00	6.99	0.21	0.27	1.13	0.00	N.A	0.44
Raghunathganj-II	9.06	7.39	8.47	5.77	13.46	8.26	0.14	1.34	0.00	0.00	0.00	0.51
Raninagar-I	0.19	0.63	0.71	0.00	0.00	0.46	0.39	0.00	0.00	0.00	N.A	0.17
Raninagar-II	0.98	1.54	1.55	3.81	N.A	1.41	0.90	0.75	0.00	N.A	N.A	0.76

(continued)

Annexure B.29 (continued)

Blocks	1995–1996						2015–2016					
	Marginal	Small	Semi-medium	Medium	Large	All classes	Marginal	Small	Semi-medium	Medium	Large	All classes
Sagardighi	2.68	1.42	2.30	2.78	N.A	2.05	0.13	0.08	0.00	0.00	N.A	0.11
Samserganj	3.01	3.88	7.26	2.58	N.A	4.11	0.00	0.00	0.00	0.00	N.A	0.00
Suti-I	4.27	3.89	5.41	5.96	0.00	4.53	0.00	0.00	0.00	0.00	N.A	0.00
Suti-II	3.56	3.35	2.79	2.88	N.A	3.36	0.00	0.00	0.00	0.00	N.A	0.00
Total	**1.88**	**1.91**	**2.03**	**2.93**	**0.87**	**1.95**	**0.23**	**0.30**	**0.22**	**0.18**	**0.00**	**0.25**

Source: Agriculture Census, Government of India, 1995–1996 and 2015–2016
N.A not available

Annexure B.30 Percentage of gross cropped area under masur crop in Murshidabad district, 1995–1996 and 2015–2016

	1995–1996						2015–2016					
Blocks	Marginal	Small	Semi-medium	Medium	Large	All classes	Marginal	Small	Semi-medium	Medium	Large	All classes
Berhampore	1.35	1.65	1.70	2.23	0.00	1.52	0.81	0.00	0.00	0.00	0.00	0.51
Burwan	2.21	1.23	1.06	3.53	N.A	1.62	0.85	6.54	1.16	0.00	0.00	2.66
Beldanga-I	0.94	0.66	1.89	0.00	N.A	0.95	0.00	0.00	0.00	0.00	N.A	0.00
Beldanga-II	1.62	1.92	1.55	4.22	3.70	1.72	0.22	0.35	0.34	0.00	N.A	0.27
Bhagwangola-I	3.52	1.43	1.88	0.99	N.A	2.55	0.82	0.55	0.00	16.17	0.00	0.74
Bhagwangola-II	5.30	3.64	6.83	4.95	N.A	5.03	1.70	4.74	1.29	0.00	N.A	2.64
Bharatpur-I	0.11	0.65	0.26	0.00	N.A	0.32	0.85	11.51	0.36	2.82	N.A	4.44
Bharatpur-II	0.02	0.00	0.00	0.00	N.A	0.01	5.25	2.97	0.00	0.00	N.A	3.97
Domkal	5.87	5.29	5.59	4.65	0.00	5.61	0.44	0.16	0.00	0.00	N.A	0.34
Farakka	5.37	2.97	2.72	0.00	0.00	3.83	2.33	0.00	0.00	0.00	0.00	0.76
Hariharpara	4.09	4.89	4.64	6.40	N.A	4.50	0.00	0.00	0.00	0.00	N.A	0.00
Jalangi	23.18	15.89	11.57	0.00	N.A	19.04	3.98	2.09	2.80	0.00	N.A	3.31
Kandi	1.58	1.36	1.02	0.32	N.A	1.43	0.00	0.00	0.00	0.00	N.A	0.00
Khargram	0.46	0.82	0.30	0.39	N.A	0.50	0.03	6.18	0.00	0.52	N.A	2.30
Lalgola	0.39	0.00	0.16	0.91	N.A	0.22	0.18	0.00	0.00	0.00	N.A	0.10
Murshidabad-Jiaganj	0.76	0.32	0.47	0.30	0.00	0.55	1.58	1.62	5.84	0.00	0.00	1.90
Nabagram	0.84	0.79	0.95	0.15	N.A	0.82	0.05	3.11	0.00	0.00	N.A	1.20
Nawda	11.34	10.07	7.68	6.92	N.A	10.09	0.00	0.00	0.00	0.88	0.00	0.01
Raghunathganj-I	3.39	6.66	3.78	6.28	0.00	4.76	6.07	3.13	1.48	0.00	N.A	3.89
Raghunathganj-II	7.83	5.81	5.62	5.12	13.46	6.78	1.05	1.34	0.00	0.00	0.00	1.00
Raninagar-I	1.67	0.93	9.04	0.00	0.00	3.05	0.12	0.00	0.00	0.00	N.A	0.06
Raninagar-II	5.95	5.58	6.98	3.14	N.A	5.81	2.19	1.51	0.00	N.A	N.A	1.77

(continued)

Annexure B.30 (continued)

Blocks	1995–1996						2015–2016					
	Marginal	Small	Semi-medium	Medium	Large	All classes	Marginal	Small	Semi-medium	Medium	Large	All classes
Sagardighi	1.54	1.06	1.77	1.88	N.A	1.40	4.09	1.08	0.00	0.00	N.A	2.85
Samserganj	8.22	5.22	6.56	2.58	N.A	6.76	2.48	22.13	14.16	0.00	N.A	13.80
Suti-I	7.73	7.96	6.91	10.17	0.00	7.69	0.00	0.00	0.00	0.00	N.A	0.00
Suti-II	11.33	7.84	13.46	0.00	N.A	10.40	0.22	0.00	0.00	0.00	N.A	0.10
Total	**3.66**	**3.05**	**2.98**	**2.90**	**1.12**	**3.31**	**1.26**	**2.64**	**1.04**	**1.07**	**0.00**	**1.70**

Source: Agriculture Census, Government of India, 1995–1996 and 2015–2016

N.A not available

Annexure B.31 Percentage of gross cropped area under maize crop in Murshidabad district, 1995–1996 and 2015–2016

Blocks	1995–1996						2015–2016					
	Marginal	Small	Semi-medium	Medium	Large	All classes	Marginal	Small	Semi-medium	Medium	Large	All classes
Berhampore	0.02	0.00	0.00	0.00	0.00	0.01	0.23	0.00	0.00	0.00	0.00	0.14
Burwan	0.06	0.00	0.00	0.00	N.A	0.02	0.00	0.12	0.00	0.00	0.00	0.04
Beldanga-I	0.10	0.05	0.00	0.00	N.A	0.08	0.00	0.00	0.00	0.00	N.A	0.00
Beldanga-II	0.29	0.73	0.14	0.00	0.00	0.38	0.03	0.00	0.00	0.00	N.A	0.02
Bhagwangola-I	0.10	0.21	0.00	0.00	N.A	0.11	1.48	3.38	0.00	0.00	0.00	1.93
Bhagwangola-II	0.19	0.00	0.65	0.00	N.A	0.21	0.56	0.00	0.00	0.00	N.A	0.28
Bharatpur-I	0.01	0.00	0.00	0.00	N.A	0.00	0.08	0.05	0.00	0.00	N.A	0.07
Bharatpur-II	0.26	0.00	0.00	0.00	N.A	0.09	0.01	0.00	0.00	0.00	N.A	0.00
Domkal	0.03	0.00	0.00	0.00	0.00	0.02	0.06	0.00	0.00	0.00	N.A	0.04
Farakka	0.36	0.00	0.00	0.00	0.00	0.15	1.26	0.67	0.00	0.00	0.00	0.78
Hariharpara	0.09	0.15	0.00	0.00	N.A	0.09	0.00	0.00	0.00	0.00	N.A	0.00
Jalangi	0.13	0.00	0.00	0.00	N.A	0.07	0.20	0.94	0.00	0.00	N.A	0.41
Kandi	0.05	0.00	0.00	0.00	N.A	0.03	0.07	0.00	0.00	0.00	N.A	0.04
Khargram	0.00	0.00	0.00	0.00	N.A	0.00	0.00	0.00	0.00	0.00	N.A	0.00
Lalgola	0.32	0.78	0.00	0.00	N.A	0.37	7.66	8.72	6.06	0.00	N.A	7.85
Murshidabad-Jiaganj	0.14	0.27	0.16	0.00	0.00	0.18	1.89	4.05	13.84	0.00	0.00	3.42
Nabagram	0.13	0.01	0.00	0.00	N.A	0.05	0.20	0.19	0.00	0.00	N.A	0.18
Nawda	0.24	0.22	0.19	0.35	N.A	0.23	0.00	0.00	0.00	0.00	0.00	0.00
Raghunathganj-I	0.00	0.00	0.00	0.00	0.00	0.00	0.00	0.11	0.00	0.00	N.A	0.04
Raghunathganj-II	0.14	0.21	0.16	1.96	0.00	0.36	4.45	3.42	0.00	0.00	0.00	3.58
Raninagar-I	0.41	0.46	0.00	0.00	0.00	0.33	0.29	0.00	0.00	0.00	N.A	0.13
Raninagar-II	1.54	1.37	0.81	1.46	N.A	1.38	0.03	0.18	0.00	N.A	N.A	0.07

(continued)

Annexure B.31 (continued)

Blocks	1995–1996						2015–2016					
	Marginal	Small	Semi-medium	Medium	Large	All classes	Marginal	Small	Semi-medium	Medium	Large	All classes
Sagardighi	0.00	0.00	0.06	0.00	N.A	0.01	0.00	0.27	2.91	0.00	N.A	0.27
Samserganj	0.92	0.30	0.00	0.00	N.A	0.51	0.06	0.00	0.00	0.00	N.A	0.02
Suti-I	0.47	0.00	0.01	0.00	0.00	0.16	0.00	0.00	0.00	0.00	N.A	0.00
Suti-II	0.02	0.15	0.13	0.00	N.A	0.07	0.00	0.00	0.00	0.00	N.A	0.00
Total	**0.19**	**0.17**	**0.07**	**0.20**	**0.00**	**0.16**	**0.58**	**0.78**	**0.99**	**0.00**	**0.00**	**0.68**

Source: Agriculture Census, Government of India, 1995–1996 and 2015–2016
N.A not available

Annexure B.32 Percentage of gross cropped area under other crops in Murshidabad district, 1995–1996 and 2015–2016

	1995–1996						2015–2016					
Blocks	Marginal	Small	Semi-medium	Medium	Large	All classes	Marginal	Small	Semi-medium	Medium	Large	All classes
Berhampore	10.86	11.70	14.98	14.62	5.99	11.97	8.16	10.73	0.00	0.00	0.00	8.34
Burwan	5.14	4.99	4.81	3.53	N.A	4.99	14.05	10.92	2.87	22.83	0.00	12.90
Beldanga-I	16.76	14.11	16.97	9.31	N.A	16.02	19.63	8.54	0.00	0.00	N.A	15.20
Beldanga-II	13.44	11.90	12.49	12.95	1.85	12.83	12.14	13.33	3.14	0.00	N.A	11.73
Bhagwangola-I	13.81	15.81	18.67	19.80	N.A	15.43	19.90	15.32	5.61	0.00	0.00	15.77
Bhagwangola-II	14.00	20.26	12.39	11.39	N.A	15.63	18.29	6.46	6.49	0.00	N.A	13.39
Bharatpur-I	2.39	2.41	2.10	1.50	N.A	2.33	19.99	9.30	1.27	5.37	N.A	15.09
Bharatpur-II	1.33	1.14	1.42	2.02	N.A	1.28	16.21	8.50	1.07	0.00	N.A	11.11
Domkal	10.75	12.03	10.61	9.88	0.00	11.05	18.02	16.92	-0.04	0.00	N.A	16.72
Farakka	12.66	14.70	32.54	1.39	0.00	16.49	6.11	0.00	5.67	0.00	0.00	2.68
Hariharpara	12.17	11.69	12.62	13.36	N.A	12.17	19.02	15.38	0.00	0.00	N.A	16.62
Jalangi	6.61	7.15	10.41	3.01	N.A	7.15	19.69	16.07	2.59	0.00	N.A	17.16
Kandi	10.76	11.87	12.30	13.27	N.A	11.36	17.54	18.19	0.05	0.00	N.A	16.44
Khargram	12.90	13.50	14.44	13.30	N.A	13.33	15.31	12.10	0.37	0.00	N.A	12.81
Lalgola	11.46	11.63	9.48	12.04	N.A	11.00	6.81	11.00	6.63	0.00	N.A	8.19
Murshidabad-Jiaganj	7.37	9.12	9.20	3.18	0.00	8.11	22.93	25.86	14.88	0.00	0.00	23.20
Nabagram	12.81	11.19	12.55	11.35	N.A	12.16	11.17	11.93	0.36	5.33	N.A	10.57
Nawda	10.69	11.40	9.05	9.76	N.A	10.65	16.14	15.09	0.00	1.54	0.00	14.57
Raghunathganj-I	12.05	14.60	9.55	8.52	0.00	12.33	11.49	10.76	0.04	0.00	N.A	8.41
Raghunathganj-II	9.86	9.32	11.55	14.60	17.31	10.56	19.06	25.90	95.13	87.67	0.00	31.00
Raninagar-I	5.51	6.71	2.50	10.36	10.00	5.33	13.67	12.91	6.03	10.96	N.A	12.23

(continued)

Annexure B.32 (continued)

Blocks	1995–1996						2015–2016					
	Marginal	Small	Semi-medium	Medium	Large	All classes	Marginal	Small	Semi-medium	Medium	Large	All classes
Raninagar-II	14.54	19.78	19.77	23.04	N.A	17.43	20.65	22.80	13.01	N.A	N.A	20.44
Sagardighi	4.08	3.08	2.51	2.18	N.A	3.19	9.46	7.46	0.00	0.00	N.A	8.19
Samserganj	29.14	28.53	26.60	14.84	N.A	28.28	19.32	16.49	11.05	0.00	N.A	16.48
Suti-I	17.08	20.62	17.70	19.70	0.00	18.56	12.35	6.62	0.11	0.00	N.A	8.87
Suti-II	18.68	21.61	17.71	3.60	N.A	18.88	13.55	1.85	0.04	0.00	N.A	6.89
Total	**11.19**	**10.81**	**11.17**	**10.95**	**3.75**	**11.05**	**15.28**	**12.43**	**5.63**	**5.08**	**0.00**	**13.43**

Source: Agriculture Census, Government of India, 1995–1996 and 2015–2016

N.A not available

Annexure B.33 List of variables of sustainable land use transformation

Variables	Description of the variables
X1	Gross cropped area (in hectares)
X2	Percentage of gross cropped area under cereal crops
X3	Percentage of gross cropped area under pulse crops
X4	Percentage of gross cropped area under spice crops
X5	Percentage of gross cropped area under fruit crops
X6	Percentage of gross cropped area under vegetable crops
X7	Percentage of gross cropped area under oil seed crops
X8	Percentage of gross cropped area under fibre crops
X9	Percentage of gross cropped area under total non-food crops
X10	Percentage of forest area to the total area
X11	Percentage of net sown area
X12	Percentage of fallow land to the total area
X13	Percentage of culturable waste land to the total area
X14	Percentage of barren land to the total area
X15	Percentage of ecological land to the total area
X16	Yield rate of paddy in kg/ha
X17	Yield rate of wheat in kg/ha
X18	Yield rate of jute in kg/ha
X19	Yield rate of mango in kg/ha
X20	Yield rate of potato in kg/ha
X21	Yield rate of mustard and rapeseed in kg/ha
X22	Yield rate of gram in kg/ha
X23	Yield rate of masur in kg/ha
X24	Yield rate of maize in kg/ha

Annexure B.34 Composite standardized score of sustainable land use index by land size classes in Murshidabad district (1995–1996 and 2015–2016)

	1995–1996						2015–2016					
Blocks	Marginal	Small	Semi-medium	Medium	Large	All classes	Marginal	Small	Semi-medium	Medium	Large	All classes
Berhampore	0.475	0.469	0.458	0.444	0.491	0.476	0.548	0.571	0.542	0.533	0.555	0.549
Burwan	0.421	0.423	0.412	0.394	N.A	0.418	0.497	0.510	0.468	0.445	0.454	0.500
Beldanga-I	0.424	0.394	0.383	0.389	N.A	0.407	0.504	0.507	0.448	0.454	N.A	0.502
Beldanga-II	0.396	0.385	0.374	0.355	0.407	0.392	0.503	0.510	0.468	0.422	N.A	0.505
Bhagwangola-I	0.414	0.410	0.399	0.371	N.A	0.414	0.463	0.473	0.425	0.428	0.402	0.466
Bhagwangola-II	0.403	0.394	0.383	0.380	N.A	0.399	0.457	0.470	0.440	0.473	N.A	0.463
Bharatpur-I	0.368	0.363	0.352	0.344	N.A	0.367	0.526	0.437	0.426	0.412	N.A	0.529
Bharatpur-II	0.439	0.441	0.430	0.410	N.A	0.438	0.463	0.479	0.481	0.503	N.A	0.467
Domkal	0.466	0.443	0.432	0.437	0.365	0.460	0.578	0.558	0.513	0.516	N.A	0.575
Farakka	0.337	0.336	0.325	0.315	0.258	0.331	0.470	0.406	0.409	0.401	0.347	0.479
Hariharpara	0.408	0.389	0.378	0.378	N.A	0.403	0.506	0.513	0.449	0.462	N.A	0.507
Jalangi	0.419	0.402	0.391	0.388	N.A	0.414	0.512	0.505	0.475	0.482	N.A	0.513
Kandi	0.407	0.390	0.379	0.369	N.A	0.406	0.489	0.493	0.433	0.449	N.A	0.488
Khargram	0.414	0.399	0.388	0.370	N.A	0.408	0.435	0.456	0.470	0.430	N.A	0.442
Lalgola	0.438	0.455	0.444	0.431	N.A	0.445	0.501	0.510	0.464	0.509	N.A	0.503
Murshidabad-Jiaganj	0.404	0.397	0.386	0.372	0.389	0.405	0.520	0.511	0.491	0.466	0.433	0.526
Nabagram	0.336	0.355	0.344	0.340	N.A	0.347	0.451	0.474	0.412	0.407	N.A	0.459
Nawda	0.432	0.422	0.411	0.403	N.A	0.434	0.499	0.512	0.445	0.478	0.479	0.502
Raghunathganj-I	0.283	0.284	0.273	0.266	0.246	0.286	0.411	0.322	0.352	0.328	N.A	0.414
Raghunathganj-II	0.462	0.453	0.342	0.351	0.375	0.455	0.476	0.474	0.419	0.428	0.398	0.472
Raninagar-I	0.388	0.389	0.378	0.364	0.311	0.388	0.560	0.523	0.447	0.443	N.A	0.567
Raninagar-II	0.425	0.411	0.400	0.414	N.A	0.421	0.490	0.490	0.456	0.481	N.A	0.490

Sagardighi	0.457	0.475	0.464	0.456	N.A	0.474	0.542	0.563	0.536	0.514	N.A	0.545
Samserganj	0.365	0.361	0.350	0.343	N.A	0.359	0.455	0.468	0.439	0.419	N.A	0.463
Suti-I	0.416	0.426	0.415	0.407	0.348	0.418	0.428	0.440	0.484	0.469	N.A	0.430
Suti-II	0.438	0.428	0.417	0.413	N.A	0.428	0.527	0.531	0.513	0.521	N.A	0.536
Total	**0.409**	**0.404**	**0.389**	**0.381**	**0.360**	**0.407**	**0.493**	**0.489**	**0.458**	**0.457**	**0.438**	**0.496**

Source: Calculated by researcher (data obtained from Agriculture Census, Government of India, 1995–1996 and 2015–2016)
N.A not available

Annexure C: Chapter 3

Annexure C.1 List of selected input indicators and variables of agricultural innovation

Variables	Description of the variables
Land	
X1	Percentage of land size operational holders
X2	Average number of parcels per holding
X3	Average size of landholding
X4	Percentage of area wholly own and self-operated
X5	Percentage of leased in area
X6	Percentage of net area sown
X7	Percentage of area located entirely within the village of residence
X8	Percentage of area located entirely outside the village of residence
Irrigation	
X9	Percentage of net irrigated area in total gross cropped area
X10	Percentage of area irrigated by canal, tank, and other sources
X11	Number of electric well and tube well in per 100 ha of GCA
X12	Number of diesel-operated and tube wells in per 100 ha of GCA
X13	Percentage of all crops irrigated in total GCA
Seeds	
X14	Percentage of households used certified seeds
X15	Percentage of households used HYV seeds
Credit	
X16	Percentage of households avail institutional credit facilities
X17	Percentage of households taken credit from PACS
X18	Percentage of households taken credit from PLDB
X19	Percentage of households taken credit from CBB
X20	Percentage of households taken credit from RRB
Agricultural machineries	
X21	Number of hand-operated machines and implements in per hectare of GCA
X22	Number of animal-operated machines and implements in per hectare of GCA
X23	Number of power-operated machines and implements in per hectare of GCA
Fertilizers and pesticides	
X24	Percentage of GCA treated with fertilizers
X25	Percentage of GCA treated with pesticides
X26	Amount of NPK mixture used, kg/ha
X27	Amount of urea, DAP, superphosphate mixture used, kg/ha
Socio-economic	
X28	Percentage of main worker to the total cultivators
X29	Percentage of marginal worker to the total cultivators
X30	Literacy rate of the operational holders
X31	Average age of the operational holders
X32	Average size of households

Annexure C.2 Percentage of landholders by land size class in Murshidabad district, 1995–1996 and 2015–2016

	1995–1996						2015–2016					
Blocks	Marginal	Small	Semi-medium	Medium	Large	All classes	Marginal	Small	Semi-medium	Medium	Large	All classes
Berhampore	77.31	16.94	5.32	0.36	0.08	100	86.11	11.79	1.92	0.16	0.03	100
Burwan	64.53	28.82	6.28	0.38	N.A	100	82.64	14.48	2.76	0.09	0.02	100
Beldanga-I	84.72	11.69	3.1	0.49	N.A	100	85.27	12.06	2.59	0.07	N.A	100
Beldanga-II	76.6	17.53	5.65	0.19	0.02	100	83.91	13.24	2.84	0.02	N.A	100
Bhagwangola-I	76.45	16.51	6.81	0.23	N.A	100	80.31	15.93	3.63	0.1	0.03	100
Bhagwangola-II	79.82	15.4	4.58	0.2	N.A	100	74.67	20.07	4.81	0.45	N.A	100
Bharatpur-I	75.61	19.82	4.19	0.38	N.A	100	80.55	16.06	3.23	0.17	N.A	100
Bharatpur-II	61.98	29.92	7.95	0.15	N.A	100	74.02	21.68	4.27	0.03	N.A	100
Domkal	78.06	15.87	5.04	1	0.02	100	88.52	9.51	1.9	0.07	N.A	100
Farakka	74.95	19.91	4.71	0.34	0.09	100	59.53	32.73	7.61	0.07	0.06	100
Hariharpara	76.63	16.48	5.9	0.99	N.A	100	82.78	14.46	2.63	0.13	N.A	100
Jalangi	80.91	14.85	3.88	0.36	N.A	100	86.67	11.17	2.16	0	N.A	100
Kandi	78.36	18.66	2.89	0.09	N.A	100	87.11	10.11	2.77	0	N.A	100
Khargram	84.45	9.52	5.6	0.43	N.A	100	82.42	14.48	3.01	0.08	N.A	100
Lalgola	75.21	16.73	7.53	0.53	N.A	100	82.95	14.61	2.44	N.A	N.A	100
Murshidabad-Jiaganj	74.6	19.71	5.14	0.52	0.03	100	85.61	12.04	2.33	N.A	0.02	100
Nabagram	66.83	22.13	9.92	1.12	N.A	100	81.2	15.78	2.88	0.13	N.A	100
Nawda	71.07	21.89	6.29	0.75	N.A	100	82.68	14.54	2.54	0.22	0.02	100
Raghunathganj-I	66.37	24.67	8.25	0.66	0.05	100	74.63	13.82	11.19	0.36	N.A	100
Raghunathganj-II	80.39	13.02	5.01	1.5	0.07	100	81.53	12.02	6.25	0.15	0.05	100
Raninagar-I	70.44	22.93	6.42	0.2	N.A	100	71.64	22.92	5.19	0.25	N.A	100
Raninagar-II	79.41	14.86	4.66	1.07	N.A	100	85.27	12.2	2.54	N.A	N.A	100

(continued)

Annexure C.2 (continued)

	1995–1996						2015–2016					
Blocks	Marginal	Small	Semi-medium	Medium	Large	All classes	Marginal	Small	Semi-medium	Medium	Large	All classes
Sagardighi	71.42	20.07	7.89	0.62	N.A	100	83.72	13.96	2.24	0.08	N.A	100
Samserganj	80.47	15.07	4.35	0.11	N.A	100	58.52	33.91	7.56	0.01	N.A	100
Suti-I	62.58	24.28	11.98	1.13	0.04	100	74.71	20.63	4.55	0.12	N.A	100
Suti-II	77.21	17.09	5.17	0.53	N.A	100	65.54	28.45	6	0.01	N.A	100
District total	**74.92**	**18.5**	**6.03**	**0.54**	**0.01**	**100**	**82.03**	**14.77**	**3.1**	**0.09**	**0.01**	**100**

Source: Agriculture Census, Government of India, 1995–1996 and 2015–2016
N.A not available

Annexure C.3 Average number of parcels per holdings by land size class in Murshidabad district, 1995–1996 and 2015–2016

	1995–1996						2015–2016					
Blocks	Marginal	Small	Semi-medium	Medium	Large	All classes	Marginal	Small	Semi-medium	Medium	Large	All classes
Berhampore	2.16	4	4.87	7.25	1	2.59	3.15	5.64	3.39	3.28	1	3.26
Burwan	2.37	3.94	4.67	6.71	N.A	3.06	3.26	5.28	3.39	2.72	1	3.47
Beldanga-I	2.72	4.04	5.26	7.06	N.A	2.63	3.01	5.57	3.43	2.9	N.A	3.21
Beldanga-II	2.49	3.86	5.72	7.77	1	2.81	3.15	5.28	3.24	4.3	N.A	3.31
Bhagwangola-I	2.46	4.23	5.18	6.91	N.A	2.87	3.16	5.56	3.36	2.83	1	3.65
Bhagwangola-II	2.12	4.19	5.44	6.88	N.A	2.48	3.49	5.57	3.39	3.54	N.A	4.28
Bharatpur-I	2.11	3.81	5.22	6.97	N.A	2.52	3.35	5.44	3.39	3.01	N.A	3.7
Bharatpur-II	2.43	3.6	5.18	6.77	N.A	3.13	3.61	5.33	3.36	6.3	N.A	4.2
Domkal	2.89	4.95	5.2	6.6	1	3.23	2.92	5.32	3.36	2.72	N.A	2.88
Farakka	2.24	4.01	5.16	8.74	1	2.72	3.34	5.69	3.38	3.15	1	5.47
Hariharpara	2.41	3.99	5.33	7.11	N.A	2.86	3.27	5.32	3.39	3.45	N.A	3.47
Jalangi	2.39	5.41	5.12	6.91	N.A	2.79	2.3	5.23	3.38	5.21	N.A	2.59
Kandi	1.99	3.6	5.01	6.89	N.A	2.22	2.77	5.25	3.37	5.21	N.A	2.9
Khargram	2.14	4.04	5.09	6.91	N.A	2.35	2.8	5.21	3.42	4.16	N.A	3.21
Lalgola	1.68	3.87	5.04	6.25	N.A	2.44	3.26	5.5	3.38	3.91	N.A	3.46
Murshidabad-Jiaganj	1.91	3.19	5.11	6.82	1	2.36	2.86	5.37	3.35	5.21	1	3.04
Nabagram	2.55	4.02	5.18	6.52	N.A	3.38	3.37	5.44	3.35	3.23	N.A	3.65
Nawda	2.09	4.11	5.24	7.07	N.A	2.87	3.13	5.49	3.31	4.01	1	3.45
Raghunathganj-I	2.79	3.98	5.3	6.98	1	3.43	2.35	5.77	3.39	2.99	N.A	3.99
Raghunathganj-II	2.12	3.98	5.3	8.41	1	2.68	2.92	5.09	3.4	2.97	1	3.51
Raninagar-I	2.3	3.96	5.13	6.85	N.A	2.88	3.49	5.49	3.38	2.74	N.A	4.37
Raninagar-II	1.97	3.98	5.34	6.44	N.A	2.46	2.89	5.44	3.4	2.58	N.A	3.1

(continued)

Annexure C.3 (continued)

Blocks	1995–1996						2015–2016					
	Marginal	Small	Semi-medium	Medium	Large	All classes	Marginal	Small	Semi-medium	Medium	Large	All classes
Sagardighi	1.35	4.55	4.82	7.12	N.A	2.57	3.24	5.11	3.31	3.24	N.A	3.32
Samserganj	1.58	4.05	5.26	8.22	N.A	2.08	3.51	5.69	3.43	5.21	N.A	5.29
Suti-I	2.5	4.04	5.07	6.48	1	3.56	3.32	5.61	3.45	2.87	N.A	4.12
Suti-II	2.4	3.91	5.33	6.97	N.A	2.73	3.55	5.56	3.37	5.21	N.A	5.01
District total	**2.24**	**4.05**	**5.18**	**7.06**	**0.31**	**2.76**	**3.13**	**5.43**	**3.38**	**3.73**	**0.27**	**3.69**

Source: Agriculture Census, Government of India, 1995–1996 and 2015–2016

N.A not available

Annexure C.4 Average size of landholdings by land size class in Murshidabad district, 1995–1996 and 2015–2016

	1995–1996						2015–2016					
Blocks	Marginal	Small	Semi-medium	Medium	Large	All classes	Marginal	Small	Semi-medium	Medium	Large	All classes
Berhampore	0.44	1.34	2.31	5.23	12.33	0.72	0.48	1.68	2.72	5.04	15.91	0.68
Burwan	0.49	1.32	2.21	4.84	N.A	0.85	0.5	1.57	2.72	4.18	10.6	0.72
Beldanga-I	0.56	1.35	2.49	5.09	N.A	0.73	0.46	1.66	2.75	4.44	N.A	0.66
Beldanga-II	0.51	1.29	2.71	5.6	10.2	0.78	0.48	1.57	2.6	6.6	N.A	0.69
Bhagwangola-I	0.5	1.42	2.45	4.98	N.A	0.8	0.48	1.65	2.69	4.35	14	0.76
Bhagwangola-II	0.44	1.4	2.57	4.96	N.A	0.69	0.53	1.66	2.72	5.44	N.A	0.89
Bharatpur-I	0.43	1.28	2.47	5.03	N.A	0.7	0.51	1.62	2.71	4.62	N.A	0.77
Bharatpur-II	0.5	1.21	2.45	4.88	N.A	0.87	0.55	1.59	2.69	9.67	N.A	0.87
Domkal	0.59	1.66	2.46	4.76	12	0.9	0.44	1.58	2.69	4.17	N.A	0.6
Farakka	0.46	1.34	2.44	6.3	10	0.76	0.61	1.69	2.71	4.83	11	1.13
Hariharpara	0.49	1.34	2.52	5.13	N.A	0.8	0.5	1.58	2.72	5.3	N.A	0.72
Jalangi	0.49	1.81	2.43	4.98	N.A	0.78	0.35	1.56	2.7	8	N.A	0.54
Kandi	0.41	1.21	2.37	4.97	N.A	0.62	0.42	1.56	2.7	8	N.A	0.6
Khargram	0.44	1.35	2.41	4.98	N.A	0.66	0.43	1.55	2.74	6.38	N.A	0.66
Lalgola	0.34	1.3	2.38	4.51	N.A	0.68	0.5	1.63	2.7	6	N.A	0.72
Murshidabad-Jiaganj	0.39	1.07	2.42	4.92	14.2	0.66	0.44	1.6	2.68	8	13.4	0.63
Nabagram	0.52	1.35	2.45	4.7	N.A	0.94	0.51	1.62	2.68	4.95	N.A	0.76
Nawda	0.43	1.38	2.48	5.1	N.A	0.8	0.48	1.63	2.65	6.16	10	0.71
Raghunathganj-I	0.57	1.33	2.51	5.03	14	0.96	0.36	1.72	2.72	4.59	N.A	0.83
Raghunathganj-II	0.44	1.33	2.51	6.07	10	0.75	0.44	1.51	2.73	4.56	10.4	0.73
Raninagar-I	0.47	1.33	2.43	4.94	N.A	0.8	0.53	1.63	2.71	4.21	N.A	0.91
Raninagar-II	0.4	1.33	2.53	4.65	N.A	0.69	0.44	1.62	2.72	N.A	N.A	0.64

(continued)

Annexure C.4 (continued)

Blocks	1995–1996						2015–2016					
	Marginal	Small	Semi-medium	Medium	Large	All classes	Marginal	Small	Semi-medium	Medium	Large	All classes
Sagardighi	0.28	1.53	2.28	5.13	N.A	0.72	0.49	1.52	2.65	4.97	N.A	0.69
Samserganj	0.32	1.36	2.49	5.93	N.A	0.58	0.53	1.69	2.75	8	N.A	1.1
Suti-I	0.51	1.35	2.4	4.67	10.8	0.99	0.51	1.67	2.76	4.41	N.A	0.85
Suti-II	0.49	1.31	2.52	5.03	N.A	0.76	0.62	1.65	2.7	8	N.A	1.04
District total	**0.45**	**1.35**	**2.42**	**4.97**	**12.08**	**0.76**	**0.47**	**1.61**	**2.7**	**5.13**	**12.67**	**0.71**

Source: Agriculture Census, Government of India, 1995–1996 and 2015–2016

N.A not available

Annexure C.5 Percentage of wholly owned and self-operated areas by land size class in Murshidabad district, 1995–1996 and 2015–2016

	1995–1996						2015–2016					
Blocks	Marginal	Small	Semi-medium	Medium	Large	All classes	Marginal	Small	Semi-medium	Medium	Large	All classes
Berhampore	95.26	89.59	88.73	80.87	100	92.05	77.61	74.25	100	100	100	78.79
Burwan	93.05	93.86	94.46	79.17	N.A	93.34	70.36	73.11	100	100	100	74.56
Beldanga-I	94.44	94.43	94.99	72.89	N.A	93.76	75.92	86.02	100	100	N.A	81.66
Beldanga-II	96.09	95.18	95.41	41.07	0	94.63	72.84	73.44	100	100	N.A	75.99
Bhagwangola-I	83.06	91.34	89.51	25.57	N.A	86	76.67	89.15	100	100	100	84.29
Bhagwangola-II	96.9	91.09	64.11	42.64	N.A	88.7	66.98	83.22	96.05	100	N.A	78.3
Bharatpur-I	86.01	91.06	84.89	79.34	N.A	87.48	74.51	76.76	100	100	N.A	78.44
Bharatpur-II	95.16	93.67	85.9	68.5	N.A	92.25	68.22	71.92	98.55	100	N.A	73.82
Domkal	89.58	76.2	75.17	65.7	0	82.13	73.89	66.54	100	100	N.A	74.4
Farakka	86.48	84.86	46.75	50	100	78.99	87.38	100	100	100	100	95.97
Hariharpara	85.36	82.7	93.42	83.54	N.A	86.01	68.68	75.49	100	100	N.A	74.25
Jalangi	91.37	91.66	77.79	0	N.A	87.72	65.19	59.91	87.74	100	N.A	65.94
Kandi	94.7	95.6	83.24	30.2	N.A	93.28	73.46	72.24	100	100	N.A	76.47
Khargram	87.99	83.96	83.76	74.8	N.A	85.9	61.29	63.18	100	100	N.A	67.04
Lalgola	88.51	69.21	86.19	74.53	N.A	81.24	88.94	100	100	100	N.A	93.64
Murshidabad-Jiaganj	92.19	84.51	90.39	82.37	0	88.32	79.09	88.14	100	100	100	84
Nabagram	92.85	90.87	94.63	82.89	N.A	92.13	67.19	72.69	98.31	100	N.A	72.52
Nawda	84.26	85.06	91.72	85.19	N.A	86.06	74.73	77.58	100	100	100	78.61
Raghunathganj-I	79.43	61.95	72.54	82.78	0	71.41	62.65	61.54	100	100	N.A	76.86
Raghunathganj-II	96.48	93.49	87.36	87.6	0	92.25	92.06	79.45	100	100	100	90.9
Raninagar-I	92.12	80.99	91.49	0	N.A	86.62	78.12	88.78	100	100	N.A	86.17
Raninagar-II	93.3	90.38	96.07	92.36	N.A	92.87	73.61	75.28	100	N.A	N.A	76.96

(continued)

Annexure C.5 (continued)

Blocks	1995–1996						2015–2016					
	Marginal	Small	Semi-medium	Medium	Large	All classes	Marginal	Small	Semi-medium	Medium	Large	All classes
Sagardighi	79.76	77.53	83.78	89.77	N.A	80.27	79.04	73.7	100	100	N.A	79.32
Samserganj	96.73	80.17	67.14	0	N.A	84.32	90.08	85.02	100	100	N.A	89.3
Suti-I	92.58	77.76	69.79	80.85	100	80.49	83.57	91.56	100	100	N.A	89.31
Suti-II	78.62	54.14	82.26	0	N.A	69.3	67.55	78.35	69.46	100	N.A	72.77
District total	**89.98**	**85.66**	**86.28**	**75.06**	**51.75**	**87.25**	**74.28**	**78.07**	**98.24**	**100**	**100**	**78.56**

Source: Agriculture Census, Government of India, 1995–1996 and 2015–2016

N.A not available

Annexure C.6 Percentage of wholly leased-in area by land size class in Murshidabad district, 1995–1996 and 2015–2016

	1995–1996						2015–2016					
Blocks	Marginal	Small	Semi-medium	Medium	Large	All classes	Marginal	Small	Semi-medium	Medium	Large	All classes
Berhampore	0.82	0	0	0	0	0.39	3.27	5.64	0	0	0	3.65
Burwan	0.83	0.42	0	0	N.A	0.49	0.14	0	0	0	0	0.08
Beldanga-I	1.48	0	0	0	N.A	0.95	0	3.99	0	0	N.A	1.2
Beldanga-II	0.06	0	0	0	0	0.03	0.94	0	0	0	N.A	0.55
Bhagwangola-I	0.28	0	0	0	N.A	0.14	2.01	0	0	0	0	1.03
Bhagwangola-II	0.47	0	0	0	N.A	0.24	7.04	0	0	0	N.A	3.15
Bharatpur-I	1.24	0	0	0	N.A	0.58	0.33	0	0	0	N.A	0.18
Bharatpur-II	1.02	0.8	0	0	N.A	0.69	0.78	0	0	0	N.A	0.36
Domkal	0.1	0	0	0	0	0.05	0.65	1.31	0	0	N.A	0.76
Farakka	1.38	0	0	0	0	0.63	0.03	0	0	0	0	0.01
Hariharpara	0.59	0	0	0	N.A	0.28	0	0	0	0	N.A	0
Jalangi	1.4	0	0	0	N.A	0.71	1.55	1.54	0	0	N.A	1.38
Kandi	0.43	0	0	0	N.A	0.22	0	0	0	0	N.A	0
Khargram	0.8	0	0	0	N.A	0.45	2.93	3.18	0	0	N.A	2.63
Lalgola	0.59	0	0	0	N.A	0.23	0.7	0	0	0	N.A	0.4
Murshidabad-Jiaganj	1.37	1.72	0	0	0	1.16	4.28	5.21	0	0	0	4.13
Nabagram	0.01	0	0	0	N.A	0	14.3	8.62	0.83	0	N.A	10.87
Nawda	1.83	0.44	0	0	N.A	0.86	0	0	0	0	0	0
Raghunathganj-I	1.07	1.03	0	0	0	0.78	0.49	0	0	0	N.A	0.16
Raghunathganj-II	0.25	0	0	0	0	0.11	0	0	0	0	0	0
Raninagar-I	0	0	0	0	N.A	0	0.04	0	0	0	N.A	0.02
Raninagar-II	0.63	0	0	0	N.A	0.3	0.05	2.43	0	N.A	N.A	0.78

(continued)

Annexure C.6 (continued)

	1995–1996						2015–2016					
Blocks	Marginal	Small	Semi-medium	Medium	Large	All classes	Marginal	Small	Semi-medium	Medium	Large	All classes
Sagardighi	3.75	2.56	0.83	0	N.A	2.34	0.31	0	0	0	N.A	0.19
Samserganj	0.2	0	0	0	N.A	0.09	0	0	0	0	N.A	0
Suti-I	0	0	0	0	0	0	0	0	0	0	N.A	0
Suti-II	0	0	0	0	N.A	0	4.06	7.16	0	0	N.A	4.83
District total	**0.85**	**0.45**	**0.09**	**0**	**0**	**0.54**	**1.9**	**1.69**	**0.04**	**0**	**0**	**1.6**

Source: Agriculture Census, Government of India, 1995–1996 and 2015–2016
N.A not available

Annexure C.7 Percentage of net sown area in total area by land size class in Murshidabad district, 1995–1996 and 2015–2016

	1995–1996						2015–2016					
Blocks	Marginal	Small	Semi-medium	Medium	Large	All classes	Marginal	Small	Semi-medium	Medium	Large	All classes
Berhampore	98.97	98.64	98.78	98.92	91.89	98.74	99.48	100	100	100	100	99.68
Burwan	99.24	98.78	99.4	98.91	N.A	99.05	99.57	100	100	100	100	99.75
Beldanga-I	99.7	99.29	99.68	99.41	N.A	99.6	98.93	100	100	100	N.A	99.36
Beldanga-II	97.18	96.93	97.25	98.21	88.24	97.1	99.45	100	100	100	N.A	99.68
Bhagwangola-I	98.75	99.21	97.87	86.08	N.A	98.52	97.92	100	100	100	100	98.94
Bhagwangola-II	96.45	93.72	96.1	98.45	N.A	95.57	97.41	100	99.95	100	N.A	98.82
Bharatpur-I	98.35	97.64	100	100	N.A	98.38	98.3	98.33	98.06	100	N.A	98.3
Bharatpur-II	100	100	100	100	N.A	100	99.04	99.85	100	100	N.A	99.49
Domkal	98.37	96.13	91.48	88.88	95	96.25	98.9	100	100	100	N.A	99.28
Farakka	98.52	99.05	96.3	97.62	94	98.29	99.81	100	100	100	100	99.94
Hariharpara	98.18	95.52	94.78	89.7	N.A	96.27	99.18	97.47	100	100	N.A	98.73
Jalangi	96.86	95.99	89.82	91.82	N.A	95.59	96.76	89.49	100	100	N.A	94.75
Kandi	99.88	99.84	100	95.97	N.A	99.85	98.85	100	100	100	N.A	99.3
Khargram	99.25	98.09	98.13	92.87	N.A	98.58	99.29	100	100	100	N.A	99.62
Lalgola	89.86	86.14	87.27	87.81	N.A	87.92	98.22	100	100	100	N.A	98.98
Murshidabad-Jiaganj	99.66	98.05	99.75	79.14	95.77	98.33	98.24	100	100	100	100	98.96
Nabagram	96.47	97.33	97.8	98.3	N.A	97.19	98.33	99.3	98.47	88.46	N.A	98.59
Nawda	98.26	96.59	95.1	89.62	N.A	96.6	99.26	98.4	100	100	100	99.06
Raghunathganj-I	99.31	99.17	95.59	92.38	95.71	98.19	98.84	100	100	100	N.A	99.62
Raghunathganj-II	99.63	98.6	93.05	97.96	52	97.63	98.47	97.06	100	100	100	98.51
Raninagar-I	98.3	97.93	97.36	100	N.A	98	98.97	98.8	100	100	N.A	99.08
Raninagar-II	98.37	97.23	93.4	96.88	N.A	97.08	98.89	98.9	100	N.A	N.A	99.01

(continued)

Annexure C.7 (continued)

	1995–1996						2015–2016					
Blocks	Marginal	Small	Semi-medium	Medium	Large	All classes	Marginal	Small	Semi-medium	Medium	Large	All classes
Sagardighi	99.51	99.14	98.36	97.36	N.A	98.97	99.1	99.16	100	100	N.A	99.2
Samserganj	100	99.96	99.93	98.8	N.A	99.96	99.17	84.28	100	100	N.A	91.53
Suti-I	98.47	99.21	98.11	98.31	90.74	98.57	99.72	100	100	100	N.A	99.88
Suti-II	93.41	82.27	86.2	80.1	N.A	88.43	99.35	100	100	100	N.A	99.75
District total	**98.27**	**97.42**	**96.77**	**93.9**	**90.43**	**97.53**	**98.82**	**98.61**	**99.85**	**99.19**	**100**	**98.88**

Source: Agriculture Census, Government of India, 1995–1996 and 2015–2016
N.A not available

Annexure C.8 Percentage of area located entirely in the village of residence by land size class in Murshidabad district, 1995–1996 and 2015–2016

	1995–1996						2015–2016					
Blocks	Marginal	Small	Semi-medium	Medium	Large	All classes	Marginal	Small	Semi-medium	Medium	Large	All classes
Berhampore	72.68	58.35	59.82	55.78	0	64.53	96.78	94.62	92.87	90.33	100	96.28
Burwan	90.49	86.66	85.86	66.67	N.A	87.51	89.7	69.53	72.83	50.34	100	79.84
Beldanga-I	69.89	56.99	43.27	27.11	N.A	62.86	95.88	85.95	87.53	91.72	N.A	93.17
Beldanga-II	70.88	56.26	46.5	20.09	0	60.97	97.84	97.78	97.66	97.25	N.A	97.8
Bhagwangola-I	61.5	43	37.62	0	N.A	50.18	95.88	87.32	43.13	94.31	100	88.33
Bhagwangola-II	81.68	59.05	83.28	83.72	N.A	74.91	79.71	72.77	39.39	N.A	N.A	73.51
Bharatpur-I	62.18	42.51	52.79	60.62	N.A	53.67	83.01	67.93	61.76	20.06	N.A	74.27
Bharatpur-II	83.79	84.1	61.7	68.5	N.A	78.87	91.9	87.07	79.49	60.55	N.A	86.93
Domkal	75.78	68.22	49.9	16.01	0	66.6	96.3	90.11	88.93	51.39	N.A	93.53
Farakka	66.65	34.03	2.07	100	100	46.64	80.68	68.63	36.94	100	100	74.86
Hariharpara	75.27	73.4	60.65	54.84	N.A	70.74	96.46	91.62	90.93	92.38	N.A	94.64
Jalangi	75.21	58.2	63.08	41.26	N.A	67.07	89.61	76.28	8.52	97.05	N.A	75.12
Kandi	66.19	54.79	19.44	15.44	N.A	56.48	94.27	91.9	95.18	92.4	N.A	93.22
Khargram	58.78	38.76	25.26	24.9	N.A	46.85	98.39	97.23	95.38	98.58	N.A	97.43
Lalgola	55.07	56.91	34.04	17.5	N.A	48.78	91.46	91.03	87.47	90.74	N.A	91.16
Murshidabad-Jiaganj	67.12	61.46	47.12	24.09	0	59.35	89.31	81.74	87.03	21.13	100	86.38
Nabagram	66.61	57.42	36.14	20.04	N.A	53.25	96.07	93.53	91.48	92.92	N.A	94.51
Nawda	98.21	90.07	84.27	84.62	N.A	91.78	87.63	79.08	75.98	86.15	100	83.02
Raghunathganj-I	56.64	33.31	32.66	0	0	41.01	70.54	67.25	79.6	92.09	N.A	71.04
Raghunathganj-II	19.47	7.66	12.98	0	0	13.07	68.23	64.97	51.91	26.59	100	64.25
Raninagar-I	70.53	54.92	36.75	0	N.A	57.16	97.5	95.72	96.9	97.35	N.A	97.19

(continued)

Annexure C.8 (continued)

Blocks	1995–1996						2015–2016					
	Marginal	Small	Semi-medium	Medium	Large	All classes	Marginal	Small	Semi-medium	Medium	Large	All classes
Raninagar-II	88.47	61.87	63.76	62	N.A	74.63	93.26	81.99	69.68	80.77	N.A	86.79
Sagardighi	54.37	39.39	42.88	31.51	N.A	44.05	71	54.15	48.93	28.58	N.A	61.41
Samserganj	77.02	51.76	73.2	0	N.A	66.57	79.89	63.1	73.15	28.54	N.A	72.31
Suti-I	79.62	60.93	62.62	47.18	100	66.88	95.02	93.02	93.55	91.92	N.A	93.78
Suti-II	64.79	44.22	65.77	42.35	N.A	58.12	76.74	46.95	28.32	17.4	N.A	60.12
District total	**70.06**	**58.65**	**49.15**	**37.34**	**13.45**	**61.02**	**90.22**	**81.48**	**75.84**	**75.02**	**100**	**85.32**

Source: Agriculture Census, Government of India, 1995–1996 and 2015–2016

N.A not available

Annexure C.9 Percentage of area located entirely outside village of residence by land size class in Murshidabad district, 1995–1996 and 2015–2016

	1995–1996						2015–2016					
Blocks	Marginal	Small	Semi-medium	Medium	Large	All classes	Marginal	Small	Semi-medium	Medium	Large	All classes
Berhampore	27.32	41.65	40.18	44.22	100	35.47	3.22	5.38	7.13	9.67	0	3.72
Burwan	9.51	13.34	14.14	33.33	N.A	12.49	10.3	30.47	27.17	49.66	0	20.16
Beldanga-I	30.11	43.01	56.73	72.89	N.A	37.14	4.12	14.05	12.47	8.28	N.A	6.83
Beldanga-II	29.12	43.74	53.5	79.91	100	39.03	2.16	2.22	2.34	2.75	N.A	2.2
Bhagwangola-I	38.5	57	62.38	100	N.A	49.82	4.12	12.68	56.87	5.69	0	11.67
Bhagwangola-II	18.32	40.95	16.72	16.28	N.A	25.09	20.29	27.23	60.61	N.A	N.A	26.49
Bharatpur-I	37.82	57.49	47.21	39.38	N.A	46.33	16.99	32.07	38.24	79.94	N.A	25.73
Bharatpur-II	16.21	15.9	38.3	31.5	N.A	21.13	8.1	12.93	20.51	39.45	N.A	13.07
Domkal	24.22	31.78	50.1	83.99	100	33.4	3.7	9.89	11.07	48.61	N.A	6.47
Farakka	33.35	65.97	97.93	0	0	53.36	19.32	31.37	63.06	0	0	25.14
Hariharpara	24.73	26.6	39.35	45.16	N.A	29.26	3.54	8.38	9.07	7.62	N.A	5.36
Jalangi	24.79	41.8	36.92	58.74	N.A	32.93	10.39	23.72	91.48	2.95	N.A	24.88
Kandi	33.81	45.21	80.56	84.56	N.A	43.52	5.73	8.1	4.82	7.6	N.A	6.78
Khargram	41.22	61.24	74.74	75.1	N.A	53.15	1.61	2.77	4.62	1.42	N.A	2.57
Lalgola	44.93	43.09	65.96	82.5	N.A	51.22	8.54	8.97	12.53	9.26	N.A	8.84
Murshidabad-Jiaganj	32.88	38.54	52.88	75.91	100	40.65	10.69	18.26	12.97	78.87	0	13.62
Nabagram	33.39	42.58	63.86	79.96	N.A	46.75	3.93	6.47	8.52	7.08	N.A	5.49
Nawda	1.79	9.93	15.73	15.38	N.A	8.22	12.37	20.92	24.02	13.85	0	16.98
Raghunathganj-I	43.36	66.69	67.34	100	100	58.99	29.46	32.75	20.4	7.91	N.A	28.96
Raghunathganj-II	80.53	92.34	87.02	100	100	86.93	31.77	35.03	48.09	73.41	100	35.75
Raninagar-I	29.47	45.08	63.25	100	N.A	42.84	2.5	4.28	3.1	2.65	N.A	2.81
Raninagar-II	11.53	38.13	36.24	38	N.A	25.37	6.74	18.01	30.32	19.23	N.A	13.21

(continued)

Annexure C.9 (continued)

Blocks	1995–1996						2015–2016					
	Marginal	Small	Semi-medium	Medium	Large	All classes	Marginal	Small	Semi-medium	Medium	Large	All classes
Sagardighi	45.63	60.61	57.12	68.49	N.A	55.95	29	45.85	51.07	71.42	N.A	38.59
Samserganj	22.98	48.24	26.8	100	N.A	33.43	20.11	36.9	26.85	71.46	N.A	27.69
Suti-I	20.38	39.07	37.38	52.82	0	33.12	4.98	6.98	6.45	8.08	N.A	6.22
Suti-II	35.21	55.78	34.23	57.65	N.A	41.88	23.26	53.05	71.68	82.6	N.A	39.88
District total	**29.94**	**41.35**	**50.85**	**62.66**	**86.55**	**38.98**	**9.78**	**18.52**	**24.16**	**24.98**	**0**	**14.68**

Source: Agriculture Census, Government of India, 1995–1996 and 2015–2016
N.A not available

Annexure C.10 Percentage of net irrigated area in total gross cropped area by land size class in Murshidabad district, 1995–1996 and 2015–2016

	1995–1996						2015–2016					
Blocks	Marginal	Small	Semi-medium	Medium	Large	All classes	Marginal	Small	Semi-medium	Medium	Large	All classes
Berhampore	40.71	39.41	38.69	39.29	40.06	39.87	36.74	39.11	44.5	50.56	100	40.71
Burwan	47.59	51.53	52.98	52.76	N.A	50.31	32.75	34.04	42.24	35.14	100	47.59
Beldanga-I	32.99	34.67	41.61	34.85	N.A	34.19	53.33	68.05	100	100	N.A	32.99
Beldanga-II	37.93	42.49	41.48	37.65	29.63	39.78	36.03	39.47	49.38	100	N.A	37.93
Bhagwangola-I	36.31	37.1	31.67	29.31	N.A	35.54	44.32	53.58	65.66	52.1	100	36.31
Bhagwangola-II	34.46	31.74	27.53	27.23	N.A	32.42	39.84	56.22	56.65	100	N.A	34.46
Bharatpur-I	61.95	65.32	64.21	85.22	N.A	64.05	31.08	31.49	65.91	45.48	N.A	61.95
Bharatpur-II	58.12	59.15	57.69	60.1	N.A	58.47	26.24	35.56	42.59	100	N.A	58.12
Domkal	29.75	31.73	37.81	38.29	23.33	31.6	47.31	43.98	81.42	100	N.A	29.75
Farakka	1.05	0	0.54	4.17	14.89	0.79	28.89	22.11	33.92	27.59	100	1.05
Hariharpara	35.24	38.93	39.85	31.05	N.A	36.75	52.26	57.87	100	100	N.A	35.24
Jalangi	16.13	25.73	32.51	17.79	N.A	21.21	37.42	41.55	57.71	100	N.A	16.13
Kandi	40.62	41.5	44.5	39.81	N.A	41.38	52.71	53.67	100	100	N.A	40.62
Khargram	42.95	46.23	50.06	45.18	N.A	45.03	34.38	31	56.92	85.57	N.A	42.95
Lalgola	42.14	39.52	42.33	40.4	N.A	41.29	52.44	54.6	49.98	100	N.A	42.14
Murshidabad-Jiaganj	35.99	38.87	35.75	33.97	0	36.68	33.77	38.56	50.35	100	100	35.99
Nabagram	44.98	44.86	43.46	39.32	N.A	44.22	43.5	36.02	52.97	54.86	N.A	44.98
Nawda	30.32	30.36	35.49	39.38	N.A	31.51	58.83	61.67	100	100	100	30.32
Raghunathganj-I	21.3	27.66	26.92	41.03	6.94	25.32	23.21	23.46	49.73	100	N.A	21.3
Raghunathganj-II	4.03	5.66	1.27	9.26	0	4.53	38.66	18.93	4.54	100	100	4.03
Raninagar-I	44.22	45.11	39.44	46.15	0	43.47	43.21	49.6	52.74	62.72	N.A	44.22

((continued))

Annexure C.10 (continued)

	1995–1996						2015–2016					
Blocks	Marginal	Small	Semi-medium	Medium	Large	All classes	Marginal	Small	Semi-medium	Medium	Large	All classes
Raninagar-II	31.06	30.1	30.29	30.77	N.A	30.64	43.15	51.22	51.06	N.A	N.A	31.06
Sagardighi	32.26	32.29	37.33	32.76	N.A	33.53	35.12	33.15	52.41	39.29	N.A	32.26
Samserganj	1.1	3.36	1.89	16.13	N.A	2.23	40.61	48.19	65.34	100	N.A	1.1
Suti-I	23.58	22.9	20.78	26.77	15.09	22.71	51.36	76.43	99.83	100	N.A	23.58
Suti-II	5.77	12.25	3.75	0	N.A	7.03	46.49	62.76	63.86	100	N.A	5.77
District total	**35.99**	**38.03**	**38.61**	**35.95**	**22.1**	**37.11**	**40.53**	**42.74**	**57.93**	**71.06**	**100**	**35.99**

Source: Agriculture Census, Government of India, 1995–1996 and 2015–2016
N.A not available

Annexure C.11 Percentage of gross cropped area irrigated by tank and other sources (except well and tube well) by land size class in Murshidabad district, 1995–1996 and 2015–2016

	1995–1996						2015–2016					
Blocks	Marginal	Small	Semi-medium	Medium	Large	All classes	Marginal	Small	Semi-medium	Medium	Large	All classes
Berhampore	18.86	14.71	13.86	41.23	0	17.05	10.7	11.06	16.21	8.73	N.A	11.35
Burwan	58.44	61.23	57.53	94.88	N.A	60.35	13.58	10.01	9.81	10.84	N.A	11.76
Beldanga-I	4.35	0.79	0.21	1.99	N.A	3.06	8.49	7.63	20.48	0	N.A	10.3
Beldanga-II	39.6	41.2	38.4	28.8	68.75	39.78	31.45	30.29	25.26	0.52	0	30.18
Bhagwangola-I	2.87	7.83	7.1	0	N.A	5.11	1.64	4.28	0	7.4	N.A	2.54
Bhagwangola-II	4.52	0	0	10.91	N.A	2.6	2.47	0.1	0	N.A	N.A	1.26
Bharatpur-I	57.95	67.45	56.67	68.42	N.A	61.51	26.56	33.25	18.23	32.41	N.A	28.3
Bharatpur-II	56.92	47.46	50.26	52.1	N.A	51.47	11.8	16.68	11.47	3.43	N.A	13.35
Domkal	5.5	5.32	7.06	12.99	100	6.14	5.7	4.38	7.19	0	0	5.3
Farakka	100	N.A	100	66.67	0	78.05	4.46	5.01	N.A	N.A	0	4.75
Hariharpara	12.18	12.46	10.86	17.8	N.A	12.25	6.35	11.59	0.67	2.2	N.A	7.25
Jalangi	0.3	1.18	1.47	0	N.A	0.85	20.19	17.49	2.71	4.36	N.A	17
Kandi	40.69	34.18	36.13	8.94	N.A	37.52	24.31	15.17	15.83	16.39	N.A	20.32
Khargram	42.72	48.43	58.65	50.36	N.A	47.5	11.05	13.27	21.39	9.16	N.A	13.06
Lalgola	13.1	8.55	3.85	2.04	N.A	8.73	5.45	12.07	12.18	26.4	N.A	8.49
Murshidabad-Jiaganj	14.78	8.92	3.27	22.44	N.A	10.98	10.3	8.73	8.19	37.33	N.A	9.87
Nabagram	19.65	24.93	23.84	19.89	N.A	22.45	5.27	5.44	6.98	8.84	N.A	5.6
Nawda	5.3	3.18	2.7	8.96	N.A	4.19	29.17	25.75	14.83	30.45	N.A	25.48
Raghunathganj-I	35.21	68.45	64.66	72.68	40	56.75	28.99	35.56	45.63	4.3	0	33.95
Raghunathganj-II	91.81	65.74	100	100	N.A	86.58	5.78	13.07	2.06	9.89	N.A	7.52
Raninagar-I	8.78	9.79	6	0	N.A	8.47	3.17	0.89	0.57	2.68	N.A	2.19

((continued))

Annexure C.11 (continued)

Blocks	1995–1996						2015–2016					
	Marginal	Small	Semi-medium	Medium	Large	All classes	Marginal	Small	Semi-medium	Medium	Large	All classes
Raninagar-II	2.44	0.04	0.63	7.47	N.A	1.79	2.23	5.18	8.69	20.44	N.A	4.18
Sagardighi	45.66	39.6	36.68	33.94	N.A	40.22	10.52	8.36	7.34	13.41	N.A	9.51
Samserganj	31.34	3.59	0	8	N.A	9.42	15.62	5.12	4.24	13.09	N.A	10.48
Suti-I	90.19	88.1	82.81	96.44	100	87.98	32.27	51.71	21.99	41.15	0	38.05
Suti-II	2	1.24	0	N.A	N.A	1.47	7.55	5.77	6.39	12.92	N.A	6.9
District total	**26.37**	**29.38**	**27.89**	**31.58**	**19.77**	**27.81**	**12.22**	**12.72**	**12.57**	**16.49**	**0**	**12.48**

Source: Agriculture Census, Government of India, 1995–1996 and 2015–2016
N.A not available

Annexure C.12 Number of electric well and tube wells in per 10 ha of GCA by land size class in Murshidabad district

1995–1996 and 2015–2016												
	1995–1996						2015–2016					
Blocks	Marginal	Small	Semi-medium	Medium	Large	All classes	Marginal	Small	Semi-medium	Medium	Large	All classes
Berhampore	0	0	0	0	1	0	0	0	0	1	0	0
Burwan	0	0	0	0	N.A	0	1	1	1	3	N.A	1
Beldanga-I	4	3	1	2	N.A	3	0	0	0	0	N.A	0
Beldanga-II	0	0	0	0	0	0	0	0	0	0	0	0
Bhagwangola-I	0	0	0	0	N.A	0	0	0	0	0	N.A	0
Bhagwangola-II	1	2	1	0	N.A	1	0	0	0	N.A	N.A	0
Bharatpur-I	8	4	2	0	N.A	5	0	0	0	0	N.A	0
Bharatpur-II	1	4	0	0	N.A	2	1	2	2	2	N.A	2
Domkal	2	1	1	0	0	1	1	0	0	0	0	0
Farakka	0	0	0	0	0	0	2	0	0	0	0	1
Hariharpara	0	0	0	0	N.A	0	0	0	0	1	N.A	0
Jalangi	1	0	0	0	N.A	1	0	0	0	2	N.A	0
Kandi	2	3	0	0	N.A	2	0	1	1	1	N.A	0
Khargram	0	0	0	0	N.A	0	0	0	0	0	N.A	0
Lalgola	0	0	0	0	N.A	0	0	0	1	0	N.A	0
Murshidabad-Jiaganj	1	1	2	2	0	1	0	0	0	0	0	0
Nabagram	1	0	0	0	N.A	1	0	0	0	0	N.A	0
Nawda	1	0	0	0	N.A	0	2	2	1	0	N.A	2
Raghunathganj-I	0	0	0	0	1	0	0	0	0	0	0	0
Raghunathganj-II	0	0	0	0	0	0	0	0	0	0	0	0
Raninagar-I	0	0	0	0	N.A	0	0	0	0	0	N.A	0

(continued)

Annexure C.12 (continued)

1995–1996 and 2015–2016												
	1995–1996						2015–2016					
Blocks	Marginal	Small	Semi-medium	Medium	Large	All classes	Marginal	Small	Semi-medium	Medium	Large	All classes
Raninagar-II	0	1	0	0	N.A	0	0	0	0	0	N.A	0
Sagardighi	1	0	0	0	N.A	0	0	0	0	0	N.A	0
Samserganj	0	0	0	1	N.A	0	0	0	0	0	N.A	0
Suti-I	0	0	0	0	0	0	2	0	0	0	0	1
Suti-II	4	0	0	0	N.A	2	4	0	13	0	N.A	4
District total	**1**	**1**	**0**	**0**	**1**	**1**	**0**	**0**	**1**	**0**	**0**	**0**

Source: Agriculture Census, Government of India, 1995–1996 and 2015–2016
N.A not available

Annexure C.13 Number of diesel-operated well and tube wells in per 10 ha of GCA by land size class in Murshidabad district

1995–1996 and 2015–2016												
	1995–1996						2015–2016					
Blocks	Marginal	Small	Semi-medium	Medium	Large	All classes	Marginal	Small	Semi-medium	Medium	Large	All classes
Berhampore	0	1	0	1	0	0	0	1	0	0	N.A	0
Burwan	0	0	0	0	N.A	0	1	0	1	0	N.A	0
Beldanga-I	15	12	11	5	N.A	13	6	8	6	4	N.A	7
Beldanga-II	1	2	1	0	2	1	0	1	0	0	0	1
Bhagwangola-I	7	4	3	0	N.A	5	0	0	0	0	N.A	0
Bhagwangola-II	5	4	3	2	N.A	4	0	0	0	N.A	N.A	0
Bharatpur-I	3	1	1	0	N.A	2	2	0	0	0	N.A	1
Bharatpur-II	1	1	1	1	N.A	1	0	0	0	0	N.A	0
Domkal	5	4	4	3	0	4	6	7	4	2	0	6
Farakka	0	0	0	0	0	0	0	0	0	0	0	0
Hariharpara	3	2	1	1	N.A	2	0	0	0	0	N.A	0
Jalangi	0	1	1	1	N.A	1	1	1	1	1	N.A	1
Kandi	0	1	0	0	N.A	0	0	0	0	0	N.A	0
Khargram	6	3	0	0	N.A	4	0	0	0	0	N.A	0
Lalgola	4	2	1	0	N.A	2	1	1	3	3	N.A	1
Murshidabad-Jiaganj	2	1	1	1	0	1	0	0	0	0	0	0
Nabagram	6	4	3	2	N.A	4	0	0	0	0	N.A	0
Nawda	15	15	3	3	N.A	12	4	1	0	0	N.A	2
Raghunathganj-I	1	1	0	1	0	1	0	0	0	0	0	0
Raghunathganj-II	0	0	0	0	0	0	0	0	0	0	0	0
Raninagar-I	3	2	2	0	N.A	3	4	5	4	4	N.A	4

(continued)

Annexure C.13 (continued)

1995–1996 and 2015–2016												
	1995–1996						2015–2016					
Blocks	Marginal	Small	Semi-medium	Medium	Large	All classes	Marginal	Small	Semi-medium	Medium	Large	All classes
Raninagar-II	9	5	4	2	N.A	7	0	0	0	0	N.A	0
Sagardighi	2	1	1	0	N.A	1	2	0	0	0	N.A	1
Samserganj	0	0	0	1	N.A	0	0	0	0	0	N.A	0
Suti-I	0	0	0	0	0	0	4	1	1	0	0	2
Suti-II	1	1	0	0	N.A	1	0	0	0	3	N.A	0
District Total	**4**	**3**	**2**	**1**	**0**	**3**	**2**	**1**	**1**	**0**	**0**	**1**

Source: Agriculture Census, Government of India, 1995–1996 and 2015–2016
N.A not available

Annexure C.14 Percentage of irrigated area in GCA under all crops by land size class in Murshidabad district, 1995–1996 and 2015–2016

	1995–1996						2015–2016					
Blocks	Marginal	Small	Semi-medium	Medium	Large	All classes	Marginal	Small	Semi-medium	Medium	Large	All classes
Berhampore	84.68	85.08	89.34	84.82	40.69	85.38	64.93	65.35	66.59	50.56	100	65.13
Burwan	77.85	81.53	83.01	76.62	N.A	80.29	59.73	57.75	56.7	59.06	100	58.87
Beldanga-I	76.99	79.62	76.34	80	N.A	77.56	94.67	92.97	100	100	N.A	94.59
Beldanga-II	68.93	73.18	72.95	43.07	40.74	70.49	65.52	67.42	65.8	100	N.A	66.14
Bhagwangola-I	67.39	68.55	59.7	51.29	N.A	66.02	82.07	75.59	80.66	100	100	80
Bhagwangola-II	65.89	62.83	65.06	42.57	N.A	64.54	71.51	79.66	98.71	100	N.A	78.65
Bharatpur-I	84.93	87.96	89.15	100	N.A	87.01	52.81	50.22	67.14	56.78	N.A	52.88
Bharatpur-II	96.75	97.91	95.27	95.96	N.A	96.9	50.36	60.99	51.02	100	N.A	54.3
Domkal	69.82	70.33	73.24	74.87	23.33	70.5	85.41	83.7	86.7	100	N.A	85.03
Farakka	1.37	0	0.76	5.56	14.89	1	48	41.27	56.48	27.59	100	45.43
Hariharpara	81.76	78.35	71.31	45.14	N.A	77.5	93.26	94.69	100	100	N.A	94.15
Jalangi	39.69	56.32	64.47	22.81	N.A	47.76	75.02	76.92	83.53	100	N.A	76.31
Kandi	81.15	84.69	91.84	92.23	N.A	83.76	87	96.72	100	100	N.A	90.65
Khargram	84.68	89.18	94.24	80.27	N.A	87.29	60.98	57.94	57.22	85.57	N.A	59.64
Lalgola	83.43	79.61	87.53	84.34	N.A	83.37	95.2	97.35	88.36	100	N.A	95.23
Murshidabad-Jiaganj	87.68	91.25	84.28	88.97	0	87.95	59.31	51.73	83.47	100	100	58.82
Nabagram	79.19	80.91	83.53	75.6	N.A	80.72	64.16	54.39	62.74	94.98	N.A	60.56
Nawda	64.46	60.24	64.31	62.21	N.A	62.76	93.79	98.95	100	100	100	96.04
Raghunathganj-I	39.11	42.93	40.28	53.81	6.94	41.06	43.31	45.81	49.73	100	N.A	46.25
Raghunathganj-II	6.37	9.74	1.19	9.15	0	6.62	67.24	36.87	4.76	100	100	49.79
Raninagar-I	82.3	85.01	81.74	100	100	83.39	81.91	85.88	90.38	100	N.A	84.88
Raninagar-II	75.85	73.29	67.37	69.11	N.A	73.39	84.56	83.45	98.42	N.A	N.A	85.72

(continued)

Annexure C.14 (continued)

Blocks	1995–1996						2015–2016					
	Marginal	Small	Semi-medium	Medium	Large	All classes	Marginal	Small	Semi-medium	Medium	Large	All classes
Sagardighi	51.66	49.96	56.25	46.41	N.A	51.82	49.19	52.67	55.5	50.45	N.A	50.71
Samserganj	1.51	5.59	2.59	18.71	N.A	3.38	64.25	63.72	74.8	100	N.A	65.94
Suti-I	43.78	36.29	32.47	38.6	20.75	37.77	74.03	98.2	99.83	100	N.A	85.84
Suti-II	11.5	24.27	8.95	1.44	N.A	14.27	91.69	97.56	88.91	100	N.A	93.7
District total	**71.37**	**71.23**	**72.04**	**64**	**24.72**	**71.18**	**70.74**	**69.58**	**74.37**	**81.91**	**100**	**70.74**

Source: Agriculture Census, Government of India, 1995–1996 and 2015–2016
N.A not available

Annexure C.15 Percentage of households used certified seeds by land size class in Murshidabad district, 1996–1997 and 2016–2017

	1996–1997						2016–2017					
Blocks	Marginal	Small	Semi-medium	Medium	Large	All classes	Marginal	Small	Semi-medium	Medium	Large	All classes
Berhampore	39.25	46.33	46.61	57.23	0.00	39.45	31.61	29.86	31.45	28.13	0.00	29.08
Burwan	43.06	45.69	44.70	53.02	N.A	46.60	32.75	27.98	31.45	23.33	0.00	30.95
Beldanga-I	49.45	46.76	50.34	55.74	N.A	40.12	30.19	29.51	31.79	24.82	N.A	28.62
Beldanga-II	45.16	44.68	54.72	61.32	0.00	42.86	31.62	27.97	30.05	36.85	N.A	29.50
Bhagwangola-I	44.69	49.06	49.61	54.57	N.A	43.71	31.75	29.43	31.15	24.29	0.00	32.54
Bhagwangola-II	38.52	48.50	52.03	54.33	N.A	37.82	35.05	29.52	31.46	30.38	N.A	38.10
Bharatpur-I	38.33	44.10	49.94	55.07	N.A	38.44	33.63	28.81	31.39	25.81	N.A	32.97
Bharatpur-II	44.21	41.77	49.54	53.49	N.A	47.76	36.23	28.22	31.11	53.98	N.A	37.39
Domkal	52.51	57.32	49.81	52.15	0.00	49.28	29.26	28.16	31.11	23.30	N.A	25.71
Farakka	40.75	46.43	49.34	68.98	0.00	41.44	40.05	30.13	31.30	26.99	0.00	48.70
Hariharpara	43.72	46.25	50.99	56.15	N.A	43.67	32.86	28.14	31.42	29.61	N.A	30.96
Jalangi	43.33	62.65	49.04	54.55	N.A	42.52	23.09	27.72	31.28	44.67	N.A	23.07
Kandi	36.12	41.72	47.96	54.38	N.A	33.81	27.83	27.80	31.21	44.67	N.A	25.86
Khargram	38.86	46.76	48.69	54.53	N.A	35.87	28.14	27.58	31.67	35.65	N.A	28.58
Lalgola	30.50	44.82	48.22	49.35	N.A	37.17	32.72	29.10	31.29	33.50	N.A	30.85
Murshidabad-Jiaganj	34.61	36.91	48.92	53.88	0.00	35.93	28.72	28.44	31.03	44.67	0.00	27.13
Nabagram	46.29	46.55	49.59	51.49	N.A	51.60	33.79	28.78	31.05	27.65	N.A	32.51
Nawda	37.98	47.63	50.14	55.82	N.A	43.81	31.39	29.07	30.68	34.41	0.00	30.73
Raghunathganj-I	50.65	46.09	50.70	55.11	0.00	52.32	23.61	30.57	31.46	25.63	N.A	35.52
Raghunathganj-II	38.54	46.10	50.72	66.43	0.00	40.90	29.27	26.94	31.54	25.48	0.00	31.25
Raninagar-I	41.70	45.91	49.11	54.12	N.A	43.90	35.01	29.07	31.36	23.48	N.A	38.98
Raninagar-II	35.74	46.09	51.14	50.86	N.A	37.53	29.02	28.81	31.48	N.A	N.A	27.62

(continued)

Annexure C.15 (continued)

Blocks	1996–1997						2016–2017					
	Marginal	Small	Semi-medium	Medium	Large	All classes	Marginal	Small	Semi-medium	Medium	Large	All classes
Sagardighi	24.48	52.71	46.17	56.20	N.A	39.15	32.48	27.07	30.69	27.76	N.A	29.61
Samserganj	28.75	46.88	50.30	64.92	N.A	31.75	35.23	30.15	31.76	44.67	N.A	47.16
Suti-I	45.36	46.77	48.52	51.15	0.00	54.34	33.34	29.70	31.97	24.63	N.A	36.71
Suti-II	43.57	45.26	51.02	55.03	N.A	41.62	40.70	29.45	31.27	44.67	N.A	44.68
District total	**39.47**	**46.55**	**48.96**	**54.47**	**0.00**	**41.43**	**31.11**	**28.65**	**31.26**	**28.65**	**0.00**	**30.75**

Source: Agriculture Census, Government of India, 1995–1996 and 2015–2016

N.A not available

Annexure C.16 Percentage of households used hybrid seeds by land size class in Murshidabad district, 1996–1997 and 2016–2017

	1996–1997						2016–2017					
Blocks	Marginal	Small	Semi-medium	Medium	Large	All classes	Marginal	Small	Semi-medium	Medium	Large	All classes
Berhampore	4.74	4.89	5.92	7.75	0.00	4.66	8.52	6.61	8.28	4.77	0.00	7.64
Burwan	5.20	4.82	5.67	7.18	N.A	5.50	8.83	6.19	8.28	3.96	0.00	8.13
Beldanga-I	5.97	4.93	6.39	7.55	N.A	4.74	8.14	6.53	8.37	4.21	N.A	7.52
Beldanga-II	5.45	4.71	6.94	8.30	0.00	5.06	8.53	6.19	7.91	6.25	N.A	7.75
Bhagwangola-I	5.39	5.18	6.30	7.39	N.A	5.16	8.56	6.51	8.20	4.12	0.00	8.55
Bhagwangola-II	4.65	5.12	6.60	7.36	N.A	4.47	9.45	6.53	8.28	5.16	N.A	10.01
Bharatpur-I	4.63	4.65	6.34	7.46	N.A	4.54	9.07	6.37	8.26	4.38	N.A	8.66
Bharatpur-II	5.34	4.41	6.29	7.24	N.A	5.64	9.77	6.24	8.19	9.16	N.A	9.82
Domkal	6.34	6.05	6.32	7.06	0.00	5.82	7.89	6.23	8.19	3.95	N.A	6.75
Farakka	4.92	4.90	6.26	9.34	0.00	4.89	10.80	6.67	8.24	4.58	0.00	12.79
Hariharpara	5.28	4.88	6.47	7.60	N.A	5.16	8.86	6.23	8.27	5.02	N.A	8.13
Jalangi	5.23	6.61	6.22	7.39	N.A	5.02	6.23	6.13	8.23	7.58	N.A	6.06
Kandi	4.36	4.40	6.09	7.36	N.A	3.99	7.51	6.15	8.21	7.58	N.A	6.79
Khargram	4.69	4.93	6.18	7.38	N.A	4.24	7.59	6.10	8.34	6.05	N.A	7.51
Lalgola	3.68	4.73	6.12	6.68	N.A	4.39	8.82	6.44	8.24	5.69	N.A	8.10
Murshidabad-Jiaganj	4.18	3.89	6.21	7.30	0.00	4.24	7.75	6.29	8.17	7.58	0.00	7.13
Nabagram	5.59	4.91	6.29	6.97	N.A	6.09	9.11	6.37	8.17	4.69	N.A	8.54
Nawda	4.58	5.02	6.36	7.56	N.A	5.17	8.47	6.43	8.07	5.84	0.00	8.07
Raghunathganj-I	6.11	4.86	6.43	7.46	0.00	6.18	6.37	6.76	8.28	4.35	N.A	9.33
Raghunathganj-II	4.65	4.86	6.44	8.99	0.00	4.83	7.89	5.96	8.30	4.32	0.00	8.21
Raninagar-I	5.03	4.84	6.23	7.33	N.A	5.18	9.44	6.43	8.25	3.99	N.A	10.24
Raninagar-II	4.31	4.86	6.49	6.89	N.A	4.43	7.83	6.37	8.29	N.A	N.A	7.26

(continued)

Annexure C.16 (continued)

Blocks	1996–1997						2016–2017					
	Marginal	Small	Semi-medium	Medium	Large	All classes	Marginal	Small	Semi-medium	Medium	Large	All classes
Sagardighi	2.95	5.56	5.86	7.61	N.A	4.62	8.76	5.99	8.08	4.71	N.A	7.78
Samserganj	3.47	4.95	6.38	8.79	N.A	3.75	9.50	6.67	8.36	7.58	N.A	12.39
Suti-I	5.48	4.93	6.16	6.93	0.00	6.42	8.99	6.57	8.41	4.18	N.A	9.64
Suti-II	5.26	4.77	6.48	7.45	N.A	4.92	10.98	6.52	8.23	7.58	N.A	11.74
District total	**4.76**	**4.91**	**6.21**	**7.38**	**0.00**	**4.89**	**8.39**	**6.34**	**8.23**	**4.86**	**0.00**	**8.08**

Source: Agriculture Census, Government of India, 1995–1996 and 2015–2016

N.A not available

Annexure C.17 Percentage of households avail institutional credit facilities by land size class in Murshidabad district, 1996–1997 and 2016–2017

	1996–1997						2016–2017					
Blocks	Marginal	Small	Semi-medium	Medium	Large	All classes	Marginal	Small	Semi-medium	Medium	Large	All classes
Berhampore	9.44	12.24	12.81	14.59	0.00	9.78	30.37	53.90	64.56	65.63	0.00	32.36
Burwan	10.36	12.07	12.28	13.52	N.A	11.56	31.46	50.52	64.55	54.44	0.00	34.43
Beldanga-I	11.90	12.35	13.83	14.21	N.A	9.95	29.00	53.28	65.26	57.91	N.A	31.85
Beldanga-II	10.86	11.80	15.04	15.64	0.00	10.63	30.38	50.50	61.68	85.99	N.A	32.82
Bhagwangola-I	10.75	12.96	13.63	13.92	N.A	10.84	30.50	53.12	63.94	56.68	0.00	36.20
Bhagwangola-II	9.27	12.81	14.30	13.85	N.A	9.38	33.67	53.29	64.57	70.89	N.A	42.39
Bharatpur-I	9.22	11.65	13.72	14.04	N.A	9.53	32.31	52.01	64.43	60.21	N.A	36.68
Bharatpur-II	10.63	11.03	13.61	13.64	N.A	11.85	34.81	50.95	63.86	125.94	N.A	41.60
Domkal	12.63	15.14	13.69	13.30	0.00	12.22	28.11	50.83	63.85	54.36	N.A	28.60
Farakka	9.80	12.27	13.56	17.59	0.00	10.28	38.48	54.39	64.25	62.97	0.00	54.19
Hariharpara	10.52	12.22	14.01	14.32	N.A	10.83	31.56	50.81	64.49	69.09	N.A	34.45
Jalangi	10.42	16.55	13.47	13.91	N.A	10.54	22.19	50.04	64.21	104.23	N.A	25.67
Kandi	8.69	11.02	13.18	13.87	N.A	8.38	26.74	50.18	64.06	104.23	N.A	28.77
Khargram	9.35	12.35	13.38	13.90	N.A	8.90	27.03	49.79	65.01	83.18	N.A	31.80
Lalgola	7.34	11.84	13.25	12.58	N.A	9.22	31.43	52.54	64.23	78.17	N.A	34.32
Murshidabad-Jiaganj	8.32	9.75	13.44	13.74	0.00	8.91	27.59	51.35	63.69	104.23	0.00	30.18
Nabagram	11.13	12.30	13.63	13.13	N.A	12.80	32.46	51.96	63.72	64.52	N.A	36.17
Nawda	9.14	12.58	13.78	14.23	N.A	10.86	30.16	52.48	62.98	80.29	0.00	34.19
Raghunathganj-I	12.18	12.18	13.93	14.05	0.00	12.97	22.68	55.19	64.58	59.81	N.A	39.52
Raghunathganj-II	9.27	12.18	13.94	16.94	0.00	10.14	28.12	48.64	64.74	59.44	0.00	34.76
Raninagar-I	10.03	12.13	13.49	13.80	N.A	10.89	33.63	52.48	64.38	54.80	N.A	43.37
Raninagar-II	8.60	12.17	14.05	12.97	N.A	9.31	27.88	52.01	64.62	N.A	N.A	30.73

(continued)

Annexure C.17 (continued)

	1996–1997						2016–2017					
Blocks	Marginal	Small	Semi-medium	Medium	Large	All classes	Marginal	Small	Semi-medium	Medium	Large	All classes
Sagardighi	5.89	13.92	12.69	14.33	N.A	9.71	31.20	48.87	63.00	64.77	N.A	32.95
Samserganj	6.92	12.38	13.82	16.55	N.A	7.87	33.84	54.42	65.19	104.23	N.A	52.47
Suti-I	10.91	12.35	13.33	13.04	0.00	13.48	32.03	53.62	65.62	57.48	N.A	40.85
Suti-II	10.48	11.96	14.02	14.03	N.A	10.32	39.10	53.16	64.18	104.23	N.A	49.72
District total	**9.49**	**12.30**	**13.45**	**13.89**	**0.00**	**10.27**	**29.89**	**51.73**	**64.16**	**66.84**	**0.00**	**34.21**

Source: Agriculture Census, Government of India, 1995–1996 and 2015–2016
N.A not available

Annexure C.18 Percentage of households taken credit from Primary Agricultural Credit Societies (PACS) by land size class in Murshidabad district, 1996–1997 and 2016–2017

	1996–1997						2016–2017					
Blocks	Marginal	Small	Semi-medium	Medium	Large	All classes	Marginal	Small	Semi-medium	Medium	Large	All classes
Berhampore	2.00	3.71	3.68	3.05	0.00	2.33	7.81	11.36	13.79	9.38	0.00	7.90
Burwan	2.19	3.66	3.53	2.83	N.A	2.75	8.10	10.65	13.79	7.78	0.00	8.41
Beldanga-I	2.51	3.75	3.98	2.97	N.A	2.37	7.46	11.23	13.94	8.27	N.A	7.78
Beldanga-II	2.30	3.58	4.33	3.27	0.00	2.53	7.82	10.64	13.18	12.28	N.A	8.01
Bhagwangola-I	2.27	3.93	3.92	2.91	N.A	2.58	7.85	11.20	13.66	8.10	0.00	8.84
Bhagwangola-II	1.96	3.88	4.11	2.90	N.A	2.23	8.66	11.23	13.80	10.13	N.A	10.35
Bharatpur-I	1.95	3.53	3.95	2.94	N.A	2.27	8.31	10.96	13.76	8.60	N.A	8.96
Bharatpur-II	2.25	3.35	3.92	2.85	N.A	2.82	8.96	10.74	13.64	17.99	N.A	10.16
Domkal	2.67	4.59	3.94	2.78	0.00	2.91	7.23	10.71	13.64	7.77	N.A	6.98
Farakka	2.07	3.72	3.90	3.68	0.00	2.44	9.90	11.46	13.73	9.00	0.00	13.23
Hariharpara	2.22	3.70	4.03	2.99	N.A	2.57	8.12	10.71	13.78	9.87	N.A	8.41
Jalangi	2.20	5.02	3.88	2.91	N.A	2.51	5.71	10.55	13.72	14.89	N.A	6.27
Kandi	1.84	3.34	3.79	2.90	N.A	1.99	6.88	10.58	13.69	14.89	N.A	7.02
Khargram	1.98	3.75	3.85	2.91	N.A	2.11	6.96	10.49	13.89	11.88	N.A	7.76
Lalgola	1.55	3.59	3.81	2.63	N.A	2.19	8.09	11.08	13.72	11.17	N.A	8.38
Murshidabad-Jiaganj	1.76	2.96	3.87	2.87	0.00	2.12	7.10	10.82	13.61	14.89	0.00	7.37
Nabagram	2.35	3.73	3.92	2.75	N.A	3.04	8.35	10.95	13.61	9.22	N.A	8.83
Nawda	1.93	3.81	3.96	2.98	N.A	2.58	7.76	11.06	13.45	11.47	0.00	8.35
Raghunathganj-I	2.58	3.69	4.01	2.94	0.00	3.08	5.84	11.63	13.80	8.54	N.A	9.65
Raghunathganj-II	1.96	3.69	4.01	3.54	0.00	2.41	7.24	10.25	13.83	8.49	0.00	8.49
Raninagar-I	2.12	3.68	3.88	2.89	N.A	2.59	8.65	11.06	13.75	7.83	N.A	10.59

(continued)

Annexure C.18 (continued)

	1996–1997						2016–2017					
Blocks	Marginal	Small	Semi-medium	Medium	Large	All classes	Marginal	Small	Semi-medium	Medium	Large	All classes
Raninagar-II	1.82	3.69	4.04	2.71	N.A	2.21	7.17	10.96	13.81	N.A	N.A	7.50
Sagardighi	1.24	4.22	3.65	3.00	N.A	2.31	8.03	10.30	13.46	9.25	N.A	8.05
Samserganj	1.46	3.75	3.98	3.46	N.A	1.87	8.71	11.47	13.93	14.89	N.A	12.81
Suti-I	2.31	3.75	3.84	2.73	0.00	3.20	8.24	11.30	14.02	8.21	N.A	9.97
Suti-II	2.22	3.63	4.03	2.94	N.A	2.45	10.06	11.21	13.71	14.89	N.A	12.14
District Total	**2.01**	**3.73**	**3.87**	**2.91**	**0.00**	**2.44**	**7.69**	**10.95**	**13.71**	**9.55**	**0.00**	**8.35**

Source: Agriculture Census, Government of India, 1995–1996 and 2015–2016
N.A not available

Annexure C.19 Percentage of households taken credit from Primary Land Development Bank (PLDB) by land size class in Murshidabad district, 1996–1997 and 2016–2017

	1996–1997						2016–2017					
Blocks	Marginal	Small	Semi-medium	Medium	Large	All classes	Marginal	Small	Semi-medium	Medium	Large	All classes
Berhampore	0.35	2.86	4.27	8.86	0.00	1.06	0.36	1.59	1.65	0.00	0.00	0.53
Burwan	0.38	2.82	4.10	8.20	N.A	1.25	0.37	1.49	1.65	0.00	0.00	0.57
Beldanga-I	0.43	2.89	4.61	8.62	N.A	1.07	0.34	1.57	1.67	0.00	N.A	0.52
Beldanga-II	0.40	2.76	5.01	9.49	0.00	1.15	0.36	1.49	1.58	0.00	N.A	0.54
Bhagwangola-I	0.39	3.03	4.55	8.45	N.A	1.17	0.36	1.56	1.64	0.00	0.00	0.60
Bhagwangola-II	0.34	3.00	4.77	8.41	N.A	1.01	0.39	1.57	1.65	0.00	N.A	0.70
Bharatpur-I	0.34	2.73	4.58	8.52	N.A	1.03	0.38	1.53	1.65	0.00	N.A	0.60
Bharatpur-II	0.39	2.58	4.54	8.28	N.A	1.28	0.41	1.50	1.64	0.00	N.A	0.68
Domkal	0.46	3.54	4.56	8.07	0.00	1.32	0.33	1.50	1.64	0.00	N.A	0.47
Farakka	0.36	2.87	4.52	10.68	0.00	1.11	0.45	1.60	1.65	0.00	0.00	0.89
Hariharpara	0.38	2.86	4.67	8.69	N.A	1.17	0.37	1.49	1.65	0.00	N.A	0.57
Jalangi	0.38	3.87	4.49	8.44	N.A	1.14	0.26	1.47	1.64	0.00	N.A	0.42
Kandi	0.32	2.58	4.39	8.42	N.A	0.90	0.31	1.48	1.64	0.00	N.A	0.47
Khargram	0.34	2.89	4.46	8.44	N.A	0.96	0.32	1.46	1.67	0.00	N.A	0.52
Lalgola	0.27	2.77	4.42	7.64	N.A	0.99	0.37	1.55	1.64	0.00	N.A	0.56
Murshidabad-Jiaganj	0.30	2.28	4.48	8.34	0.00	0.96	0.32	1.51	1.63	0.00	0.00	0.50
Nabagram	0.41	2.88	4.54	7.97	N.A	1.38	0.38	1.53	1.63	0.00	N.A	0.59
Nawda	0.33	2.94	4.59	8.64	N.A	1.17	0.35	1.54	1.61	0.00	0.00	0.56
Raghunathganj-I	0.45	2.85	4.65	8.53	0.00	1.40	0.27	1.62	1.65	0.00	N.A	0.65
Raghunathganj-II	0.34	2.85	4.65	10.28	0.00	1.09	0.33	1.43	1.66	0.00	0.00	0.57
Raninagar-I	0.37	2.84	4.50	8.38	N.A	1.17	0.39	1.54	1.65	0.00	N.A	0.71

(continued)

Annexure C.19 (continued)

Blocks	1996–1997						2016–2017					
	Marginal	Small	Semi-medium	Medium	Large	All classes	Marginal	Small	Semi-medium	Medium	Large	All classes
Raninagar-II	0.31	2.85	4.69	7.87	N.A	1.00	0.33	1.53	1.65	N.A	N.A	0.51
Sagardighi	0.22	3.26	4.23	8.70	N.A	1.05	0.36	1.44	1.61	0.00	N.A	0.54
Samserganj	0.25	2.90	4.61	10.05	N.A	0.85	0.40	1.60	1.67	0.00	N.A	0.86
Suti-I	0.40	2.89	4.45	7.92	0.00	1.45	0.37	1.58	1.68	0.00	N.A	0.67
Suti-II	0.38	2.80	4.68	8.52	N.A	1.11	0.46	1.56	1.64	0.00	N.A	0.82
District total	**0.35**	**2.88**	**4.49**	**8.43**	**0.00**	**1.11**	**0.35**	**1.52**	**1.64**	**0.00**	**0.00**	**0.56**

Source: Agriculture Census, Government of India, 1995–1996 and 2015–2016
N.A not available

Annexure C.20 Percentage of households taken credit from commercial bank branch (CBB) by land size class in Murshidabad district, 1996–1997 and 2016–2017

Blocks	1996–1997						2016–2017					
	Marginal	Small	Semi-medium	Medium	Large	All classes	Marginal	Small	Semi-medium	Medium	Large	All classes
Berhampore	1.92	2.50	2.17	1.24	0.00	1.96	21.13	38.57	45.18	46.88	0.00	22.66
Burwan	2.11	2.47	2.08	1.15	N.A	2.31	21.90	36.15	42.70	38.89	0.00	24.11
Beldanga-I	2.42	2.52	2.34	1.21	N.A	1.99	20.18	38.13	44.69	41.36	N.A	22.30
Beldanga-II	2.21	2.41	2.54	1.33	0.00	2.13	21.14	36.14	44.26	61.42	N.A	22.98
Bhagwangola-I	2.19	2.65	2.31	1.18	N.A	2.17	21.23	38.02	44.70	40.48	0.00	25.35
Bhagwangola-II	1.89	2.62	2.42	1.18	N.A	1.88	23.43	38.13	44.60	50.64	N.A	29.68
Bharatpur-I	1.88	2.38	2.32	1.19	N.A	1.91	22.48	37.22	44.21	43.01	N.A	25.68
Bharatpur-II	2.17	2.25	2.30	1.16	N.A	2.37	24.22	36.46	44.68	89.96	N.A	29.13
Domkal	2.57	3.09	2.32	1.13	0.00	2.45	19.56	36.38	44.20	38.83	N.A	20.03
Farakka	2.00	2.51	2.29	1.50	0.00	2.06	26.78	38.93	44.48	44.98	0.00	37.94
Hariharpara	2.14	2.50	2.37	1.22	N.A	2.17	21.97	36.36	44.65	49.35	N.A	24.12
Jalangi	2.12	3.38	2.28	1.18	N.A	2.11	15.44	35.81	44.45	74.45	N.A	17.97
Kandi	1.77	2.25	2.23	1.18	N.A	1.68	18.61	35.91	44.35	74.45	N.A	20.14
Khargram	1.90	2.52	2.26	1.18	N.A	1.78	18.81	35.63	45.01	59.42	N.A	22.26
Lalgola	1.49	2.42	2.24	1.07	N.A	1.85	21.88	37.60	44.46	55.84	N.A	24.03
Murshidabad-Jiaganj	1.70	1.99	2.27	1.17	0.00	1.78	19.20	36.75	44.09	74.45	0.00	21.13
Nabagram	2.27	2.51	2.31	1.12	N.A	2.56	22.59	37.19	44.11	46.09	N.A	25.32
Nawda	1.86	2.57	2.33	1.21	N.A	2.17	20.99	37.56	43.59	57.35	0.00	23.94
Raghunathganj-I	2.48	2.49	2.36	1.20	0.00	2.60	15.78	39.50	44.70	42.72	N.A	27.67
Raghunathganj-II	1.89	2.49	2.36	1.44	0.00	2.03	19.57	34.81	44.82	42.46	0.00	24.34
Raninagar-I	2.04	2.48	2.28	1.17	N.A	2.18	23.40	37.56	44.57	39.14	N.A	30.37

(continued)

Annexure C.20 (continued)

Blocks	1996–1997 Marginal	Small	Semi-medium	Medium	Large	All classes	2016–2017 Marginal	Small	Semi-medium	Medium	Large	All classes
Raninagar-II	1.75	2.49	2.38	1.10	N.A	1.86	19.40	37.22	44.73	N.A	N.A	21.52
Sagardighi	1.20	2.85	2.15	1.22	N.A	1.94	21.71	34.98	43.61	46.27	N.A	23.07
Samserganj	1.41	2.53	2.34	1.41	N.A	1.58	23.55	38.95	45.13	74.45	N.A	36.73
Suti-I	2.22	2.52	2.26	1.11	0.00	2.70	22.29	38.37	45.43	41.06	N.A	28.60
Suti-II	2.13	2.44	2.37	1.19	N.A	2.07	27.21	38.05	44.43	74.45	N.A	34.81
District total	**1.93**	**2.51**	**2.28**	**1.18**	**0.00**	**2.06**	**20.80**	**37.02**	**44.42**	**47.74**	**0.00**	**23.95**

Source: Agriculture Census, Government of India, 1995–1996 and 2015–2016

N.A not available

Annexure C.21 Percentage of households taken credit from regional rural bank (RRB) by land size class in Murshidabad district, 1996–1997 and 2016–2017

	1996–1997						2016–2017					
Blocks	Marginal	Small	Semi-medium	Medium	Large	All classes	Marginal	Small	Semi-medium	Medium	Large	All classes
Berhampore	5.18	3.16	2.67	1.38	0.00	4.44	1.07	2.91	4.42	9.38	0.00	1.34
Burwan	5.68	3.11	2.57	1.27	N.A	5.25	1.10	2.72	4.42	7.78	0.00	1.43
Beldanga-I	6.52	3.19	2.89	1.34	N.A	4.52	1.02	2.87	4.46	8.27	N.A	1.32
Beldanga-II	5.96	3.04	3.14	1.47	0.00	4.83	1.07	2.72	4.22	12.28	N.A	1.36
Bhagwangola-I	5.89	3.34	2.85	1.31	N.A	4.92	1.07	2.86	4.37	8.10	0.00	1.50
Bhagwangola-II	5.08	3.31	2.99	1.31	N.A	4.26	1.18	2.87	4.42	10.13	N.A	1.76
Bharatpur-I	5.06	3.01	2.87	1.32	N.A	4.33	1.13	2.80	4.41	8.60	N.A	1.52
Bharatpur-II	5.83	2.85	2.84	1.29	N.A	5.38	1.22	2.75	4.37	17.99	N.A	1.72
Domkal	6.93	3.91	2.86	1.25	0.00	5.55	0.99	2.74	4.37	7.77	N.A	1.19
Farakka	5.37	3.16	2.83	1.66	0.00	4.66	1.35	2.93	4.40	9.00	0.00	2.24
Hariharpara	5.77	3.15	2.93	1.35	N.A	4.92	1.11	2.74	4.41	9.87	N.A	1.43
Jalangi	5.71	4.27	2.81	1.31	N.A	4.79	0.78	2.70	4.39	14.89	N.A	1.06
Kandi	4.76	2.84	2.75	1.31	N.A	3.81	0.94	2.71	4.38	14.89	N.A	1.19
Khargram	5.13	3.19	2.79	1.31	N.A	4.04	0.95	2.68	4.45	11.88	N.A	1.32
Lalgola	4.02	3.05	2.77	1.19	N.A	4.18	1.10	2.83	4.39	11.17	N.A	1.42
Murshidabad-Jiaganj	4.56	2.52	2.81	1.29	0.00	4.04	0.97	2.77	4.36	14.89	0.00	1.25
Nabagram	6.10	3.17	2.85	1.24	N.A	5.81	1.14	2.80	4.36	9.22	N.A	1.50
Nawda	5.01	3.25	2.88	1.34	N.A	4.93	1.06	2.83	4.31	11.47	0.00	1.42
Raghunathganj-I	6.68	3.14	2.91	1.32	0.00	5.89	0.80	2.98	4.42	8.54	N.A	1.64
Raghunathganj-II	5.08	3.14	2.91	1.60	0.00	4.60	0.99	2.62	4.43	8.49	0.00	1.44
Raninagar-I	5.50	3.13	2.82	1.30	N.A	4.94	1.18	2.83	4.40	7.83	N.A	1.80
Raninagar-II	4.71	3.14	2.93	1.22	N.A	4.23	0.98	2.80	4.42	N.A	N.A	1.27

(continued)

Annexure C.21 (continued)

Blocks	1996–1997						2016–2017					
	Marginal	Small	Semi-medium	Medium	Large	All classes	Marginal	Small	Semi-medium	Medium	Large	All classes
Sagardighi	3.23	3.59	2.65	1.35	N.A	4.41	1.09	2.64	4.31	9.25	N.A	1.37
Samserganj	3.79	3.19	2.89	1.56	N.A	3.57	1.19	2.93	4.46	14.89	N.A	2.17
Suti-I	5.98	3.19	2.78	1.23	0.00	6.12	1.12	2.89	4.49	8.21	N.A	1.69
Suti-II	5.75	3.08	2.93	1.32	N.A	4.69	1.37	2.87	4.39	14.89	N.A	2.06
District total	**5.21**	**3.17**	**2.81**	**1.31**	**0.00**	**4.66**	**1.05**	**2.79**	**4.39**	**9.55**	**0.00**	**1.42**

Source: Agriculture Census, Government of India, 1995–1996 and 2015–2016

N.A not available

Annexure C.22 Number of hand-operated machines and implements in per 10 ha of GCA by land size class in Murshidabad district, 1996–1997 and 2016–2017

	1996–1997						2016–2017					
Blocks	Marginal	Small	Semi-medium	Medium	Large	All classes	Marginal	Small	Semi-medium	Medium	Large	All classes
Berhampore	59	27	18	15	0	40	30	10	8	7	0	22
Burwan	66	35	27	20	N.A	43	25	9	7	7	1	19
Beldanga-I	37	23	17	15	N.A	33	43	17	16	17	N.A	34
Beldanga-II	46	30	19	18	0	37	29	11	9	11	N.A	22
Bhagwangola-I	49	26	19	18	N.A	36	35	14	11	9	0	26
Bhagwangola-II	60	28	20	19	N.A	45	33	15	9	13	N.A	23
Bharatpur-I	92	46	28	25	N.A	65	27	11	13	8	N.A	20
Bharatpur-II	64	38	24	20	N.A	42	25	13	11	8	N.A	19
Domkal	40	22	22	20	0	32	40	12	14	18	N.A	31
Farakka	105	47	28	21	0	67	25	9	11	15	1	14
Hariharpara	46	29	21	20	N.A	36	40	15	16	14	N.A	31
Jalangi	51	21	22	20	N.A	38	41	12	10	9	N.A	30
Kandi	65	31	19	15	N.A	49	46	14	17	9	N.A	35
Khargram	60	29	21	19	N.A	47	32	9	10	10	N.A	22
Lalgola	86	32	21	18	N.A	48	39	13	8	12	N.A	28
Murshidabad-Jiaganj	56	31	18	17	0	40	35	11	10	9	1	26
Nabagram	63	35	22	21	N.A	39	34	11	11	10	N.A	23
Nawda	56	27	22	21	N.A	38	47	15	17	12	1	33
Raghunathganj-I	58	33	26	20	0	40	37	9	12	16	N.A	21
Raghunathganj-II	69	36	26	17	0	50	39	11	16	16	1	26
Raninagar-I	61	31	17	16	0	39	31	13	9	11	N.A	20

(continued)

Annexure C.22 (continued)

	1996–1997						2016–2017					
Blocks	Marginal	Small	Semi-medium	Medium	Large	All classes	Marginal	Small	Semi-medium	Medium	Large	All classes
Raninagar-II	57	25	19	17	N.A	40	38	14	8	N.A	N.A	28
Sagardighi	118	31	27	19	N.A	53	35	12	11	11	N.A	26
Samserganj	91	30	21	14	N.A	56	34	17	11	9	N.A	21
Suti-I	54	29	22	18	0	32	43	19	16	17	N.A	31
Suti-II	57	36	23	21	N.A	45	28	15	11	9	N.A	20
District total	**60**	**30**	**22**	**19**	**0**	**42**	**35**	**12**	**11**	**11**	**1**	**25**

Source: Agriculture Census, Government of India, 1995–1996 and 2015–2016
N.A not available

Annexure C.23 Number of animal-operated machines and implements in per 10 ha of GCA by land size class in Murshidabad district, 1996–1997 and 2016–2017

Blocks	1996–1997						2016–2017					
	Marginal	Small	Semi-medium	Medium	Large	All classes	Marginal	Small	Semi-medium	Medium	Large	All classes
Berhampore	14	11	7	6	0	12	14	5	4	4	0	10
Burwan	15	14	11	8	N.A	13	12	4	4	4	1	9
Beldanga-I	9	9	7	6	N.A	10	20	8	8	9	N.A	16
Beldanga-II	11	12	8	7	0	11	13	5	4	6	N.A	10
Bhagwangola-I	11	10	8	7	N.A	10	16	7	5	5	0	12
Bhagwangola-II	14	11	8	7	N.A	13	15	7	5	8	N.A	10
Bharatpur-I	21	18	12	10	N.A	19	12	5	6	4	N.A	9
Bharatpur-II	15	15	10	7	N.A	12	11	6	6	4	N.A	9
Domkal	9	9	9	8	0	9	18	5	7	10	N.A	14
Farakka	24	18	12	8	0	20	12	4	6	8	1	7
Hariharpara	11	11	9	7	N.A	11	18	7	8	8	N.A	14
Jalangi	12	8	9	8	N.A	11	19	6	5	5	N.A	14
Kandi	15	12	8	6	N.A	14	21	7	8	5	N.A	16
Khargram	14	12	9	7	N.A	14	15	4	5	5	N.A	10
Lalgola	20	13	9	7	N.A	14	18	6	4	7	N.A	13
Murshidabad-Jiaganj	13	12	7	6	0	12	16	5	5	5	1	12
Nabagram	15	14	9	8	N.A	11	16	5	5	5	N.A	11
Nawda	13	11	9	8	N.A	11	21	7	8	7	1	15
Raghunathganj-I	14	13	11	8	0	12	17	4	6	9	N.A	9
Raghunathganj-II	16	14	11	7	0	14	18	5	8	9	1	12
Raninagar-I	14	12	7	6	0	11	14	6	4	6	N.A	9

(continued)

Annexure C.23 (continued)

Blocks	1996–1997						2016–2017					
	Marginal	Small	Semi-medium	Medium	Large	All classes	Marginal	Small	Semi-medium	Medium	Large	All classes
Raninagar-II	13	10	8	6	N.A	12	17	6	4	N.A	N.A	13
Sagardighi	27	12	11	7	N.A	15	16	6	6	6	N.A	12
Samserganj	21	12	9	5	N.A	16	16	8	5	5	N.A	10
Suti-I	13	12	9	7	0	9	20	9	8	9	N.A	14
Suti-II	13	14	9	8	N.A	13	13	7	5	5	N.A	9
District total	**14**	**12**	**9**	**7**	**0**	**12**	**16**	**6**	**5**	**6**	**1**	**11**

Source: Agriculture Census, Government of India, 1995–1996 and 2015–2016
N.A not available

Annexure C.24 Number of power-operated machines and implements in per 10 ha of GCA by land size class in Murshidabad district, 1996–1997 and 2016–2017

Blocks	1996–1997						2016–2017					
	Marginal	Small	Semi-medium	Medium	Large	All classes	Marginal	Small	Semi-medium	Medium	Large	All classes
Berhampore	1	2	2	2	0	2	17	7	6	6	1	13
Burwan	2	2	3	2	N.A	2	14	6	5	6	1	11
Beldanga-I	1	2	2	2	N.A	1	24	11	12	13	N.A	20
Beldanga-II	1	2	2	2	0	2	16	7	6	8	N.A	13
Bhagwangola-I	1	2	2	2	N.A	2	20	9	8	7	1	15
Bhagwangola-II	1	2	2	2	N.A	2	18	10	7	10	N.A	13
Bharatpur-I	2	3	3	3	N.A	3	15	7	10	6	N.A	12
Bharatpur-II	2	2	2	2	N.A	2	14	8	9	6	N.A	11
Domkal	1	1	2	2	0	1	22	8	10	13	N.A	18
Farakka	3	3	3	2	0	3	14	6	8	12	1	8
Hariharpara	1	2	2	2	N.A	2	22	10	12	11	N.A	18
Jalangi	1	1	2	2	N.A	2	23	8	7	7	N.A	17
Kandi	2	2	2	2	N.A	2	26	9	12	7	N.A	21
Khargram	1	2	2	2	N.A	2	18	6	7	7	N.A	13
Lalgola	2	2	2	2	N.A	2	22	9	6	9	N.A	16
Murshidabad-Jiaganj	1	2	2	2	0	2	19	7	8	7	1	15
Nabagram	2	2	2	2	N.A	2	19	7	8	7	N.A	14
Nawda	1	2	2	2	N.A	2	26	10	13	9	1	19
Raghunathganj-I	1	2	3	2	0	2	21	6	9	12	N.A	12
Raghunathganj-II	2	2	3	2	0	2	22	7	12	12	1	15
Raninagar-I	1	2	2	2	0	2	17	8	7	8	N.A	12

(continued)

Annexure C.24 (continued)

Blocks	1996–1997						2016–2017					
	Marginal	Small	Semi-medium	Medium	Large	All classes	Marginal	Small	Semi-medium	Medium	Large	All classes
Raninagar-II	1	2	2	2	N.A	2	21	9	6	N.A	N.A	17
Sagardighi	3	2	3	2	N.A	2	19	8	9	9	N.A	15
Samserganj	2	2	2	1	N.A	2	19	11	8	7	N.A	12
Suti-I	1	2	2	2	0	1	24	12	12	13	N.A	18
Suti-II	1	2	2	2	N.A	2	16	10	8	7	N.A	12
District total	**1**	**2**	**2**	**2**	**0**	**2**	**19**	**8**	**8**	**8**	**1**	**15**

Source: Agriculture Census, Government of India, 1995–1996 and 2015–2016

N.A not available

Annexure C.25 Percentage of gross cropped area treated with fertilizers by land size class in Murshidabad district, 1996–1997 and 2016–2017

	1996–1997						2016–2017					
Blocks	Marginal	Small	Semi-medium	Medium	Large	All classes	Marginal	Small	Semi-medium	Medium	Large	All classes
Berhampore	72.42	89.47	92.84	97.61	0.00	82.28	91.97	93.73	82.93	94.39	100.00	92.57
Burwan	69.34	86.45	89.01	96.76	N.A	80.91	93.14	94.02	84.90	94.33	100.00	93.79
Beldanga-I	82.80	91.03	92.90	97.60	N.A	85.40	88.40	89.33	66.06	87.42	N.A	88.63
Beldanga-II	78.29	88.46	92.28	97.08	0.00	83.60	92.10	93.25	82.27	91.53	N.A	92.73
Bhagwangola-I	77.18	90.12	92.08	97.03	N.A	84.08	90.45	91.17	77.25	93.31	100.00	91.50
Bhagwangola-II	71.77	89.13	91.88	96.88	N.A	80.30	91.09	90.64	80.56	89.73	N.A	92.44
Bharatpur-I	56.99	82.24	88.62	95.90	N.A	71.32	92.67	93.26	73.35	94.16	N.A	93.22
Bharatpur-II	70.20	85.38	90.31	96.82	N.A	81.57	93.35	92.03	76.06	94.22	N.A	93.71
Domkal	81.28	91.54	91.15	96.80	0.00	85.87	89.21	92.68	71.24	86.60	N.A	89.74
Farakka	51.00	82.09	88.54	96.64	0.00	70.33	93.15	94.05	76.56	88.43	100.00	95.23
Hariharpara	78.41	88.96	91.35	96.86	N.A	83.95	89.27	90.31	65.65	89.46	N.A	89.87
Jalangi	76.03	91.89	91.19	96.72	N.A	83.06	88.90	92.36	80.09	93.01	N.A	90.08
Kandi	69.48	87.88	92.23	97.65	N.A	78.57	87.52	91.08	65.42	93.01	N.A	88.26
Khargram	71.93	88.84	91.26	96.89	N.A	79.18	91.31	94.06	79.98	92.51	N.A	92.77
Lalgola	59.97	87.70	91.49	97.17	N.A	78.75	89.38	91.61	82.76	90.68	N.A	90.84
Murshidabad-Jiaganj	73.67	87.89	92.86	97.29	0.00	82.38	90.59	92.92	78.60	93.01	100.00	91.39
Nabagram	70.56	86.50	91.03	96.70	N.A	82.87	90.74	93.26	78.12	92.64	N.A	92.27
Nawda	73.93	89.48	90.99	96.58	N.A	83.47	87.47	90.23	64.83	90.93	0.00	88.91
Raghunathganj-I	72.79	87.24	89.22	96.74	0.00	82.55	90.02	94.37	75.16	87.82	N.A	93.19
Raghunathganj-II	67.62	86.25	89.47	97.23	0.00	78.15	89.55	93.29	66.07	87.75	100.00	91.26
Raninagar-I	71.40	87.91	92.88	97.49	N.A	82.67	91.72	92.05	81.85	91.66	N.A	93.43

(continued)

Annexure C.25 (continued)

Blocks	1996–1997						2016–2017					
	Marginal	Small	Semi-medium	Medium	Large	All classes	Marginal	Small	Semi-medium	Medium	Large	All classes
Raninagar-II	73.45	90.38	92.07	97.29	N.A	82.48	89.75	91.28	82.50	N.A	N.A	90.59
Sagardighi	44.95	88.06	89.02	96.86	N.A	76.67	90.70	92.28	76.28	91.27	N.A	91.44
Samserganj	57.67	88.56	91.63	97.81	N.A	75.14	90.82	89.53	77.80	93.01	N.A	93.09
Suti-I	74.56	88.83	91.07	97.08	0.00	85.72	88.44	88.22	66.25	87.33	N.A	89.86
Suti-II	73.20	86.15	90.77	96.60	N.A	80.01	92.48	90.47	77.96	93.01	N.A	93.44
District total	**71.95**	**88.26**	**91.19**	**96.98**	**0.00**	**81.44**	**90.64**	**92.30**	**77.36**	**91.78**	**100.00**	**91.71**

Source: Agriculture Census, Government of India, 1995–1996 and 2015–2016
N.A not available

Annexure C.26 Percentage of gross cropped area treated with pesticides by land size class in Murshidabad district, 1996–1997 and 2016–2017

	1996–1997						2016–2017					
Blocks	Marginal	Small	Semi-medium	Medium	Large	All classes	Marginal	Small	Semi-medium	Medium	Large	All classes
Berhampore	37.63	52.87	57.32	62.05	0.00	45.10	54.33	64.04	79.07	77.72	100.00	58.76
Burwan	30.65	39.33	34.48	48.49	N.A	40.88	60.96	65.72	81.49	77.48	100.00	65.53
Beldanga-I	61.10	59.87	57.66	61.93	N.A	54.78	33.97	38.84	58.39	50.06	N.A	36.90
Beldanga-II	50.91	48.33	54.01	53.67	0.00	49.19	55.07	61.32	78.26	66.37	N.A	59.66
Bhagwangola-I	48.39	55.78	52.79	52.79	N.A	50.70	45.68	49.37	72.12	73.42	100.00	52.82
Bhagwangola-II	36.17	51.36	51.58	50.50	N.A	38.98	49.30	46.33	76.17	59.21	N.A	58.06
Bharatpur-I	2.74	20.51	32.16	34.84	N.A	11.18	58.26	61.37	67.33	76.80	N.A	62.34
Bharatpur-II	32.60	34.56	42.24	49.50	N.A	42.91	62.17	54.32	70.66	77.04	N.A	65.10
Domkal	57.66	62.14	47.24	49.16	0.00	56.25	38.60	58.00	64.75	46.81	N.A	43.02
Farakka	10.81	19.82	31.67	46.59	0.00	8.11	61.02	65.86	71.26	54.08	100.00	73.52
Hariharpara	51.17	50.59	48.44	50.05	N.A	50.30	38.92	44.42	57.90	58.15	N.A	43.76
Jalangi	45.79	63.68	47.47	47.96	N.A	47.52	36.86	56.19	75.60	72.26	N.A	44.91
Kandi	30.99	45.77	53.72	62.67	N.A	33.62	28.97	48.83	57.62	72.26	N.A	34.80
Khargram	36.52	50.04	47.93	50.57	N.A	35.50	50.55	65.96	75.46	70.26	N.A	59.83
Lalgola	9.48	44.94	49.29	54.98	N.A	34.19	39.54	51.90	78.87	63.01	N.A	49.14
Murshidabad-Jiaganj	40.46	45.80	57.46	57.02	0.00	45.41	46.44	59.42	73.77	72.26	100.00	52.21
Nabagram	33.42	39.58	46.53	47.52	N.A	46.96	47.29	61.36	73.18	70.78	N.A	57.09
Nawda	41.04	52.92	46.29	45.71	N.A	48.79	28.70	43.98	56.88	63.98	0.00	38.45
Raghunathganj-I	38.46	42.87	35.73	48.27	0.00	45.94	43.21	67.73	69.55	51.66	N.A	62.18
Raghunathganj-II	26.77	38.45	37.23	56.02	0.00	32.34	40.55	61.53	58.41	51.36	100.00	51.49
Raninagar-I	35.33	45.90	57.54	60.18	N.A	46.33	52.88	54.40	77.75	66.90	N.A	63.53
Raninagar-II	39.95	56.96	52.76	56.89	N.A	45.72	41.68	49.97	78.54	N.A	N.A	47.77

(continued)

Annexure C.26 (continued)

Blocks	1996–1997						2016–2017					
	Marginal	Small	Semi-medium	Medium	Large	All classes	Marginal	Small	Semi-medium	Medium	Large	All classes
Sagardighi	24.49	46.56	34.56	50.16	N.A	27.74	47.09	55.76	70.93	65.32	N.A	52.47
Samserganj	4.27	48.79	50.11	65.27	N.A	22.99	47.78	39.96	72.78	72.26	N.A	61.61
Suti-I	42.46	50.02	46.75	53.57	0.00	55.78	34.23	32.47	58.62	49.69	N.A	43.71
Suti-II	39.40	38.02	45.02	46.05	N.A	38.09	57.23	45.35	72.98	72.26	N.A	63.60
District total	**36.56**	**47.43**	**47.50**	**51.95**	**0.00**	**42.51**	**46.74**	**55.86**	**72.25**	**67.38**	**100.00**	**54.00**

Source: Agriculture Census, Government of India, 1995–1996 and 2015–2016

N.A not available

Annexure C.27 Amount of NPK mixture used (kg/ha) by land size class in Murshidabad district, 1996–1997 and 2016–2017

	1996–1997						2016–2017					
Blocks	Marginal	Small	Semi-medium	Medium	Large	All classes	Marginal	Small	Semi-medium	Medium	Large	All classes
Berhampore	29	30	34	34	0	32	78	31	36	33	0	60
Burwan	27	31	36	37	N.A	27	66	29	32	33	0	50
Beldanga-I	23	30	32	35	N.A	31	117	55	91	79	N.A	96
Beldanga-II	26	32	29	32	0	29	77	33	38	51	N.A	59
Bhagwangola-I	26	29	32	36	N.A	29	94	44	52	40	0	70
Bhagwangola-II	30	29	31	36	N.A	33	87	47	43	63	N.A	61
Bharatpur-I	30	32	32	36	N.A	33	71	33	64	34	N.A	55
Bharatpur-II	26	34	32	37	N.A	26	64	40	56	34	N.A	50
Domkal	22	25	32	38	0	26	108	36	72	85	N.A	86
Farakka	28	30	32	29	0	30	66	29	54	72	0	38
Hariharpara	26	31	31	35	N.A	29	107	49	93	65	N.A	85
Jalangi	27	23	32	36	N.A	30	111	38	44	41	N.A	83
Kandi	32	34	33	36	N.A	37	127	45	94	41	N.A	100
Khargram	30	30	33	36	N.A	35	85	29	44	45	N.A	59
Lalgola	38	31	33	40	N.A	34	106	42	37	57	N.A	76
Murshidabad-Jiaganj	33	38	32	37	0	35	93	35	48	41	0	71
Nabagram	25	30	32	38	N.A	24	91	33	50	44	N.A	63
Nawda	30	30	32	35	N.A	29	128	49	96	55	0	94
Raghunathganj-I	23	31	31	36	0	24	99	27	59	76	N.A	55
Raghunathganj-II	30	31	31	30	0	31	104	33	91	77	0	72
Raninagar-I	28	31	32	36	N.A	29	81	39	39	50	N.A	53

(continued)

Annexure C.27 (continued)

Blocks	1996–1997						2016–2017					
	Marginal	Small	Semi-medium	Medium	Large	All classes	Marginal	Small	Semi-medium	Medium	Large	All classes
Raninagar-II	32	31	31	39	N.A	34	102	44	38	N.A	N.A	78
Sagardighi	47	27	34	35	N.A	32	92	38	55	53	N.A	70
Samserganj	40	30	32	30	N.A	40	90	53	51	41	N.A	56
Suti-I	25	30	33	39	0	23	117	61	90	80	N.A	85
Suti-II	27	31	31	36	N.A	30	73	48	50	41	N.A	53
District total	**29**	**30**	**32**	**36**	**0**	**30**	**92**	**38**	**52**	**49**	**0**	**68**

Source: Agriculture Census, Government of India, 1995–1996 and 2015–2016

N.A not available

Annexure C.28 Amount of urea, DAP, and superphosphate mixture used (kg/ha) by land size class in Murshidabad district, 1996–1997 and 2016–2017

	1996–1997						2016–2017					
Blocks	Marginal	Small	Semi-medium	Medium	Large	All classes	Marginal	Small	Semi-medium	Medium	Large	All classes
Berhampore	25	26	30	31	0	27	76	29	35	30	0	59
Burwan	22	27	31	34	N.A	23	64	28	30	30	0	48
Beldanga-I	20	26	28	32	N.A	27	115	52	87	72	N.A	94
Beldanga-II	21	27	25	29	0	25	75	32	36	46	N.A	57
Bhagwangola-I	22	25	28	33	N.A	25	92	43	50	36	0	68
Bhagwangola-II	25	25	27	33	N.A	29	85	45	41	57	N.A	60
Bharatpur-I	25	28	28	33	N.A	28	69	32	61	31	N.A	53
Bharatpur-II	22	29	28	34	N.A	23	62	38	53	31	N.A	49
Domkal	18	21	28	35	0	22	106	35	68	78	N.A	84
Farakka	24	26	28	26	0	26	64	28	52	66	0	37
Hariharpara	22	26	27	32	N.A	25	105	47	89	59	N.A	82
Jalangi	22	19	28	33	N.A	25	109	36	42	38	N.A	80
Kandi	27	29	29	33	N.A	32	125	43	89	38	N.A	97
Khargram	25	26	29	33	N.A	30	83	28	42	41	N.A	57
Lalgola	32	27	29	36	N.A	29	104	40	35	52	N.A	74
Murshidabad-Jiaganj	28	33	28	33	0	30	91	33	46	38	0	69
Nabagram	21	26	28	35	N.A	21	89	32	47	40	N.A	61
Nawda	25	25	28	32	N.A	25	125	48	92	50	0	91
Raghunathganj-I	19	26	27	33	0	21	97	26	56	70	N.A	53
Raghunathganj-II	25	26	27	27	0	26	102	32	87	70	0	70
Raninagar-I	23	26	28	33	N.A	25	79	38	38	46	N.A	51
Raninagar-II	27	26	27	35	N.A	29	100	42	36	N.A	N.A	76

(continued)

Annexure C.28 (continued)

	1996–1997						2016–2017					
Blocks	Marginal	Small	Semi-medium	Medium	Large	All classes	Marginal	Small	Semi-medium	Medium	Large	All classes
Sagardighi	40	23	30	32	N.A	28	90	37	53	48	N.A	68
Samserganj	34	26	28	28	N.A	34	88	51	48	38	N.A	54
Suti-I	21	26	29	35	0	20	114	59	86	73	N.A	82
Suti-II	22	27	27	33	N.A	26	71	46	48	38	N.A	51
District total	**25**	**26**	**28**	**33**	**0**	**26**	**90**	**37**	**50**	**45**	**0**	**66**

Source: Agriculture Census, Government of India, 1995–1996 and 2015–2016
N.A not available

Annexure C.29 Percentage of main cultivators to the total cultivators by land size class in Murshidabad district, 1996–1997 and 2016–2017

	1996–1997						2016–2017					
Blocks	Marginal	Small	Semi-medium	Medium	Large	All classes	Marginal	Small	Semi-medium	Medium	Large	All classes
Berhampore	93.04	92.34	96.25	98.00	100.00	94.05	88.79	89.18	92.72	100.00	100.00	90.60
Burwan	93.46	93.21	95.32	96.24	N.A	94.83	76.92	79.94	88.27	95.61	100.00	78.37
Beldanga-I	94.21	95.53	92.82	97.25	N.A	96.92	90.21	92.62	90.94	95.75	N.A	91.91
Beldanga-II	92.57	94.72	96.28	100.00	100.00	96.79	93.03	89.15	86.76	92.91	N.A	92.49
Bhagwangola-I	93.94	92.49	97.37	96.89	N.A	93.77	90.87	92.38	94.13	93.56	100.00	91.18
Bhagwangola-II	90.47	91.18	92.04	93.12	N.A	96.56	86.79	88.35	91.58	92.00	N.A	87.21
Bharatpur-I	92.89	90.34	92.36	85.45	N.A	91.98	92.96	94.46	94.57	94.38	N.A	93.85
Bharatpur-II	93.12	90.76	91.26	89.95	N.A	91.23	88.39	87.47	93.24	96.25	N.A	89.71
Domkal	90.51	93.75	89.27	98.45	100.00	93.94	91.19	92.38	94.56	92.11	N.A	92.31
Farakka	89.17	93.27	94.58	95.55	100.00	90.47	83.28	85.34	88.72	90.27	100.00	84.69
Hariharpara	97.56	93.82	94.26	100.00	N.A	96.89	89.02	87.37	84.25	94.22	N.A	88.39
Jalangi	94.35	92.75	93.21	86.74	N.A	93.12	90.53	92.71	95.48	94.82	N.A	91.79
Kandi	90.83	89.62	91.11	88.71	N.A	90.51	90.89	93.74	89.57	94.17	N.A	91.24
Khargram	88.92	90.29	90.83	86.55	N.A	89.17	86.93	83.39	85.58	90.36	N.A	85.42
Lalgola	92.79	90.22	94.71	90.81	N.A	91.56	82.73	84.38	88.59	91.23	N.A	83.13
Murshidabad-Jiaganj	93.77	85.23	85.54	89.27	100.00	91.62	93.08	93.94	96.33	100.00	100.00	94.40
Nabagram	91.56	89.04	85.17	90.64	N.A	90.95	90.04	92.83	92.45	94.24	N.A	90.59
Nawda	91.98	95.86	93.20	94.67	N.A	93.35	94.34	95.88	97.53	96.49	100.00	95.00
Raghunathganj-I	96.23	95.26	96.08	100.00	100.00	97.38	86.91	90.07	92.73	93.56	N.A	88.59
Raghunathganj-II	84.94	86.89	92.11	100.00	100.00	85.97	96.74	95.21	100.00	100.00	100.00	97.37
Raninagar-I	95.47	94.82	93.06	96.00	N.A	95.75	91.02	93.29	95.48	94.22	N.A	92.04
Raninagar-II	93.39	94.51	95.63	92.17	N.A	94.47	92.29	91.47	96.10	98.16	N.A	93.94

(continued)

Annexure C.29 (continued)

Blocks	1996–1997						2016–2017					
	Marginal	Small	Semi-medium	Medium	Large	All classes	Marginal	Small	Semi-medium	Medium	Large	All classes
Sagardighi	92.57	94.86	95.37	96.26	N.A	94.26	87.08	90.41	90.30	92.37	N.A	87.61
Samserganj	86.11	92.40	93.17	100.00	N.A	90.69	87.36	88.94	95.23	92.11	N.A	88.41
Suti-I	96.68	95.19	98.72	100.00	100.00	97.28	86.23	89.62	85.34	90.27	N.A	87.32
Suti-II	91.16	92.57	93.28	95.68	N.A	92.71	91.85	91.63	95.33	94.15	N.A	92.48
District total	**92.37**	**92.34**	**93.19**	**94.55**	**100.00**	**93.56**	**89.21**	**90.24**	**92.15**	**94.35**	**100.00**	**90.08**

Source: Agriculture Census, Government of India, 1995–1996 and 2015–2016
N.A not available

Annexure C.30 Percentage of marginal cultivators to the total cultivators by land size class in Murshidabad district, 1996–1997 and 2016–2017

	1996–1997						2016–2017					
Blocks	Marginal	Small	Semi-medium	Medium	Large	All classes	Marginal	Small	Semi-medium	Medium	Large	All classes
Berhampore	6.96	7.66	3.75	2.00	0.00	5.95	11.21	10.82	7.28	0.00	0.00	9.40
Burwan	6.54	6.79	4.68	3.76	N.A	5.17	23.08	20.06	11.73	4.39	0.00	21.63
Beldanga-I	5.79	4.47	7.18	2.75	N.A	3.08	9.79	7.38	9.06	4.25	N.A	8.09
Beldanga-II	7.43	5.28	3.72	0.00	0.00	3.21	6.97	10.85	13.24	7.09	N.A	7.51
Bhagwangola-I	6.06	7.51	2.63	3.11	N.A	6.23	9.13	7.62	5.87	6.44	0.00	8.82
Bhagwangola-II	9.53	8.82	7.96	6.88	N.A	3.44	13.21	11.65	8.42	8.00	N.A	12.79
Bharatpur-I	7.11	9.66	7.64	14.55	N.A	8.02	7.04	5.54	5.43	5.62	N.A	6.15
Bharatpur-II	6.88	9.24	8.74	10.05	N.A	8.77	11.61	12.53	6.76	3.75	N.A	10.29
Domkal	9.49	6.25	10.73	1.55	0.00	6.06	8.81	7.62	5.44	7.89	N.A	7.69
Farakka	10.83	6.73	5.42	4.45	0.00	9.53	16.72	14.66	11.28	9.73	0.00	15.31
Hariharpara	2.44	6.18	5.74	0.00	N.A	3.11	10.98	12.63	15.75	5.78	N.A	11.61
Jalangi	5.65	7.25	6.79	13.26	N.A	6.88	9.47	7.29	4.52	5.18	N.A	8.21
Kandi	9.17	10.38	8.89	11.29	N.A	9.49	9.11	6.26	10.43	5.83	N.A	8.76
Khargram	11.08	9.71	9.17	13.45	N.A	10.83	13.07	16.61	14.42	9.64	N.A	14.58
Lalgola	7.21	9.78	5.29	9.19	N.A	8.44	17.27	15.62	11.41	8.77	N.A	16.87
Murshidabad-Jiaganj	6.23	14.77	14.46	10.73	0.00	8.38	6.92	6.06	3.67	0.00	0.00	5.60
Nabagram	8.44	10.96	14.83	9.36	N.A	9.05	9.96	7.17	7.55	5.76	N.A	9.41
Nawda	8.02	4.14	6.80	5.33	N.A	6.65	5.66	4.12	2.47	3.51	0.00	5.00
Raghunathganj-I	3.77	4.74	3.92	0.00	0.00	2.62	13.09	9.93	7.27	6.44	N.A	11.41
Raghunathganj-II	15.06	13.11	7.89	0.00	0.00	14.03	3.26	4.79	0.00	0.00	0.00	2.63
Raninagar-I	4.53	5.18	6.94	4.00	N.A	4.25	8.98	6.71	4.52	5.78	N.A	7.96

(continued)

Annexure C.30 (continued)

Blocks	1996–1997						2016–2017					
	Marginal	Small	Semi-medium	Medium	Large	All classes	Marginal	Small	Semi-medium	Medium	Large	All classes
Raninagar-II	6.61	5.49	4.37	7.83	N.A	5.53	7.71	8.53	3.90	1.84	N.A	6.06
Sagardighi	7.43	5.14	4.63	3.74	N.A	5.74	12.92	9.59	9.70	7.63	N.A	12.39
Samserganj	13.89	7.60	6.83	0.00	N.A	9.31	12.64	11.06	4.77	7.89	N.A	11.59
Suti-I	3.32	4.81	1.28	0.00	0.00	2.72	13.77	10.38	14.66	9.73	N.A	12.68
Suti-II	8.84	7.43	6.72	4.32	N.A	7.29	8.15	8.37	4.67	5.85	N.A	7.52
District total	**7.63**	**7.66**	**6.81**	**5.45**	**0.00**	**6.44**	**10.79**	**9.76**	**7.85**	**5.65**	**0.00**	**9.92**

Source: Agriculture Census, Government of India, 1995–1996 and 2015–2016
N.A not available

Annexure C.31 Literacy rate of the operational holders by land size class in Murshidabad district, 1996–1997 and 2016–2017

	1996–1997						2016–2017					
Blocks	Marginal	Small	Semi-medium	Medium	Large	All classes	Marginal	Small	Semi-medium	Medium	Large	All classes
Berhampore	75.88	85.41	91.38	97.69	100.00	79.59	82.24	92.60	94.49	95.57	100.00	85.28
Burwan	73.54	85.61	91.73	97.86	N.A	75.88	81.60	93.07	94.49	96.32	100.00	84.34
Beldanga-I	69.61	85.28	90.69	97.75	N.A	79.24	83.04	92.69	94.43	96.09	N.A	85.52
Beldanga-II	72.25	85.93	89.87	97.52	100.00	77.82	82.24	93.07	94.74	94.19	N.A	85.07
Bhagwangola-I	72.54	84.55	90.82	97.79	N.A	77.38	82.16	92.71	94.54	96.17	100.00	83.53
Bhagwangola-II	76.33	84.73	90.37	97.80	N.A	80.43	80.31	92.69	94.49	95.21	N.A	80.72
Bharatpur-I	76.45	86.12	90.76	97.77	N.A	80.11	81.11	92.86	94.50	95.93	N.A	83.32
Bharatpur-II	72.84	86.85	90.83	97.84	N.A	75.28	79.64	93.01	94.55	91.49	N.A	81.08
Domkal	67.74	81.95	90.78	97.89	100.00	74.50	83.56	93.02	94.55	96.33	N.A	86.99
Farakka	74.96	85.38	90.87	97.21	100.00	78.56	81.19	92.54	94.52	95.75	100.00	75.35
Hariharpara	73.14	85.44	90.57	97.73	N.A	77.40	81.54	93.03	94.50	95.33	N.A	84.33
Jalangi	73.38	80.27	90.93	97.79	N.A	78.00	87.03	93.13	94.52	92.96	N.A	88.32
Kandi	77.81	86.86	91.13	97.80	N.A	82.50	84.36	93.11	94.53	92.96	N.A	86.91
Khargram	76.12	85.28	90.99	97.80	N.A	81.44	84.19	93.17	94.45	94.38	N.A	85.54
Lalgola	81.26	85.89	91.08	98.00	N.A	80.76	81.62	92.79	94.52	94.72	N.A	84.39
Murshidabad-Jiaganj	78.74	88.38	90.95	97.82	100.00	81.41	83.86	92.95	94.56	92.96	100.00	86.27
Nabagram	71.56	85.34	90.82	97.92	N.A	73.30	81.02	92.87	94.56	95.64	N.A	83.55
Nawda	76.67	85.00	90.72	97.74	N.A	77.33	82.36	92.80	94.63	94.58	100.00	84.45
Raghunathganj-I	68.88	85.49	90.62	97.77	100.00	72.93	86.74	92.43	94.49	95.96	N.A	82.02
Raghunathganj-II	76.32	85.49	90.61	97.31	100.00	78.84	83.55	93.33	94.47	95.99	100.00	84.19
Raninagar-I	74.38	85.54	90.91	97.81	N.A	77.28	80.33	92.80	94.51	96.30	N.A	80.27
Raninagar-II	78.04	85.49	90.54	97.94	N.A	80.58	83.70	92.86	94.49	N.A	N.A	86.02

(continued)

Annexure C.31 (continued)

Blocks	1996–1997						2016–2017					
	Marginal	Small	Semi-medium	Medium	Large	All classes	Marginal	Small	Semi-medium	Medium	Large	All classes
Sagardighi	84.96	83.40	91.46	97.73	N.A	79.74	81.75	93.29	94.62	95.63	N.A	85.01
Samserganj	82.33	85.24	90.69	97.38	N.A	83.57	80.21	92.53	94.44	92.96	N.A	76.14
Suti-I	72.13	85.28	91.02	97.93	100.00	71.88	81.27	92.64	94.40	96.12	N.A	81.42
Suti-II	73.23	85.75	90.56	97.78	N.A	78.46	80.01	92.70	94.52	92.96	N.A	77.39
District total	**75.75**	**85.34**	**90.94**	**97.80**	**100.00**	**78.56**	**82.52**	**92.90**	**94.52**	**95.49**	**100.00**	**84.44**

Source: Agriculture Census, Government of India, 1995–1996 and 2015–2016
N.A not available

Annexure C.32 Average age of the operational holders by land size class in Murshidabad district, 1996–1997 and 2016–2017

Blocks	1996–1997						2016–2017					
	Marginal	Small	Semi-medium	Medium	Large	All classes	Marginal	Small	Semi-medium	Medium	Large	All classes
Berhampore	49	51	53	55	52	50	48	54	58	59	51	52
Burwan	46	50	52	52	N.A	47	47	52	56	57	46	48
Beldanga-I	44	50	52	54	N.A	45	45	52	56	57	N.A	47
Beldanga-II	44	50	51	54	49	46	45	52	56	57	N.A	48
Bhagwangola-I	46	49	51	52	N.A	47	47	51	56	57	44	48
Bhagwangola-II	45	48	51	52	N.A	47	47	51	56	57	N.A	48
Bharatpur-I	42	49	52	52	N.A	45	44	50	56	57	N.A	48
Bharatpur-II	43	50	52	50	N.A	46	44	52	56	54	N.A	48
Domkal	43	50	52	50	46	46	45	52	56	57	N.A	48
Farakka	45	50	54	56	46	47	47	52	57	59	45	48
Hariharpara	41	48	52	52	N.A	45	43	51	56	57	N.A	46
Jalangi	42	50	50	52	N.A	46	45	52	56	57	N.A	48
Kandi	44	50	50	52	N.A	47	47	52	56	57	N.A	48
Khargram	46	50	50	52	N.A	47	47	52	54	57	N.A	48
Lalgola	49	53	54	57	N.A	50	51	55	56	60	N.A	53
Murshidabad-Jiaganj	49	52	54	57	49	49	50	52	56	60	48	50
Nabagram	50	50	52	52	N.A	50	51	52	55	57	N.A	52
Nawda	51	50	52	54	N.A	48	52	52	56	57	47	48
Raghunathganj-I	43	46	53	54	45	47	44	49	56	57	N.A	48
Raghunathganj-II	42	50	52	52	45	44	43	52	56	57	46	46
Raninagar-I	47	50	49	52	N.A	47	49	52	53	57	N.A	48
Raninagar-II	48	48	49	52	N.A	47	49	52	54	57	N.A	48

(continued)

Annexure C.32 (continued)

Blocks	1996–1997						2016–2017					
	Marginal	Small	Semi-medium	Medium	Large	All classes	Marginal	Small	Semi-medium	Medium	Large	All classes
Sagardighi	45	50	52	52	N.A	47	47	52	56	54	N.A	48
Samserganj	49	50	52	52	N.A	47	51	52	56	55	N.A	48
Suti-I	43	50	52	55	48	45	46	52	56	57	N.A	46
Suti-II	48	49	51	52	N.A	47	49	51	54	56	N.A	48
District total	**46**	**50**	**52**	**52**	**48**	**47**	**47**	**52**	**56**	**57**	**46**	**48**

Source: Agriculture Census, Government of India, 1995–1996 and 2015–2016
N.A not available

Annexure C.33 Average family size by land size class in Murshidabad district, 1996–1997 and 2016–2017

	1996–1997						2016–2017					
Blocks	Marginal	Small	Semi-medium	Medium	Large	All classes	Marginal	Small	Semi-medium	Medium	Large	All classes
Berhampore	4	4	5	5	3	4	4	4	5	5	5	4
Burwan	5	5	6	6	N.A	5	5	5	6	6	5	5
Beldanga-I	5	5	6	6	N.A	5	5	5	6	6	N.A	5
Beldanga-II	5	5	6	6	4	5	5	6	5	6	N.A	5
Bhagwangola-I	5	5	6	6	N.A	5	5	5	6	6	5	5
Bhagwangola-II	5	5	6	6	N.A	5	5	5	6	6	N.A	5
Bharatpur-I	6	6	7	6	N.A	6	6	5	7	6	N.A	6
Bharatpur-II	5	5	6	6	N.A	5	5	5	6	7	N.A	5
Domkal	5	5	6	6	5	5	5	5	6	6	N.A	5
Farakka	5	5	5	6	4	5	5	4	5	5	6	5
Hariharpara	6	5	6	6	N.A	6	6	5	6	6	N.A	6
Jalangi	5	5	6	6	N.A	5	5	5	6	6	N.A	5
Kandi	5	6	7	7	N.A	6	5	6	7	6	N.A	5
Khargram	5	5	6	6	N.A	5	5	5	6	6	N.A	5
Lalgola	5	5	6	6	N.A	5	5	5	6	6	N.A	5
Murshidabad-Jiaganj	4	4	5	4	4	4	4	5	4	5	5	4
Nabagram	5	5	6	6	N.A	5	5	5	6	6	N.A	5
Nawda	5	5	6	6	N.A	5	5	5	6	6	4	5
Raghunathganj-I	5	5	6	6	4	5	5	5	6	6	N.A	5
Raghunathganj-II	5	5	6	6	4	5	5	5	6	6	5	5
Raninagar-I	7	6	6	7	N.A	7	6	6	6	7	N.A	7
Raninagar-II	6	5	6	6	N.A	5	5	5	6	6	N.A	5

(continued)

Annexure C.33 (continued)

Blocks	1996–1997						2016–2017					
	Marginal	Small	Semi-medium	Medium	Large	All classes	Marginal	Small	Semi-medium	Medium	Large	All classes
Sagardighi	5	5	6	6	N.A	5	5	5	6	6	N.A	5
Samserganj	5	5	6	6	N.A	5	5	5	6	6	N.A	5
Suti-I	5	5	6	6	4	5	5	5	6	6	N.A	5
Suti-II	5	5	5	6	N.A	5	5	5	5	6	N.A	5
District total	**5**	**5**	**6**	**6**	**4**	**5**	**5**	**5**	**6**	**6**	**5**	**5**

Source: Agriculture Census, Government of India, 1995–1996 and 2015–2016

N.A not available

Annexure C.34 Composite standardized score of agricultural innovation inputs by land size classes in Murshidabad district (1995–1996 and 2015–2016)

Blocks	1995–1996						2015–2016					
	Marginal	Small	Semi-medium	Medium	Large	All classes	Marginal	Small	Semi-medium	Medium	Large	All classes
Berhampore	0.394	0.355	0.424	0.386	0.328	0.397	0.505	0.482	0.458	0.433	0.479	0.493
Burwan	0.405	0.389	0.397	0.401	N.A	0.403	0.536	0.477	0.444	0.451	0.501	0.514
Beldanga-I	0.428	0.440	0.421	0.419	N.A	0.435	0.551	0.562	0.632	0.569	N.A	0.557
Beldanga-II	0.369	0.345	0.346	0.348	0.354	0.380	0.475	0.489	0.443	0.502	N.A	0.491
Bhagwangola-I	0.401	0.377	0.403	0.340	N.A	0.407	0.478	0.506	0.473	0.444	0.512	0.496
Bhagwangola-II	0.375	0.374	0.375	0.372	N.A	0.382	0.481	0.450	0.444	0.467	N.A	0.473
Bharatpur-I	0.388	0.379	0.445	0.377	N.A	0.419	0.521	0.501	0.526	0.464	N.A	0.535
Bharatpur-II	0.375	0.444	0.397	0.456	N.A	0.437	0.479	0.522	0.497	0.502	N.A	0.530
Domkal	0.425	0.349	0.406	0.377	0.328	0.404	0.518	0.541	0.557	0.528	N.A	0.514
Farakka	0.386	0.349	0.364	0.384	0.246	0.350	0.464	0.428	0.434	0.483	0.535	0.419
Hariharpara	0.396	0.388	0.406	0.390	N.A	0.410	0.525	0.483	0.593	0.522	N.A	0.515
Jalangi	0.361	0.356	0.324	0.365	N.A	0.362	0.484	0.420	0.484	0.512	N.A	0.459
Kandi	0.432	0.412	0.427	0.395	N.A	0.417	0.547	0.542	0.614	0.518	N.A	0.540
Khargram	0.381	0.348	0.386	0.380	N.A	0.386	0.460	0.441	0.453	0.467	N.A	0.467
Lalgola	0.450	0.422	0.405	0.400	N.A	0.433	0.506	0.496	0.472	0.545	N.A	0.524
Murshidabad-Jiaganj	0.369	0.316	0.388	0.395	0.330	0.373	0.452	0.518	0.441	0.444	0.445	0.471
Nabagram	0.392	0.368	0.428	0.408	N.A	0.425	0.496	0.474	0.471	0.433	N.A	0.510
Nawda	0.441	0.422	0.390	0.426	N.A	0.420	0.599	0.522	0.570	0.488	0.499	0.565
Raghunathganj-I	0.381	0.337	0.437	0.405	0.336	0.406	0.435	0.427	0.513	0.543	N.A	0.457
Raghunathganj-II	0.344	0.289	0.360	0.307	0.283	0.343	0.436	0.403	0.492	0.479	0.537	0.422
Raninagar-I	0.435	0.427	0.384	0.332	N.A	0.427	0.505	0.492	0.475	0.504	N.A	0.512
Raninagar-II	0.388	0.393	0.387	0.410	N.A	0.398	0.514	0.459	0.441	0.483	N.A	0.515

(continued)

Annexure C.34 (continued)

	1995–1996						2015–2016					
Blocks	Marginal	Small	Semi-medium	Medium	Large	All classes	Marginal	Small	Semi-medium	Medium	Large	All classes
Sagardighi	0.366	0.327	0.397	0.339	N.A	0.376	0.448	0.447	0.494	0.419	N.A	0.452
Samserganj	0.392	0.388	0.368	0.311	N.A	0.407	0.488	0.493	0.498	0.461	N.A	0.520
Suti-I	0.399	0.392	0.428	0.418	0.335	0.394	0.577	0.593	0.608	0.543	N.A	0.565
Suti-II	0.390	0.370	0.378	0.400	N.A	0.395	0.471	0.481	0.467	0.484	N.A	0.501
Total	**0.395**	**0.375**	**0.395**	**0.382**	**0.317**	**0.399**	**0.498**	**0.487**	**0.500**	**0.488**	**0.501**	**0.501**

Source: Calculated by researcher (data extracted from Agriculture Census and Input Survey, District Statistical Handbook and District Census Handbook)
N.A not available

Annexure C.35 Yang's crop yield index by land size category in Murshidabad district, 1995–1996 and 2015–2016

	1995–1996						2015–2016					
Blocks	Marginal	Small	Semi-medium	Medium	Large	All classes	Marginal	Small	Semi-medium	Medium	Large	All classes
Berhampore	91.62	103.31	97.5	109.22	108.32	94.2	93.98	99.62	99.12	103	120.07	95.72
Burwan	117.35	97.53	102.04	111.6	N.A	114.12	121.17	105.73	107.46	126.65	104.9	120.02
Beldanga-I	115.41	115.02	111.57	104.16	N.A	114.82	110.9	108.27	106.27	112.62	N.A	109.01
Beldanga-II	87.31	95.64	96.51	93.58	104.87	89.63	88.11	98.28	99.05	89.33	N.A	90.54
Bhagwangola-I	88.49	81.99	91.41	83.21	N.A	86.85	81.84	83.82	95.42	71.35	101.12	82.59
Bhagwangola-II	105.15	111.59	113.96	110.24	N.A	107.67	106.92	102.36	101.33	104.69	N.A	104.75
Bharatpur-I	134.34	103.29	125.32	111.33	N.A	129.6	129.16	114.28	111.67	120.84	N.A	126.09
Bharatpur-II	107.82	113.58	113.46	125.11	N.A	109.04	114.94	127.75	123.01	135.74	N.A	118.53
Domkal	97.37	85.85	94.89	90.5	101.78	95.57	91.85	88.88	96.33	86.85	N.A	91.09
Farakka	89.66	106.41	86.28	85.71	80.94	91.64	87.55	98.52	86.2	88.01	84.21	90.85
Hariharpara	102.94	108.55	88.75	99.57	N.A	103.58	104.87	101.23	89.35	103.78	N.A	102.06
Jalangi	100.46	96.57	107.81	85.08	N.A	99.39	97.11	90.43	103.37	86.72	N.A	95.29
Kandi	110.23	102	104.29	106.52	N.A	108.88	114.89	106.31	99.17	112.14	N.A	111.1
Khargram	116.21	107.91	102.85	107.24	N.A	113.21	120.35	112.14	108.34	122.42	N.A	117.2
Lalgola	104.96	111.59	106.47	91.25	N.A	106.17	99.33	106.86	100.53	94.85	N.A	102.84
Murshidabad-Jiaganj	99.28	97.24	96.99	89.6	126.28	98.52	106.48	99.55	100.72	91.5	116.92	104.08
Nabagram	97.75	95.12	91.39	99.01	N.A	96.91	95.04	90.47	94.78	95.09	N.A	93.51
Nawda	83.86	88.71	96.95	88.12	N.A	85.02	85.81	89.58	93.97	88.1	80.52	87.91
Raghunathganj-I	95.67	108.09	104.55	111.42	98.22	97.97	97.6	103.6	109.28	102.71	N.A	99.37
Raghunathganj-II	84.84	93.04	87.6	96.6	84.23	86.1	85.3	94.56	89.61	88.89	92.23	88.28
Raninagar-I	110.09	117.07	103.95	109.74	N.A	112.05	106.42	104.02	111.68	99.46	N.A	105.61
Raninagar-II	86.61	85.43	96.56	83.51	N.A	86.1	87.69	85.31	91.01	84.66	N.A	87.07

(continued)

Annexure C.35 (continued)

Blocks	1995–1996						2015–2016					
	Marginal	Small	Semi-medium	Medium	Large	All classes	Marginal	Small	Semi-medium	Medium	Large	All classes
Sagardighi	92.04	104.11	105.24	104.07	N.A	93.65	92.93	110.49	105.97	99.07	N.A	96.61
Samserganj	96.15	82.05	81.32	110.83	N.A	94.48	95.19	89.57	84.2	101.6	N.A	92.8
Suti-I	89.1	96.22	93.33	93.91	95.33	90.84	87.82	93.73	98.25	94.99	N.A	90.83
Suti-II	95.17	92.17	99.11	98.77	N.A	94.09	96.67	94.67	93.79	94.95	N.A	96.28

Source: Calculated by researcher (data extracted from Agriculture Census and Input Survey, District Statistical Handbook)

N.A not available

Annexure C.36 Crop diversification index by land size category in Murshidabad district, 1995–1996 and 2015–2016

	1995–1996						2015–2016					
Blocks	Marginal	Small	Semi-medium	Medium	Large	All classes	Marginal	Small	Semi-medium	Medium	Large	All classes
Berhampore	0.615	0.602	0.522	0.653	0.224	0.611	0.639	0.654	0.583	0.706	0.567	0.642
Burwan	0.383	0.302	0.307	0.347	N.A	0.379	0.568	0.536	0.535	0.683	0.471	0.565
Beldanga-I	0.694	0.535	0.44	0.633	N.A	0.686	0.632	0.59	0.327	0.546	N.A	0.628
Beldanga-II	0.621	0.581	0.534	0.585	0.433	0.613	0.724	0.802	0.627	0.562	N.A	0.729
Bhagwangola-I	0.708	0.684	0.688	0.701	N.A	0.703	0.832	0.777	0.826	0.778	0.443	0.821
Bhagwangola-II	0.701	0.702	0.478	0.502	N.A	0.702	0.823	0.808	0.717	0.647	N.A	0.82
Bharatpur-I	0.44	0.321	0.228	0.192	N.A	0.432	0.64	0.636	0.336	0.355	N.A	0.638
Bharatpur-II	0.369	0.284	0.259	0.285	N.A	0.366	0.621	0.558	0.396	0.402	N.A	0.615
Domkal	0.718	0.707	0.52	0.744	0.206	0.715	0.733	0.756	0.695	0.658	N.A	0.739
Farakka	0.592	0.589	0.64	0.231	0.112	0.59	0.643	0.531	0.555	0.4	0.348	0.634
Hariharpara	0.723	0.575	0.381	0.588	N.A	0.719	0.615	0.618	0.563	0.623	N.A	0.616
Jalangi	0.696	0.682	0.71	0.312	N.A	0.693	0.834	0.845	0.874	0.435	N.A	0.839
Kandi	0.376	0.308	0.375	0.447	N.A	0.371	0.64	0.716	0.305	0.506	N.A	0.644
Khargram	0.455	0.428	0.439	0.45	N.A	0.449	0.486	0.586	0.356	0.241	N.A	0.488
Lalgola	0.637	0.623	0.567	0.515	N.A	0.634	0.766	0.778	0.735	0.576	N.A	0.767
Murshidabad-Jiaganj	0.624	0.633	0.647	0.643	N.A	0.629	0.837	0.833	0.751	0.749	0.363	0.836
Nabagram	0.449	0.404	0.407	0.358	N.A	0.442	0.533	0.521	0.528	0.277	N.A	0.53
Nawda	0.635	0.719	0.503	0.62	N.A	0.641	0.736	0.606	0.611	0.686	0.422	0.731
Raghunathganj-I	0.586	0.549	0.437	0.525	N.A	0.582	0.551	0.644	0.404	0.689	N.A	0.556
Raghunathganj-II	0.65	0.574	0.582	0.618	0.21	0.644	0.829	0.781	0.602	0.716	0.523	0.822
Raninagar-I	0.679	0.653	0.692	0.663	N.A	0.677	0.818	0.846	0.801	0.758	N.A	0.823
Raninagar-II	0.741	0.735	0.646	0.797	N.A	0.739	0.844	0.881	0.789	0.872	N.A	0.847

(continued)

Annexure C.36 (continued)

	1995–1996						2015–2016					
Blocks	Marginal	Small	Semi-medium	Medium	Large	All classes	Marginal	Small	Semi-medium	Medium	Large	All classes
Sagardighi	0.511	0.419	0.416	0.407	N.A	0.504	0.641	0.46	0.43	0.421	N.A	0.633
Samserganj	0.724	0.679	0.676	0.569	N.A	0.718	0.781	0.723	0.732	0.497	N.A	0.778
Suti-I	0.57	0.535	0.461	0.571	0.107	0.563	0.613	0.612	0.302	0.682	N.A	0.612
Suti-II	0.757	0.694	0.754	0.733	N.A	0.751	0.788	0.648	0.835	0.658	N.A	0.779
Total	0.602	0.558	0.512	0.526	0.162	0.598	0.699	0.682	0.585	0.582	0.448	0.697

Source: Calculated by researcher (data extracted from Agriculture Census, Government of India, 1995–1996 and 2015–2016)
N.A not available

Annexure C.37 Cropping intensity by land size category in Murshidabad district, 1995–1996 and 2015–2016

	1995–1996						2015–2016					
Blocks	Marginal	Small	Semi-medium	Medium	Large	All classes	Marginal	Small	Semi-medium	Medium	Large	All classes
Berhampore	199.83	204.33	150.42	163.86	107.09	201.05	247.48	238.84	200.95	197.79	131.52	245.66
Burwan	163.84	160.95	156.50	154.17	N.A	161.14	279.39	263.29	227.22	235.90	126.35	274.42
Beldanga-I	154.32	207.76	145.04	138.43	N.A	156.72	188.22	242.06	198.25	189.32	N.A	193.16
Beldanga-II	220.64	193.27	182.13	148.21	115.88	213.85	251.48	236.98	202.51	172.17	N.A	248.55
Bhagwangola-I	202.09	205.65	165.67	163.43	N.A	203.06	227.14	251.03	182.29	191.95	134.57	231.25
Bhagwangola-II	198.95	189.14	181.96	156.59	N.A	192.37	241.07	224.27	216.44	204.58	N.A	235.87
Bharatpur-I	161.21	127.30	135.30	117.35	N.A	154.03	254.54	230.39	178.99	207.02	N.A	246.52
Bharatpur-II	164.17	163.23	160.19	155.91	N.A	162.82	260.66	198.91	184.89	187.25	N.A	253.20
Domkal	210.06	205.59	174.41	158.84	121.30	205.89	252.89	216.85	220.64	175.34	N.A	245.23
Farakka	108.34	119.85	135.95	114.29	100.00	116.60	179.36	187.45	147.05	167.89	135.66	180.86
Hariharpara	179.13	195.25	174.38	150.14	N.A	184.57	208.03	233.92	202.34	184.32	N.A	213.06
Jalangi	208.26	196.09	137.93	148.33	N.A	203.99	245.02	211.15	173.29	169.57	N.A	240.16
Kandi	196.24	197.20	206.50	207.38	N.A	197.81	237.72	218.62	225.17	228.34	N.A	234.50
Khargram	198.28	190.99	180.79	156.22	N.A	193.88	256.79	273.12	205.06	179.60	N.A	258.41
Lalgola	177.20	180.80	187.44	189.53	N.A	181.49	232.16	213.16	200.06	205.37	N.A	228.21
Murshidabad-Jiaganj	217.43	213.03	177.27	161.83	122.34	216.35	256.24	242.14	224.51	197.20	158.37	254.76
Nabagram	158.75	158.66	172.86	155.83	N.A	160.20	200.63	230.58	191.85	208.63	N.A	206.05
Nawda	218.46	198.96	170.20	138.94	N.A	211.91	259.63	257.45	208.62	186.52	128.97	254.99
Raghunathganj-I	156.91	169.43	140.68	147.68	112.86	158.97	266.50	239.76	164.28	188.41	N.A	259.32
Raghunathganj-II	173.33	157.25	143.96	144.11	124.11	168.23	205.32	217.39	174.34	179.88	145.64	207.66
Raninagar-I	181.39	179.62	169.81	195.38	N.A	180.43	216.61	193.44	209.59	212.34	N.A	214.19
Raninagar-II	197.93	174.91	159.74	192.03	N.A	195.64	232.18	217.14	195.83	209.68	N.A	229.26

(continued)

Annexure C.37 (continued)

Blocks	1995–1996						2015–2016					
	Marginal	Small	Semi-medium	Medium	Large	All classes	Marginal	Small	Semi-medium	Medium	Large	All classes
Sagardighi	160.53	158.42	151.71	150.35	N.A	156.95	207.91	214.10	178.24	178.74	N.A	208.22
Samserganj	177.72	185.85	182.65	186.75	N.A	181.60	194.23	221.66	213.04	221.59	N.A	201.78
Suti-I	187.43	190.88	177.39	177.32	108.96	188.77	234.29	227.85	229.73	204.92	N.A	233.09
Suti-II	185.28	159.04	163.38	141.84	N.A	182.29	205.23	199.33	196.59	168.73	N.A	202.75
Total	**182.99**	**180.13**	**164.78**	**158.26**	**114.07**	**181.95**	**232.33**	**226.96**	**198.15**	**194.35**	**137.30**	**230.81**

Source: Calculated by researcher (data extracted from Agriculture Census, Government of India, 1995–1996 and 2015–2016)

N.A not available

Annexure C.38 Composite standardized score of agricultural output by land size category in Murshidabad district, 1995–1996 and 2015–2016

	1995–1996						2015–2016					
Blocks	Marginal	Small	Semi-medium	Medium	Large	All classes	Marginal	Small	Semi-medium	Medium	Large	All classes
Berhampore	0.455	0.577	0.400	0.527	0.425	0.480	0.535	0.544	0.524	0.549	0.546	0.536
Burwan	0.466	0.390	0.395	0.457	N.A	0.442	0.680	0.605	0.621	0.730	0.417	0.685
Beldanga-I	0.504	0.637	0.433	0.445	N.A	0.518	0.487	0.578	0.503	0.522	N.A	0.482
Beldanga-II	0.516	0.505	0.487	0.405	0.524	0.485	0.535	0.563	0.536	0.402	N.A	0.536
Bhagwangola-I	0.465	0.474	0.449	0.412	N.A	0.469	0.482	0.525	0.508	0.449	0.244	0.481
Bhagwangola-II	0.524	0.611	0.555	0.486	N.A	0.541	0.614	0.553	0.580	0.562	N.A	0.592
Bharatpur-I	0.540	0.388	0.426	0.357	N.A	0.528	0.698	0.607	0.486	0.562	N.A	0.677
Bharatpur-II	0.426	0.483	0.448	0.512	N.A	0.420	0.647	0.560	0.571	0.559	N.A	0.652
Domkal	0.516	0.501	0.455	0.445	0.510	0.516	0.556	0.464	0.579	0.423	N.A	0.534
Farakka	0.320	0.444	0.332	0.229	0.222	0.346	0.353	0.373	0.321	0.332	0.146	0.365
Hariharpara	0.496	0.581	0.400	0.439	N.A	0.510	0.498	0.553	0.473	0.495	N.A	0.479
Jalangi	0.541	0.536	0.450	0.310	N.A	0.525	0.583	0.476	0.497	0.361	N.A	0.562
Kandi	0.494	0.495	0.559	0.578	N.A	0.491	0.604	0.540	0.527	0.627	N.A	0.584
Khargram	0.538	0.539	0.493	0.461	N.A	0.538	0.635	0.646	0.536	0.463	N.A	0.632
Lalgola	0.492	0.577	0.554	0.469	N.A	0.499	0.553	0.542	0.539	0.520	N.A	0.557
Murshidabad-Jiaganj	0.538	0.600	0.499	0.427	0.585	0.561	0.645	0.587	0.623	0.520	0.457	0.633
Nabagram	0.393	0.392	0.413	0.398	N.A	0.396	0.425	0.456	0.468	0.470	N.A	0.405
Nawda	0.461	0.505	0.450	0.366	N.A	0.466	0.545	0.532	0.521	0.465	0.301	0.538
Raghunathganj-I	0.409	0.518	0.388	0.477	0.383	0.406	0.571	0.602	0.452	0.517	N.A	0.565
Raghunathganj-II	0.408	0.411	0.349	0.402	0.457	0.384	0.452	0.496	0.414	0.454	0.284	0.457
Raninagar-I	0.508	0.612	0.518	0.599	N.A	0.531	0.562	0.496	0.633	0.591	N.A	0.551
Raninagar-II	0.473	0.438	0.447	0.494	N.A	0.458	0.519	0.473	0.511	0.360	N.A	0.502

(continued)

Annexure C.38 (continued)

	1995–1996						2015–2016					
Blocks	Marginal	Small	Semi-medium	Medium	Large	All classes	Marginal	Small	Semi-medium	Medium	Large	All classes
Sagardighi	0.376	0.446	0.418	0.425	N.A	0.367	0.452	0.537	0.474	0.424	N.A	0.459
Samserganj	0.475	0.431	0.448	0.567	N.A	0.463	0.462	0.472	0.448	0.573	N.A	0.456
Suti-I	0.434	0.494	0.445	0.463	0.359	0.430	0.478	0.483	0.533	0.540	N.A	0.481
Suti-II	0.491	0.434	0.490	0.446	N.A	0.469	0.492	0.428	0.536	0.429	N.A	0.475
Total	**0.472**	**0.501**	**0.450**	**0.446**	**0.433**	**0.471**	**0.541**	**0.527**	**0.516**	**0.496**	**0.342**	**0.534**

Source: Calculated by researcher (data extracted from Agriculture Census and Input Survey, District Statistical Handbook)

Annexure C.39 Karl Pearson's correlation of coefficient of agricultural input and output dimensions by land size category in Murshidabad district (1995–1996 and 2015–2016)

Crop diversification												
Indicators	1995–1996						2015–2016					
	Marginal	Small	Semi-medium	Medium	Large	All classes	Marginal	Small	Semi-medium	Medium	Large	All classes
Land	−0.298	0.117	−0.236	0.257	−0.160	−0.273	−0.017	0.042	0.291	0.123	0.211	0.006
Irrigation	0.602**	0.554**	0.542**	0.395*	0.712*	0.614***	0.656***	0.538**	0.542**	0.613***	0.869*	0.650***
Seeds	0.388	0.450*	0.253	0.187	−0.006	0.326	0.448*	0.079	0.057	−0.254	0.036	0.437*
Credits	0.255	0.322	0.157	0.059	0.000	0.179	0.301	0.165	−0.131	−0.153	0.000	0.289
Agricultural machinery	0.635***	0.508**	0.685***	0.483*	0.000	0.621***	0.519**	0.571**	0.374	0.675***	0.942**	0.533**
Fertilizers and pesticides	0.417*	0.055	0.252	0.501**	0.412	0.399*	0.523**	0.458*	0.309	0.548**	0.464	0.518**
Socio-economic aspects	0.332	0.130	−0.291	0.056	0.563	0.185	0.371	0.306	0.552**	0.213	0.167	0.351
Productivity												
Indicators	**1995–1996**						**2015–2016**					
	Marginal	**Small**	**Semi-medium**	**Medium**	**Large**	**All classes**	**Marginal**	**Small**	**Semi-medium**	**Medium**	**Large**	**All classes**
Land	0.593**	0.287	0.327	0.018	0.806*	0.567**	0.670***	0.553**	0.713***	0.466*	0.796*	0.579**
Irrigation	0.468*	0.565**	0.532**	0.267	0.076	0.482*	0.702***	0.444*	0.684***	0.548**	0.685	0.683***
Seeds	0.442*	0.628***	0.525**	0.427*	0.065	0.501**	0.539**	0.172	0.557**	0.719***	0.400	0.522
Credits	−0.130	0.099	0.126	0.463*	0.000	−0.098	0.404	0.245	−0.133	0.343	0.000	0.387
Agricultural machinery	0.643***	0.493*	0.497**	0.604**	0.000	0.617***	0.689***	0.505**	0.701***	0.405*	0.901**	0.667***

(continued)

Annexure C.39 (continued)

Crop diversification												
Fertilizers and pesticides	0.410*	0.547**	0.370	0.379	0.162	0.428*	0.511**	0.446*	0.538**	0.524**	0.023	0.509**
Socio-economic aspects	0.257	0.030	0.037	0.183	0.709*	0.246	0.396*	0.079	0.012	0.412	0.511	0.391*
Cropping intensity												
Indicators	**1995–1996**						**2015–2016**					
	Marginal	**Small**	**Semi-medium**	**Medium**	**Large**	**All classes**	**Marginal**	**Small**	**Semi-medium**	**Medium**	**Large**	**All classes**
Land	0.413*	0.465*	0.593**	0.369	0.242	0.438*	0.520**	0.565**	0.339	0.118	0.044	0.531**
Irrigation	0.530**	0.415*	0.461*	0.092	−0.193	0.521**	0.608***	0.642***	0.352	0.453*	0.681	0.616***
Seeds	−0.033	0.192	0.282	0.459*	0.350	0.062	0.217	0.366	0.093	−0.164	0.227	0.267
Credits	0.331	0.409*	0.154	0.547**	0.000	0.347	0.402*	0.227	0.573**	0.361	0.000	0.459*
Agricultural machinery	0.363	0.390*	−0.300	0.173	0.000	0.372	0.378	0.280	0.316	0.622***	−0.204	0.382
Fertilizers and pesticides	0.306	0.239	0.061	0.484*	0.568	0.303	0.298	0.152	−0.107	0.066	−0.219	0.276
Socio-economic aspects	0.545**	0.518**	0.201	0.266	0.051	0.537**	0.680***	0.493*	0.245	0.393*	0.684	0.654***
Cropping pattern												
Indicators	**1995–1996**						**2015–2016**					
	Marginal	**Small**	**Semi-medium**	**Medium**	**Large**	**All classes**	**Marginal**	**Small**	**Semi-medium**	**Medium**	**Large**	**All classes**
Land	−0.171	0.241	0.168	0.495*	0.429	−0.104	0.093	0.100	0.239	0.173	0.007	0.082
Irrigation	0.475*	0.172	0.165	0.184	0.720*	0.447*	0.505**	0.466*	0.104	0.447*	0.010	0.410*

Seeds	0.278	0.389*	0.420*	0.307	0.062	0.318	0.260	0.143	−0.113	0.078	0.758*	0.241
Credits	0.229	0.236	0.387	0.243	0.000	0.035	0.263	0.075	0.063	0.391*	0.000	0.351
Agricultural machinery	0.453*	0.405*	0.310	0.029	0.000	0.408*	0.669***	0.508**	0.530**	0.262	0.091	0.632***
Fertilizers and pesticides	0.327	−0.110	0.093	0.450*	0.036	0.290	0.515**	0.258	0.529**	0.321	0.488	0.436*
Socio-economic aspects	0.369	0.142	0.320	0.143	0.372	0.322	0.222	0.147	0.325	−0.065	0.097	0.168

Source: Calculated by author (based on data obtained from Agriculture Census and Input Survey, 1995–1996 and 2015–2016)
Note: * $p < 0.05$, ** $p < 0.01$, *** $p < 0.001$

Annexure C.40 Composite standardized score of agricultural innovation input–output by land size category in Murshidabad district, 1995–1996 and 2015–2016

Blocks	1995–1996						2015–2016					
	Marginal	Small	Semi-medium	Medium	Large	All classes	Marginal	Small	Semi-medium	Medium	Large	All classes
Berhampore	0.425	0.466	0.412	0.457	0.376	0.438	0.52	0.513	0.491	0.491	0.513	0.515
Burwan	0.436	0.39	0.396	0.429	N.A	0.423	0.608	0.541	0.532	0.59	0.459	0.599
Beldanga-I	0.466	0.538	0.427	0.432	N.A	0.477	0.519	0.57	0.568	0.546	N.A	0.52
Beldanga-II	0.443	0.425	0.417	0.377	0.439	0.433	0.505	0.526	0.49	0.452	N.A	0.513
Bhagwangola-I	0.433	0.425	0.426	0.376	N.A	0.438	0.48	0.516	0.491	0.447	0.378	0.489
Bhagwangola-II	0.45	0.493	0.465	0.429	N.A	0.461	0.548	0.502	0.512	0.515	N.A	0.533
Bharatpur-I	0.464	0.384	0.435	0.367	N.A	0.474	0.61	0.554	0.506	0.513	N.A	0.606
Bharatpur-II	0.401	0.464	0.422	0.484	N.A	0.429	0.563	0.541	0.534	0.53	N.A	0.591
Domkal	0.471	0.425	0.431	0.411	0.419	0.46	0.537	0.503	0.568	0.476	N.A	0.524
Farakka	0.353	0.396	0.348	0.307	0.234	0.348	0.408	0.401	0.378	0.407	0.341	0.392
Hariharpara	0.446	0.485	0.403	0.415	N.A	0.46	0.511	0.518	0.532	0.508	N.A	0.497
Jalangi	0.451	0.446	0.387	0.337	N.A	0.444	0.533	0.448	0.491	0.437	N.A	0.511
Kandi	0.463	0.454	0.493	0.486	N.A	0.454	0.576	0.541	0.571	0.572	N.A	0.562
Khargram	0.459	0.444	0.44	0.42	N.A	0.462	0.547	0.543	0.495	0.465	N.A	0.549
Lalgola	0.471	0.499	0.48	0.435	N.A	0.466	0.529	0.519	0.506	0.532	N.A	0.541
Murshidabad-Jiaganj	0.454	0.458	0.443	0.411	0.457	0.467	0.549	0.553	0.532	0.482	0.451	0.552
Nabagram	0.392	0.38	0.42	0.403	N.A	0.411	0.46	0.465	0.47	0.452	N.A	0.458
Nawda	0.451	0.464	0.42	0.396	N.A	0.443	0.572	0.527	0.546	0.477	0.4	0.552
Raghunathganj-I	0.395	0.428	0.413	0.441	0.359	0.406	0.503	0.515	0.482	0.53	N.A	0.511
Raghunathganj-II	0.376	0.35	0.355	0.355	0.37	0.364	0.444	0.449	0.453	0.467	0.411	0.44
Raninagar-I	0.472	0.52	0.453	0.466	N.A	0.479	0.533	0.494	0.554	0.548	N.A	0.532
Raninagar-II	0.43	0.415	0.417	0.452	N.A	0.428	0.516	0.466	0.476	0.421	N.A	0.509

Sagardighi	0.371	0.386	0.407	0.382	N.A	0.371	0.45	0.492	0.484	0.422	N.A	0.455
Samserganj	0.434	0.41	0.408	0.439	N.A	0.435	0.475	0.482	0.473	0.517	N.A	0.488
Suti-I	0.416	0.443	0.437	0.441	0.347	0.412	0.528	0.538	0.57	0.541	N.A	0.523
Suti-II	0.441	0.402	0.434	0.423	N.A	0.432	0.482	0.454	0.501	0.456	N.A	0.488
Total	**0.433**	**0.438**	**0.423**	0.414	**0.375**	**0.435**	**0.519**	**0.507**	**0.508**	0.492	**0.251**	**0.517**

Source: Calculated by researcher (data extracted from Agriculture Census and Input Survey, District Statistical Handbook)
N.A not available

Annexure D: Chapter 4

Annexure D.1 Sustainable land use and agricultural innovation under marginal land size class in Murshidabad district, 1995–1996 to 2015–2016

			Agricultural innovation											
			Inputs							Outputs				
S. no	Blocks	SLU	Land	Irrigation	Seeds	Credit	Agri. machinery	Fertiliser & pes.	Socio-eco	CDI	CYI	CI	CP
1	Berhampore	0.073	0.074	0.055	0.331	0.076	0.399	0.029	0.170	0.357	0.651	0.314	0.082
2	Burwan	0.075	0.011	0.106	0.367	0.071	0.487	0.328	0.155	0.788	0.756	1.000	0.055
3	Beldanga-I	0.080	0.051	0.013	0.215	0.054	0.436	0.381	0.028	0.123	0.155	0.176	0.054
4	Beldanga-II	0.107	0.024	0.025	0.317	0.063	0.225	0.344	0.066	0.567	0.538	0.145	0.113
5	Bhagwangola-I	0.049	0.061	0.036	0.246	0.061	0.170	0.170	0.029	0.625	0.000	0.086	0.104
6	Bhagwangola-II	0.053	0.059	0.053	0.309	0.064	0.446	0.295	0.038	0.619	0.608	0.259	0.105
7	Bharatpur-I	0.158	0.076	0.352	0.361	0.090	0.390	0.486	0.227	0.829	0.106	0.776	0.165
8	Bharatpur-II	0.024	0.003	0.142	0.297	0.068	0.466	0.404	0.021	0.969	0.994	0.208	0.054
9	Domkal	0.112	0.097	0.165	0.186	0.046	0.455	0.174	0.053	0.330	0.082	0.266	0.104
10	Farakka	0.133	0.036	0.055	0.276	0.070	0.156	0.313	0.042	0.425	0.328	0.550	0.106
11	Hariharpara	0.098	0.055	0.007	0.254	0.063	0.386	0.286	0.103	0.000	0.619	0.525	0.075
12	Jalangi	0.093	0.072	0.149	0.159	0.020	0.574	0.315	0.069	0.662	0.238	0.205	0.054
13	Kandi	0.082	0.028	0.146	0.274	0.059	0.439	0.303	0.063	1.000	0.817	0.252	0.054
14	Khargram	0.021	0.017	0.048	0.290	0.046	0.476	0.252	0.026	0.373	0.779	0.424	0.032
15	Lalgola	0.063	0.109	0.008	0.332	0.076	0.218	0.011	0.036	0.636	0.074	0.388	0.055
16	Murshidabad-Jiaganj	0.117	0.168	0.114	0.248	0.048	0.409	0.199	0.026	0.864	1.000	0.225	0.095
17	Nabagram	0.115	0.055	0.161	0.270	0.054	0.358	0.346	0.085	0.516	0.284	0.256	0.087
18	Nawda	0.067	0.013	0.161	0.226	0.054	0.645	0.342	0.029	0.563	0.621	0.249	0.104
19	Raghunathganj-I	0.127	0.079	0.017	0.161	0.033	0.104	0.366	0.035	0.198	0.619	0.940	0.055

20	Raghunathganj-II	0.014	0.030	0.023	0.220	0.041	0.149	0.221	0.012	0.770	0.513	0.156	0.029
21	Raninagar-I	0.172	0.150	0.362	0.254	0.062	0.103	0.065	0.045	0.666	0.215	0.389	0.134
22	Raninagar-II	0.065	0.071	0.076	0.343	0.081	0.472	0.518	0.032	0.567	0.558	0.179	0.065
23	Sagardighi	0.085	0.115	0.139	0.329	0.067	0.237	0.305	0.024	0.639	0.544	0.312	0.080
24	Samserganj	0.091	0.098	0.175	0.328	0.073	0.043	0.118	0.031	0.443	0.411	0.000	0.054
25	Suti-I	0.012	0.001	0.216	0.200	0.053	0.580	0.373	0.062	0.405	0.388	0.106	0.016
26	Suti-II	0.088	0.062	0.249	0.309	0.087	0.315	0.028	0.024	0.373	0.588	0.435	0.063

Source: Calculated by authors

Note: SLU sustainable land use index, *Agri. machinery* agricultural machinery, *Fertilizers & pes.* fertilizer and pesticides, *Socio-eco.* socioeconomic, *CDI* crop diversification index, *CYI* crop yield index, *CI* cropping intensity, *CP* cropping pattern

Annexure D.2 Sustainable land use and agricultural innovation under small land size class in Murshidabad district, 1995–1996 to 2015–2016

			Agricultural innovation										
			Inputs							Outputs			
S. no.	Blocks	SLU	Land	Irrigation	Seeds	Credit	Agr. Mch	Fert & pres	Socio-eco	CDI	CYI	CI	CP
1	Berhampore	0.102	0.110	0.142	0.402	0.073	0.389	0.267	0.056	0.315	0.644	0.253	0.034
2	Burwan	0.087	0.065	0.164	0.433	0.083	0.313	0.230	0.114	0.666	0.581	0.992	0.034
3	Beldanga-I	0.113	0.116	0.204	0.365	0.071	0.453	0.142	0.041	0.322	0.631	0.251	0.034
4	Beldanga-II	0.125	0.130	0.363	0.394	0.064	0.282	0.405	0.209	0.641	0.876	0.353	0.133
5	Bhagwangola-I	0.063	0.043	0.105	0.411	0.067	0.347	0.208	0.053	0.394	0.247	0.372	0.033
6	Bhagwangola-II	0.077	0.038	0.050	0.416	0.076	0.051	0.230	0.081	0.419	0.140	0.260	0.034
7	Bharatpur-I	0.074	0.042	0.164	0.368	0.089	0.597	0.332	0.021	0.821	0.883	1.000	0.033
8	Bharatpur-II	0.038	0.029	0.132	0.425	0.095	0.391	0.163	0.124	0.743	1.000	0.266	0.033
9	Domkal	0.115	0.088	0.242	0.438	0.098	0.624	0.418	0.048	0.310	0.591	0.000	0.034
10	Farakka	0.070	0.081	0.144	0.406	0.081	0.224	0.206	0.106	0.105	0.190	0.614	0.033
11	Hariharpara	0.124	0.106	0.225	0.369	0.066	0.167	0.115	0.089	0.299	0.711	0.299	0.084
12	Jalangi	0.103	0.079	0.123	0.403	0.044	0.056	0.064	0.052	0.530	0.254	0.041	0.034
13	Kandi	0.102	0.053	0.206	0.454	0.077	0.456	0.263	0.058	1.000	0.738	0.111	0.034
14	Khargram	0.057	0.042	0.185	0.420	0.067	0.250	0.189	0.042	0.520	0.335	0.772	0.033
15	Lalgola	0.055	0.054	0.081	0.370	0.070	0.070	0.088	0.070	0.514	0.306	0.230	0.033
16	Murshidabad-Jiaganj	0.115	0.044	0.140	0.450	0.092	0.668	0.366	0.275	0.601	0.564	0.195	0.034
17	Nabagram	0.119	0.109	0.270	0.415	0.076	0.313	0.297	0.098	0.440	0.709	0.661	0.084
18	Nawda	0.090	0.043	0.204	0.354	0.067	0.307	0.087	0.042	0.000	0.511	0.514	0.034
19	Raghunathganj-I	0.038	0.073	0.086	0.439	0.085	0.139	0.163	0.070	0.398	0.314	0.643	0.033
20	Raghunathganj-II	0.021	0.017	0.092	0.416	0.075	0.128	0.168	0.099	0.613	0.135	0.532	0.033

21	Raninagar-I	0.134	0.183	0.246	0.364	0.087	0.618	0.349	0.240	0.586	0.000	0.028	0.134
22	Raninagar-II	0.080	0.027	0.109	0.378	0.068	0.209	0.061	0.098	0.497	0.475	0.337	0.034
23	Sagardighi	0.088	0.098	0.064	0.410	0.073	0.251	0.203	0.080	0.296	0.714	0.484	0.034
24	Samserganj	0.107	0.084	0.208	0.373	0.074	0.079	0.113	0.041	0.300	0.756	0.267	0.034
25	Suti-I	0.014	0.032	0.019	0.402	0.049	0.030	0.072	0.041	0.364	0.088	0.280	0.033
26	Suti-II	0.103	0.064	0.146	0.398	0.075	0.001	0.089	0.051	0.127	0.571	0.316	0.034

Source: Calculated by authors

Note: SLU sustainable land use index, *Agri. Machinery* agricultural machinery, *Fertilizers & pes.* fertilizer and pesticides, *Socio-eco.* socioeconomic, *CDI* crop diversification index, *CYI* crop yield index, *CI* cropping intensity, *CP* cropping pattern

Annexure D.3 Sustainable land use and agricultural innovation under semi-medium land size class in Murshidabad district, 1995–1996 to 2015–2016

			Agricultural innovation										
			Inputs							Outputs			
S. No.	Blocks	SLU	Land	Irrigation	Seeds	Credit	Agr. Mch	Fert & pres	Socio-eco	CDI	CYI	CI	CP
1	Berhampore	0.085	0.094	0.030	0.487	0.110	0.161	0.310	0.141	0.553	0.658	0.661	0.029
2	Burwan	0.056	0.029	0.139	0.515	0.146	0.259	0.198	0.130	0.970	0.822	1.000	0.023
3	Beldanga-I	0.065	0.117	0.159	0.445	0.106	0.592	0.289	0.132	0.116	0.360	0.706	0.161
4	Beldanga-II	0.095	0.134	0.035	0.385	0.079	0.054	0.198	0.086	0.632	0.698	0.856	0.083
5	Bhagwangola-I	0.026	0.026	0.014	0.441	0.104	0.059	0.028	0.170	0.744	0.761	0.093	0.098
6	Bhagwangola-II	0.057	0.065	0.003	0.441	0.107	0.007	0.127	0.157	1.000	0.044	0.392	0.018
7	Bharatpur-I	0.074	0.038	0.336	0.450	0.137	0.147	0.431	0.042	0.671	0.000	0.546	0.044
8	Bharatpur-II	0.051	0.056	0.226	0.438	0.085	0.185	0.362	0.118	0.745	1.000	0.228	0.066
9	Domkal	0.081	0.135	0.119	0.439	0.105	0.127	0.176	0.130	0.838	0.650	0.589	0.144
10	Farakka	0.084	0.162	0.007	0.484	0.080	0.340	0.083	0.046	0.186	0.585	0.000	0.078
11	Hariharpara	0.072	0.086	0.120	0.435	0.102	0.407	0.361	0.128	0.857	0.614	0.483	0.235
12	Jalangi	0.084	0.113	0.100	0.448	0.097	0.403	0.448	0.234	0.810	0.397	0.704	0.057
13	Kandi	0.054	0.072	0.058	0.454	0.113	0.451	0.129	0.135	0.225	0.368	0.127	0.049
14	Khargram	0.081	0.095	0.138	0.418	0.093	0.107	0.051	0.101	0.193	0.825	0.221	0.070
15	Lalgola	0.020	0.016	0.133	0.454	0.109	0.322	0.134	0.050	0.823	0.332	0.025	0.041
16	Murshidabad-Jiaganj	0.106	0.159	0.128	0.398	0.085	0.214	0.052	0.040	0.660	0.749	0.806	0.505
17	Nabagram	0.068	0.134	0.162	0.439	0.091	0.139	0.026	0.062	0.702	0.734	0.132	0.029
18	Nawda	0.034	0.009	0.166	0.489	0.085	0.518	0.370	0.118	0.671	0.460	0.158	0.193
19	Raghunathganj-I	0.079	0.161	0.020	0.388	0.106	0.409	0.127	0.090	0.315	0.792	0.410	0.039
20	Raghunathganj-II	0.077	0.123	0.181	0.410	0.080	0.276	0.386	0.112	0.449	0.675	0.323	0.098
21	Raninagar-I	0.069	0.163	0.101	0.416	0.090	0.019	0.154	0.101	0.673	0.922	0.481	0.144

22	Raninagar-II	0.056	0.025	0.010	0.376	0.107	0.219	0.022	0.154	0.759	0.349	0.419	0.067
23	Sagardighi	0.072	0.031	0.064	0.515	0.122	0.079	0.007	0.260	0.434	0.620	0.259	0.043
24	Samserganj	0.089	0.045	0.282	0.479	0.092	0.129	0.001	0.118	0.540	0.713	0.723	0.402
25	Suti-I	0.069	0.108	0.211	0.348	0.089	0.490	0.153	0.117	0.000	0.800	0.492	0.090
26	Suti-II	0.096	0.091	0.250	0.432	0.106	0.294	0.059	0.065	0.601	0.359	0.771	0.522

Source: Calculated by authors

Note: SLU sustainable land use index, *Agri. Machinery* agricultural machinery, *Fertilizers & pes.* fertilizer and pesticides, *Socio-eco.* socioeconomic, *CDI* crop diversification index, *CYI* crop yield index, *CI* cropping intensity, *CP* cropping pattern

Annexure D.4 Sustainable land use and agricultural innovation under medium land size class in Murshidabad district, 1995–1996 to 2015–2016

			Agricultural innovation										
			Inputs							Outputs			
S. No.	Blocks	SLU	Land	Irrigation	Seeds	Credit	Agr. Mch	Fert & pres	Socio-eco	CDI	CYI	CI	CP
1	Berhampore	0.089	0.111	0.125	0.449	0.090	0.020	0.013	0.019	0.481	0.209	0.245	0.057
2	Burwan	0.051	0.009	0.049	0.438	0.140	0.483	0.443	0.004	0.046	0.095	0.092	0.056
3	Beldanga-I	0.066	0.199	0.023	0.366	0.089	0.474	0.236	0.042	0.223	0.752	0.475	0.056
4	Beldanga-II	0.067	0.259	0.191	0.508	0.146	0.036	0.143	0.059	0.342	0.282	0.110	0.057
5	Bhagwangola-I	0.057	0.350	0.165	0.367	0.084	0.285	0.131	0.012	0.125	0.000	0.172	0.059
6	Bhagwangola-II	0.093	0.032	0.173	0.479	0.098	0.056	0.086	0.010	0.650	0.234	0.435	0.057
7	Bharatpur-I	0.068	0.074	0.240	0.545	0.111	0.056	0.564	0.099	0.682	0.790	1.000	0.058
8	Bharatpur-II	0.093	0.019	0.101	0.571	0.207	0.392	0.137	0.066	0.599	0.832	0.210	0.057
9	Domkal	0.079	0.149	0.099	0.398	0.098	0.225	0.207	0.072	0.226	0.304	0.609	0.056
10	Farakka	0.086	0.066	0.042	0.301	0.055	0.081	0.571	0.117	0.694	0.524	0.512	0.072
11	Hariharpara	0.083	0.124	0.252	0.436	0.102	0.004	0.142	0.003	0.448	0.594	0.248	0.056
12	Jalangi	0.093	0.256	0.394	0.430	0.169	0.307	0.108	0.034	0.610	0.499	0.732	0.058
13	Kandi	0.080	0.267	0.217	0.399	0.175	0.011	0.135	0.077	0.492	0.647	0.069	0.056
14	Khargram	0.060	0.216	0.044	0.539	0.142	0.201	0.068	0.013	0.000	1.000	0.102	0.057
15	Lalgola	0.078	0.223	0.335	0.545	0.141	0.015	0.134	0.027	0.496	0.472	0.000	0.056
16	Murshidabad-Jiaganj	0.094	0.022	0.009	0.444	0.211	0.069	0.124	0.018	0.778	0.709	0.825	0.056
17	Nabagram	0.067	0.190	0.054	0.440	0.081	0.305	0.176	0.025	0.235	0.294	0.501	0.067
18	Nawda	0.075	0.028	0.130	0.511	0.131	0.184	0.064	0.053	0.506	0.438	0.430	0.057
19	Raghunathganj-I	0.062	0.263	0.049	0.413	0.090	0.082	0.201	0.054	0.685	0.116	0.337	0.057
20	Raghunathganj-II	0.077	0.085	0.183	0.297	0.065	0.392	0.465	0.013	0.563	0.154	0.270	0.097
21	Raninagar-I	0.079	0.373	0.246	0.358	0.066	0.109	0.067	0.037	0.558	0.059	0.415	0.057
22	Raninagar-II	0.067	0.053	0.045	0.401	0.099	0.158	0.008	0.011	0.520	0.481	0.124	0.444

23	Sagardighi	0.058	0.008	0.020	0.437	0.108	0.148	0.283	0.077	0.408	0.254	0.017	0.057
24	Samserganj	0.076	0.180	0.258	0.470	0.165	0.123	0.103	0.092	0.252	0.097	0.257	0.057
25	Suti-I	0.062	0.140	0.156	0.353	0.097	0.318	0.107	0.099	0.288	0.279	0.159	0.066
26	Suti-II	0.108	0.069	0.388	0.491	0.225	0.434	0.094	0.045	1.000	0.798	0.850	0.056

Source: Calculated by authors

Note: SLU sustainable land use index, Agri. Machinery agricultural machinery, Fertilizers & p*es*. fertilizer and pesticides, *Socio-eco*. socioeconomic, *CDI* crop diversification index, *CYI* crop yield index, *CI* cropping intensity, *CP* cropping pattern

Annexure D.5 Sustainable land use and agricultural innovation under large land size class in Murshidabad district, 1995–1996 to 2015–2016

			Agricultural innovation											
			Inputs							Outputs				
S. no	Blocks	SLU	Land	Irrigation	Seeds	Agr. Mch	Fert & pres	Socio-eco	Input	CDI	CYI	CI	CP	Output
1	Berhampore	0.064	0.106	0.267	0.263	0.367	0.317	0.124	0.241	0.839	1.000	0.200	0.000	0.510
10	Farakka	0.089	0.140	0.618	0.289	0.696	0.555	0.115	0.402	0.000	0.598	0.974	1.000	0.643
16	Murshidabad-Jiaganj	0.044	0.053	0.400	0.472	0.217	0.289	0.211	0.274	1.000	0.000	1.000	0.500	0.625
20	Raghunathganj-II	0.023	0.008	0.205	0.439	0.113	0.222	0.083	0.178	0.606	0.822	0.000	0.500	0.482

Source: Calculated by authors

Note: SLU sustainable land use index, Agri. Machinery agricultural machinery, Fertilizers & pes. fertilizer and pesticides, *Socio-eco*. socioeconomic, *CDI* crop diversification index, *CYI* crop yield index, *CI* cropping intensity, *CP* cropping pattern

Annexure D.6 Sustainable land use and agricultural innovation under large land size class in Murshidabad district, 1995–1996 to 2015–2016

			Agricultural innovation										
			Inputs							Outputs			
S. no	Blocks	SLU	Land	Irrigation	Seeds	Credit	Agr. Mch	Fert & pres	Socio-eco	CDI	CYI	CI	CP
1	Berhampore	0.073	0.080	0.067	0.350	0.078	0.390	0.071	0.154	0.362	0.652	0.319	0.074
2	Burwan	0.083	0.019	0.116	0.385	0.076	0.460	0.315	0.148	0.777	0.734	1.000	0.051
3	Beldanga-I	0.095	0.064	0.044	0.246	0.059	0.448	0.347	0.034	0.153	0.235	0.207	0.055
4	Beldanga-II	0.113	0.044	0.075	0.335	0.065	0.228	0.349	0.087	0.583	0.593	0.196	0.116
5	Bhagwangola-I	0.052	0.061	0.046	0.279	0.064	0.194	0.172	0.037	0.597	0.057	0.128	0.095
6	Bhagwangola-II	0.064	0.057	0.053	0.333	0.068	0.378	0.281	0.048	0.609	0.528	0.267	0.093
7	Bharatpur-I	0.162	0.071	0.328	0.370	0.093	0.413	0.469	0.193	0.830	0.220	0.810	0.144
8	Bharatpur-II	0.029	0.008	0.144	0.325	0.074	0.451	0.371	0.039	0.937	1.004	0.218	0.052
9	Domkal	0.115	0.098	0.175	0.233	0.056	0.471	0.211	0.056	0.344	0.173	0.244	0.096
10	Farakka	0.148	0.047	0.066	0.303	0.072	0.171	0.297	0.053	0.380	0.321	0.548	0.096
11	Hariharpara	0.103	0.065	0.044	0.280	0.066	0.356	0.266	0.102	0.072	0.638	0.495	0.081
12	Jalangi	0.098	0.077	0.148	0.206	0.027	0.499	0.285	0.072	0.654	0.250	0.204	0.052
13	Kandi	0.083	0.035	0.154	0.309	0.065	0.442	0.293	0.065	0.982	0.799	0.229	0.052
14	Khargram	0.034	0.025	0.070	0.318	0.052	0.435	0.238	0.031	0.389	0.728	0.468	0.034
15	Lalgola	0.059	0.101	0.025	0.346	0.078	0.201	0.027	0.041	0.630	0.119	0.355	0.052
16	Murshidabad-Jiaganj	0.122	0.151	0.118	0.286	0.057	0.440	0.219	0.061	0.829	0.939	0.247	0.099
17	Nabagram	0.112	0.067	0.177	0.300	0.059	0.348	0.332	0.086	0.513	0.360	0.314	0.085
18	Nawda	0.068	0.018	0.168	0.257	0.058	0.596	0.308	0.034	0.493	0.605	0.288	0.098
19	Raghunathganj-I	0.128	0.083	0.027	0.211	0.043	0.119	0.332	0.042	0.237	0.583	0.886	0.052
20	Raghunathganj-II	0.017	0.032	0.039	0.256	0.048	0.153	0.223	0.027	0.744	0.467	0.217	0.033
21	Raninagar-I	0.179	0.158	0.340	0.277	0.067	0.174	0.108	0.075	0.661	0.207	0.345	0.135

22	Raninagar-II	0.069	0.064	0.079	0.353	0.081	0.429	0.439	0.045	0.568	0.545	0.210	0.065
23	Sagardighi	0.071	0.110	0.126	0.350	0.070	0.236	0.284	0.039	0.589	0.573	0.334	0.073
24	Samserganj	0.103	0.097	0.185	0.343	0.075	0.052	0.115	0.036	0.429	0.469	0.062	0.063
25	Suti-I	0.013	0.010	0.190	0.237	0.055	0.504	0.325	0.062	0.390	0.361	0.144	0.021
26	Suti-II	0.108	0.064	0.239	0.330	0.088	0.275	0.038	0.029	0.355	0.587	0.437	0.073

Source: Calculated by authors

Note: *SLU* sustainable land use index, *Agri. Machinery* agricultural machinery, *Fertilizers & pes.* fertilizer and pesticides, *Socio-eco.* socioeconomic, *CDI* crop diversification index, *CYI* crop yield index, *CI* cropping intensity, *CP* cropping pattern

Annexure D.7 Changes of agricultural innovation per unit sustainable land use change by land size category in Murshidabad district, 1995–1996 to 2015–2016

Blocks	Input dimension						Output dimension						Overall agricultural innovation					
	Marginal	Small	Semi-medium	Medium	Large	All classes	Marginal	Small	Semi-medium	Medium	Large	All classes	Marginal	Small	Semi-medium	Medium	Large	All classes
Berhampore	2.41	2.1	2.64	2.26	3.47	2.4	1.65	1.23	2.37	2.57	2.42	1.62	2.03	1.67	2.51	2.42	2.95	2.01
Burwan	3.01	2.22	3.23	1.18	N.A	2.86	1.98	0.88	2.2	0.86	N.A	1.8	2.5	1.55	2.72	1.02	N.A	2.33
Beldanga-I	2.68	2.14	3.22	2.67	N.A	2.59	1.03	1.83	2.97	1.92	N.A	1.18	1.86	1.99	3.1	2.3	N.A	1.88
Beldanga-II	1.8	2.49	3.48	2.68	N.A	1.92	1.08	2.79	3.01	1.38	N.A	1.35	1.44	2.64	3.25	2.03	N.A	1.64
Bhagwangola-I	2.38	2.3	5.18	2.45	N.A	2.4	2.42	0.69	2.78	0.83	N.A	2.14	2.4	1.5	3.98	1.64	N.A	2.27
Bhagwangola-II	3.53	2.25	3.09	3.11	N.A	3.3	2.08	0.83	1.96	1.91	N.A	1.88	2.81	1.54	2.53	2.51	N.A	2.59
Bharatpur-I	1.85	2.62	2.95	2.02	N.A	1.96	1.72	0.72	2.24	2.29	N.A	1.58	1.79	1.67	2.6	2.16	N.A	1.77
Bharatpur-II	5.81	3.14	4.26	2.91	N.A	5.46	2.09	0.99	2.35	2.6	N.A	1.93	3.95	2.07	3.31	2.76	N.A	3.69
Domkal	1.5	2.32	2.87	2.26	N.A	1.63	1.32	2.1	2.66	1.31	N.A	1.44	1.41	2.21	2.77	1.79	N.A	1.54
Farakka	1.34	2.47	2.78	2.78	4.99	1.58	0.66	0.67	2.24	2.21	3.13	0.73	1	1.57	2.51	2.5	4.06	1.16
Hariharpara	1.74	2.34	3.26	2.47	N.A	1.84	1.11	2.3	2.9	2.17	N.A	1.31	1.43	2.32	3.08	2.32	N.A	1.58
Jalangi	2.09	1.99	3.44	3.13	N.A	2.09	1.16	1.59	2.13	2.82	N.A	1.24	1.63	1.79	2.79	2.98	N.A	1.67
Kandi	2.19	1.9	3.52	2.43	N.A	2.16	1.64	1.44	2.38	1.55	N.A	1.61	1.92	1.67	2.95	1.99	N.A	1.88
Khargram	7.18	2.5	2.64	2.06	N.A	6.93	3.31	0.77	2.02	1.39	N.A	2.88	5.25	1.64	2.33	1.73	N.A	4.9
Lalgola	1.89	2.26	5.35	2.56	N.A	2	1.38	0.75	2.56	1.55	N.A	1.3	1.64	1.51	3.96	2.06	N.A	1.65
Murshidabad-Jiaganj	1.53	2.46	3.31	3.1	4.64	1.74	1.55	1.99	3.36	2.83	3.75	1.68	1.54	2.23	3.34	2.97	4.2	1.71
Nabagram	1.72	2.15	2.6	2.81	N.A	1.79	1.39	2.19	1.8	2.11	N.A	1.5	1.56	2.17	2.2	2.46	N.A	1.65
Nawda	3.05	2	4	2.17	N.A	2.88	1.52	1.16	2.95	1.56	N.A	1.48	2.29	1.58	3.48	1.87	N.A	2.18
Raghunathganj-I	1.21	3.31	2.9	2.35	N.A	1.54	1.23	0.72	1.29	2.07	N.A	1.16	1.22	2.02	2.1	2.21	N.A	1.35
Raghunathganj-II	7	4.64	2.76	2.62	4.12	6.53	4.55	1.05	1.83	1.44	4.36	3.97	5.78	2.85	2.3	2.03	4.24	5.25
Raninagar-I	0.86	2.63	2.89	2.05	N.A	1.15	1.8	3.36	2	1.62	N.A	2	1.33	3	2.45	1.84	N.A	1.58

Raninagar-II	3.55	1.97	3.03	1.86	N.A	3.27	1.18	0.84	1.98	3.04	N.A	1.16	2.37	1.41	2.51	2.45	N.A	2.21
Sagardighi	1.87	2	3.41	1.33	N.A	1.89	1.3	0.95	1.94	1.32	N.A	1.25	1.59	1.48	2.68	1.33	N.A	1.57
Samserganj	1.32	2.03	2.97	2.54	N.A	1.45	1.16	1.78	3.36	1.51	N.A	1.28	1.24	1.91	3.17	2.03	N.A	1.37
Suti-I	5.42	5.6	2.89	2.66	N.A	4.51	2.61	1.31	2.46	1.51	N.A	2.39	4.02	3.46	2.68	2.09	N.A	3.45
Suti-II	1.58	2.01	3.33	3.72	N.A	1.68	1.44	1.7	3.91	3.12	N.A	1.53	1.51	1.86	3.62	3.42	N.A	1.6
Total	2.71	2.53	3.31	2.47	4.31	2.68	1.71	1.41	2.45	1.9	3.42	1.67	2.21	1.97	2.88	2.19	3.86	2.17

Source: Calculated by researcher
N.A not available

Annexure E: Chapter 5

Annexure E.2 Characteristics of soil (nitrogen, phosphorus, and potassium) in the sample villages, 2016–2017

Annexure E.1 General information of the sample villages, 2011 (area in hectares)

Blocks	Sample villages	Total area	Total households	Total population	Total workers	Total cultivators
Berhampore	Chhayghari	1020.25	3143	12,388	4741	901
Burwan	Jhikarhati	958.63	1933	8163	2579	781
Beldanga-I	Mirzapur	754.73	5380	27,364	8420	1341
Beldanga-II	Andul Baria	1547.07	3230	13,929	4515	1438
Bhagwangola-I	Dakshin Hanumanta Nagar	2133.09	4345	19,813	6218	704
Bhagwangola-II	Nasipur	1544.91	6047	25,644	8009	1302
Bharatpur-I	Bharatpur	1021.99	3445	15,812	5191	860
Bharatpur-II	Kagram	1101.12	3481	14,878	5673	623
Domkal	Garaimari	2356.7	7205	29,260	10,022	1748
Farakka	Kuli	1435.42	5511	28,092	13,359	495
Hariharpara	Rukunpur	2327.77	4429	18,663	6037	2149
Jalangi	Debipur	1872.26	4091	17,877	6221	2114
Kandi	Hijal	3680.22	2713	11,480	3970	2450
Khargram	Khargram	965.14	2648	11,209	3660	703
Lalgola	Lalgola	585.06	6260	31,698	11,918	706
Murshidabad-Jiaganj	Hasenpur	549.87	2014	8093	2898	1121
Nabagram	Panchgram	1107.14	4004	17,137	5611	767
Nawda	Bali	1741.05	5343	22,779	6990	2009
Raghunathganj-I	Ramchandrabati	393.51	1159	5648	2488	314
Raghunathganj-II	Bahara	301.51	2296	12,862	5094	355
Raninagar-I	Bishnupur	936.61	2059	8320	2596	732
Raninagar-II	Katlamari	1610.05	7555	33,290	10,864	2632
Sagardighi	Bokhara	584.38	1789	8000	3147	688
Samserganj	Malancha	551.21	3136	16,918	6016	355
Suti-I	Bahutali	1051.46	3076	15,963	6791	485
Suti-II	Mahesail	983.29	3057	16,178	7434	989

Source: Village Directory, Census of India, 2011

Sample villages	Nitrogen (%)							Phosphorus (%)					Potassium (%)		
	Very low	Low	Medium	High	Very high	Very low	Low	Medium	High	Very high	Very low	Low	Medium	High	Very high
Chhayghari	0.29	19.29	75.36	5.06	0.00	0.00	0.00	37.28	49.78	12.93	0.58	22.54	64.52	12.28	0.07
Jhikarhati	0.15	8.73	81.73	9.17	0.22	0.00	0.00	28.97	52.77	18.27	2.77	17.47	67.69	11.94	0.15
Mirzapur	0.16	15.76	76.69	7.40	0.00	0.16	0.00	27.33	65.11	7.40	0.48	9.49	78.62	11.09	0.32
Andul Baria	0.93	28.70	70.37	0.00	0.00	0.00	0.00	50.00	46.30	3.70	2.78	24.07	71.30	1.85	0.00
Dakshin Hanumanta Nagar	0.00	1.33	75.48	23.19	0.00	0.00	0.00	28.52	65.40	6.08	0.00	3.04	86.50	10.46	0.00
Nasipur	0.00	24.03	72.51	3.46	0.00	0.22	0.00	23.11	74.08	2.59	0.00	1.73	97.62	0.65	0.00
Bharatpur	0.00	0.00	0.00	100.00	0.00	0.00	0.00	0.00	100.00	0.00	0.00	8.06	69.35	22.58	0.00
Kagram	0.00	0.00	0.00	100.00	0.00	0.00	0.00	0.00	100.00	0.00	0.00	0.00	100.00	0.00	0.00
Garaimari	11.41	26.09	62.50	0.00	0.00	0.00	0.54	42.39	51.63	5.43	0.54	16.85	81.52	1.09	0.00
Kuli	0.00	35.23	64.77	0.00	0.00	0.00	10.23	51.14	21.59	17.05	0.00	30.68	54.55	14.77	0.00
Rukunpur	0.00	2.00	96.00	0.00	2.00	0.00	0.00	14.00	67.00	19.00	0.00	0.00	34.00	66.00	0.00
Debipur	1.14	17.95	80.91	0.00	0.00	0.00	0.57	17.38	42.17	39.89	7.41	10.26	66.38	15.67	0.28
Hijal	0.35	11.01	69.58	18.88	0.17	0.09	0.44	31.38	50.26	17.83	1.14	14.86	66.35	17.31	0.35
Khargram	0.44	7.93	74.01	16.74	0.88	0.00	0.00	18.94	35.68	45.37	0.00	32.60	60.79	6.61	0.00
Lalgola	0.77	10.62	85.33	3.28	0.00	0.00	0.00	17.18	69.88	12.93	0.19	7.72	33.98	58.11	0.00
Hasenpur	1.18	36.69	62.13	0.00	0.00	0.00	0.00	35.29	61.18	3.53	1.18	20.00	77.65	0.59	0.59
Panchgram	0.00	51.55	47.42	1.03	0.00	0.00	1.03	5.15	36.08	57.73	4.12	15.46	78.35	2.06	0.00
Bali	0.14	30.74	68.13	0.99	0.00	0.00	0.00	31.16	51.56	17.28	1.27	14.02	30.59	54.11	0.00
Ramchandrabati	0.00	100.00	0.00	0.00	0.00	0.00	0.00	0.00	0.00	100.00	0.00	100.00	0.00	0.00	0.00
Bahara	0.00	2.44	97.56	0.00	0.00	0.00	0.00	14.63	68.29	17.07	0.00	0.00	14.63	85.37	0.00
Bishnupur	0.49	10.78	79.41	9.31	0.00	0.00	0.00	27.94	56.86	15.20	0.98	3.43	83.33	12.25	0.01

(continued)

Annexure E.2 (continued)

Sample villages	Nitrogen (%)							Phosphorus (%)					Potassium (%)		
Katlamari	0.00	12.61	84.24	3.15	0.00	0.00	0.00	28.37	56.45	15.19	1.15	54.44	41.55	2.87	0.00
Bokhara	0.21	14.99	80.32	4.42	0.06	0.00	1.10	51.05	37.10	10.75	0.12	25.04	66.89	7.34	0.61
Malancha	0.29	33.24	65.03	1.45	0.00	0.00	3.47	79.77	14.16	2.60	0.00	3.47	13.87	70.52	12.14
Bahutali	0.00	8.55	85.53	5.92	0.00	0.00	0.00	21.05	25.66	53.29	1.32	36.84	31.58	28.95	1.32
Mahesail	0.29	8.46	83.08	8.17	0.00	0.10	0.48	52.66	43.16	3.61	0.00	2.28	34.03	62.55	1.14
Village total	0.70	19.95	66.85	12.37	0.13	0.02	0.69	28.26	51.62	19.41	1.00	18.24	57.91	22.19	0.65

Source: Soil Health Card, Department of Agriculture, Cooperation, and Welfare, Govt of India, 2016–2017

Annexure E.3 Sufficiency of macro- and micro-soil nutrients in the sample villages (in percentage), 2016–2017

Sample villages	Macro-soil nutrients				Micro-soil nutrients					
	Nitrogen (N)	Organic carbon (OC)	Phosphorus (P)	Potassium (K)	Copper (Cu)	Iron (Fe)	Manganese (Mn)	Zinc (Zn)	Boron (B)	Sulfur (S)
Chhayghari	80.42	66.04	100.00	76.88	97.47	99.64	88.24	11.34	0.36	65.15
Jhikarhati	91.12	64.41	100.00	79.76	99.93	99.78	89.03	30.20	2.98	99.49
Mirzapur	84.08	60.45	99.84	90.03	99.68	99.84	96.95	22.19	0.48	98.07
Andul Baria	70.37	72.22	100.00	73.15	97.22	99.81	53.77	1.85	0.93	91.67
Dakshin Hanumanta Nagar	98.67	69.58	100.00	96.96	100.00	99.78	99.24	25.10	0.00	100.00
Nasipur	75.97	36.50	99.78	98.27	99.57	100.00	99.78	45.14	0.22	98.27
Bharatpur	96.77	82.26	100.00	91.94	100.00	100.00	100.00	15.23	0.00	100.00
Kagram	96.77	82.26	100.00	91.94	100.00	100.00	100.00	0.00	0.00	100.00
Garaimari	69.01	66.14	97.12	78.74	100.00	100.00	35.71	1.63	0.00	98.91
Kuli	73.68	55.59	96.17	92.50	100.00	100.00	90.70	40.91	26.14	40.91
Rukunpur	83.02	75.19	99.92	63.98	100.00	98.00	46.46	99.00	2.00	3.00
Debipur	78.81	78.12	98.51	77.64	100.00	99.91	79.65	68.66	31.31	87.39
Hijal	88.64	69.67	99.47	84.00	100.00	99.97	84.44	24.48	3.76	98.78
Khargram	91.71	49.25	99.82	83.71	99.56	100.00	91.03	19.82	0.44	90.75
Lalgola	88.61	81.85	100.00	92.09	100.00	100.00	84.85	77.80	0.00	30.50
Hasenpur	82.26	61.91	99.69	76.49	99.41	100.00	98.82	28.24	0.00	94.64
Panchgram	60.25	80.37	99.74	75.12	100.00	100.00	100.00	100.00	34.02	86.60
Bali	69.12	71.67	100.00	84.71	99.10	100.00	66.16	51.70	0.28	17.14
Ramchandrabati	86.53	57.56	99.42	57.17	100.00	100.00	100.00	100.00	0.00	100.00
Bahara	87.79	83.97	100.00	96.18	100.00	100.00	31.71	100.00	2.44	43.90
Bishnupur	88.78	44.04	99.83	96.31	99.02	100.00	98.53	25.49	7.84	99.51

(continued)

Annexure E.3 (continued)

Sample villages	Macro-soil nutrients				Micro-soil nutrients					
	Nitrogen (N)	Organic carbon (OC)	Phosphorus (P)	Potassium (K)	Copper (Cu)	Iron (Fe)	Manganese (Mn)	Zinc (Zn)	Boron (B)	Sulfur (S)
Katlamari	89.03	50.85	99.96	44.88	100.00	100.00	100.00	21.49	0.57	94.54
Bokhara	84.80	61.28	98.90	74.84	100.00	100.00	96.80	29.79	5.73	79.44
Malancha	66.47	26.59	96.53	96.53	100.00	100.00	100.00	24.57	0.00	92.49
Bahutali	84.30	61.10	99.93	89.60	100.00	100.00	67.76	20.39	0.66	100.00
Mahesail	91.25	57.89	99.42	97.72	99.90	100.00	99.33	25.76	1.71	90.46
Village total	**83.01**	**64.11**	**99.39**	**83.12**	**99.65**	**99.87**	**84.58**	**38.88**	**4.69**	**80.83**

Source: Soil Health Card, Department of Agriculture, Cooperation, and Welfare, Govt of India, 2016–2017

Annexure E.4 Soil fertility index of sample villages, 2016–2017

Sample villages	Macro-soil nutrients	Micro-soil nutrients	Overall soil nutrients
Chhayghari	0.808	0.604	0.706
Jhikarhati	0.838	0.703	0.771
Mirzapur	0.836	0.695	0.766
Andul Baria	0.789	0.575	0.682
Dakshin Hanumanta Nagar	0.913	0.707	0.810
Nasipur	0.776	0.738	0.757
Bharatpur	0.927	0.692	0.810
Kagram	0.927	0.667	0.797
Garaimari	0.778	0.560	0.669
Kuli	0.795	0.664	0.730
Rukunpur	0.805	0.581	0.693
Debipur	0.833	0.778	0.805
Hijal	0.854	0.686	0.770
Khargram	0.811	0.669	0.740
Lalgola	0.906	0.655	0.781
Hasenpur	0.801	0.702	0.751
Panchgram	0.789	0.868	0.828
Bali	0.814	0.557	0.686
Ramchandrabati	0.752	0.833	0.793
Bahara	0.920	0.630	0.775
Bishnupur	0.822	0.717	0.770
Katlamari	0.712	0.694	0.703
Bokhara	0.800	0.686	0.743
Malancha	0.715	0.695	0.705
Bahutali	0.837	0.648	0.743
Mahesail	0.866	0.695	0.780
Village Total	0.824	0.681	0.752

Source: Calculated by author (data obtained from Soil Health Card, Govt. of India 2016–2017)

Annexure E.5 List of indicators and variables of agricultural development

Indicators	Variables	Description of the variables
Land	X1	Average size of landholdings
	X2	Average number of parcels per holdings
	X3	Percentage of wholly own and self-operated area
	X4	Percentage of wholly leased-in area
	X5	Percentage of net sown area to the total area
	X6	Percentage of area located entirely in the village of residence
Irrigation	X7	Percentage of net irrigated area in total GCA
	X8	Percentage of area irrigated by tank and other sources
	X9	Number of electric well and tube well per 5 ha of GCA
	X10	Number of diesel well and tube well in 10 ha of GCA
Seeds	X11	Percentage of households used certified seeds
	X12	Percentage of households used hybrid seeds
Income and credits	X13	Number of households earn more than 3000 per month from agriculture
	X14	Number of households take institutional credit
	X15	Number of households take credit from money lender
	X16	Number of households have Kisan Credit Card (KCC)
Machinery	X17	Number of hand-operated machines used in per hectare of GCA
	X18	Percentage of households use power-operated machines
	X19	Percentage of households use tractor and harvester
Soil and fertilizers	X20	Number of households have Soil Health Card
	X21	Percentage of the area treated with insecticides and pesticides
	X22	Percentage of the area treated with chemical fertilizers
	X23	Amount of NPK mixture (kg) use per hectare of GCA
	X24	Amount of urea, DAP, superphosphate mixture used (kg/ha)
Market and cold storage	X25	Percentage of products sold directly in the market
	X26	Percentage of product sold directly from home and farm
	X27	Number of households benefitted from cold storage facility
Socio-economic	X28	Farming experience (in number of years)
	X29	Percentage of main cultivators to the total cultivators
	X30	Percentage of marginal cultivators to the total cultivators
	X31	Literacy rate of the farmers
	X32	Number of farmers ever been participated in capacity building program
	X33	Number of farmers ever been participated in farming agitation
	X34	Average size of households
	X35	The average age of household 15–60 years age group

Source: Fixed by authors based on literature review

Annexure E.6 Land size class-wise score sheet of variables of agricultural innovation in sample villages, 2021

Variables	Marginal	Small	Semi-medium	Medium	All classes
X1	0.47	1.49	2.58	5.42	0.88
X2	2.69	4.74	4.28	5.4	3.22
X3	82.54	81.51	90.8	100	83.58
X4	1.24	0.89	0.03	0	0.54
X5	98.47	97.72	98.04	95.04	98.01
X6	79.08	67.8	60.73	54.66	65.57
X7	36.46	39.25	47.49	57.93	45.28
X8	21.93	19.74	19.72	23.59	21.18
X9	2	2	2	1	2
X10	4	3	3	4	5
X11	36.26	37.84	40.41	44.25	37.47
X12	6.75	5.66	7.26	6.55	6.8
X13	179	38	47	26	290
X14	46	47	24	16	133
X15	129	37	30	9	204
X16	97	32	34	17	180
X17	5	2	2	1	3
X18	71.84	64.1	92.31	100	72.93
X19	72.75	74.45	100	100	76.57
X20	68	14	13	10	105
X21	40.58	50.19	58.58	58.96	47.26
X22	80.29	90	83.43	94.09	86.26
X23	369	211	271	266	296
X24	248	304	277	269	236
X25	11.88	16.99	20.59	33.23	14.46
X26	2.83	2.04	4.14	3.93	2.74
X27	0.35	0.65	0.88	0.27	0.50
X28	39	35	38	17	37
X29	97.55	98.99	98.36	100.00	97.79
X30	33.34	32.66	29.19	41.54	35.94
X31	66.66	67.34	70.81	58.46	64.06
X32	18	8	11	7	44
X33	26	34	28	26	26
X34	91.74	90.24	100	100	93.16
X35	8.26	9.76	0	0	6.84
X36	63.53	61.54	73.08	76.92	64.82
X37	45	15	11	8	79
X38	69	23	12	6	110
X39	5	6	6	6	5
X40	47	52	55	57	48

Source: Field Survey, 2021

Annexure E.7 Percentage of GCA under major crops in sample villages of Murshidabad district, 2021

Sample villages	Paddy	Wheat	Jute	Mango	Potato	Mustard	Gram	Masur	Maize	Other crops	Total
Chhayghari	53.77	9.57	19.23	0.75	0.93	4.21	0.31	1.02	0.08	10.16	100.00
Jhikarhati	65.56	8.96	0.16	1.27	5.75	5.05	2.15	2.14	0.03	8.95	100.00
Mirzapur	53.67	13.66	8.70	1.73	2.23	2.75	1.14	0.48	0.04	15.61	100.00
Andul Baria	49.60	10.81	16.14	0.80	2.03	6.79	0.38	1.00	0.20	12.28	100.00
Dakshin Hanumanta Nagar	37.10	11.62	21.12	0.89	1.75	8.38	0.90	1.65	1.02	15.60	100.00
Nasipur	30.47	14.35	27.04	1.50	1.28	5.43	1.35	3.84	0.25	14.51	100.00
Bharatpur	72.03	7.00	0.82	0.87	2.15	5.17	0.85	2.38	0.04	8.71	100.00
Kagram	76.74	5.60	3.33	0.67	0.92	4.03	0.50	1.99	0.05	6.20	100.00
Garaimari	37.67	17.95	19.91	1.48	0.98	4.46	0.67	2.98	0.03	13.89	100.00
Kuli	58.93	4.08	9.46	8.10	0.88	5.71	0.51	2.30	0.47	9.59	100.00
Rukunpur	49.62	16.22	9.50	1.88	0.47	4.73	0.91	2.25	0.05	14.40	100.00
Debipur	31.91	18.20	20.85	1.08	0.67	3.04	0.69	11.18	0.24	12.16	100.00
Hijal	66.51	10.89	0.46	1.56	2.02	2.81	1.11	0.72	0.04	13.90	100.00
Khargram	69.05	10.17	0.28	0.74	0.62	3.43	1.26	1.40	0.00	13.07	100.00
Lalgola	30.44	15.43	32.51	1.27	0.44	5.63	0.45	0.16	4.11	9.60	100.00
Hasenpur	38.82	14.73	18.73	2.09	1.20	4.92	0.85	1.23	1.80	15.66	100.00
Panchgram	71.90	8.49	1.62	0.85	0.87	2.95	0.86	1.01	0.12	11.37	100.00
Bali	48.39	17.52	10.31	2.00	0.54	2.00	1.48	5.05	0.12	12.61	100.00
Ramchandrabati	65.96	7.09	1.97	1.85	0.34	4.37	3.72	4.33	0.02	10.37	100.00
Bahara	40.06	9.80	13.09	0.80	0.45	4.80	4.39	3.89	1.97	20.78	100.00
Bishnupur	21.41	23.09	31.45	0.91	2.85	9.43	0.32	1.56	0.23	8.78	100.00
Katlamari	26.61	18.34	22.30	1.77	1.19	5.28	1.09	3.79	0.73	18.94	100.00
Bokhara	70.91	9.65	3.15	0.32	0.69	6.27	1.08	2.13	0.14	5.69	100.00
Malancha	25.32	6.05	28.11	0.03	1.13	4.38	2.06	10.28	0.27	22.38	100.00
Bahutali	67.68	6.45	1.39	2.06	0.37	2.16	2.27	3.85	0.08	13.72	100.00
Mahesail	19.88	9.54	32.67	1.81	2.70	13.56	1.68	5.25	0.04	12.89	100.00
Total	**51.34**	**12.10**	**12.60**	**1.21**	**1.46**	**5.05**	**1.10**	**2.51**	**0.42**	**12.24**	**100.00**

Source: Field Survey, 2021

Annexure E.8 Composite standardized score of sustainable land use and agricultural innovation in the sample villages, 2021

	Sustainable land use					Agricultural innovation				
Villages	Marginal	Small	Semi-medium	Medium	All classes	Marginal	Small	Semi-medium	Medium	All classes
Chhayghari	0.531	0.501	0.531	0.419	0.525	0.520	0.497	0.507	0.456	0.514
Jhikarhati	0.490	0.497	0.471	0.450	0.488	0.517	0.479	0.466	0.472	0.508
Mirzapur	0.497	0.499	0.497	0.502	0.497	0.535	0.547	0.573	0.540	0.539
Andul Baria	0.481	0.478	0.412	0.460	0.475	0.468	0.463	0.441	0.471	0.466
Dakshin Hanumanta Nagar	0.470	0.473	0.443	0.430	0.467	0.486	0.487	0.484	0.438	0.484
Nasipur	0.461	0.463	0.443	0.458	0.460	0.474	0.458	0.455	0.465	0.471
Bharatpur	0.478	0.431	0.520	0.409	0.474	0.501	0.456	0.531	0.466	0.497
Kagram	0.482	0.491	0.486	0.488	0.483	0.473	0.529	0.493	0.525	0.482
Garaimari	0.513	0.491	0.503	0.508	0.510	0.517	0.491	0.528	0.499	0.515
Kuli	0.434	0.402	0.398	0.389	0.427	0.471	0.435	0.445	0.479	0.466
Rukunpur	0.488	0.482	0.445	0.451	0.483	0.506	0.481	0.546	0.502	0.506
Debipur	0.496	0.485	0.424	0.466	0.489	0.469	0.434	0.450	0.485	0.464
Hijal	0.479	0.503	0.437	0.440	0.477	0.536	0.523	0.567	0.502	0.535
Khargram	0.455	0.459	0.460	0.431	0.455	0.466	0.440	0.466	0.470	0.464
Lalgola	0.500	0.514	0.485	0.501	0.501	0.524	0.505	0.485	0.519	0.519
Hasenpur	0.493	0.485	0.469	0.450	0.489	0.456	0.463	0.461	0.466	0.458
Panchgram	0.425	0.446	0.409	0.405	0.425	0.490	0.467	0.495	0.463	0.487
Bali	0.497	0.498	0.459	0.472	0.493	0.566	0.518	0.526	0.503	0.556
Ramchandrabati	0.348	0.334	0.344	0.328	0.346	0.454	0.428	0.521	0.420	0.455
Bahara	0.400	0.441	0.411	0.420	0.405	0.436	0.392	0.472	0.439	0.434
Bishnupur	0.505	0.487	0.443	0.434	0.497	0.516	0.506	0.476	0.464	0.511
Katlamari	0.488	0.482	0.459	0.479	0.485	0.497	0.472	0.460	0.492	0.492

(continued)

Annexure E.8 (continued)

Villages	Sustainable land use					Agricultural innovation				
	Marginal	Small	Semi-medium	Medium	All classes	Marginal	Small	Semi-medium	Medium	All classes
Bokhara	0.531	0.550	0.531	0.516	0.532	0.453	0.533	0.491	0.525	0.466
Malancha	0.441	0.446	0.456	0.412	0.441	0.486	0.487	0.479	0.432	0.484
Bahutali	0.453	0.464	0.480	0.469	0.456	0.534	0.539	0.564	0.527	0.536
Mahesail	0.513	0.511	0.496	0.498	0.512	0.476	0.471	0.468	0.488	0.476
Total	**0.475**	**0.473**	**0.458**	**0.449**	**0.473**	**0.493**	**0.481**	**0.494**	**0.481**	**0.492**

Source: Calculated by authors (data obtained from Field Survey, 2021)

Annexure E.9 Principal component analysis (PCA)

Component loading															
Indicators	Marginal			Small			Semi-medium			Medium			All classes		
	1	2	3	1	2	3	1	2	3	1	2	3	1	2	3
Land	0.969			0.907			0.957			0.91			0.925		
Irrigation	0.485		−0.629		0.737				−0.742		−0.462	0.564	0.46	0.54	0.524
Seeds	0.968			0.925			0.953			0.908			0.896		
Income and credits		0.53		0.342	0.414	0.658		0.78			0.505				0.389
Agricultural machinery	−0.399	−0.513		−0.487		0.537		0.721			0.714		−0.334		
Soil and fertilizers	−0.303	0.677	−0.319	−0.373	0.629			0.488			0.73		−0.444	0.672	
Market and storage facilities		0.76				0.672			0.658			0.789		0.8	−0.342
Socio-economic aspects			0.871		−0.641	0.384		−0.447	0.386	0.352		0.649			−0.894
Component summary															
SS loadings	**2.77**	**1.91**	**1.46**	**2.24**	**1.6**	**1.39**	**2.08**	**1.64**	**1.33**	**2.33**	**1.78**	**1.51**	**2.59**	**1.75**	**1.43**
% of variance	**34.6**	**23.9**	**18.2**	**28**	**20**	**17.4**	**26**	**20.5**	**16.6**	**29.1**	**22.2**	**18.8**	**32.3**	**21.9**	**17.9**
Cumulative %	**34.6**	**58.5**	**76.7**	**28**	**48**	**65.4**	**26**	**46.5**	**63.1**	**29.1**	**51.3**	**70.1**	**32.3**	**54.2**	**72.1**

Source: Computed by authors (based on data collected from Field Survey, 2021)
Note: 'Varimax' rotation was used

Annexure E.10 Principal component analysis (PCA)

Component loading															
Indicators	Marginal			Small			Semi-medium			Medium			All classes		
	1	2	3	1	2	3	1	2	3	1	2	3	1	2	3
Land	0.969			0.907			0.957			0.91			0.925		
Irrigation	0.485		−0.629		0.737				−0.742		−0.462	0.564	0.46	0.54	0.524
Seeds	0.968			0.925			0.953			0.908			0.896		
Income and credits		0.53		0.342	0.414	0.658		0.78			0.505				0.389
Agricultural machinery	−0.399	−0.513		−0.487		0.537		0.721			0.714		−0.334		
Soil and fertilizers	−0.303	0.677	−0.319	−0.373	0.629			0.488			0.73		−0.444	0.672	
Market and storage facilities		0.76				0.672			0.658			0.789		0.8	−0.342
Socio-economic aspects			0.871		−0.641	0.384		−0.447	0.386	0.352		0.649			−0.894
Component summary															
SS loadings	**2.77**	**1.91**	**1.46**	**2.24**	**1.6**	**1.39**	**2.08**	**1.64**	**1.33**	**2.33**	**1.78**	**1.51**	**2.59**	**1.75**	**1.43**
% of variance	**34.6**	**23.9**	**18.2**	**28**	**20**	**17.4**	**26**	**20.5**	**16.6**	**29.1**	**22.2**	**18.8**	**32.3**	**21.9**	**17.9**
Cumulative %	**34.6**	**58.5**	**76.7**	**28**	**48**	**65.4**	**26**	**46.5**	**63.1**	**29.1**	**51.3**	**70.1**	**32.3**	**54.2**	**72.1**

Source: Computed by authors (based on data collected from Field Survey, 2021)
Note: ‘Varimax’ rotation was used

Annexure F: Chapter 6

Annexure F.1 SWOT analysis for sustainable agricultural innovation in Murshidabad district

Strengths	Weaknesses
Intensity of cropping increased	Huge concentration of small landholdings
Improvisation of transportation facilities	Lacking in advanced technology
Relatively low labour cost	Degradation of soil health
Availability of fertile soil	Low return
Good climate for agriculture	Lack of farmer awareness and training program
Increasing agricultural productivity	Limited infrastructure and marketing
Increase in diversification of cropping system	Limited agro-based industries
Availability of land for agriculture	Low institutional credit facilities
Opportunities	**Threats**
Organic farming	Flood and soil erosion
Agro-based industries	Increasing crop diseases
Export facilities	Overdose of chemical fertilizers
Shifting to commercial agriculture	Increasing fertilizer cost
Involvement of female participation	Variability of monsoon rain
Farm mechanization	Degradation of soil fertility
Intensification of high value crop	Hailstorm and Kalbaishakhi
Introduction of more hybrid seeds	Sudden price fall of agro-products

Source: Field Survey, 2021

Annexure F.2 Listing of SWOT with sustainable agriculture in the sample villages of Murshidabad district, 2021

Sl	Strengths	Sl	Weaknesses	Sl	Opportunities	Sl	Threats
1	Agricultural productivity increased	1	Deteriorating soil quality	1	Agro-based industries	1	Decreasing wetlands
2	Availability of fertile soil	2	Engagement of child labour	2	Bringing fallow land to the cultivation	2	Degradation of soil fertility
3	Available organic waste	3	Heavy doses of pesticides	3	Crop diversification	3	Dependent on rainfall-based farming
4	Conservation of rainwater	4	High cost of cultivation	4	Demand-based production	4	Erratic rainfall
5	Crop productivity increased	5	Lack of new technology	5	Efficient irrigation systems	5	Fluctuation in price
6	Demand-driven farming	6	Lack of industries	6	Export-oriented cultivation	6	Frequent occurrence of flood
7	Easy access to groundwater	7	Lack of goods and cold storage	7	Extension of fodder cultivation	7	Hailstorm and Kalbaishakhi
8	Fodder cultivation	8	Lack of interest in new farming	8	Extension of jute factory	8	Increasing crop diseases
9	Good number of dealers	9	Land fragmentation	9	Fair and regulated market system	9	Increasing migration
10	Good number of cattle	10	Large number of small landholdings	10	Increase productivity	10	Inorganic farming
11	Good road network	11	Low institutional credit facilities	11	Intensification of high value crop	11	Increasing temperature
12	High-value crop farming	12	Low returns	12	Intensive and mixed cultivation	12	Land encroachment
13	Intensity of cropping increased	13	Low yield	13	Introduction of more HYV seeds	13	Land insecurity
14	Outsourcing increasing	14	Lowering water tables	14	Involvements of NGOs	14	Overdose of chemical fertilizers
15	Plenty of labour	15	Overdoses of fertilizers	15	Nearness to urban market	15	Soil erosion
16	Presence of market	16	Poor transport	16	Organic farming	16	Sudden fall of price crops
17	Presence of water bodies	17	Rain-fed agriculture	17	Plantation-based agriculture	17	Unexpected rain
18	Self-help group (SHG)	18	Reduction in government incentives	18	Road and railway connectivity	18	Unpredictable weather
19	Small agro-industry	19	Slow adaptation of new change	19	Soil test nutrition management	19	Water shortage

(continued)

Annexure F.2 (continued)

Sl	Strengths	Sl	Weaknesses	Sl	Opportunities	Sl	Threats
20	Use of organic manure	20	Traditional crop farming	20	Vertical diversification of agriculture	20	Waterlogging
21	Water retention is high	21	Unscientific method of cultivation	21	Wetland-based cultivation		
22	Women participation	22	Migration of workforce				

Source: Field Survey, 2021

Annexure F.3 Arrangement of strengths, weaknesses, opportunities, and threats (SWOT) in agriculture of the sample villages (2021)

Villages	Strengths					Weakness					Opportunities					Threats				
Chhayghari	2	13	16	18	22	3	4	9	14	20	1	4	9	13	14	5	8	12	13	18
Jhikarhati	1	9	12	14	16	1	3	5	9	12	2	3	9	13	18	2	3	7	12	16
Mirzapur	2	5	8	17	22	2	5	9	12	15	5	7	8	9	13	3	5	9	13	15
Andul Baria	1	7	11	15	20	4	7	10	11	21	1	2	5	6	10	2	4	6	9	15
Dakshin Hanumanta Nagar	5	8	10	16	20	5	7	9	13	16	1	2	3	4	5	6	9	10	11	16
Nasipur	2	8	14	15	18	5	7	10	11	22	3	5	6	8	14	2	3	4	4	11
Bharatpur	7	10	13	14	15	4	5	9	11	13	3	7	8	10	12	3	5	7	9	10
Kagram	5	6	9	12	19	4	7	8	10	15	2	7	8	9	10	7	8	9	10	15
Garaimari	1	3	7	14	15	3	4	6	8	11	3	4	9	13	17	5	6	7	10	19
Kuli	4	7	15	17	18	3	6	11	13	15	1	2	3	13	17	2	5	7	9	11
Rukunpur	2	5	15	19	22	2	4	6	13	14	1	6	7	8	18	2	3	4	5	17
Debipur	2	5	8	13	16	2	12	13	16	18	2	3	6	9	15	3	4	7	11	16
Hijal	3	5	6	12	13	8	10	14	15	16	2	3	10	17	18	2	3	6	15	16
Khargram	8	10	16	18	19	3	4	6	8	12	2	3	4	10	13	2	4	5	10	11
Lalgola	5	6	10	14	16	2	3	10	11	13	2	4	6	7	11	3	4	12	14	15
Hasenpur	5	11	16	17	18	4	6	9	19	14	2	4	6	9	11	1	4	10	14	19
Panchgram	5	8	10	15	18	3	10	11	13	15	3	5	10	13	17	2	3	7	9	11
Bali	12	14	17	19	21	6	9	11	14	17	2	3	13	17	20	3	6	14	15	16
Ramchandrabati	7	8	9	11	15	3	5	7	9	22	6	7	8	18	19	4	5	10	11	17
Bahara	4	5	6	9	10	9	10	12	15	17	3	6	9	15	17	4	10	11	12	16
Bishnupur	2	3	5	8	10	2	7	13	16	18	3	10	17	18	21	4	9	10	14	20
Katlamari	1	5	11	13	20	6	7	10	11	14	3	4	10	13	17	3	7	9	11	14
Bokhara	5	6	9	12	19	6	8	9	15	18	3	4	6	7	8	5	6	7	13	15
Malancha	2	5	7	11	16	3	5	7	12	21	8	9	13	17	23	2	5	12	14	16
Bahutali	2	7	13	17	21	4	9	12	15	20	2	4	15	16	18	2	5	8	13	16
Mahesail	2	11	13	16	21	1	5	7	9	21	3	7	10	15	19	3	5	13	14	16

Source: Field Survey, 2021 (see Annexure F.2 for description)

Annexure F.4 List of major problems in agricultural development in sample villages of Murshidabad district, 2021

Problems		Severity (mean score)	Problems		Severity (mean score)
A. Farmer and Farm-related constraints		3.242	**B. Inputs of cultivation-related constraints**		3.569
1	Illiteracy	2.953	1	Low use of advanced technology	3.639
2	Lack of technical knowledge of scientific cultivation	2.549	2	Lack of supply of HYV seeds	3.598
3	Increasing costs of irrigation	3.681	3	Lack of supply of fertilizers	3.512
4	Unscientific use of water	3.805	4	High cost of fertilizers	3.949
5	Low farm mechanization	3.362	5	Unscientific use of pesticides	3.713
6	Degradation of soil fertility	3.112	6	Hired inputs from others	3.342
7	Natural hazards and soil erosion	3.232	7	Unskilled labours	3.234
C. Institutional and infrastructural-related constraints		3.968	**D. Production-related constraints**		3.775
1	Lack of marketing facilities	3.766	1	Low productivity	3.475
2	Lack of cold storage and warehousing facilities	4.138	2	Decreasing cropping area	3.558
3	Poor transport facilities	4.012	3	Small landholding	3.593
4	Lack of credit and loan facilities	4.182	4	High production costs	4.364
5	Lack of seed storage facilities	3.964	5	Lack of export facilities	3.427
6	Lack of fair price systems	4.13	6	Price instability	3.789
7	High transportation costs	3.584	7	Limited marketing information	4.231

Source: Field Survey, 2021

Annexure F.5 Constraints of agricultural innovation in the sample villages of Murshidabad district, 2021

Sample villages	Farmers and farm-related constraints	Input-related constraints	Institutional-related constraints	Production-related constraints	Overall problem score
Chhayghari	3.134	3.235	3.678	3.781	3.457
Jhikarhati	3.108	3.657	4.231	3.673	3.667
Mirzapur	3.105	3.124	4.126	4.124	3.62
Andul Baria	3.269	3.223	4.245	4.183	3.73
Dakshin Hanumanta Nagar	3.112	3.799	3.879	4.184	3.744
Nasipur	3.163	3.511	4.134	3.689	3.624
Bharatpur	3.608	3.794	4.152	3.142	3.674
Kagram	3.228	3.567	3.812	3.672	3.57
Garaimari	3.019	3.264	4.162	3.481	3.482
Kuli	3.452	3.99	4.376	3.589	3.852
Rukunpur	3.163	3.897	3.781	4.289	3.783
Debipur	3.219	3.571	4.141	3.69	3.655
Hijal	2.987	3.589	4.341	3.678	3.649
Khargram	3.551	3.452	3.909	3.698	3.653
Lalgola	2.964	3.674	3.348	3.895	3.47
Hasenpur	3.398	3.568	4.162	3.808	3.734
Panchgram	3.006	3.167	3.572	3.342	3.272
Bali	3.142	3.167	4.032	3.673	3.504
Ramchandrabati	3.384	3.143	4.134	4.163	3.706
Bahara	3.109	4.235	4.671	3.581	3.899
Bishnupur	3.385	3.901	3.897	3.652	3.709
Katlamari	3.213	3.675	3.789	4.172	3.712
Bokhara	3.234	3.105	4.154	3.149	3.411
Malancha	3.109	3.359	3.584	3.781	3.458
Bahutali	3.871	4.231	3.178	3.879	3.79
Mahesail	3.384	3.908	3.672	4.176	3.785
Total	**3.243**	**3.569**	**3.968**	**3.775**	**3.639**

Source: Calculated by authors

Annexure F.6 Various ongoing programs, policies, and schemes in Murshidabad district

S. no	Ongoing programs and schemes
1	Rashtriya Krishi Vikas Yojana (RKVY)
2	National Mission for Sustainable Agriculture (NMSA)
3	National Food Security Mission (NFSM)
4	National Mission on Agricultural Extension and Technology (NMAET)
5	Integrated Scheme for Farmers Income Security (ISFIS)
6	Integrated Scheme on Agriculture Marketing (ISAM)
7	Integrated Scheme on Agriculture Cooperation (ISAC)
8	Agricultural Technology Management Agency (ATMA)
9	Integrated Watershed Management Programme (IWMP)
10	Kisan Credit Card (KCC)
11	National Horticulture Mission (NHM)
12	Mission for Integrated Development of Horticulture (MIDH)
13	National Mission on Oilseed and Oil Palm (NMOOP)
14	National Crop Insurance Programme (NCIP)
15	National Mission on Food Processing (NMFP)
16	Sub-mission on Agriculture Mechanization (SMAM)
17	Paramparagat Krishi Vikas Yojana (PKVY)
18	National Agriculture Market (eNAM)
19	Pradhan Mantri Krishi Sinchai Yojana (PMKSY)
20	Pradhan Mantri Fasal Bima Yojana (PMFBY)
21	Pradhan Mantri Kisan Maandhan Yojana (PMKMY)
22	Soil Health Card Scheme

Source: State Agriculture Infrastructure Development Plan, West Bengal, 2020

Summary, Conclusion, and Suggestions

Summary and Conclusion

The pattern of agricultural land use directly or indirectly reflects the level of agricultural innovation, growth, and development in a particular region. Agriculture, the most crucial sector of the nation's economy in terms of employment opportunities, must be taken care of by the government through different policies and programs. This research work's finding will undoubtedly help design and formulate policies and programs for agricultural innovation in the study area. The summary of this study from the ongoing analytical chapters has been described in the following sequence.

Chapter 1 presents the different aspects of the physical and cultural set-up of the Murshidabad district. The district is one of the backward districts of West Bengal regarding the social and economic situation. Agriculture plays a significant role in the district as half of the population is directly or indirectly (as cultivators and agricultural labours) engaged in this sector alone. The engagement of people in the agriculture sector reports more than 80% in terms of the rural population. Murshidabad district is endowed with a fertile alluvial track as it is located at the lower Gangetic Plain and has excellent climatic conditions, which are favourable for agricultural activities. After going through the brief overview of the study area, it can be stated that the physio-cultural and socio-economic conditions of the district are highly diversified in a spatial context which leads to the variations of lifestyle and living conditions of its inhabitants. Since agriculture and its allied sectors are still the main ways for the economy of the people in the study area, the development of this sector with the pace of population growth is a major concern for policymakers.

The dynamics of sustainable land use and cropping pattern have been presented in Chap. 2. It is found that the total area, net sown area, and net cultivated area of the district have increased from 1995–1996 to 2015–2016, but their share of

H. Rahaman, S. Ansari, *Sustainable Land Use Practices and Agricultural Innovations in India*, https://doi.org/10.1007/978-3-031-83246-8

increases is almost negligible. On the plus side, the uncultivated land and the land not available for cultivation have decreased during the same time periods in the district. The scenario of gross cropped area of major crops reveals that paddy, the most dominant crop, is still catching areas under cultivation in 2015–2016 over 1995–1996, while the remaining crops have lost area significantly across land size category. A significant amount of gross cropped area has increased in the district during 1995–1996 and 2015–2016. The total area under food crops is growing over non-food crops under the different land classes except for semi-medium and medium landholdings due to the massive dominance of cereal crops. The sustainable land use index is prepared for 1995–1996 and 2015–2016 at the block level across land size categories to show the spatial variation of land use. The changes and variations of index value have been found in addition to agricultural land use intensity. The tested hypothesis has confirmed that the amount of change in agricultural land use significantly differs across land size class categories from 1995–1996 to 2015–2016. Primary survey shows that profit opportunities, consumption, higher productivity, climate change, and soil fertility are the prime reasons for the agricultural land use pattern changes in the sample villages.

Chapter 3 discusses the overall agricultural innovation concerning the input and output aspects of a spatial context under different land size classes at two points in time. There are a total of thirty-two (32) variables under the heads of seven indicators that have been selected for input dimension. In contrast, four important variables (productivity, crop diversification, cropping intensity, and cropping pattern) are chosen for the output dimension. It is found that the input variables are strongly positively related to output variables in 1995–1996, and even this relation becomes stronger in 2015–2016. Therefore, it rejects the assumption that agricultural input is negatively related to agricultural output. The spatial and temporal analyses of overall (inputs–outputs) agricultural innovation suggest an improvement in the overall agricultural situation in the district. It can be easily pointed out from the above discussion that the pace of agricultural innovation over the years is not the same; rather, it differs significantly at the block level as well as in different land size classes.

A relationship between sustainable land use and agricultural innovation is established in Chap. 4 to check the impact of these two dimensions on each other under different land size classes in the Murshidabad district. The relationship between these two domains is statistically measured through multiple linear regression (MLR) in the first section of this chapter. A positive and significant correlation between sustainable land use and agricultural innovation has proved that sustainable land use is increased in the district along with the pace of agriculture innovation. Thus, sustainable land use is significantly related to agricultural innovation under different land size categories. It is essential to point out that this relationship is most robust under the marginal land size class. The following section deals with a unit-wise comparison of input and output dimensions of agricultural innovation with respect to sustainable land use under different land size categories. Here, one unit change in sustainable land use implies any amount of change in agricultural innovation. The unit-wise change comparison reveals that the input dimension records the maximum change per unit of sustainable land use change in the district. Therefore, it is again confirmed that agricultural innovation gets altered positively with the improvement of various elements of sustainable land use intensity.

The last three analytical chapters are devoted to sustainable land use, agricultural innovation, and their relationship, mainly based on secondary data sources. Chapter 5 analyses the agricultural innovation and sustainable land use, along with factors of development, based on household survey data in selected sample villages. It is found in this chapter that there is a causal relationship existing between sustainable land use and agricultural innovation. The eight indicators under three components have explained 65.80% in total variance in marginal land size class, 65.40% in small, 63.10% in semi-medium, 63.15% in medium, and 65.15% in all land size classes in the district. Agriculture innovation in sampled villages differs within and between land size categories.

After going through extensive analyses and explanations, an attempt is made to look into the contemporary issues and challenges of agriculture innovation in Chap. 6. The first section of this chapter starts with the analysis of strengths, weaknesses, opportunities, and threats (SWOT) to identify key factors of agriculture in particular and the development process in general. The SOWT analysis reveals that massive concentration of marginal and small farmers, rare use of advanced technology, a large number of unskilled labour, low yield of crops, low price, lack of market information, limited agro-based industries, and inadequate institutional credit facilities are the major problems prevailing in the district. At the same time, the district has several potential opportunities, including the availability of fertile soil, crop diversification, agriculture diversification, organic farming, agro-industry set-up, and wetland-based cultivation. A highlight of area-specific (village-level in this case) constraints faced by the farmers using a rating scale is discussed in the next section. First of all, a list of major problems prescribed by the farmers during the field survey has been noted under four broad constraints, i.e., farmer and farm-related constraints, inputs of cultivation-related constraints, institutional and infrastructural constraints, and production-related constraints. Then, the severity of these problems is calculated on a 5-point scale rating as rated by the respondents. The result shows that out of the four major constraints, the severity scores of institutional and infrastructural-related constraints emerge to be the maximum in the sample villages.

Suggestions

1. It is almost an established fact that the horizontal expansion of agricultural land is not possible in future. Therefore, planners and different agencies should incorporate the vertical expansion of agriculture through proper land utilization and management. Vertical diversification of agriculture linked with agro-based industries will be significant beneficiaries for the farmers.
2. The change in agricultural land use pattern regularly is a healthy activity. Since it is dominated by marginal land size class, other land size categories should be encouraged to change from conventional to non-conventional cropping systems.

The contributions of district agricultural agencies, extension service centers, and agricultural development officers are particularly crucial in this context.

3. Though the cropping pattern of the district has seen changes during the study period, most of the gross cropped area is still dominated by cereal crops (more specifically by a single crop, i.e., paddy, which occupies more than 50% of the GCA). Therefore, the district agencies should encourage the farmers through different awareness and discussion of the benefits to come out of their traditional cropping pattern and cultivate trending high-value crops for the sake of their own benefit.
4. Input variables, especially credit facilities, tank and sprinkle irrigation, HYV seeds, cold storage and market, and crop insurance, are also suggested to the district administrator to take action and implement these with the Government of India schemes.
5. Agriculture innovation and its forward linkage with sustainable land use have also been highlighted in the study. It is found from the analyses of this study that different aspects of sustainable land use are directly related to the innovation of agriculture in the study area. Therefore, the focused sustainable land use aspects of development should be taken into consideration with respect to ongoing programmes.
6. Low farm mechanization and infrastructural restrictions are found to be the most common problem in the district for the extension of cultivation. Huge smallholdings and the low capital base of the farmers are the main reason for such low farm mechanization. Therefore, instead of large implements and machinery, government should provide small machinery and implements that the smallholders can use.
7. One of the serious concerns in the district is that there is an increasing trend of uncontrolled and unscientific use of chemical fertilizers, pesticides, insecticides, and herbicides. This leads to the deterioration of soil health. To avoid and minimise these issues, widespread farmers' training, demonstrations, and awareness programmes must be carried out across the district.
8. Price fluctuation is another significant problem reported by the respondents. Therefore, a strengthened marketing system is needed to overcome this problem. The administrators of the marketing boards should regulate and monitor information regarding the price of different crops. The circulation of information regarding the market system must reach out to each farmer.
9. It was noticed during the field survey that although there are many schemes and policies, most farmers do not know about the government's policies, programs, and schemes related to agricultural innovation. Therefore, it is suggested that government agencies must provide awareness to the farmers through capacity-building programs, more soil testing, frequent organization of focus group discussions (FGD), and, more importantly, bottom-up approaches. Because without this, any programs, policies, and schemes will shine only on the documents, not the actual targets.

Index

H. Rahaman, S. Ansari, *Sustainable Land Use Practices and Agricultural Innovations in India*, https://doi.org/10.1007/978-3-031-83246-8

The manufacturer's authorised representative in the EU is Springer Nature Customer Service Centre GmbH, Europaplatz 3, 69115 Heidelberg, Germany. If you have any concerns regarding our products, please contact ProductSafety@springernature.com

Printed and bound by CPI Group (UK) Ltd, Croydon, CR0 4YY
01/10/2026
02225949-0019